THERMOPHOTOVOLTAIC GENERATION OF ELECTRICITY

THERMOPHOTOVOLTAIC GENERATION OF ELECTRICITY

Third NREL Conference

Colorado Springs, CO May 1997

EDITORS
Timothy J. Coutts
Carole S. Allman
John P. Benner
National Renewable Energy Laboratory

American Institute of Physics

**AIP CONFERENCE
PROCEEDINGS 401**

Woodbury, New York

L.C. Catalog Card No. 97-74374
ISBN 1-56396-734-0
ISSN 0094-243X
DOE CONF- 9705119

Printed in the United States of America

CONTENTS

SESSION 1: OVERVIEWS OF TPV

SESSION 2: TPV DEVICES BASED ON GaSB AND RELATED MATERIALS I

SESSION 3: SELECTIVE RADIATORS I

SESSION 4: TPV DEVICES BASED ON InGaAs

SESSION 7: NOVEL CONCEPTS

SESSION 8: MODELING AND CHARACTERIZATION OF TPV SYSTEMS

ADDENDUM TO SESSION 1

This volume contains a collection of most of the papers presented at the Third NREL Conference on Thermophotovoltaic Generation of Electricity. In addition, and subsequent to this introductory section, there are overviews of the individual sessions from the Session Chairs. It is hoped that these overviews will provide the casual reader with enough information about the contents of the book to ascertain whether specific papers are potentially useful.

The objectives of the Third Conference remained as they were at the time of the First, namely to provide:

1. A meeting place at which enthusiasts in the field can discuss their latest work.
2. A venue for industry/universities/federal labs to communicate research findings and needs to one another.
3. An opportunity for partnerships to form.
4. A means of informing potentially new funding agencies and organizations that there is a vigorous R&D community and an embryonic industry forming, that ultimately could be of great importance to the Nation, in both military and non-military projects.
5. A means of archiving the most recent work of the TPV community.

Funding for TPV has increased since the last Conference almost two years ago, but the number of registered attendees remained at approximately 120. The additional funding has been used to support a few groups in the field, rather than being equally distributed amongst all. Consequently, some groups reported significant success and progress while others were struggling to survive. Interest appeared to grow outside the USA and, at this Conference, there were more attendees from Europe and, for the first time, Japan was represented. Within the USA, there was increased industrial interest, including such large organizations as Babcock and Wilcox, Coleman, and the Propane Gas Research Institute. Other established companies also appeared to be doing well and it appeared that there would be at least one non-military product on the market by the end of 1997.

At this Conference, there were more papers on GaSb and its related ternary and quaternary alloys than at the other two Conferences. The back surface reflector technique for recirculating sub-bandgap photons was more widely discussed, but less attention was paid to multilayer dielectric stacks and to transparent conducting oxides. The back surface reflector approach offers significant advantages when a semi-insulating substrate is available, although it may be more problematic with less well-developed substrate materials.

A website was developed for this Conference which contained details of registration, the program, details of the Advisory Group, links to the American Institute of Physics and the previous Proceedings, links to other TPV sites, and to other sites of relevance to the attendees. Our intention is to develop the site into a user group so that the community can share their latest results and information. Input, however, on how best to use the website, would be welcome.

The Conference Chairs received considerable help from the Advisory Group that was convened to provide a more balanced view from the community. The members of the Advisory Group consisted of Dennis Flood from NASA Lewis Research Center; Robert Rosenfeld of the Defense Advanced Research Projects Agency; Frank Rose of the Space Power Institute at Auburn University; David Riley of the Westinghouse Bettis Atomic Power Laboratory; and Richard Paur of the Army Research Office. These individuals helped with comments on the program, suggested invited speakers, and generally helped with other issues. They also acted as Chairs of the individual Sessions and provided an overview of the material presented in each of the Sessions, indicating their sense of key results presented and areas for future development.

Although the Conference was again sponsored by NREL/DOE, additional support was provided by DARPA, ARO, and NASA Lewis Research Center. This was an important development from the first two conferences because it made the point that TPV is of relevance to more than simply the Department of Energy.

There also appears to be some interesting competition developing between various sub-technologies. For example, performance of both strained InGaAs films on InP substrates and lattice-matched InGaAsSb films on GaSb substrates still appears to be comparable. Hence, the added complexity of the quaternary alloy has not yet been justified. Also, work on both selective radiator-based systems and on broadband-based systems continues and neither yet appears to have established dominance. Arguments have been made in favor of the low cost of diffusion-based device fabrication versus organometallic vapor phase epitaxy (OMVPE). However, costing studies performed elsewhere on the OMVPE process do not support the notion that epitaxy is unavoidably a high-cost process. None of the outcomes is yet guaranteed.

There is still very little work on characterization of the materials and devices of the PV converters at a fundamental level. The immediate consequence of this is that there is a shortage of reliable information for device modelers. There is also some work on the modeling of system performance and the predictions are much more modest and probably more realistic than was previously the case. We hope to see an increasing amount of work on systems modeling in the future.

TPV appears to be developing rapidly and the sense amongst the community seems to be that the real need is for systems demonstration rather than research and development into the individual components of a TPV system. TPV systems have potentially outstanding advantages in a number of military and non-military markets and demonstration of systems is likely to encourage prospective users to support the field more enthusiastically. We look forward to developments from the R&D community during the next year or two and are confident that we are observing a rapidly developing electricity generation technology.

Based on comments received from the attendees after the Conference, it seems virtually certain that a Fourth Conference in the series will be held. The time, date and venue of this have not yet been decided and those interested are encouraged to contact one of the Conference Chairs directly.

The general chairs would like to express their thanks to the entire Advisory Group for their constructive comments and help during the planning of the conference: Heather Bulmer of NREL's Conferences Group, and the entire team that coordinated the conference logistics: and to Mark Fitzgerald of Science Communications, Inc. for establishing the website and helping to keep the speakers to good times. These individuals have helped to make this conference the success it was, and we look forward to their continued involvement.

Timothy J. Coutts
Carole S. Allman
John P. Benner

Overviews

The following eight short contributions were prepared by the Chairs of the Sessions. The General Chairmen asked these individuals to introduce and summarize their sessions and to provide short written versions of their thoughts. The individual sessions are listed below. It is hoped that this summary will provide the casual reader with an overview of the entire conference without needing to delve deeply into the main body of the proceedings.

Session Number	Session Chair	Session Title
1	Timothy J. Coutts	Overviews of TPV
2	David M. Wilt	TPV Devices Based on GaSb and Related Materials
3	Robert Rosenfeld	Selective Radiators I
4	David R. Riley	TPV Devices Based on InGaAs
5	Robert E. Nelson	Selective Radiators II
6	M. Frank Rose	TPV Systems
7	John S.Kruger	Novel Concepts
8	Daniel J. Krommenhoek	Modeling and Characterization of TPV Systems

Session 1: Overviews of Thermophotovoltaics

Timothy J. Coutts, National Renewable Energy Laboratory

The present resurgence of interest in TPV has extended over the last five or six years and has been primarily militarily-driven. Although it is probably true to say that 95% of R&D is being done in the United States of America, interest in Europe and Japan is undoubtedly growing. So far as we are aware, this interest is primarily commercial rather than military, with gas utilities, and energy agencies being the market forces. During this conference, a proposal was made to the chairman (T.J.C.) to hold a TPV conference in Europe, which is clearly an indication of growing interest. With this in mind, we invited Dr. Markus Schubnell of the Paul Scherrer Institute to review European Activities in Thermophotovoltaics. There are already projects in Sweden on wood-fueled TPV systems in which co-generation of heat and electricity is the aim: in Germany where a TPV generator is being considered to ensure reliability of off-grid PV systems: and in Switzerland, where there are also interests in co-generation. There are also interests in the UK where British Gas is considering a self-powered gas appliance.

Despite interest from industries with a non-military interest, the military organizations in the United States of America are still easily the largest customers. This was taken into account in formulating this opening session by inviting Dr. Jack Kruger of the Army Research Office to discuss the findings of a group of scientists and engineers who attended a Workshop organized on behalf of ARO in the summer of 1996. The objective of the Workshop was to establish whether or not TPV could benefit the operational needs of the Soldier of the Future and, if so, for which applications. The conclusions were very positive and it was collectively thought that TPV systems would be capable of meeting the needs of applications from a few Watts to greater than 500 Watts. Interestingly, many of the considerations apply equally to non-military applications.

There is little doubt that the 'new' TPV R&D has benefited considerably from the success of the High Efficiency Solar Cell program of the Department of Energy, particularly in regard to the PV converter. The availability of high-efficiency PV cells based on III-V semiconductors is one of the main reasons for the resurgence of interest in TPV. The PV Program should be given credit for this. The latter has been in existence for about twenty years now and the embryonic TPV community can learn much from it. The centralized Measurements and Characterization capabilities at NREL have historically been, and still are, one of the major strengths of the PV program and we invited Dr. Lawrence L. Kazmerski to speak on the capabilities of this Center and on some of the lessons to be learned. The clear message to emerge is that neglect of adequate component and system characterization, comes at a price! Obviously, this conclusion applies equally to all market segments, whether military or non-military.

Session 2: TPV Devices Based on GaSb and Related Materials

Chair: David Wilt, NASA Lewis Research Center

This session encompassed TPV photovoltaic (PV) cells composed of binary, ternary and quaternary compounds of the gallium antimonide (GaSb) family, and spanned the TPV spectrum from moderate temperature, "low cost" GaSb cells to quaternary devices designed for low emitter temperature systems. This session was the largest of the conference, consisting of eleven (11) papers from industry and academia, including one international paper. The majority of the presentations reflected the interest which the lattice-matched quaternary material (~0.55eV) has generated. It is believed that this material will demonstrate higher performance (i.e. lower dark currents and better bandedge photoresponse) than lattice-mismatched indium gallium arsenide (InGaAs) devices.

JX Crystals presented a path whereby they believe photovoltaic modules of GaSb (0.72eV) cells may be delivered for ~\$1/W. They claim that the critical technologies have been demonstrated, and all that is required to meet the cost goal is increased production volume. Several groups were involved in the investigation of lattice mismatched gallium indium antimonide (GaInSb) on GaSb, for low bandgap (0.55eV) TPV devices. Theoretically these devices should have the same disadvantages as InGaAs devices, although researchers from RPI reported low dislocation densities in the GaInSb material system. Their data indicate that the lattice stress was relieved by mechanisms other than threading dislocation formation. This result holds promise for these lattice-mismatched devices.

Gallium indium arsenide antimonide (GaInAsSb) devices fabricated by liquid phase epitaxy (LPE), organo metallic vapor phase epitaxy (OMVPE), molecular beam epitaxy (MBE) and diffusion into LPE produced layers were reported. Given the complexity of this material system, most researchers reported little difficulty in producing layers and devices with reasonable electronic properties. At this point, the performance differential between the lattice-matched quaternary devices and lattice-mismatched InGaAs devices is negligible, although the GaInAsSb material system has not been developed to the same level of maturity as the InGaAs system. Continued development may realize improved electrical performance of the quaternary devices. Questions concerning the stability of meta-stable alloys in the high temperature TPV environment and the ability to use these materials in the monolithic interconnected module (MIM) design currently being developed in the InGaAs material system need to be addressed in the development of GaInAsSb TPV devices.

Session 3: Selective Radiators

Chair: R. Rosenfeld, DARPA

This session discussed the source of photons for TPV. Selective emitters have frequently been important for TPV systems operating with Silicon TPV cells and high temperature emitters. Two papers described emitters for moderate temperatures and one paper addressed the measurement of temperature in TPV emitters. Luke Ferguson of JX Crystals, Inc., started by describing the performance of a selective emitter matched to GaSb thermophotovoltaic cells. The emitter can be a Co doped spinel or Co doped MgO. The spectrum of the emitter is broad, thereby permitting high power densities within the GaSb band. This emitter greatly enhances the perfor mance of TPV systems operating in the 1400°C temperature range. Zheng Chen of Auburn University described an emitter consisting of rare earth fibers supported by an alumina fiber matrix. He described an experimental setup and experiments with Erbia, Holmia, and a mixed oxide with both Erbia and Holmia. The mixed oxide gave the best results for a 0.65 volt TPV cell. Robert Nelson of Tecogen described a technique for measuring the temperture of TPV emitters. The problem is that emissivity is unknown, but Nelson uses short wavelengths and he showed that the brightness temperature measured by a pyrometer is only slightly affected by emissivity at short wavelengths.

Session 4: TPV Devices Based on InGaAs

Chair: D. R. Riley, Westinghouse / Bettis Atomic Power Laboratory

Overview:

The indium gallium arsenide (InGaAs) system is being investigated for applications in thermophotovoltaic (TPV) power applications. InGaAs is lattice-matched to indium phosphide (InP) with a bandgap of 0.74 eV. InP is available in semi-insulating form, and therefore serves as an excellent candidate for series interconnected TPV cells, also referred to as monolithic interconnected modules (MIMs). However, several important tradeoffs must be considered when designing a TPV cell system, including electrical losses due to joule heating, optical losses caused by free carrier absorption, cell obscuration incurred with metallization, and spectral control.

Key Results:

Most of the presentations either directly or indirectly compared the performance of MIMs with conventional single junction devices. Primary differences between the two include the inclusion of a semiconducting lateral conduction layer in the MIM below the active diode layers to transport current to grid metallization on the device, and the use of a semi-insulating substrate to minimize current leakage between adjacent cells. As was pointed out, if one considers only the electrical losses (i^2R), then the key tradeoff is series resistance versus additional power gleaned by lower current. MIM devices in the past suffered from high series resistance, and in fact, electrically were shown to have performed worse than single junction devices; however, recent MIM structures grown at several laboratories, have demonstrated significantly lower series resistance and enhanced performance. Considering the gains in optical performance due to decreased free carrier absorption in semi-insulating substrates, the InGaAs MIMs performed exceedingly well.

Two primary designs for InGaAs MIMs have been developed, a *conventional design* using metal interconnects to transport current between adjacent cells and an *interdigitated design* which utilizes the actual grid fingers to transport the current. In the former, metal interconnects separate each individual cell on the wafer and grid fingers traverse the top surface to collect current. Devices of this type have been fabricated with demonstrated performance of $V_{oc} > 3.3$ volts, $J_{sc} = 590$ mAmps/cm^2 and a fill factor of 68%. The latter design (interdigitated) has been developed to minimize series resistance in the MIMs and to increase the flexibility of cell design and output. At the expense of minimal increases in optical losses, grid fingers are used to transport current on both the top surface and on the lateral conduction layer. Accordingly, the current no longer has to flow the length of the cell to get to a metal contact, but only the width of the cell to get to a grid finger. This represented a significant decrease in series resistance; this design could further be exploited by thinning the lateral conduction layer and subsequently decreasing free carrier absorption. Devices of this type have exhibited $V_{oc} > 3.6$ volts, $J_{sc} = 0.9 A/cm^2$, and a fill-factor of 72%.

Materials development efforts have concentrated on characterizing the optical properties of degenerate InGaAs layers, and InP window and back surface field layers. Degenerate InGaAs layers were shown to cause the bandgap to shift due to the Moss-Burstein effect and to effectively serve as a plasma reflector. In single junction devices, and potentially in MIMs, a degenerate emitter coupled to a bandpass filter can provide effective spectral control. With a bandgap of 1.35 eV, large band discontinuities, and effective band bending at the heterojunction, InP was shown to create a significant potential barrier to minority carriers, thereby, increasing quantum efficiency and output performance.

Final Considerations:
InGaAs devices operating with radiators at low temperatures (i.e., $< 1000^oC$) have to be fabricated lattice-mismatched to obtain reasonable power lever levels. This causes defects in the devices and magnifies fabrication issues. Procedures are readily available to fabricate small quantities of high quality devices of this nature, but scale-up may prove challenging when it comes to reproducibility and performance. The second option is to utilize lattice-matched low bandgap devices, probably incorporating gallium antimonide substrates, which currently can not be fabricated semi-insulating. This introduces additional concerns including increased free carrier absorption and current leakage. Recent advances in MIM technology and cell material development may offer solutions which minimize most of these adverse effects. The incorporation of unique grid finger designs, spectral control strategies and cell architectures have already significantly improved performance of lattice-mismatched InGaAs devices; continued efforts will only serve to further optimize these high voltage / low current devices.

xvii

Session 5: Selective Radiators II

Chair: R. E. Nelson, Quantum Group

"Influence of Ytterbium Concentration on the Emissive Properties of Yb:YAG and Yb:Y_2O_3", J.C. Panitz and M. Schubnell, Paul Scherrer Institute

Powder preparation of Yb:YAG AND Yb:Y_2O_3 was accomplished by coprecipitation and by combustion synthesis. Thermal emission measurements of calcined samples indicated better performance for Yb:YAG than Yb:Y_2O_3

A simple TPV demonstrator was described. A conventional camping lantern was fitted with an ytterbia mantle and a cylindrical array of Si photovoltaic cells surrounded the mantle. The cells were water cooled. An I-V characteristic of the cell was presented.

"TPV Power Generation Prototype Using Composite Selective Emitters," P. Adair and M.F. Rose, Auburn Space Power Institute

A bonded alumina fibrous array is used to support short staple lengths of Er_2O_3 in a cylindrical emitter to illuminate an array of InGaAs cells. A single cell was used, however, to evaluate the emitter exitance. Compressed air and propane were the air/fuel premix. A water-cooled collar stabilized the flame at the base of the cylindrical emitter. A pressed and sintered plug of Erbia closed off the top of the cylinder and forced the combustion products through the porous emitter. A quartz sleeve served to protect the PV array from the combustion products. Radiation efficiencies of the emitter and conversion efficiencies of the PV cell were used to calculate system efficiencies. The role of thermal recuperation, which was not incorporated in the prototype, was projected and its effects on system efficiency was estimated.

"Effect of Temperature Gradient on Thick Film Selective Emitter Emittance," D.L. Chubb, B.S. Good, E.B. Clark, and Z. Chen, NASA Lewis Research Center and Auburn Space Power Institute

Don Chubb presented an analysis of a thick film (100µm) emitter (selective rare earth oxide) attached to a thermal source. A vacuum condition was assumed on the emitting side. He calculated the volumetric emission. He assumed all reflections from the interfaces to be diffuse, and he calculated the fraction of total internal reflection arising from Snell's law. He neglected scattering. A significant conclusion of his model is a substantial reduction of emittance if there is a large thermal gradient in the thick film.

"Superemissive Light Pipe for TPV Applications," M.K. Goldstein, L.G. DeShazer and A.S. Kushch, Quantum Group, Inc.

Single crystal emitters of Yb, Er, and Ho in garnets were diffusion bonded to YAg light pipes. The selective emission normally attributed to these rare earth's was not compromised in these structures and off-band emission was very low.

Data on the low thermal transfer along the light pipe was also presented.

Session 6: TPV Systems

Chair: M. F. Rose,
Space Power Institute, Auburn University

The papers given in this session constitute the current level of development in TPV systems. Both system studies and experimental prototype units were described. In general, the experimental units have efficiencies less than 5 %, however. Two applications, electricity plus space heating and electricity plus hot water heating, do not demand the ultimate in the conversion of the heat of combustion of the fuel to electricity. These applications are easier technically than those associated with applications demanding only electrical output. It is expected that both of these applications could penetrate niche markets within a year or so.

System studies conducted by Babcock and Wilcox and JX Crystals indicate that a 500 watt TPV system could be built which would have an overall efficiency on the order of 8%. This unit would have recuperation and be, in principle, multifuel capable. Similarly, the Tecogen Division of Thermopower Corporation described the first high temperature TPV module in the 150 watt range. The significance of this work is the use of a ceramic recuperator. Within the present configuration, the efficiency is on the order of 4.5%. It is estimated that with optimization, the efficiency could approach 8%. This system would use the readily available silicon cell technology.

One of the key issues which makes TPV attractive is the possibility that TPV power systems could be built to be multifuel capable and therefore require no exotic fuels or storage techniques as is the case for many fuel cell types. A system configuration using a gray body emitter was described in detail. The optimistic projection for efficiency for this system is on the order of 9%. This systems study emphasized spectral control at the emitter by using high temperature coatings. The intended application is for a Army battery charger.

Based on the systems described it appears to be difficult to achieve a total system efficiency greater than 10%. If this efficiency is to be reached, it will be necessary to improve the emitter, PV cell, and the recuperators within a system context.

There is substantial research and development on the emitters and the PV cells. There is a decided lack of similar development of the compact, high temperature recuperators necessary for efficient systems.

Session 7: Novel Concepts

Chair: J.S. Kruger, Army Research Office

This session dealt with ideas that were sufficiently different from the other aspects of TPV technology to warrant their own showcase.

The first paper presented by Ray Martinelli of Sarnoff Corp. discussed excellent minority carrier transport in InGaAsSb/GaSb TPV diodes making them very suitable for TPV applications. A p-on-n mesa test structure was grown by molecular beam epitaxy, with a flat response over the 1-2.2 μm range. A model was fitted to spectral quantum efficiency showing a sharp rise from the bandgap and radiation absorbed in the GaSb cap layer. For a highly doped emitter there is a lower series resistance and a higher open-circuit voltage. Martinelli also discussed radiation absorption and reradiation effects.

The second talk was given by Alex Kushch for his colleague Vadim Unishkov from Quantum Group. Unishkov was not able to return to the U.S. in time for the TPV Conference. This talk presented characteristics of Ge vertical multijunction (VMJ) PV cells. Kushch noted that lots of cells were tested, and discussed cost effectiveness, competitiveness, challenges for cells and TPV systems. He also dealt with the motivation for using germanium and the VMJ cells. This work demonstrated the feasibility of converting sunlight, thermal radiation and laser emissions in a single cell.

Next Miguel Contreras from NREL talked about thin film polycrystalline materials and their potentially lower cost. He mentioned that GaSb is the material of choice because of the ease in vaporizing its constituents. He then evaluated how easy or difficult it is to grow such devices, mentioned optical characterization and the effect of alloying GaSb with In to reduce the bandgap, and questioned the utility for TPV systems.

Paul Griffin from Imperial College presented advantages of quantum well solar cells for TPV. The key here is to insert quantum wells into the intrinsic region of a p-i-n cell. The result is a dramatically lower dark current, higher open-circuit voltage, and a band edge which can easily be varied to match the emission. Griffin discussed dark and light IV measurements, using a 3000-K tungsten lamp to represent a blackbody and interference filters to simulate Er and Yb emitters. There is still a need to optimize for TPV devices and to include a mirrored backing to increase the efficiency.

The fifth paper by Neelkanth Dhere of the Florida Solar Energy Center again treated polycrystalline thin-film TPV cells. He discussed the renewed interest in materials. One motivation for looking at such cells is to respond to selective radiators over a wide range. The thin-film cells based on InGaAs, GaSb, GaInSb, and InGaAsSb, as well as solar cell such as CdTe, and $CuInGaSe_2$, and HgCdTe and PbCdTe alloys cover a wide range of cutoff wavelengths. He mentioned a large number of deposition techniques. Dhere noted that polycrystalline thin films show about 2/3 of the conversion efficiency of the single crystal cells. He concluded that in order to bring the cost down large consumer and residential applications, such as a hybrid car, must be found. It might then be feasible to expect 35¢/W or even 17¢/W

The last talk was given by Bill Biter of Sensortex. He detailed work on a TPV system using a gold filter and $CuInSe_2$ solar cells. The Au/dielectric stack filter is a simple 3-layer induced transmission design, wherein an Au layer is sandwiched between two dielectric layers. Tradeoffs are possible with layer thickness. The concept is to provide high transmission at the bandpass, while still being lower cost. Biter reported 90% transmission at 1 μm with 99% reflectivity further out in the infrared. This ought to reclaim energy which otherwise would be thrown away as waste heat. A distinct advantage of this approach is the use of existing solar cells for the energy conversion.

Session 8: Modeling and Characterization of TPV Systems

Chair: D.J. Krommenhoek,Lockheed-Martin

Foremost Observation

Modeling is now more representative of the real situation. Simulation of TPV systems includes detailed descriptions of the components that comprise the heat input, radiator photon source, cavity geometry between the radiator and TPV semiconductor cells, spectral control and the TPV cell array. System models serve as the integrators for the solid state physicists, electrical engineers, materials scientists, chemists and design engineers to discern the impact of distinct technology efforts on overall performance. Predictions for proof of principle system tests performed with today's technology are realistically in the 1 to 6% efficiency range - this agrees with test data.

Key System Issues

Thermal management dominates the system efficiencies. Hence, thermal issues emerged as the dominant area to improve system efficiency. These issues are in response to the following physical effects. First, photon radiator birth temperatures turn out much less than combustion flame temperatures due to combustion,thermal convection and radiation losses. Second, photon losses after birth in the cavity and before reaching the cells show the significant impact of cavity optical properties and geometry. Third, incomplete spectral control recuperation with today's selective emitter and filter technology show the need to improve photon recycling. Finally, nonlinear and unexpected cell electrical outputs show the important feedback between non-uniform radiator temperatures and reflective TPV cells. TPV cell quantum efficiencies, voltage factors and fill factors are >65% - however, inefficient thermal management for the combustor, cavity, spectral control devices and radiators limits system efficiency to single digits with today's technology.

Key Analysis Finding

A key thermal system ingredient is the photon tumult in the cavity, specifically, the tracking of photon interactions from birth at the radiation source until they escape or are absorbed. Absorption depends on the energy dependent reflection for every geometrical surface within the cavity. Modeling the "cavity efficiency" included: (1) detailed Monte Carlo photon transport based on photon energy, location, direction and geometry; (2) view factors with wavelength dependent optical properties of components; (3) photon ray tracking formulations with configuration factors; and (4) comparisons of data for selective or broad band photon radiators and for cylindrical or flat geometries. Photon effects are non-intuitive - small changes in optical properties and geometric photon leakage paths result in significant variations in electrical output.

Future Work

Based on the thermal issues, future work will concentrate on combustion, cavity, spectral control, geometry and cell modeling to optimize electrical output. TPV systems analysis should capitalize on its small size compared to solar systems at the same power. New efforts should be initiated to include photon wave effects in thin films, diffuse photon radiation, cell grid photon feedback, and economic modeling to determine tradeoffs in spectral control methods, cost of cells and hardware. For the first time, important system parameters have been articulated - this will focus technology and should drive system efficiencies into the teens to yield cost competition TPV systems.

TPV Market Review

Steve Johnson, Sabrina, Inc

A market survey was performed to identify and quantify potential markets for Thermophotovoltaic (TPV) devices.

In order to assess potential markets it was necessary to develop a product specification for a hypothetical TPV device. These specifications were as follows:

1. Variable power output capability 500 to 5000 watts.

2. Efficiency equal to approximately 10%.

3. Cost equal to approximately $1 to $2 per watt.

4. The product was defined as having the following dimensions:

    Width = two feet

    Length = two feet

    Height = three feet

Market Segments

Four primary markets were identified for a TPV device possessing these properties. These markets are as follows:

1. Recreational Vehicles (RV's)

2. Remote Homes

    - USA

    - Canada

3. Uninterruptible Power

4. Military

Recreational Vehicles

According to the National Recreational Vehicle Association, there are 100,000 generators sold per year into this market. Discussions with the various manufacturers indicated that a moderately priced TPV device could anticipate capturing approximately 20% of this market. These sales would be primarily into the most expensive class of RV's which appeal to those consumers with the largest dispensable income. This volume could potentially represent approximately $80 million in annual sales.

Remote Homes

Throughout North America, there are large numbers of homes, which are not currently connected to the utility electrical grid and due to their remoteness, most likely never will be. The Solar Energy Industry Association estimates that there are approximately 150,000 homes in the US, which fall into this category. Competitors to TPV in this category are gasoline generators and photovoltaics. The quiet operation and low maintenance of the TPV device makes it preferable to the gasoline generator, while its lower capital costs will make it preferable to some over the Photovoltaic system. In the US market, it is anticipated that 50% to 80% market share can be achieved representing approximately $280 million in potential sales.

In Canada, due to its size and remoteness, the total number of remote dwellings is greater. Estimates of the number of remote homes range as high as 350,000. However, due to the fact that many of these remote dwellings are not used during the winter and that many are used by Canadian Indians, who most likely could not afford a TPV device, the total market Penetration will only be approximately 20% or $280 million in potential sales.

Uninterruptible Power Supplies

Internationally, reliable power represents a very significant market. Sales into this market currently are at one billion dollars annually. Unlike the US electrical grid, much of the world experiences frequent brownouts and blackouts that can last for hours, or even days. These extended blackouts are beyond the capacity of batteries and require supplemental power generation. Currently, this power is being supplied by diesel and gasoline generators. The low noise and low maintenance characteristics of TPV will make it attractive to this segment. However, due to its higher cost versus conventional generation, it is anticipated that initial market share will only be .05%. Due to the large size of this market segment, this still represents $50 million in annual sales. Additionally, this market is a rapidly growing one and offers significant sales potential well into the future.

Military

Several specific applications have been identified by the military for a liquid fuel powered TPV device. A conservative estimate of this market is $25 million in annual sales.

Competitive Technologies

Base Load

Due to the theoretical limitations of TPV, the applications for this technology will generally be in applications of 5 kW's or less. Above that size, such technologies as diesels, combined cycle gas turbines, and fuel cells, have efficiency or lower capital cost advantages over TPV.

Internal Combustion Engines

Below 5 kW's, small diesel or gas internal combustion engines (ICE) and thermoelectrics represent TPV's primary competition. Although TPV is more expensive than conventional ICE's, its quiet operation, lack of vibration, and low maintenance characteristics, offer significant advantages.

Thermoelectrics

Thermoelectrics are highly developed and have niches where reliability is an issue and efficiency is not. As the TPV device has been defined, the thermoelectrics would be quickly displaced by TPV in most applications.

Summary of an Open Discussion: A Coordinated National Program for TPV
Led by: Holly Thomas
National Renewable Energy Laboratory

As part of earlier conferences, NREL conducted discussion groups to review the need and possibilities for a national TPV program and recommendations for such a program. Since the 1995 Conference, considerable additional interest in has been expressed in TPV, spurred by technical advancements, product development and new military requirements. Reflecting this growing interest, NREL established a Business Development Team [1] to review the status of the technology as a whole and asses the potential for a coordinated national plan to focus additional funds into TPV and raise the general awareness of the potential for this technology. The team incorporated the comments of industry to develop a proposed structure for a program, and the Third Conference provided a forum to present ideas and discuss opportunities.

The proposed objective for a coordinated, dual-use national program would be the establishment of a nationally coordinated TPV program to facilitate cooperation among researchers, system developers, component manufacturers and funding sources to rapidly develop a robust U.S. industry. It would be conceived from the needs of the market, with development focussed to support the market-identified requirements, so advancements would be rapidly utilized by industry.

There are several key aspects for the program. Performance standards are required, as well as system and component testing and characterization. Additional technical development of systems and components that incorporate both commercial and military requirements are another key element. And finally, a vigorous program for subcontracted work, to include universities, and a plan for manufacturing development, are needed. These elements could be modeled after the approach used in terrestrial PV. In this program, a national laboratory research program works in partnership through subcontracts to universities and companies. Manufacturing improvements and process developments are advanced through a cost-shared program with industry to develop and maintain a U.S. commercial market share. The basis for all the work in a national program would be the TPV market. Those applications and related requirements would determine the developments required to rapidly develop a strong, internationally competitve industry.

The proposed technical elements of the program include the following:

Standards and characterization
- reliability and stability
- testing and performance assessment
- performance standards

Energy Supply and Recovery
- burner development
- alternative fuels
- heat management

Device (Converter) Development
- new designs and materials for commercial applications
- next generation converter design and process development

Photon Management
- selective filters
- back surface reflectors

System Design and Development

Manufacturing Technology
- cost-shared development of manufacturing methods and processes to result in increased capacity and lower costs

Market Research and Development

The intention of the discussion was to obtain comments and feedback from the industry about the content and next steps. Generally, those comments can be summarized as follows:

1. Incorporate commercial needs, as these are different from military requirements. The combined effort of meeting military and commercial needs will speed development and utilization of TPV.

2. TPV is where some other technologies were 15 -20 years ago, with a matrix of problems. A program grouped to address specific issues might yield the most rapid progress. A key element to the program should be an assessment of the market for TPV. Markets include military applications, satellite/non-terrestrial uses and commercial markets for small power supplies for uninterruptable power and off-grid uses.

3. Systems development support is an important area that is and should be included in any plan. Demonstrations of operational systems and bulk purcneglected hases would help establish the industry.

4. Solar PV is part of a Solar Energy Association that promotes the needs and interests of solar- related businesses. A similar organization for TPV interests may be warranted.

5. Focus on particular issues related to TPV in order to rapidly achieve resolution.
6. More participation from companies who will be marketing TPV products would be useful in future discussions.

Combined industrial, educational and military participants comments are currently under review. NREL has concluded that there is a strong interest in exploring support for testing, characterization and standards. Other areas generated a high degree of interest as well. In the coming months a coordinated effort to develop additional funding for TPV, supported by combined industry and government interest, is planned.

---

[1]Team members were Tim Coutts, John Benner, Holly Thomas, Ed Witt, Dave Christansen, Mark Fitzgerald, Steve Johnson and Gary Smith.

SESSION 1:
OVERVIEWS OF TPV

Overview of European Activities in Thermophotovoltaics

Markus Schubnell*, Hansjoerg Gabler† and Lars Broman‡

*Paul Scherrer Institute, CH-5232 Villigen, Switzerland
†Fraunhofer Institute for Solar Energy Systems ISE, D-79100 Freiburg, Germany
‡Solar Energy Research Center, Högskolan Dalarna, S-78188 Borlänge, Sweden

Abstract.

In Europe several research institutes and private companies are currently working or have been working in thermophotovoltaics. At the Solar Energy Research Center SERC in Sweden, the possibility of using wood as a fuel for a TPV-system emphasizing the aspect of co-generation of heat and electricity is investigated. SERC also does work on characterization of TPV cells and of TPV generator optics. The Fraunhofer Institute for Solar Energy Systems in Germany plans to develop a TPV-generator to assure the supply reliability for small off grid photovoltaic systems and in Switzerland the research foundation of the Swiss gas industry has approved a project proposed by the Paul Scherrer Institute to develop a residential TPV-gas-burner for co-generation of heat and electricity. We also report on activities in the United Kingdom as well as on a past research program conducted in the Netherlands.

INTRODUCTION

Thermophotovoltaics (TPV) as an option to convert thermal radiation into electricity has been suggested some 35 years ago [1] as a result of the fundamental limitation of the conversion efficiency of broadband radiation into electricity by means of photovoltaic cells. The basic principle consists in utilizing only a selected part of the thermal radiation that is matched to the bandgap of the cell. The remaining radiation is kept within the system which is therefore heated up and kept at a generally rather high equilibrium temperature. Selective filtering of the thermal radiation is achieved by a combination of a selective emitter and a filter. Thus, a thermophotovoltaic generator consists of a heat source, a selective emitter/filter system and photovoltaic cells. An overview on the work done in the field of thermophotovoltaics can be found in [2].

On purely thermodynamic grounds, i.e. assuming ideal performance of the

CP401, *Thermophotovoltaic Generation of Electricity: Third NREL Conference,*
edited by Benner/Coutts

system and its components, TPV is a very attractive concept allowing thermal to electricity conversion efficiencies close to the Carnot efficiency. However, real life components detract the estimated TPV performance dramatically. As an example we mention the thermodynamical limit of an ideal solar powered TPV-system of which the conversion rate of the input concentrated radiation into electricity has been calculated to 85 % [3]. However, considering available components the expected conversion efficiency of a solar TPV converter using Si-cells is expected to be around 30 %. If this is compared with the most advanced Si-cells, which convert about 20 % of the solar radiation into electricity, the advantage of a solar TPV-system over flat Si-cell modules is no longer evident.

Nevertheless there is presently a considerable renewed interest in TPV mainly by two reasons. First, several niches have been identified, where TPV-power generators might be a viable option. This includes applications such as low electric power systems (10 - 50 W, e.g. for space power systems [4]) or self powering residential heating systems, where the aspect of cogeneration of heat and electricity is most important [5,6]. Second, there has been considerable technical progress in the development of low bandgap PV-cells [7], and in the design of efficient selective emitter and filter systems [8,9].

In this paper we present an overview on recent thermophotovoltaic activities in Europe.

TPV IN NORTHERN EUROPE

Introduction

A joint project between US National Renewable Energy Laboratory NREL, Swedish University of Agricultural Sciences SLU, and Solar Energy Research Center SERC at Högskolan Dalarna, Sweden aims at building a wood powder fuelled TPV generator. Swedish participators are Jorgen Marks SLU, and Lars Broman and Kenneth Jarefors SERC. The work started with a thorough literature search [2].

Sweden's most promising renewable energy source for substantial growth during the next decade is biomass, especially from trees. It is believed that the present bioenergy ratio of 15 % of the country's total energy use may double within 10-15 years. Several different wood fuels are used including processed fuels such as chunks, chips, pellets, etc. The most refined fuel, wood powder, has a high energy density, 20 MJ/kg dry matter, and it is combustible also in existing oil furnaces (in the range 1-15 MW) with little alteration [10].

Plants for co-generation of heat and electricity in the multi-MW range are increasingly being built in Sweden, feeding hot water into city heating systems as well as electricity into the electric grid. A wide variety of fuels are being employed, and the most common electricity generating process is a generator driven by a Rankine turbine.

Wood powder might have the prospects of becoming an excellent renewable energy source for co-generation of heat and electricity also in the 10-1000 kW range, utilizing thermophotovoltaic conversion in water cooled TPV cells. Since a TPV converter works without moving parts (except water and air pumps), an automatically working system is envisioned.

Thus, much of the initial work in the present project has dealt with the problem of combusting wood powder in such a way that a stable flame temperature of about 1200 °C is maintained [11,6]. In a prototype pilot-scale burner that includes feeder and combustion chamber, this goal was achieved. In co-operation with Borlange Energy, Ltd., a new project has started, aimed at installing and testing wood powder burners, rated in the order of 1-10 MW, in the company's existing plants, which produce hot water for the city's district heating systems.

Biomass Fuel

There are some different biomass fuels capable of producing a temperature of about 1200 °C in the flame, especially gas and liquid fuels of different kinds. When wood is the basis of the fuel, gasification or liquefying however consumes a higher percentage of the initial energy content than grinding and drying. This makes making wood powder, the only sufficiently uniform solid wood fuel, interesting for high temperature combustion [10]. When upgrading wood fuel to a very fine and dry fraction, like wood powder, the result is a uniformed fuel, easy to handle and with a high energy content. Wood powder has a low heat value around 18.3 MJ/kg. Wood powder will be burned in the same way as oil or gas and at a combustion temperature over 1100 °C. In principle, all heating done by oil can be replaced by wood powder even in small boilers in limited space and in densely built-up areas.

The energy spent in upgrading is fully compensated for through more efficient handling and combustion. As with the energy balance, the costs of production are recovered in the combustion. Higher combustion efficiency also results in reduced fuel consumption. Other factors, such as the use of a cheap raw material and a large market with high annual utilization, also have a positive influence on the economy.

Outline of a TPV Generator Fueled with Wood Powder

Burner - Emitter

When used in connection with TPV cells, the powder must be burned in such a way that most of the energy is released as radiation rather than as hot exhaust gases, and that the temperature is sufficiently high. The temperature goal is 1200 °C, a compromise between most useful emitter spectrum and low NOX production.

The development of a TPV converter that uses combustion of wood powder as energy source has started with development of the combustion source. SLU/SERC's plans for continued development of this and other components of the converter include the following: Reliable feeding, and flame ignition and control, are just some of the mechanisms that have to be developed. A feeding mechanism and a combustion chamber that seem very promising have been constructed. A 10 kW flame can be kept steadily burning for several minutes at the time and about 1200 °C has been reached in one of the two prototype burners. The next step in the development process is to further increase the radiation fraction, and direct the hot exhausts through a radiative emitter.

Converters

Simultaneously, development work is being carried out on other parts of the TPV generator [12]. Four converters for testing have been built at NREL. They are 0.6 eV $In_{0.65}Ga_{0.35}As$ epitaxially grown lattice mismatched on an InP substrate using MOVPE. The metallizations and sawing were done at ASEC and the 2-layer antireflection coatings were applied at NREL. The converters are mounted onto 50*50*2 mm^3 copper plates (electroplated with gold) by means of electrically conductive epoxy.

Modeling is in progress; internal quantum efficiency data have been measured at NREL and reflection losses have been measured at SERC, typically being about 10 %. The diagram in Figure 1 shows absorption, internal quantum efficiency and external quantum efficiency for one of the cells. It could be worth noting that the reflectance does not increase at wavelengths over that corresponding to the bandgap (2.0 μm). An efficient edge filter that reflects back the radiation over 2 μm to the emitter (or a selective emitter) is therefore necessary for high converter efficiency.

Optics

Calculations on ideal filters [13] have shown that optimum bandgap E_g of the converter is strongly dependent not only on the (greybody) emitter

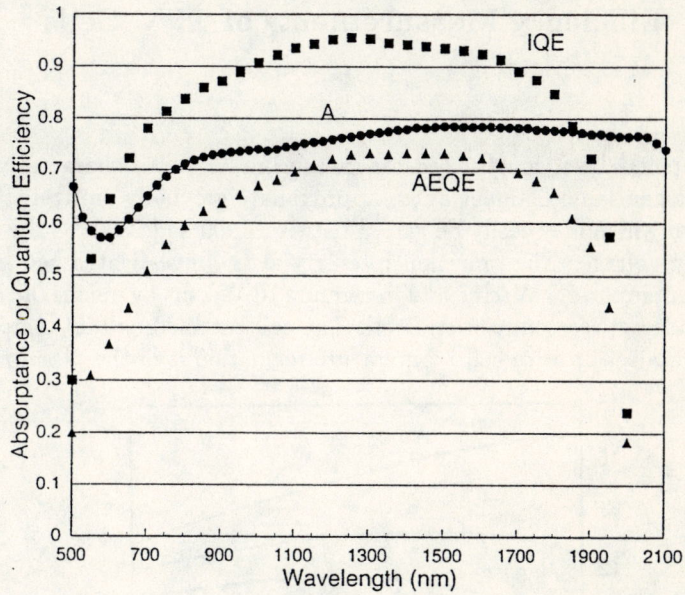

FIGURE 1. Absorption, internal quantum efficiency and external quantum efficiency for one of the four cells that are being studied.

temperature, but also on the efficiency of the edge filter between the emitter and the converter. As an example, $E_g = 0.50$ eV is optimum if the filter's reflectivity above the corresponding wavelength is 90 %, while $E_g = 0.38$ is optimum if the filter reflectivity is only 70 % for an emitter temperature of 1200 °C.

A setup for testing the chain IR emitter - selectively reflecting filter - TPV cell is in the process of being designed. During the coming year, making and investigation of multiple layer dielectric filters is in the planning. In order to limit the region of filter incident angles - which will make the edge filter act more efficiently - a special geometry of the internally reflecting tube that transmits the radiation is considered: a tube in the shape of two V-cones joined at their large apertures and with the filter perpendicular to the cones' axes between the cones. As an example, $C = 4$ cones will make almost all rays reach the filter at incidence angles less than 30 °C. An experimental setup for studying such geometries has recently been constructed.

Efficiency Measurements of TPV Cells

A set-up has been constructed for measuring TPV cell characteristics. It uses a halogen lamp filament as an approximate greybody emitter, the temperature of which may easily be varied between 1200 and 2300 °C by varying the supply voltage. The radiation intensity at the investigated cell is varied between minimum 0.3 W/cm² and maximum 10 W/cm² by means of changing the distance between emitter and cell. The cell is cooled and its temperature monitored; a set-up with cell temperature regulation is in the planning.

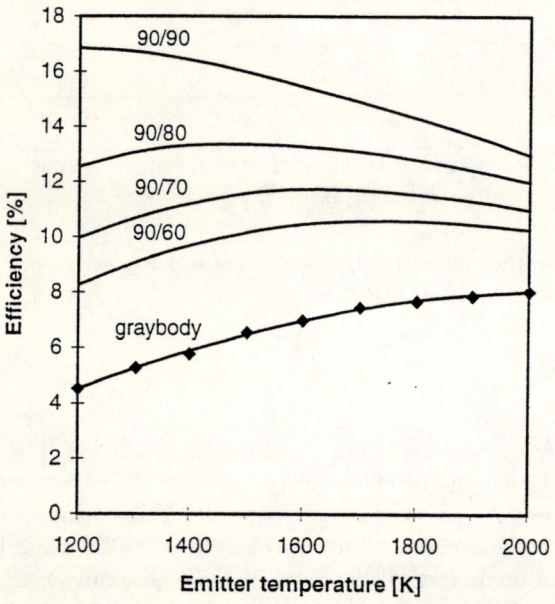

FIGURE 2. Efficiency of a typical TPV cell measured for graybody spectrum and calculated for filtered spectra.

The first four cells that have been characterized are the 0.6 eV $In_{0.65}Ga_{0.35}As$ thin film cells grown at NREL. One series of efficiency measurements on one of the cells is shown in the diagram, Figure 2. Incident radiation intensity is 1 W/cm², and the lowest curve shows how the cell efficiency varies with emitter temperature. The top four curves show the efficiency of the cell combined with a hypothetical edge filter. All filters are assumed to have 90 % transmittance of photons above 0.6 eV. They reflect back 90, 80, 70 and 60 percent of photons below 0.6 eV respectively to the emitter. This work will be reported at this conference [14].

Concluding Remarks

A Swedish TPV R&D program is in the planning, aiming at co-generation of electricity and heat from refined wood fuel. Many components of the TPV converter; combustion chamber, emitter, optical path design, dichroic filter, and TPV cell mounting and cooling, are considered. Close co-operation with researchers at NREL, working on thin film TPV cell development, is an important aspect of the program.

Work at SERC is so far not appropriately financed, since it has not yet been able to convince Swedish National Research Councils about the usefulness of TPV for power production. To be of interest, the R&D efforts have to be directed towards plants in the size of 1 MW_e or larger. The probability of m^2-quantities of TPV cells, with high efficiency and long lifetime, being produced and marketed within the next 5-10 years must be shown to be large if any substantial funds shall be provided.

TPV IN THE UNITED KINGDOM

Cogeneration of heat and electricity by TPV

Together with various partners (Morgan Material Technology, Gwent Electronic Materials and The University of Warwick) the British Gas Research & Technology Centre is currently developing a radiant burner system incorporating rare earth oxides to achieve selective emission in a three years research program. The project commenced in October 1995 and its total budget amounts to 656'000 £ of which 45 % are funded by the UK Department of Trade & Industry. To estimate the system performance a detailed computer based TPV-burner and TPV-system model have been constructed. According to a system design and costing study a sub 1 \$/W capital costs potential has been identified for the system. The experimental work is clearly focused on realizing a selective emitter, assuming that PV-cell technology is available or will be made available in the near future by others. Actually, the work is not restricted to a special cell technology although at the present stage cost arguments favour Si-cells. Several coating technologies and burner materials are under development. Work is underway to design and demonstrate the burner system. By the end of the project the feasibility of a 300 W electrical output module will be demonstrated. By commercial confidentiality reasons no more technical details can be given at this stage.

Low bandgap TPV-cells

In October 1996 a joint project between the Clarendon Laboratory (University of Oxford), the Northumbria University and the Surrey University has started to develop a TPV cell based on relaxed InGaAs layers grown on GaAs. It is expected that InGaAs grown mismatched on GaAs substrate is capable of providing efficient TPV cells at low costs and at a potentially more manufacturable volume than e.g. InGaAs on InP. Three different surface lattice constants are aimed for different wavelength cut-off for operation with and without spectrally narrow emission that can be obtained from selective emitters (1000 nm, 1500 nm and 1800 nm). So far the plastic relaxation of strained $In_{<0.25}Ga_{>0.75}As$/GaAs grown by MOCVD has been verified. In a next step p-n junction cells on a GaAs substrate to work at 1000 nm shall be fabricated.

TPV IN THE NETHERLANDS

Introduction

In 1993/1994 the Dutch Energy Research Foundation (ECN) has conducted a study by order of the Dutch Agency of Energy and Environment (NOVEM) on the potential of thermophotovoltaic cogeneration of heat and electricity. The main goal of the study was to be conclusive on the technical and economic feasibility of a thermophotovoltaic central heating unit using Si-cells and ceramic foam radiant burners as developed at ECN. The design goal of such a system has been set to a thermal input of about 30 kW at a share of about 10 % electrical output.

At ECN radiant burners have been developed on the basis of highly porous ceramic foam materials with a typical porosity of about 90 %. Apart from surface radiant burners delivering about 25 % radiant output also multilayer burners have been developed, where the combustion of the premixed air/gas-mixture occurs within the top layer, which is a black coated ceramic foam with relatively large pores. To prevent flashback a second foam layer with smaller pores is used as a flame arrestor. With multilayer burners radiant outputs of up to 50 % have been achieved. Even higher radiant outputs (up to 70 %) have been achieved in a multistep multilayer burner as shown in Fig. 1 [15,16].

Selective radiant burners for TPV

Experimental work on TPV has been focused on tests with variously modified ECN standard burners. ECN's standard ceramic foam burner is mainly made of SiO_2 and Al_2O_3. To protect the burner plate from overheating it is coated with a black layer, thus radiating in a broad wavelength band. To achieve a better spectral match to the Si-cells standard foams coated with Yb_2O_3, foams made of SiO_2, Al_2O_3 and Yb_2O_3 (at various ratios), foams made of clay and Yb_2O_3, foams made of yttria stabilized ZrO_2 and standard foams with a highly emissive coating each at different pore sizes (0.3 mm up

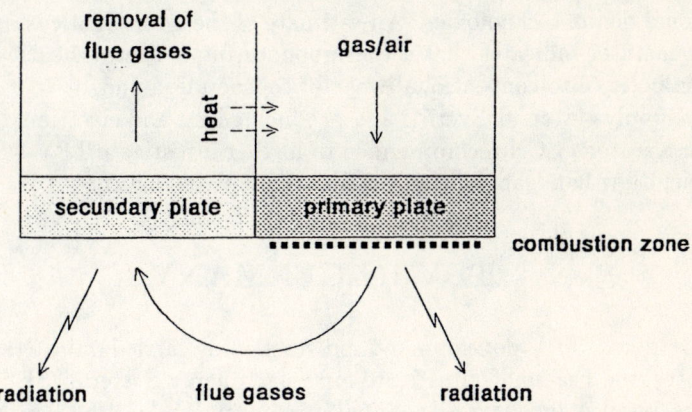

FIGURE 3. Operating principle of a multistep burner. The flue gases are used for extra radiation output from a secondary ceramic foam plate and to preheat the gas/air mixture at the inlet of the burner.

to 2 mm) have been tested. The burners have been operated with methane at about 5 % to 10 % air excess, and at loads between 100 kW/m² and 600 kW/m² yielding surface temperatures between 1000 °C and 1400 °C. In all experiments the radiant output in the wavelength band relevant for Si-cell operation (i.e. at wavelengths < 1.1 μm) was below 5% of the input fuel enthalpy which resulted in a maximum electrical conversion efficiency of about 0.5 %. For the conversion of the radiant output ECN used common Si-cells with a solar efficiency of about 12 % as well as concentrator Si-cell with an efficiency around 20 % at 20 suns.

The influence of Yb_2O_3 on the emissive properties of the foam could clearly be demonstrated: as compared to a standard ECN ceramic foam burner an Yb_2O_3 coated standard burner, an Yb_2O_3/clay burner and a $Yb_2O_3/SiO_2/Al_2O_3$ burner yielded a 3, 22 and 22 times higher radiant power output below 1.1 μm. Apart from filters protecting the PV-cells from the IR-part of the radiation no further effort has been made to increase the selectivity of the burner by using dedicated selective transmitting/reflecting filters.

Conclusion

In conclusion ECN found that using Si-cell technology a 10 % electrical conversion efficiency can only be reached if emitter temperatures of about 2100 °C can be achieved. However, this high emitter temperature implies severe mechanical, chemical and thermal problems to the emitter, and the ceramic foam technique as developed at ECN cannot be used in this temperature range. Furthermore, the expected high NO_x-output at this temperature requires additional denox-technologies. An estimate of the additional costs due to the TPV-generator indicates that a thermophotovoltaic central heating system is not competitive to conventional ways of cogeneration and/or to the common energy supply system delivering and producing heat and electricity separately. By these reasons ECN recommended to further investigate TPV-systems only if efficient low-bandgap cells are available at reasonable costs.

TPV IN GERMANY

In Germany, TPV development and research is carried out at two research institutes, the Fraunhofer Institute for Solar Energy Systems ISE in Freiburg and the Institut für solare Energieversorgungstechnik (ISET) in Kassel. As ISET does not publish reports on TPV development upto now, the remainder of this section reports on the work at FraunhoferISE.

Introduction

Remote electrical loads, located far away from the public electricity grid, may be supplied by means of photovoltaics. Especially where solar radiation varies significantly throughout the year, these PV power supplies should be combined with a fossil fueled generator. Such PV hybrid systems have the advantages of higher reliability and reduced installation costs. Due to their projected properties (high reliability and low costs for maintenance), TPV generators have an excellent potential to serve as auxiliary generators in small stand-alone PV systems. Therefore, the work at Fraunhofer ISE aims at the development of a small (50-100 W) generator which can be used for this purpose.

Description of the demonstration system

A laboratory demonstration system [17] has been set up and is currently being tested. The system consists of the following parts:

- A propane burner with a thermal power between 1 and 2 kW. Two different layouts have been developed and tested at Fraunhofer ISE, a radiation burner and a burner with a combustion chamber. Propane and air are premixed in both cases. Emitter temperatures of more than 1100 °C have been reached with both designs.

- An emitter plate. The currently investigated materials are silicon carbide (SiC) and high temperature-resistant metals.

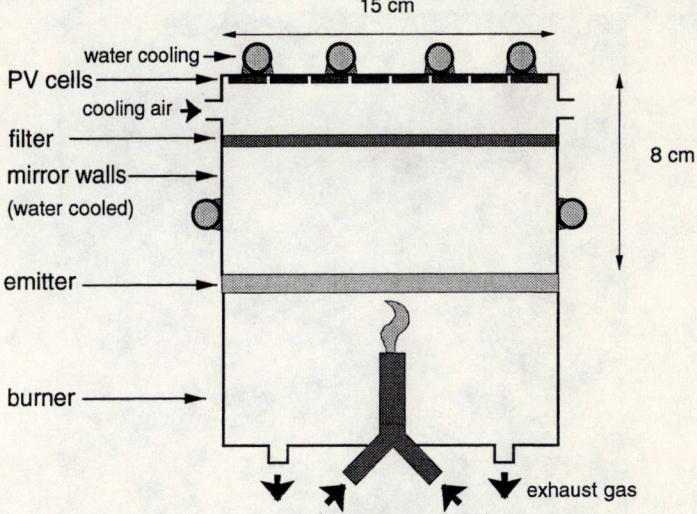

FIGURE 4. Vertical cross section of the TPV demonstration experiment at Fraunhofer ISE.

- A quartz window with a selective plasma-interference filter coating, reflecting the radiation with $\lambda > 2\mu m$ back to the emitter. The coating has been developed and is currently fabricated at Fraunhofer ISE.

- A PV module consisting of 32 GaSb cells fabricated at Fraunhofer ISE with an active area of 1.66 cm^2 per cell, mounted on a water cooled copper plate.

Emitter, filter and module are circular plates with 15 cm of diameter. The lateral walls are reflective in order to minimize radiation losses. The whole system is water cooled, additional air cooling is used for filter and PV module. A schematic view of the system layout is shown in Fig. 4.

Fig. 5 shows the demonstration device during a test measurement. For an emitter temperature of approximately 1100 °C, an electrical power density PV of 150 mW/cm^2 cell area has been reached.

Modelling of the Energy flows in a TPV system

To predict energy flows and temperatures in the TPV device, two simulation tools have been developed [18].

A quick, approximative response is given by a one-dimensional model which takes into account radiation and heat conduction inside the cavity. A simplified geometry without side effects due to the lateral walls is assumed. The

13

FIGURE 5. Photograph of the TPV demonstration system at Fraunhofer ISE

spectrum is divided into two ranges separated by the wavelength corresponding to the bandgap E_g of the GaSb cells ($E_g = 0.7$ eV). The calculated net radiation and conduction flows between emitter, filter and PV array are shown in Fig. 6.

To get a more realistic idea of the radiation inside the cavity in three dimensions, a ray tracing program is used, closely modelling the real geometry of the test device and calculating the distribution of the radiation intensity on the cell array, the absorption in the cavity walls and in the filter and the reflection back to the emitter. The results are combined with those from the one-dimensional model.

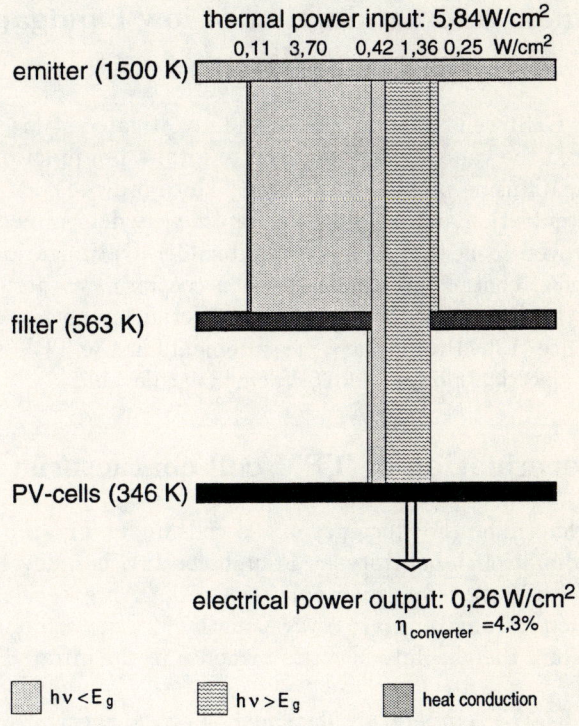

FIGURE 6. Net energy flows in the TPV device at Fraunhofer ISE as calculated by a one-dimensional model. The spectral range is divided into two parts ($h\nu > E_g$ and $h\nu < E_g$). A greybody emitter ($\epsilon = 0.85$), an actively cooled filter with a reflectivity of 0.8 for $h\nu < E_g$ and a transmission of 0.8 for $h\nu > E_g$ and PV cells with $\eta=20\%$ for radiation with $h\nu > E_g$ are assumed.

Emissivity measurement

The Fraunhofer ISE has developed a setup to measure directly the angle- and wavelength-dependent emissivity of emitter materials at temperatures of interest for TPV.

The samples (sheets of 2.5×2.5cm^2) are placed in an evacuated environment. They are radiation-heated from the back by a tungsten heating foil. The radiation from the front side is coupled into a Fourier interference spectrometer which gives high spectral resolution. The temperature is determined with a tungsten-tantalum thermocouple. Absolute measurements are possible by comparison to a built-in blackbody radiator.

With this setup the emissivity can be measured within a temperature range from 700 °C to 1200 °C, a spectral range from 1 μm to 20 μm, and an angular range from 0° to 60°.

Development and production of low bandgap GaSb cells

2.1 x 0.9 cm^2 GaSb cells were developed and investigated at Fraunhofer ISE. Zn diffusion from the vapor phase was used to form a p-n junction in n-GaSb substrates. The diffusion process parameters, which provide good control over Zn surface concentration and p-n junction depth, were determined. Thus, this method has proved to be simple and reproducible. Optimization of the cell design (Zn profile, contact grid, antireflection coating) was performed. The parameters of the GaSb cells are described in detail in another contribution to this conference [19]. Preliminary measurements in the TPV system have shown that an electrical power of 0.25W/cm^2 is achievable.

Determination of TPV cell characteristics

In order to determine the efficiency of PV cells under TPV operating conditions, the calibration laboratory at Fraunhofer ISE has developed a new method of I-V-calibration for this application [20]. An extension of an existing grating monochromator up to a wavelength of 2 μm was done in order to be able to measure the absolute spectral response in the infrared region with high accuracy.

For GaSb cells, the temperature dependency of the spectral response and the cell parameters were modelled using a semi-empirical description of the absorption coefficient.

TPV IN SWITZERLAND

Introduction

In Switzerland, 62 % of the electricity is generated by hydropower, 36 % by nuclear and 2 % by fossil fuels. Due to the non acceptance of nuclear power the existing nuclear power plants in Switzerland will probably not be renewed which leads to an electricity generation deficit of at least 3 GW by the year 2020. Electricity generation by an extended use of natural gas is the most probable scenario. In this scenario the use of a thermophotovoltaic central heating systems might contribute to the Swiss electricity supply system. This concept is attractive by several reasons. Usually residential need for heat is much higher than for electricity (on an annual basis households in Switzerland spend about 215 PJ of heat and 13 PJ of electricity). Therefore, a system delivering electricity at a low conversion efficiency and keeping the rest of the energy input as heat matches the actual residential energy demands. Furthermore, electricity is available when heat is required, i.e. during day time and in winter and when, accordingly, electricity utilities are usually tendentially short of supply (this holds at least for mid European countries). Finally the fuel input in a cogenerating system is revalued from e.g. in Switzerland 5 cts/kWh gas to 20 cts/kWh electricity which is an attractive economic impetus if the system can be designed at appropriate costs.

Thermophotovoltaic Cogeneration of heat and electricity

In 1996 Paul Scherrer Institute (PSI) has elaborated a proposal to investigate the technical and economic potential of a thermophotovoltaic central heating system (TCHS). This proposal has found the interest of official institutions as well as of private companies. Financial support has finally been approved by the Research Foundation of the Swiss Gas Industry (FOGA) and by a private company (Hovalwerk AG). The total project budget amounts to 920'000 sfr of which 30 % are funded by the partners. The program is of three years duration and has commenced in October 1996. The objective of the TPV-project at PSI is to demonstrate the feasibility of a TCHS with a thermal power output of about 20 kW and some 500 W electricity share. Thus, the main interest is not to reach a high electrical conversion rate but to build a complete TCHS, to identify possibilities to improve the electrical share at the output and to learn about the technical and economic potential of such a system. Due to mainly cost reasons our TPV-generator will relay on Si-cells which implies the use of Yb_2O_3-selective emitters.

Surface temperature measurements

At PSI a radiometric multiwavelength temperature measurement technique has recently been developed to measure the temperature of highly irradiated surfaces (e.g. in a solar furnace) [21,22]. The method bases on measuring the brightness of an absorber and of a reference target with known reflectivity at various wavelengths with a carefully calibrated CCD-camera. This is done with and without an additional light source of known intensity such as a powerful flash. From these images the reflectivities of the absorber at the wavelengths considered as well as the surface temperature and the irradiance can be evaluated. The same technique has been successfully applied to highly emissive hot emitters as well. As an example we show in Fig. 2 the temperature distribution on the mantle of a butene fueled lamp as it is widely used by campers. The temperature distribution clearly shows the mesh structure of the glowing mantle, the cold spots being the holes of the mantle.

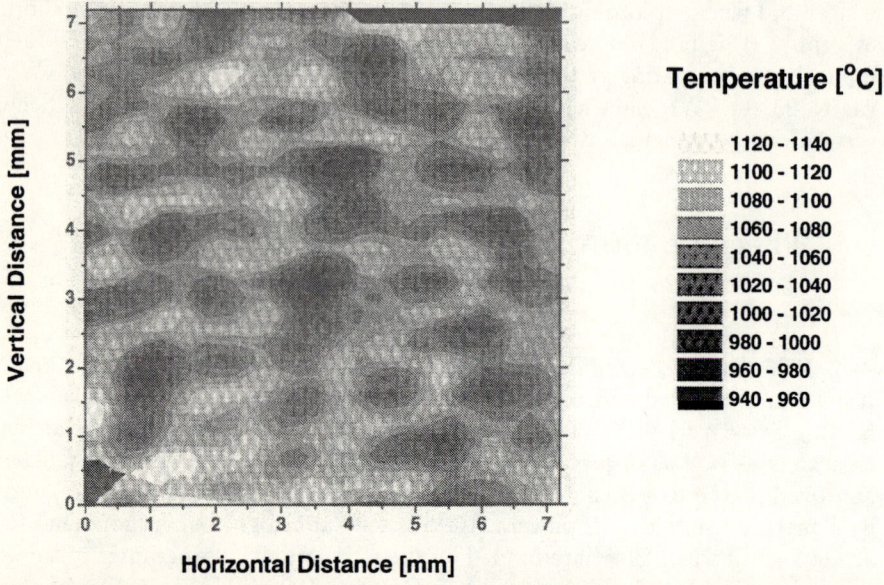

Temperature [°C]

- 1120 - 1140
- 1100 - 1120
- 1080 - 1100
- 1060 - 1080
- 1040 - 1060
- 1020 - 1040
- 1000 - 1020
- 980 - 1000
- 960 - 980
- 940 - 960

FIGURE 7. Temperature distribution on the glowing mantle of a butene fueled lamp; spatial resolution 0.225 mm per pixel.

Characterisation of TPV-cells

To monitor the performance of either single solar cells (down to sizes of 3 mm by 3 mm) or modules under realistic conditions a PC based measurement system has been developed [23]. The system allows to measure current-voltage characteristics, ambient and cell/module temperatures and irradiances (direct and global normal) onto the cell. Based on these data the maximum power point, the conversion efficiency and the temperature coefficient are calculated. To use this system for thermophotovoltaic purposes it has to be slightly modified. The modifications mainly concern the electronic loads and the irradiation measurements. Instead of using spectrally not selective radiometers we plan to include the acquisition of spectrally relevant irradiation data extending to wavelengths up to 3 μm.

System design considerations

With a simple model and assuming optimistic selective emitter/filter properties we have estimated an upper limit of the electricity conversion efficiency of about 5 % (c. f. table 1). Even this rather low efficiency can only be reached if a recuperative burner system is used and if emitter temperatures of about

Thermal input power density	150 kW/m^2
Combustion gas temperature	1500 °C
Emitter temperature	1400 °C
Inlet temperature of air/fuel mixture	350 °C
Broadband absorptivity of the filter	0.1
Transmittance between 700 nm and 1100 nm	0.9
Off-band Reflectivity	0.9
Fuel/air ratio	1.05
Cell efficiency	0.56
View factor	0.8
Conversion Efficiency	5.1 %

TABLE 1. Characteristic data of a Si-based TCHS. Losses due to convection and conduction are not included.

1400 °C can be maintained. Utilizing low bandgap TPV-cells would approximately double the electrical share. Experimental work has been focused so far on investigating the optical properties of ytterbia emitters. For this purpose the influence of ytterbium concentration on the emissivity of ytterbium doped YAG and yttria has been studied. Results of this work will be reported at this conference [24]. Future experimental work will include the development of a low thermal load burner (typically 150 kW/m^2) with a high radiative output that can be operated at temperatures up to 1400 °C.

SUMMARY

TPV-projects in Europe mainly focus on small scaled heat and electricity cogenerating systems. As a summary we have compiled an overview on the European TPV activities in table 2.

ACKNOWLEDGEMENTS

This work has been supported by the Research Foundation of the Swiss Gas Industry (FOGA). M. Schubnell wishes to thank A. van der Drift and G. J. J. Beckers from ECN (Netherlands), A. Jickells from British Gas Research&Technology Centre (UK) and D. J. Dunstan from Queen Mary Westfield College for information on their TPV-activities.

Country	Institutions	Duration	Objectives	Status
Sweden	SERC / SLU / NREL	not defined	Co-generation of heat and electricity from refined wood in the 10 to 1000 kW range.	10 kW prototype wood powder burners have been built and successfully operated.
Germany	Fraunhofer ISE	not defined	Small TPV generator (50 - 100 W_e) to provide backup electricity complementing small PV installations with high reliability in remote areas.	A setup to measure TPV cell characteristics has been constructed and cell efficiencies for T = 1200 - 2300 °C and P = 3 - 20 W/cm^2 have been measured. A setup to test the chain emitter - filter - TPV cell is currently being built. A laboratory 2 kW_e TPV demonstration system with a metallic emitter, a selective plasma interference filter and GaSb cells has been built and tested. GaSb cells have been fabricated. Facilities to measure TPV cell characteristics as well as spectral emissivities of materials up to 1500 °C are available.
United Kingdom	British Gas Morgan Materials Gwent Electronic Materials Univ. Warwick	1995 - 1998	Development of a radiant burner system which incorporates rare earth oxides to achieve selective emission. The system will be demonstrated at the 300 W_e output scale.	A detailed computer based TPV burner and TPV system model has been constructed. System design and costing study carried out. Opportunity to achieve sub-1$/W capital costs has been identified. Coating technologies are under development to realise a selective emitter. System components have been successfully coated. Burner system materials have been identified and are being produced.
	Clarendon Laboratories Northumbria University Surrey University	1996 - 1998	Low bandgap TPV-cells based on relaxed InGaAs layers grown on GaAs at various lattice constants	Plastic relaxation of strained $In_{0.53}Ga_{0.47}As$/GaAs has been verified. Next step is to fabricate p-n junction cells on a GaAs substrate.
Switzerland	PSI	1996 - 1999	Co-generation of heat and electricity from natural gas. Key elements: Si-cells, Yb_2O_3-emitter, ceramic foam burner. Project goal: demonstration of the principle (10 - 20 kW_e range) and estimation of the technical and economical potential.	A methane burner with a thermal power of about 2 kW_{th} is available. Emitter temperatures above 1100 °C have been reached. Selective emitter materials on Yb_2O_3 basis have been evaluated. Dedicated measurement techniques (surface temperatures, TPV-cell characterisation) are available.
Netherlands	ECN	1993 / 1994	Technical and economic feasibility study on the potential of a thermophotovoltaic central heating unit at about 30 kW_e, and an eletricity share of 10 %.	Conclusion: From a technical point of view the design goals can in principle be reached. However, the investment costs in such a system are far from being competitive. A necessary prerequisite for a TPV-co-generating system is the availability of low bandgap TPV cell at low costs.

TABLE 2. Summary of European activities in TPV

REFERENCES

1. J. Blair D. C. White, B. D. Wedlock. Recent advance in thermal energy conversion. In *Proc. 15th Power Source Conf.*, pages 125 – 132, 1961.

2. L. Broman. Thermophotovoltaics bibliography. *Prog. in Photovoltaics*, 3:65 – 74, 1995.

3. H. Hoefler. *Thermophotovoltaische Konversion der Sonnenergie*. PhD thesis, University of Karlsruhe, 1984.

4. R. E. Nelson. Grid-independent residential power systems. In *2nd NREL Conf. On Thermophotovoltaic Generation of Electricity*, 221 – 237, 1995.

5. V. Kumar A. Schock, C. Or. Small radioisotope thermophotovoltaic generators. In *2nd NREL Conf. On Thermophotovoltaic Generation of Electricity*, 81 – 97, 1995.

6. L. Broman and J. Marks. Development of a TPV converter for co-generation of electricity and heat from combustion of wood powder. *Proc. 1st World Conference on Photovoltaic Energy Conversion*, 1995.

7. D. M. Wilt et al. InGaAs PV device development for TPV power systems. In *1st NREL Conf. On Thermophotovoltaic Generation of Electricity*, 210 – 220, 1994.

8. R. E. Nelson. Thermophotovoltaic emitter development. In *1st NREL Conference on Themophotovoltaic Generation of Electricity*, 80 – 96, 1994.

9. V. S. Sundaram, W. E. Horne, M. D. Morgan. Ir filters for TPV converter modules. In *2nd NREL Conf. On Thermophotovoltaic Generation of Electricity*, pages 35 – 51, 1995.

10. J. Marks. Wood powder: An upgraded wood fuel. *Forest Prod. J.*, 42:52 – 56, 1992.

11. L. Broman and J. Marks. Co-generation of electricity and heat from combustion of wood powder utilizing thermophotovoltaic conversion. *Proc. 1st NREL Conference on Thermophotovoltaic Generation of Electricity*, 133 – 138, 1995.

12. L. Broman, K. Jarefors, J. Marks, and M. Wanlass. Electricity from wood powder. Report on a TPV generator in progress. *Proc. 2nd NREL Conference on Thermophotovoltaic Generation of Electricity*, 177 – 180, 1996.

13. L. Broman. Calculations of optimum bandgaps of TPV devices for blackbody emitters and selective mirrors. *To be published*, 1997.

14. L. Broman, K. Jarefors, J. Marks, and M. Wanlass. Efficiency measurements of TPV cells. *Proc. 3rd NREL Conference on Thermophotovoltaic Generation of Electricity*, 1997.

15. A. Van der Drift, G. J. J. Beckers, K. Smit, and J. Beesteheerde. An experimental and numerical study of porous radiant burners. Technical Report ECN-RX–94-061, Netherlands Energy Research Foundation ECN, 1994.

16. S. L. Tjeng, A. Van der Drift, G. J. J. Beckers, and L. P. L. M. Rabou. High efficiency radiant burners greatly enhance heat transfer. Technical Report ECN-I-95-058, Netherlands Energy Research Foundation ECN, 1995.

17. H. Gabler, M. Hein, and M. Zenker. Thermophotovoltaic generation of electricity. In *Proceedings of the EuroSun Conference*, Freiburg, 1996.

18. H. Gabler, M. Hein, and M. Zenker. Modellierung der Energiefluesse in einem Thermophotovoltaik-Generator. In *12. Symposium Photovoltaische Solarenergie*, 1997.

19. A. Bett, S. Keser, G. Stollwerck, and O. Sulima. Large area GaSb photovoltaic cells. In *Proceedings of the 3rd NREL Conference on Thermophotovoltaic Generation of Electricity*, 1997.

20. R. H. Beckert and K. Buecher. Investigation of temperature effects in GaSb-based (thermo-)photovoltaic cells based on infrared spectral response measurements. In *Proceedings of the 14th European Photovoltaic Solar Energy Conference and Exhibition*, Barcelona, 1997.

21. H. R. Tschudi and M. Schubnell. Simultaneous measurement of irradiation, temperature and reflectivity on hot irradiated surfaces. *Appl. Phys. A*, 60:581 – 587, 1995.

22. M. Schubnell, H. R. Tschudi, and Chr. Mueller. Temperature measurement under concentrated radiation. *Solar Energy*, 58:69 – 75, 1996.

23. W. Durisch, J. Urban, and G. Smestad. Characterisation of solar cells and modules under actual operatiung conditions. In *Proceedings of World Renewable Energy Congress*, Vol. 1, p. 359 f. Pergamon, 1996.

24. J.-C. Panitz, M. Schubnell, and W. Duriusch. Influence of ytterbium concentration on the emissive properties of Yb:YAG and Yb:Y_2O_3. *Proc. 3rd NREL Conference on Thermophotovoltaic generation of Electricity*, 1997.

Review of a Workshop on Thermophotovoltaics Organized for the Army Research Office

John S. Kruger

Research and Technology Integration
U.S. Army Research Office
P.O. Box 12211
Research Triangle Park, North Carolina 27709-2211

Abstract. Under the sponsorship of the U.S. Army Research Office (ARO), a workshop dealing with thermophotovoltaic (TPV) power technology was held in Durham, NC, in July 1996. This was one of a series of highly focused workshops on key issues associated with the science and technology of advanced power systems.

The focus of this workshop was to assess whether TPV is capable of meeting some of the needs of the dismounted soldier as a fighting platform, the Army's "Soldier System."

To accomplish the objectives of the workshop, a group of scientists and engineers, active in the field, from government laboratories, industry, and academia were invited to lecture on a wide range of topics pertinent to TPV. The technical program consisted of plenary, state-of-the-art technology updates and working group sessions covering as wide a range of topics as time permitted.

This review is planned to share some of the results and ideas which emerged.

INTRODUCTION

Recent advances in thermophotovoltaic (TPV) technology, especially as presented in previous NREL Conferences, have suggested that TPV might play a key role in future power systems for the Dismounted Soldier. As the Army becomes more mobile, reliable and minimal mass systems are required. Improvements in power technology and systems, in terms of reliability, cost and maximum energy/power density, translate to increased capability and reduced cost.

The idea for a series of highly focused workshops dealing with key issues associated with the science and technology of advanced power systems had its genesis in many conversations with outstanding technologists from the wide world. Power systems offer unique challenges to engineers and application specialists. While a mission or capability may be feasible, more often than not, the power technology available determines the total mission profile. As the mission expands, the associated power demand quickly increases beyond the state-of-the-art.

Often promising concepts are abandoned due to the lack of a foreseeable power technology that could move it beyond being a mere laboratory curiosity. For many advanced concepts, power technology is totally enabling.

CP401, *Thermophotovoltaic Generation of Electricity: Third NREL Conference,*
edited by Benner/Coutts
© 1997 The American Institute of Physics 1-56396-734-0/97/$10.00

In 1990 the first workshop on Mobile Battlefield Power concluded that research and development is needed to improve the Army's power technology up to about 500 W, especially in man-portable systems. The second power workshop concentrated on the particular challenges for power needs associated with the Soldier as a System and the ability of electrochemical power sources to meet the requirements. Other related workshops dealt with radioisotope thermal generators, small combustion-engine-driven generator systems, micromachines, and small fuels cells.

Motivation for a TPV Workshop

Since TPV is emerging and could use battlefield logistic fuels, a workshop was organized for ARO by the Dr. Frank Rose and the staff of the Space Power Institute, Auburn University. This TPV workshop, reported here, was the eighth in the continuing series dealing directly with advanced technologies applied to the needs of the individual soldier. The intention was to assess the potential of TPV and to put its attributes in perspective with respect to batteries, fuel cells and small motor-generator sets. The workshop addressed requirements, key research issues, and projected capabilities and development opportunities.

Goals

Specific goals articulated for this TPV workshop were to:

• Assess the state-of-the-art of TPV systems, characterizing innovative photoconversion techniques and determining their suitability to power applications;
• Identify key research issues pacing the development (or limiting full development) of efficient, mobile, fieldable, high-power TPV cells/systems with acceptable life;
• Determine major limiting factors which must be addressed as part of overall system design;
• Prioritize research issues, indicating the impact if research is successful;
• Provide milestones for research teams to assure significant improvements over the near term and the longer term .
• Identify operational and environmental constraints such as materials, signatures, manufacturing capability and pollution effects.
• Establish scaling laws and, whenever possible, compare TPV innovations with other power sources.

To accomplish these goals, a group of scientists and engineers, active or interested in the field, were invited to present their perspectives on the current state of TPV research and technology. The workshop was patterned after the highly successful Gordon Conferences that have formal morning and evening sessions, leaving afternoons free for small group discussions, laboratory tours, or recreation at the participants discretion.

PUTTING TPV IN PERSPECTIVE

The challenge to the participants was to: tell the Army what TPV is good for; give an honest picture of remaining technical barriers; decide where to use TPV or hybrid; consider cost and whether TPV is affordable; and to treat the fuel issue so that we can live with.

Workshop Structure

The 50 participants were drawn from Industry (26), Academia (10), and Government/National Laboratories (14) and represented an adequate cross-section of scientists and engineers working, or interested, in the TPV field. The workshop was organized into a plenary session, technology updates and working group deliberations. The following is a synopsis of the key determinations from the participants.

Plenary Session

A plenary session was organized to present government and industrial perspectives on the potential of TPV. Talks were presented on:

- the Army's motivation to look at TPV;
- what ARO is sponsoring;
- an industry perspective;
- why NASA should care about TPV; and
- competing technologies.

As confirmed by the plenary speakers, there is much military, civil, and space interest in TPV. Many of the applications appear to be possible in the near term if cost is not a factor.

Technology Updates

To accurately determine the applicability of TPV, it was necessary to define the state-of-the-art in the relevant technologies. To assist the workshop process invited speakers presented summaries about where various technologies stand and gave their opinions of ultimate limitations with some consideration for practicality. TPV topics summarized at the workshop included:

- liquid fuel combustion;
- recuperators;
- system aspects of energy conversion;
- small radioisotopes as heat sources;
- NREL activities and capabilities;
- JPL research;
- advanced detectors;
- line emitters and selective filters;
- selective emitters;

- Auburn University, Space Power Institute research;
- emitter fabrication;
- system performance modeling; and
- cell fabrication.

Working Group Discussions

The remainder of the workshop centered on small working groups or discussion panels. At the end of the workshop, the findings of these deliberations were summarized by the respective chairmen of each group. The groups and their output are discussed below.

Customer Requirements, Specific Mission Needs, State-of-the-Art. This group examined the state-of-the-art in TPV devices, and attempted to present cost and technological limits. The members of this working group were drawn from military, industrial and academic organizations allowing a broad perspective of the issues involved. The group was heavily weighted by industrial producers of TPV technologies and systems. Thus insights into the technological issues and drivers were available. Of significant benefit were the thoughts of a major supplier of commercial equipment to be fielded in difficult locations.

Several Army applications were used were set the stage for discussions of TPV technologies and state-of-the-art. Initial discussions focused on Army requirements for some potential TPV applications. The applications include a 300-500 W battery charging station using liquid or gaseous fuel, weighing roughly 10-15 lbs. Another was a replacement for the hand-cranked generator, capable of supplying 20-30W at night (while photovoltaics could provide the power from sunlight in the daytime). A substitute for the BA5590 battery that could supply 5W continuously and 50 W peak was also examined. Cogeneration possibilities were dealt with in lesser detail. These included shower, laundry, and kitchen equipment

Subsequent comparisons were made of TPV systems with competing technologies including thermoelectric generators, fuel cells, solar photovoltaics, battery hybrids, Stirling engines, motor generators, batteries, and small turbines for a range of applications. These comparisons allowed clarification of TPV attributes and limitations.

It was concluded that the absence of system-level demonstrations has not allowed a downselection of the strongest technology options and system concepts and the industrial teams. Nor has it allowed potential users to ensure themselves of TPV system viability and competitiveness. Thus credible, government-sponsored demonstrations conducted through consortia with assured DoD customers were proposed in order to validate this technology.

It was also observed that many of the separate elements of the TPV system may be independently marketable to provide cash flow. Applications such as radiant heat heating (using the emitters' narrow spectral emissivity to more efficiently dry things like crops and laundry) was mentioned as a specific application to highlight that possibility. It was noted that the international marketplace should be a strong consideration. An optimistic cost goal for TPV of 35¢/W was suggested.

Key Research Issues, Major Limiting Factors, Constraints. The second panel started with a detailed discussion of the requirements and applications which were envisioned for the military. In terms of energy, the "technology to beat" was judged to be the battery, with representative specific energies on the order of 175 Whrs/kg and specific powers on the order of 20-50 W/kg. It was pointed out that battlefield fuels have energy densities on the order of 13,000 Whrs/kg, so that conversion of this stored energy to electricity at an efficiency of less than 2% would beat batteries in terms of storage density. This is noteworthy since TPV technologists are trying desperately to raise system efficiencies from the 3-7% level to possibly 15-20%.

This group discussed the positive and negative attributes of TPV so they could compare with alternative technologies. The positive attributes considered include:

- inherently quiet operation;
- intrinsically light weight;
- multifuel operation, capable of using logistic fuel;
- low emission; no moving parts in main power stream;
- cogeneration compatible (convenient for incidental electric generation on heating systems;
- tolerant of low temperatures;
- moderate efficiency;
- simple to start;
- excellent dormancy; and
- throttleable.

Negative attributes consist of:

- need to operate at elevated (combustion) temperatures;
- may have thermal signature;
- smaller industrial base so far;
- poor system experience; and
- limited efficiency.

The panel then did compare the projected performance parameters with some competing technologies. TPV has the promise of being far more efficient than thermoelectric (TE) power generation. The power and energy density is also greater than for thermoelectrics. Further, the cost of the TPV elements should be less than the TE ones.

When measured against fuel cells, the TPV is multifuel capable, as well as being able to use logistic fuels. It is, however, less efficient by more than a factor of two. Both fuel cells and TPV are airbreathers and should have comparable power densities.

TPV technology is far more energy dense than batteries, and has a low life-cycle cost. Further, the ability to refuel and the "infinite" shelf life are attributes that are not achievable for battlefield batteries. The TPV power density is also far greater than that of batteries.

It was the unanimous opinion of this working group that there were no "fundamental problems" standing in the way of the development of systems with overall 10% efficiency. Further, the panel thought that the upper limit on efficiency is on the order of 30%. This is governed mainly by the physics of the photovoltaic (PV) materials and the wavelengths at which energy can be efficiently converted.

The panel also agreed that there is a need for standard methods and approaches to defining the performance of TPV components. The participants realize that this is difficult to do because there is a coupling of components. It is necessary to develop standards for emitters, both blackbody and selective, which allow for the many systems choices. In order to facilitate standard measurements, there also needs to be an "agreed upon" set of definitions.

The panel closed its deliberations with the observation that the development of <u>integrated systems</u> has the highest priority. The general idea is to learn the design rules by conceiving systems and trying to build them. In this mode, demonstrators will emerge, as well as key R&D items, which will extend the systems state-of-the-art. The approach should be to develop teams to build systems, which may have flaws and mistakes, but will stimulate rapid learning.

By developing working systems, the next level of component research needed to improve systems will be identified. By this process, poor approaches will be eliminated, the reliability of components will be established and demonstrated, efficiency will be determined and refined, poorly performing subcomponents will be redesigned, and better methods of fabrication will be established. All this will lead to sample fieldable units within a couple of years and eventually to production.

Strategies & Technologies, Priorities, Near & Long Term Developments, Milestones to Achieve Priorities. The third working group focused on the direction for the nation's TPV program. They also ensured that the Army's priorities would drive the direction of any major TPV development. The kinds of questions considered are quite fundamental:

- Why should there be TPV program?
- What does TPV bring to the playing field that other technologies lack?
- What price is paid for TPV over other competitors?
- What is the cost to convert to TPV?
- What are the limitations of TPV?
- Are there other markets for TPV, should it become technologically achievable at reasonable cost?
- What breakthroughs are required to make TPV viable?
- How should the Army proceed toward a TPV program?
- What recommendations can be given to Army developers?
- In what time frame does TPV appear achievable?

This working group recommended that the Army establish a development program, consisting of a technology demonstration phase to address near-term systems issues, and a production (engineering) phase that would treat the longer term to meet specific customer requirements. Thus the panel identified a strategy, cost goals, and time frame to arrive at some near- and longer-term applications for the TPV development program. This strategy would be effective if there were early successes, visible to potential users of the technology (perhaps through demonstrations), and if a "niche market" were identified so as to ensure a successful introduction for the TPV capability.

The group further advocated that the Army establish a consortium of DoD, industry and academic institutions. The Army should look at opportunities or unique ways in which cost sharing could be accomplished during the first phase. This is generally not a popular approach during periods of fiscal constraint among industrial teams, so definition of cost-sharing arrangements needs to be reviewed quite carefully.

The establishment of a mechanism for information exchange is highly desirable, so as to promote a "common language" to discuss TPV technologies, facilitate the sharing of data, and more accurately and efficiently describe new developments in publications.

SUMMARY

This technology for photoconverting energy from an incandescent source (which can be heated by any number of heat sources) to electricity is quite unique and has great promise for the development of portable power sources for the Army's dismounted soldier, as well as the consumer market. Recent advances in the technology associated with TPV suggest that power systems could be built in the range of a few watts to greater than 500 W which would impact the requirements for the Dismounted Soldier.

Improvements in photovoltaics and emitters have been made in terms of increased reliability and energy efficiency, and reduced size and weight. But fieldable technology rarely equals theoretical or even laboratory prototype capability. This workshop sought to uncover obstacles which might be overcome through appropriate research, innovative development techniques, and skillful engineering.

TPV is by its very nature multidisciplinary. Solid-state converters must be combined with a radiant element, which is heated from a combustion source. And recuperation is key for overall efficiency. While the ARO, DARPA and others are investing in technologies which tend to define the state-of-the-art, much must be learned to make usable, affordable systems. Questions such as: whether TPV devices can be mass produced from affordable materials; whether systems can be made robust enough, and reliable enough to function in a hostile battlefield environment; whether embodiments can be engineered with minimal (acoustic and thermal) signatures; and whether TPV materiel will provide enhanced functionality and capability to the soldier in the field. Many of these same questions apply to the civilian segment too.

As the workshop progressed, several pacing ideas emerged which were used to guide the discussion process:

• There is a potentially large civilian market for TPV;
• It is only necessary to convert less than 2% of the energy in diesel fuel to produce a power supply that is more energetic than existing batteries (such as the Army BA5590 battery, the most common electrical sources for man-portable radios);
• There do not appear to be any fundamental physical reasons that TPV systems cannot be built with modest efficiencies;
• There is a definite lack of engineering experience with TPV systems;
• There is also a scarcity of standards and "agreed upon" test procedures;
• A concerted program could field a device within 3-5 years.

It is insufficient to think only in terms of system energy density or power density. Due to environmental or safety concerns, the power technologist may be forced to employ non-optimum power systems that drastically limit the performance envelope. Clearly, there is a set of tradeoff parameters that must be manipulated to provide an optimum system for a given mission, and of course, it is impossible to optimize all parameters simultaneously.

ARO hopes that its sponsorship of workshops such as this is of benefit to the technologist. . We are indebted to Auburn University, the Space Power Institute, and especially Dr. M. Frank Rose for the success of this workshop, for the detailed proceedings reporting the findings, and for facilitating the other power workshops.

SESSION 2:
TPV DEVICES BASED ON GaSb
AND RELATED MATERIALS I

Low Cost High Power GaSb Photovoltaic Cells

Lewis M. Fraas, Han X. Huang, Shi-Zhong Ye, She Hui, James Avery, and
Russell Ballantyne

JX Crystals Inc.
Issaquah, WA 98027

Abstract. High power density and high capacity factor are important attributes of a
thermophotovoltaic (TPV) system and GaSb cells are enabling for TPV systems. A TPV
cogeneration unit at an off grid site will compliment solar arrays producing heat and
electricity on cloudy days with the solar arrays generating electricity on sunny days.
Herein, we project that GaSb cells generating 2 Watts each can be made in 1 MW
quantities at $4 per cell. This will allow TPV circuits to be made at $2 per Watt. At this
cost, the off-grid cogeneration and self-powered furnace markets will be viable.

Introduction

Over the last two decades, the photovoltaic community has focused its
attention on producing low cost flat plate solar arrays. This has proved to be a
very formidable task for two reasons. First, the solar intensity is very small
corresponding to only 0.1 Watts per square centimeter and, second, sunshine is
only available during daylight hours. A 10% efficient traditional solar cell,
therefore, will only produce 0.01 Watts per sq. cm. for approximately 5 hours per
day. It then takes 20,000 days or approximately 50 years for this 1 sq. cm. cell to
produce I kWhr of electricity. If a solar cell were to be made for 10 cents per sq.
cm., it would then take approximately 50 years for it to pay for itself, yet 10 cents
per sq. cm. is a very small number compared to the cost of all other semiconductor
devices. These facts have led to a focus on thin film solar cells for low cost
electricity produced from photovoltaic devices.

There are, however, alternatives based on higher power density. In a first
alternative, lenses are used with solar cells to concentrate the sun's energy onto the
cell. For example, a concentration ratio of 100 in a desert location where one
could expect 10 hours per day of sunshine would reduce the time to produce 1
kWhr to 100 days. Unfortunately, this alternative has received much less attention
than the thin film option perhaps because it is considered to be a much more
geographically limited option.

The truth is that all solar cells, even thin film cells, will be limited to sunny
locations, and furthermore, people do live in colder climates as well as warmer
climates and even in warmer climates, winters are colder with less sunshine than
summers. These facts may be limiting for solar cells but they create an opportunity
for thermophotovoltaic cells, a second high power density alternative. Referring to
figure 1, infrared sensitive thermophotovoltaic cells can be combined with man

CP401, *Thermophotovoltaic Generation of Electricity: Third NREL Conference,*
edited by Benner/Coutts
© 1997 The American Institute of Physics 1-56396-734-0/97/$10.00

made heat sources to cogenerate electricity along with heat in colder climates and during winter months.

Perhaps the most interesting thing about the thermophotovoltaic application is that the electric power that a 1 sq. cm. cell can produce has been measured at 2 Watts. Furthermore, the generator can run day or night. This implies that a TPV cell operating 10 hours per day will produce 1 kWhr of electricity in 50 days..

From the above considerations, the economic advantages of high power density and higher capacity factor for the concentrator and thermophotovoltaic approaches are considerable. Geographically, these approaches are complimentary. Furthermore, both approaches are well founded on single crystal semiconductor device technology. So, no miraculous breakthroughs in thin film solar cell technology are required.

For the past several years, JX Crystals has been fabricating GaSb thermophotovoltaic cells. We have the equipment, facility, and know-how to grow the semiconductor crystals, convert the crystals to wafers, process the wafers into cells, and mount the cells into power producing circuits. This paper will focus on a quantitative analysis of the cost of thermophotovoltaic crystals, cells, and circuits as well as the implications of these resultant costs for various potential thermophotovoltaic market sectors.

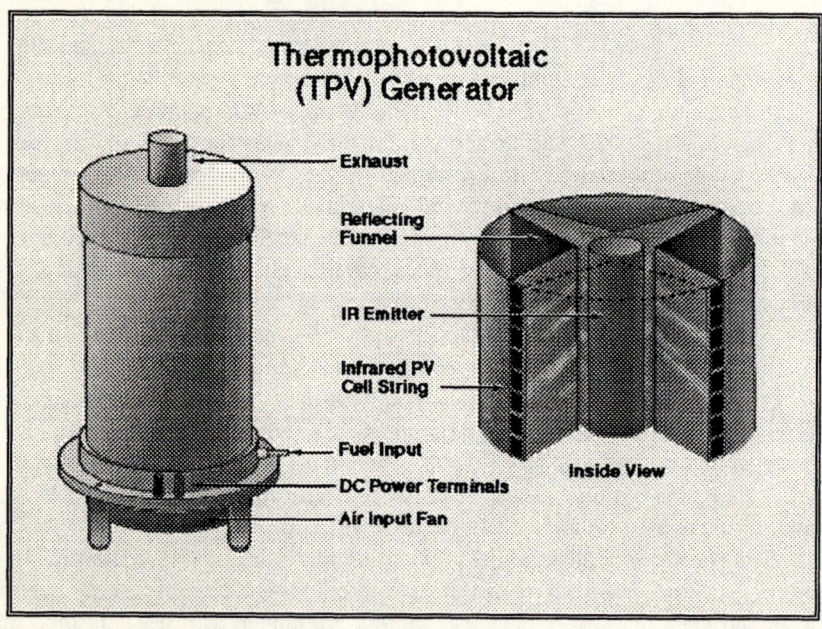

FIGURE 1: Thermophotovoltaic generator concept

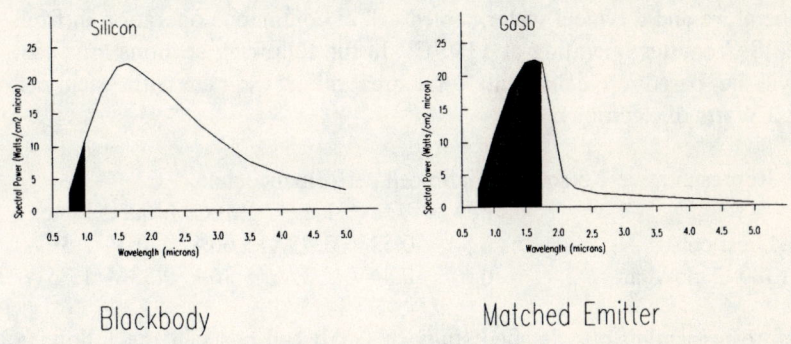

Blackbody   Matched Emitter

FIGURE 2: Silicon vs GaSb with a blackbody emitter and matched emitter.

Gallium Antimonide Thermophotovoltaic Cells

The idea of thermophotovoltaic generators originated in the 1970's. However until recently, there was a mismatch between the power spectrum emitted by man-made heat sources and the response band of the available photovoltaic cells. This problem and its solution are illustrated in figure 2. The temperature of the heated and radiating emitter in a TPV generator is limited to the 1400 C to 1700 C range by the nature of hydrocarbon combustion and by the thermal durability of appropriate ceramic materials. This means that nearly all of the energy emitted is in the infrared at wavelengths longer than 1 micron. Unfortunately, the only available photovoltaic cell until recently, the silicon cell, responds to wavelengths shorter than 1.1 microns. Referring to figure 2a, the power available to a Si cell from a 1700 C emitter, the black sliver under the power curve, is very small while the wasted infrared power, the white region under the power curve, is very large.

The solution to this problem is shown in figure 2b. In 1989, Fraas, Avery, and Girard invented the GaSb infrared cell while at Boeing. Referring to figure 2b, this GaSb cell which responds out to 1.8 microns, dramatically increases the convertible power. In 1993, JX Crystals Inc. obtained from Boeing an exclusive license to this cell technology. Subsequently in 1996, JX Crystals Inc. discovered a ceramic emitter material which emits strongly below 1.8 microns and weakly at longer wavelengths. Figure 2b shows the results of combining the GaSb cell with the new matched emitter. JX Crystals Inc. has filed a patent application on this new matched emitter and has been recently informed that the appropriate claims have now been allowed. As figure 2b shows, the combination of these two new key components now provides for higher power density and higher conversion efficiency than prior art TPV systems.

JX Crystals Inc. now has GaSb cells in low volume production Table 1 summarizes illuminated current vs voltage test data for a best cell flash tested at room temperature and a typical water cooled cell in continuous operation in front of a heated SiC emitter operating at 1380 C. In the following sections, our cost analysis will be based on cells with total area of 1.5 sq. cm.with each cell producing 2 Watts of electricity.

TABLE 1: Representative 1.5 sq. cm. GaSb cell performance data.

	FF	Voc	Isc	Imax	Vmax	Pmax
Flash tested best cell	0.84	0.52V	5.95A	5.65A	0.46V	2.59W
Cell with 1380 C glowbar	0.67	0.44V	5.87A	5.36A	0.33V	1.76W

Before presenting our detailed study of GaSb cell costs in the following section, some overview observations are appropriate. The readers initial question may be: GaSb, is the cost of Ga and Sb raw material prohibitive? The answer to this question is that the total cost of Ga and Sb in a 1.5 sq. cm. GaSb cell is 15 cents. For a flat plate solar cell, this cost would be prohibitive, but for a 2 Watt TPV cell, this cost is nearly insignificant. The real cost of a single crystal semiconductor chip is not the material cost but the processing cost. This point can be easily made by noting that carbon is cheap but diamonds are very expensive.

Given that processing cost dominate, we have made an effort to copy low cost silicon solar cell processing in fabricating GaSb cells. Specifically, we use converted silicon pullers for crystal growth and diffusions for junction formation. Nearly all of the equipment at our facility is used silicon processing equipment. Epitaxy, generally used in GaAs space solar cell fabrication, is an example of a process we have avoided. This process uses toxic gases creating a huge safety expense. Furthermore, it is very capital intensive and has low wafer throughput. The key to low cost devices is high throughput with minimal capital and labor.

Projected Wafer, Cell, and Circuit Fabrication Costs

Table 2 summarizes our projected costs at annual production volumes of 100 kW and 1 MW per year. These costs are broken down into three areas: Wafco by which we mean wafer fabrication; Cellco by which we mean cell fabrication; and Circo by which we mean power circuit fabrication. Wafco operations include crystal growth, rounding, slicing, and etching. Cellco operations include diffusion, photolithography, front and back metal deposition, filter deposition, testing, scribing, and dicing. And finally, Circo operations include solder dispense, pick and place, die attach, inner and outer lead bonding, and testing. The finished product is a 20 cell circuit that produces 40 Watts of electricity.

The data in table 2 are based on the present equipment, facility and know-how at JX Crystals. For example, referring to Wafco, we presently have two crystal pullers which together can produce one 3" diameter crystal per day where each crystal when sliced will produce 110 wafers per day. Assuming 5 days per week and 52 weeks per year and 90% yield, this implies that we can produce approximately 25,000 wafers per year. Given that there are 20 cells per wafer at 2 Watts each, then Wafco's full capacity is 1 MW. Similarly, we chose the 100 kW throughput number because this represents the single shift present capacity for Circo as well as the probable market demand for the first year of production.

Summing the cost in table 2 gives a power circuit cost of $4.82 per Watt at a production volume of 100 kW per year dropping to $1.97 per Watt at a volume of 1 MW per year. These cost translate to a per cell cost in a circuit of $9.64 and $3.94 respectively. Note that at a production volume of 100 kW, capital and labor cost are still dominate.

Discussion of Costs

While space does not allow a complete discussion of all of the inputs in table 2, we will discuss some specific examples to provide a flavor of the manufacturing assumptions.

As a first example, we describe the front and back metallization in the Cellco process. We presently use an electron beam evaporation machine with an 18" bell jar capable of coating eighteen 3" diameter wafers at a time. A 100 kW cell annual production rate given 40 Watts per wafer is equivalent to 50 wafers per week or 10 wafers per day. Fifty four wafers can be coated per week in 3 afternoon runs for front metal and 3 morning runs for back metal. This implies a single shift equipment utilization of 60%. In order to produce 1 MW per year, we would use two machines all five days per week on three shifts.

As a second example, refer to the Ga and Sb materials cost in the 100 kW Wafco column in table 2. Note that 30 cents per Watt corresponds to 60 cents per cell which is 4 times larger than the cost of Ga and Sb in a cell. Why? It turns out that for the 4 kg that goes into a crystal, only 1 kg is incorporated in cells for our current process. This is because 1 kg is lost in rounding and 1.5 kg is lost in slicing and 0.5 kg is lost at the edge of a completely processed wafer. While this yield is poor, the total Ga and Sb costs are still insignificant at these production rates. However, this does not mean that improvements in utilization efficiency are not possible. For example, automatic diameter control during crystal growth will reduce rounding losses, a wire saw rather than an ID saw will reduce kerf losses, and larger 4" or 6" wafers will reduce the edge loss percent.

Turning now to Circo costs, we note that 100 kW per year can be accomplished in batches primarily by manual labor. However, when it comes to 1 MW per year, we plan to invest in automated equipment for the pick and place and lead bonding operations. Also note that the purchased substrate direct material cost is as large as the Ga and Sb materials cost.

For experts in semiconductor wafer fabrication, it is instructive to state costs in per wafer costs. Thus, in the 100 kW production column, we are fabricating 3" wafers at a cost of approximately $160 each, and in the 1 MW column, we are fabricating 3" wafers at $56 each. These numbers can be compared with the cost of a 4" single crystal silicon solar cell produced at approximately $5 each in much higher volumes.

Clearly, processed wafer costs decrease with increasing volume implying that still further cost reduction can occur for TPV cells beyond the projections in table 2. However, on a cost per Watt bases, note that the $5 per Watt circuit cost projection is comparable to the cost of a flat plate solar module today and note that the $2 per Watt cost projection is comparable to the target set for solar modules for the last several years, a target still not achieved. Also note that $2 per Watt for a TPV circuit will really be more cost effective than $2 per Watt peak for a flat plate solar module because of the higher capacity factor for the TPV generator. The TPV generator can generate more Watt-hrs per day.

Applications

The first commercial market for a TPV generator is as a compliment to flat plate solar panels in off-grid dwellings. Our market research indicates that there are at least 20 thousand potential customers per year purchasing solar panels in off-grid environments. These environments include mountain cabins, sailboats, and recreational vehicles. Many of these customers are accessible through solar panel distributors. Figure 3 shows a Midnight Sun Micro-Cogeneration unit designed and built by JX Crystals to address this market (patent pending). This unit produces approximately 500 Watts-hrs per day of electricity and 8,000 BTU per hour of heat.

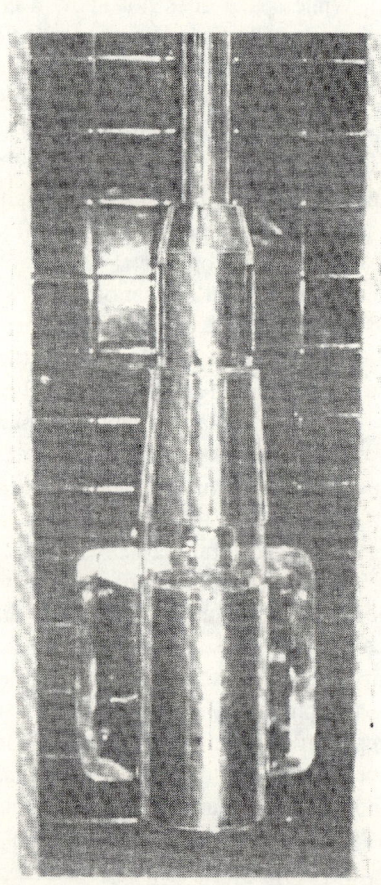

FIGURE 3: Midnight Sun Micro-Cogenerator produces 500 W-hrs / day of electricity and 8,000 BTU per hour of heat.

We believe that our Midnight Sun Micro-Cogeneration unit will sell well in this market because, while solar panels work well during warm summer months, when winter comes around and clouds, rain, or snow come, there is a need for heat and a need to provide for the electricity no longer provided by the solar panel. The midnight Sun unit meets these needs perfectly. Solar panels and the Midnight Sun both are designed to provide 12 Volts of DC to a battery bank. Another positive selling point will be that the customer is already familiar with and sympathetic to photovoltaic panels. Finally, there will be an additional benefit for new solar panel customers in that they can down size their solar panel and battery bank requirements since there will no longer be a need for excess capacity for several days of bad weather. This market should be accessible at the $10 per cell level.

The next largest TPV market and the first grid connected market is the self powered furnace market. A Gas Research Institute study concluded that among the 500,000 to 1,000,000 households in the Northeast and North Central US which experienced outages greater than 4 hours in duration each year, there exist a niche markets willing to pay a premium of $500 to maintain heater service during outages. This need can be met with a TPV system in which a heating furnace is redesigned to include an infrared emitter and a 200 W TPV array wired to a DC blower for forced air heating or a water pump for radiator heating. An Arthor D. Little survey indicated that 20% of new family dwellings with natural gas (70% of total) or approximately 150,000 customers per year would be willing to pay $400 extra for a self powered furnace. Note that $400 for 200 Watts corresponds to $2 per Watt and that 150,000 x 200 W is equal to 30 MW per year. This appears to be an exciting and viable market.

Finally, the biggest market for TPV is probably the home cogen market in which more TPV output is added to the home furnace for colder climates to provide the majority of electricity needed by the home. However, there will be hurdles to overcome to enter this market in that AC will probably be required and the TPV cost will need to be below $2 per Watt. Still, the following calculation indicates why this market may be interesting. At $2 per Watt and 10 hours per day, the TPV cogen unit will cost 20 cents per Watt hour per day. If electricity costs 10 cents per kWhr, the system will pay for itself in 2000 days or 5.5 years.

Conclusions

High power density and high capacity factor are important attributes of a TPV system and GaSb cells are enabling for TPV systems. It is very reasonable to project that GaSb cells generating 2 Watts each can be made in 1 MW or larger quantities at $4 per cell. This will allow TPV circuits to be made at $2 per Watt. At this cost, the off-grid cogeneration and self-powered furnace markets will be viable. It would be appropriate for the photovoltaic community to target the home cogeneration market as a huge potential market in the next century and to expand their horizon to beyond the solar PV silicon cell.

Table 2: Projected Manufacturing Costs in $ / Watt

	100 kW per year	1 MW per year
WAFCO		
Capital	0.54	0.07
Lease	0.14	0.01
Leasehold Improvement	0.07	0.02
Power	0.01	0.01
Direct Material	0.30	0.25
Indirect Material	0.07	0.05
Maintenance	0.14	0.06
Direct Labor	0.48	0.19
Benefits @ 35%	0.17	0.07
Wafco Total	**$ 1.92**	**$ 0.73**
CELLCO		
Capital	0.58	0.09
Lease	0.19	0.02
Leasehold Improvement	0.07	0.02
Power	--	--
Direct Material	0.11	0.08
Indirect Material	0.21	0.15
Maintenance	0.15	0.11
Direct Labor	0.50	0.17
Benefits @ 35%	0.17	0.06
Cellco Total	**$ 1.98**	**$ 0.70**
CIRCO		
Capital	0.07	0.10
Lease	--	--
Leasehold Improvement	--	--
Power	--	--
Direct Material	0.37	0.24
Indirect Material	--	--
Maintenance	0.03	0.02
Direct Labor	0.33	0.13
Benefits @ 35%	0.12	0.05
Circo Total	**$ 0.92**	**$ 0.54**
GRAND TOTAL	**$ 4.82**	**$ 1.97**

Large-Area GaSb Photovoltaic Cells

Andreas W. Bett, Silke Keser, Gunther Stollwerck
and Oleg V. Sulima[*]

*Fraunhofer Institute for Solar Energy Systems, Oltmannsstr. 5, D-79100 Freiburg, Germany
Phone: (+49) 761-4588257, Fax: (+49) 761-4588250, E-mail: bett@ise.fhg.de
[*]Ioffe Institute, Polytechnicheskaya 26, 194021 St. Petersburg, Russia
Phone: (+7) 812-2479394, Fax: (+7) 812-2471017*

Abstract. GaSb photovoltaic (PV) cells with different designs for thermophotovoltaic generators were developed and investigated. Zn diffusion was used to form a p-n junction in n-GaSb substrates. Two diffusion technologies were studied: From the vapor and the liquid phase. Both diffusion technologies resulted in good PV cell parameters. However, diffusion from the vapor phase was found to be more simple and reproducible. The diffusion process parameters, which provide good control over Zn surface concentration and p-n junction depth, were determined. Photovoltaic parameters of the GaSb cells were measured under various illumination conditions and temperatures. The cell power output at maximum efficiency reached 952 mW for 1.89 cm^2 cells. The developed cell fabrication technology have shown good reproducibility of cell parameters.

INTRODUCTION

GaSb and related low band gap compounds (InGaSb, InGaAsSb) are of large interest for photovoltaic (PV) cells [1,2]. GaSb cells exhibit high efficiency of photovoltaic conversion of infrared radiation with wavelengths up to 1750 nm. Moreover, the fabrication process of GaSb cells is relatively simple.

The main goal of this work was to study various types of GaSb PV cells for different illumination levels. Therefore, GaSb PV cells with different areas and contact grids were developed.

To meet the demands of a large-scale production one should develop a GaSb-cell technology which is simple, reproducible and harmless to the environment. Nowadays, most of the III-V structures for optoelectronic application are produced by different epitaxial methods: metalorganic chemical vapour deposition (MOCVD), molecular beam epitaxy (MBE) or liquid phase epitaxy (LPE). In order to avoid problems arising from a high surface recombination, a heterostructure multi-layer design is commonly used. However, GaSb seems to have a relatively low surface recombination. Therefore, a more simple, pure diffusion process can be used successfully to fabricate PV cells. In this work the pseudo-closed box method was used to perform the diffusion. This method

CP401, *Thermophotovoltaic Generation of Electricity: Third NREL Conference,*
edited by Benner/Coutts
© 1997 The American Institute of Physics 1-56396-734-0/97/$10.00

avoids the inconvenience of sealed ampoules and proved to be quite simple and reproducible. As an alternative method, the Zn diffusion into n-GaSb substrate from a Ga-Sb-Zn melt was studied.

P-N JUNCTION FORMATION BY ZINC DIFFUSION INTO n-GaSb

Two methods of Zn-diffusion were experimentally investigated: (a) diffusion from the vapor phase in a pseudo-closed box system and (b) diffusion from the liquid phase using a Ga-Sb-Zn melt. Secondary ion mass spectroscopy (SIMS), Raman scattering spectroscopy and electrochemical capacity-voltage (C-V) profiling were used to study p-GaSb diffused layers. A precise thinning of diffused GaSb layers was performed through anodic oxidation and selective etching.

Zn-Diffusion from the Vapor Phase

Zn-diffusion from the vapor phase was performed into 2.5 x 2.5 cm^2 n-GaSb:Te (n$\approx$3x10^{17} cm^{-3}) (100) substrates in a purified H$_2$ atmosphere. A multi-wafer graphite boat shown in Fig. 1 was designed for the diffusion process. This boat allows (i) to place 4x4 cm^2 or smaller wafers in a substrate holder, (ii) to vary the distance between the wafers with a minimum distance of 0.5 mm, (iii) to use several separated Zn vapor sources (pure Zn or Zn-Ga melt) and Sb vapor sources. This design of the graphite boat ensures the uniformity of the Zn vapor pressure across the wafer surface, and thus the uniformity of the p-GaSb layer depth.

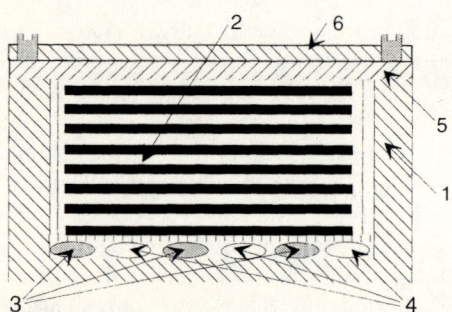

FIGURE 1. Scheme of a graphite diffusion boat:
1 - body of the boat, 2 - GaSb wafers in a holder, 3 - Zn or Zn+Ga as sources of Zn vapor, 4 - Sb as a source of Sb vapor, 5 - cover, 6 - lid with screws

Fig. 2 shows Zn concentration profiles obtained from the vapor phase at different temperatures. It is noteworthy, that there is almost no difference in the Zn surface concentration at 450-500°C after 1h - diffusion. This shows, that in this time-temperature range within the measurement error of SIMS, the saturation of Zn concentration in GaSb takes place at the level of $1.0\text{-}1.7\text{x}10^{20}$ cm^{-3}. This corresponds to the GaSb:Zn solubility at 500°C ($1.5\text{x}10^{20}$ cm^{-3}) published in Ref. [3].

The fact that the surface concentration can be saturated is very important for the reproducibility and simplicity of the p-layer formation process. If one fixes the surface concentration at the saturation level, the p-layer thickness can be regulated either by diffusion temperature or diffusion time with high accuracy. Fig. 3 shows that in the case of pure Zn as a vapor source, the Zn surface concentration slowly increases with diffusion time and reaches a saturation level after 1-2 hours. Afterwards, the p-layer thickness changes according to the square root time dependence only. If one needs to decrease the surface concentration/p-layer thickness after diffusion, a precise anodic oxidation of the corresponding part of the diffused layer succeeded by selective etching is applied.

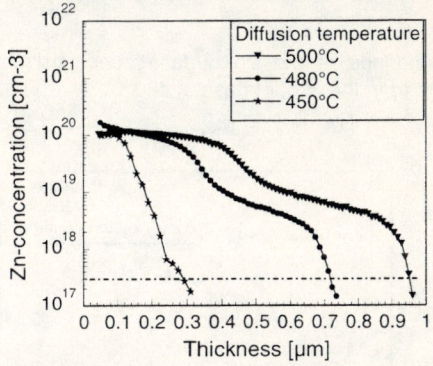

FIGURE 2. SIMS profiles of Zn concentration after diffusion of Zn from the vapor phase at different temperatures for pure Zn and Sb vapor sources. The diffusion time was 1 hour. The dash dot line shows the doping level in the n-GaSb:Te substrate.

We investigated both diluted Zn (3-10wt.% Zn + 97-90wt.% Ga) and pure Zn vapor sources. The p-layer thickness dependence on diffusion time is not accurately predictable, if diluted Zn sources are used. The effect can be explained by a Zn depletion in the Zn-Ga melt and a Zn out-diffusion from the diffused layer. This can not happen with the pure Zn source which is practically infinite. The details of this investigation are published elsewhere [4].

In order to investigate the electrical activity of Zn-atoms in GaSb, free-hole profiles in p-GaSb were measured by Raman-spectroscopy and electrochemical C-V profiling methods. In the case of Raman-spectroscopy measurements, the depth resolution was provided by a precise layer-by-layer anodic oxidation of GaSb and corresponding selective etching of the oxide. Details of these measurements are described in Ref. [5]. A Raman scattering free-carrier profile in comparison with a SIMS measured Zn profile is shown in Fig. 4. The agreement of the SIMS and Raman spectroscopy measurements for the Zn (hole) concentrations between 10^{19}-10^{20} cm^{-3} is quite good. That means, that almost all Zn atoms are electrically active in that part of the diffused layer.

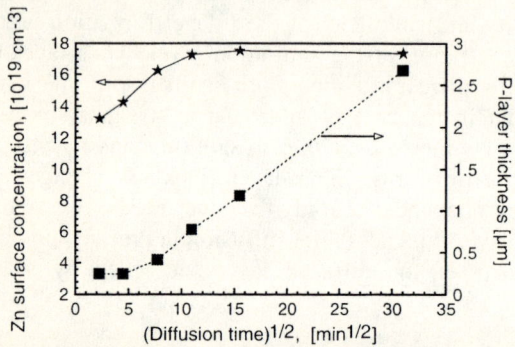

FIGURE 3. Time dependence of the Zn surface concentration and of the p-layer thickness for Zn diffusion from the vapor phase at 480°C.

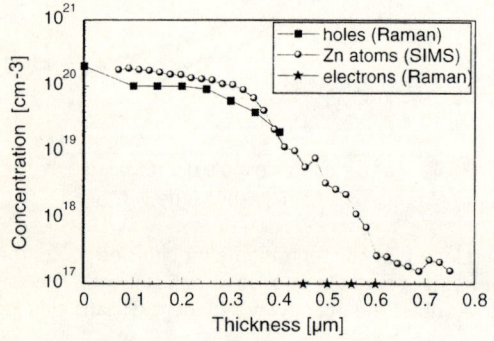

FIGURE 4. Raman scattering free-carrier profiles and SIMS Zn profile of a GaSb structure after diffusion of Zn from the vapor phase at 450°C.

The electrochemical C-V profiling was performed with a 0.1 M tiron solution $[C_6H_2(OH)_2(SO_3Na)_2 \cdot H_2O]$ as an electrolyte. Fig. 5 shows again a relatively good agreement of free-hole and Zn-atom profiles.

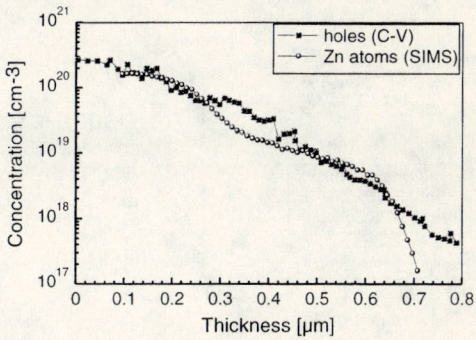

FIGURE 5. Electrochemical C-V carrier profiles and SIMS Zn profile of the GaSb structure after diffusion of Zn from the vapor phase at 480°C.

Zn-Diffusion from the Liquid Phase

Zn-diffusion from the liquid phase was performed into 1.5 x 1.5 cm² n-GaSb:Te ($n \approx 3 \times 10^{17}$ cm⁻³) (100) substrates from a Ga-Sb-Zn melt. A simple graphite one-melt one-substrate LPE boat was used. This boat can also be considered as a „pseudo-closed-box". The substrate was placed in horizontal position. The melt depth was 4 mm. The process was performed in a H_2 atmosphere purified by a Pd cell.

It was found that the wetting of the substrate by the melt was always successful if the melt and the substrate had been baked at 600°C during 1 hour prior to the diffusion process. The baking helped to reduce native oxides on the GaSb surface and to dissolve Sb and Zn in Ga homogeneously. The process temperature was 550°C. The Ga-Sb-Zn melt was undersaturated with respect to Sb. The undersaturation level was chosen to cause a dissolution of roughly 20 µm of the GaSb substrate. Afterwards the Zn diffusion into GaSb took place. The melt-substrate contact time was 1h. The process was terminated by removing the melt from the substrate.

Minor changes took place on the GaSb surface after the etchback-diffusion process and it retained almost mirror like.

The experiments on Zn-diffusion into GaSb from vapor and liquid phases were started simultaneously and were performed in parallel. From the very beginning we have observed that the diffusion from the liquid phase was less reproducible with respect to the surface concentration and the emitter thickness. One can assume, that the evaporation of Zn from the Ga-Sb-Zn melt is the main reason of the lower reproducibility. Figure 6 shows SIMS profiles of Zn concentration after Zn-diffusion (550°C, 1h) from two Ga-Sb-Zn melts with different Zn content. One can see that the Zn diffusion from the liquid phase results in lower surface concentration values compared to the diffusion from the vapour phase and in "no-kink" profiles.

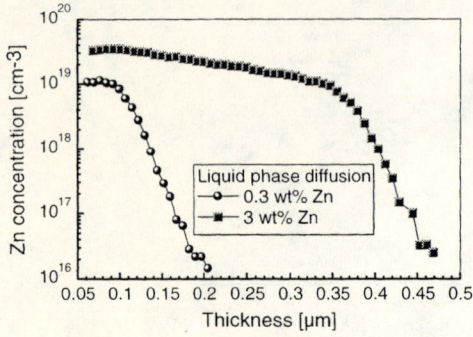

FIGURE 6. SIMS profiles of Zn concentration after Zn-diffusion (550°C, 1h) from two Ga-Sb-Zn melts with different Zn content.

PV - CELL FABRICATION

Different GaSb PV cell designs were developed for various illumination levels. Cells with the following total areas were investigated: 0.16 cm^2 , 1.05 cm^2 , 1.44 cm^2, 1.89 cm^2 and 4.00 cm^2. The GaSb PV cell structure investigated in this work is shown schematically in Fig.7.

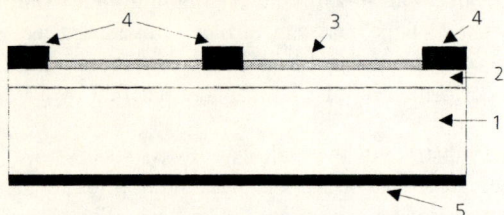

FIGURE 7. Scheme of a GaSb PV cell:

1 - n-GaSb base (substrate), 2 - diffused p-GaSb emitter, 3 - antireflection coating (anodic oxide, SiN or their combination), 4 - front contacts, 5 - back contact

A p-layer (emitter) was formed by Zn diffusion as described before. In the case of our standard diffusion from the vapor phase, a post-diffusion thinning of the p-layer by anodic oxidation was usually applied. The resulting thickness of the emitter is approximately 0.5 µm. This value is typical for the cells used in TPV module [6]. The anodization technique used was essentially that reported for GaAs in Ref. [7]. The oxidation depth rate of GaSb in the range of an applied voltage of 10-100 V was determined as 2.0 nm/V [5]. Diluted HCl or HF were used as a selective etchant of the anodic oxide on GaSb.

SiN (n≈2.2), GaSb anodic oxide (n≈2.2) or their combination were used as antireflection coatings for the cells.

As mentioned before, a constant value of the emitter thickness (0.5 µm) was used in this work. Therefore, the optimization of the cells for different illumination intensities was performed only through contact grid design. Different types of contact grids used in this work are shown in Fig.8. Their parameters are presented in Table 1. One should note that the contact grids were designed for different photocurrent densities corresponding to various illumination conditions.

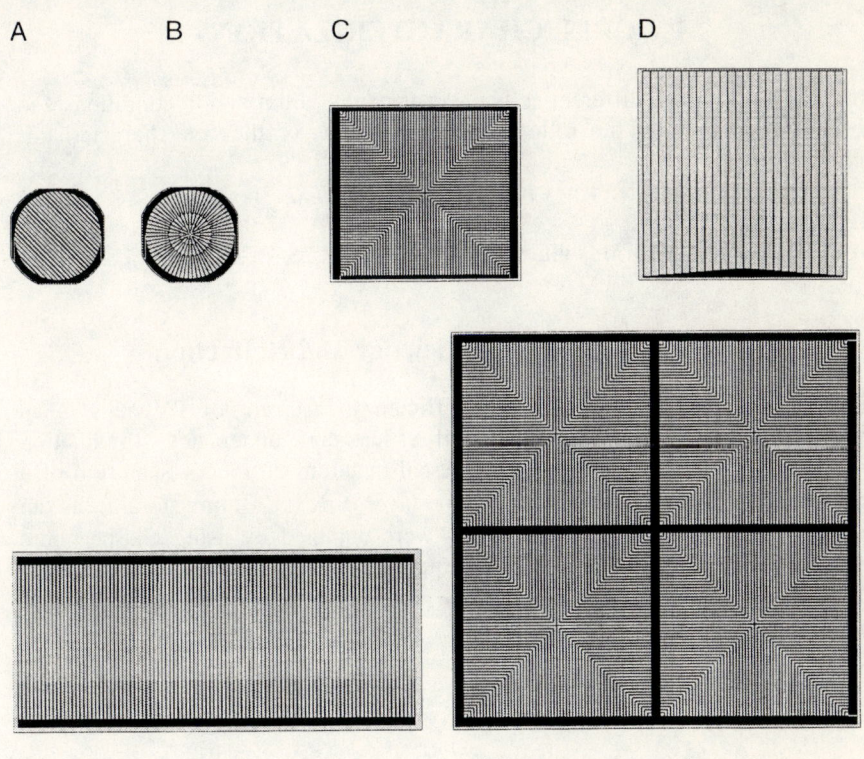

FIGURE 8. Different grid designs developed and investigated in this work (for the details see Table 1)

TABLE 1. Parameters of contact grids

Type	Total area (cm²)	Area without busbar (cm²)	Shadowing (%)	Shadowing without busbar (%)	Nominal current density (mA/cm²)
A	0.16	0.13	23.3	9.0	6 000
B	0.16	0.13	20.9	6.0	5 500
C	1.05	0.92	15.4	4.9	500
D	1.44	1.41	7.0	4.6	30
E	1.89	1.66	18.0	9.0	1200
F	4.00	3.61	16.8	9.0	150

PV CELL CHARACTERIZATION

In order to compare different cells under the same illumination conditions and the same temperature, the following measurements of the cell characteristics were performed:
(1) External quantum efficiency measurement,
(2) Reflectivity,
(3) I-V curves under various illumination intensity at 25-75°C.

External Quantum Efficiency and Reflection

Determination of external quantum efficiency spectrum of PV cells is an important issue for calculation of cell efficiencies under any illumination spectrum. Fig. 9 shows a measured external quantum efficiency spectrum of a standard GaSb cell (25°C) with a one-layer SiN (130 nm, n≈2.2) as an antireflection coating. The same results were obtained by using GaSb anodic oxide as an antireflection coating. The equality of the refractive indexes of both materials makes them interchangeable and allows to use them in any combination.

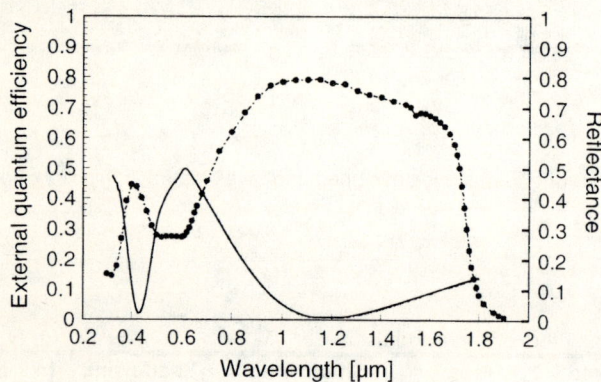

FIGURE 9. Measured external quantum efficiency spectrum of a standard GaSb cell (25°C) with a one-layer SiN (130 nm, n≈2.2) as an antireflection coating. The reflectance of the cell is shown by the solid line.

I-V Curves and Efficiencies versus Illumination Intensity and Temperature

It is common believe that TPV cells operate at high power density levels and at enhanced temperatures. Therefore, it is important to study TPV cell characteristics in a range of intensities and temperatures. However, it is not clear yet how to compare TPV cells for different applications.

We decided to measure voltage and FF of TPV cells in dependence of current. We have proved that I-V curves of the cells do not depend on illumination spectra if the same short-circuit currents are generated. Therefore, one can measure I-V curves with the necessary short-circuit current at any illumination spectrum (for example AM 1.5) by choosing the right intensity. In combination with the precise external quantum efficiency measurements it makes possible to calculate efficiencies for any spectrum, for example blackbody radiation.

Fig. 10 shows dependencies of open-circuit voltages (V_{oc}) of PV cells with different areas and grid designs (see Fig. 8 and Table 1) on short-circuit current (I_{sc}). The strong increase of V_{oc} of the small cell (type B) at low current values can be explained by the influence of the perimeter recombination (recombination component of dark current). The perimeter recombination is saturated at higher illumination levels and therefore, the V_{oc} value is determined by the diffusion component of dark current. Therefore, cells with larger area/perimeter ratio and lower recombination currents achieve high values of V_{oc} already at low illumination levels. An exception is a cell of type E which exhibits nearly the same dependence as the small cell at low current values. However, in this case such behaviour was caused by a relatively low shunt resistance of the cell. The influence of the shunt resistance decreases with the current.

Fig. 11 shows dependencies of the fill-factor (FF) of PV cells with different areas and grid designs (see Fig. 8 and Table 1) on I_{sc}. The influence of the cell series resistance at high current values is typical for all type of cells. However, the exact value of maximum in FF(I_{sc}) curves depends on the grid design.

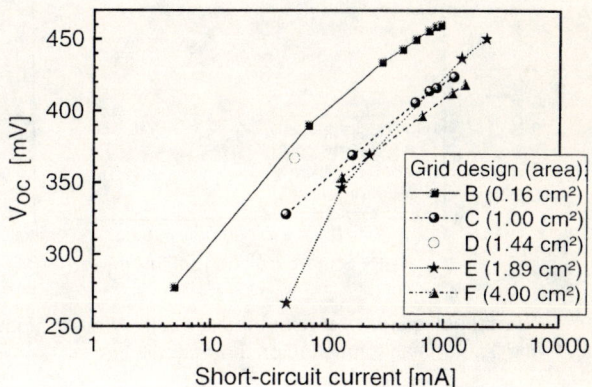

FIGURE 10. V_{oc} of PV cells with different areas and grid designs (see Fig. 8 and Table 1) as a function of short-circuit current.

49

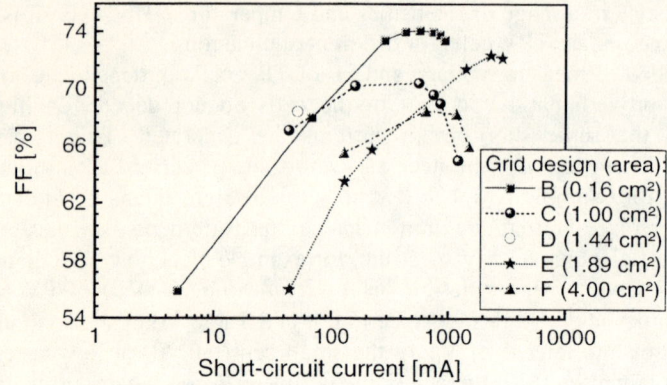

FIGURE 11. FF of PV cells with different areas and grid designs (see Fig. 8 and Table 1) as a function of short-circuit current.

Figure 12 shows temperature coefficients of V_{oc} measured between 25°C and 75 °C as a function of I_{sc}. The cells of type B, C and F were tested. All types of cell exhibit the following dependence: the higher the I_{sc} value the lower the temperature coefficient of V_{oc}. The values of temperature coefficients of V_{oc} were in the range between -1.42 mV/K and -1.86 mV/K.

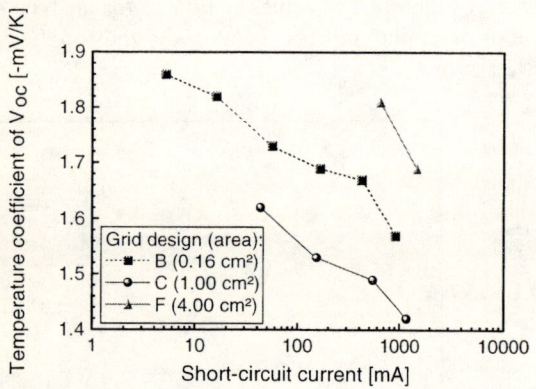

FIGURE 12. Temperature coefficients of V_{oc} for the cells with different areas and grid designs (see Fig. 8 and Table 1) as a function of short-circuit current.

Figure 13 shows temperature coefficients of FF measured between 25°C and 75 °C as a function of I_{sc}. The cells of type B, C and F were tested. For the cells with low series resistance (types B,C) the decrease of the temperature coefficient of FF takes place at low I_{sc} values due to the temperature dependence of V_{oc}. At higher currents the influence of series resistance becomes stronger and leads to the increase of the temperature coefficient. The values of temperature coefficients of FF were in the range between -0.099 abs.%/K and -0.179 abs.%/K.

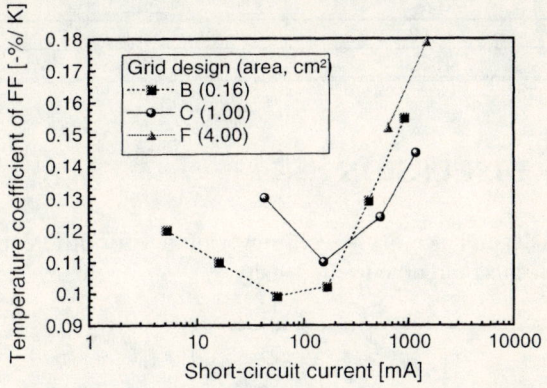

FIGURE 13. Temperature coefficients of FF for the cells with different areas and grid designs (see Fig. 8 and Table 1) as a function of short-circuit current.

We have performed calculations of a PV cell efficiency for blackbody temperatures 1000-1750 K. The calculations were based on the measured external quantum efficiency (Fig. 9), open circuit voltage and fill factor of the type B cell (Figs.10,11). It was assumed that 60% of the blackbody irradiation reaches the cell. Fig. 14 shows the results of these calculations for two different conditions: (i) cut-off of the not absorbed IR part of the illumination spectrum ($\lambda > 1800$ nm), (ii) the whole blackbody illumination spectrum. The efficiency for a blackbody illumination (T=1250-1750K) is as high as 20-25%, if one assumes an illumination spectrum with a cut-off wavelength at $\lambda = 1800$ nm

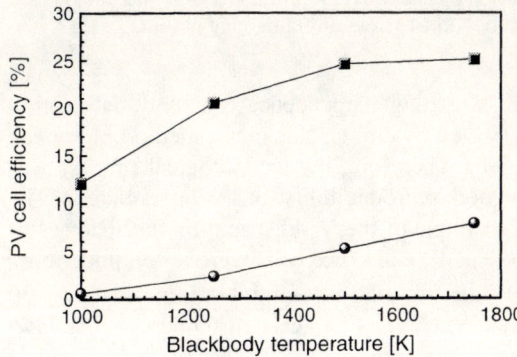

FIGURE 14. Efficiency for a blackbody illumination as a function of the blackbody temperature.
The upper curve: An illumination spectrum with a cut-off wavelength at $\lambda = 1800$ nm is assumed.
The lower curve: The whole blackbody spectrum.

Table 2 shows the output power of the GaSb cell illuminated with blackbody spectra at different temperatures together with its efficiency.

TABLE 2. Output power and efficiency of the GaSb cell illuminated with blackbody spectra at different temperatures

Blackbody temperature (K)	Output power (mW/cm^2)	Efficiency (%)
1000	27	12.1
1250	328	20.5
1500	1443	24.6
1750	4083	25.1

DISCUSSION

Fig. 15 shows the measured dependence of the cell power output for different types of cells in dependence on short-circuit current density.

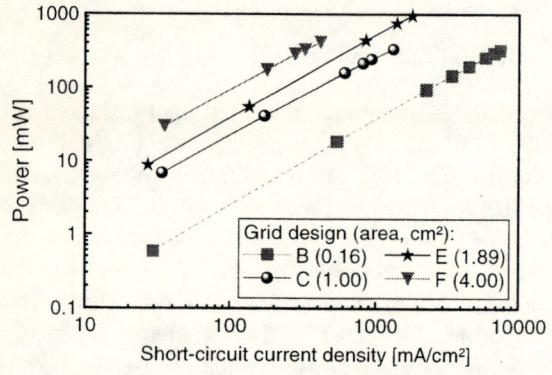

FIGURE 15. Measured power output of PV cells with different areas and grid designs (see Fig. 8 and Table 1) as a function of short-circuit current density at 25°C.

PV cell parameters exhibited a strong dependence on cell fabrication technology. For example, V_{oc} is affected by emitter thickness, method of contact fabrication and the surface treatment. Therefore, the technological process was optimized in order to obtain a good reproducibility of PV parameters. The histograms on Figure 16 show an example of the V_{oc} distribution in 2 batches of 1.44 cm^2 (Type D) and 1.89 cm^2 (Type E) cells. The cells were tested under one-sun illumination conditions (AM1.5g) with the short-circuit current density $\approx$ 30 mA/cm^2. One should mention that V_{oc} values of the cells, operating at low-illumination conditions (e.g. one sun), are especially sensitive to emitter thickness, method of contact fabrication and the surface treatment. Therefore, we consider the V_{oc} distribution to be quite good.

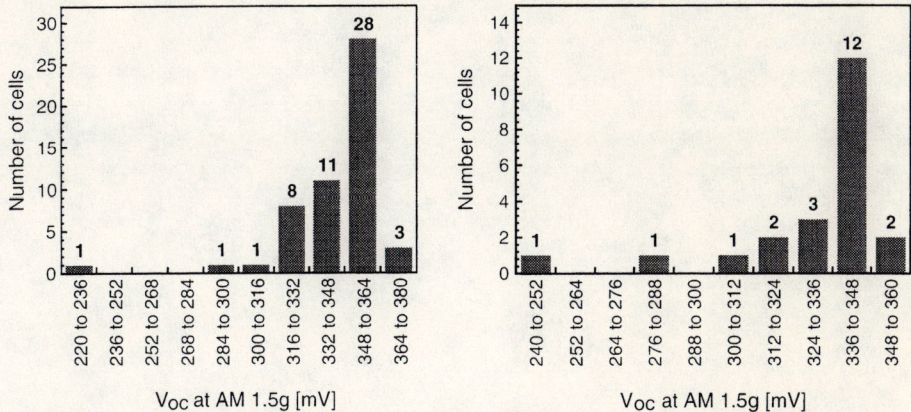

FIGURE 16. V_{oc} distribution of 50 cells of type D (left) and 22 cells of type E (right) at one-sun illumination (AM1.5g) corresponding to the short-circuit current density $\approx 30 mA/cm^2$.

ACKNOWLEDGEMENTS

The authors wish to thank Dr. B. Ber for SIMS measurements, Dr. A. Mintairov for Raman spectroscopy measurements, R. Beckert for the solar cells parameter measurement, Prof. J. Luther and Prof. W. Wettling for fruitful discussions. This work was partly financed by the Federal Ministry of Education, Science, Research and Technology (BMBF), Germany (contract 0328554D). The authors are responsible for the content of this publication.

REFERENCES

1. Fraas, L., Ballantyne, R., Samaras, J., and Seal, M., "A thermophotovoltaic electric generator using GaSb cells with a hydrocarbon burner", in Proceedings of the 1-st World Conference on Photovoltaic Energy Conversion, Waikoloa, Hawaii, December 5-9, 1994, pp. 1713-1716.
2. Shellenbarger, Z.A., Mauk, M.G., DiNetta, L.C., and Charache, G.W., " Recent progress in InGaAs/GaSb TPV Devices", in the Proceedings of the 25th IEEE Photovoltaic Specialists Conference, Washington, DC, May 13-17, 1996, pp. 81-84.
3. DaCunha, S.F., and Bougnot, J., Phys. Stat. Sol. **22**, 205 (1974).
4. Bett, A.W., Keser, S., and Sulima, O.V., Journal of Crystal Growth, 1997, (in press).
5. Sulima, O.V., Faleev, N.N., Kazantsev, A.B., Mintairov, A.M., and Namazov, A., "Low-temperature Zn diffusion for GaSb solar cell structures fabrication", in Proceedings of the 4-th European Space Power Conference, Poitiers, France, September 5-8, 1995, pp. 641-644.
6. Schubnell, M., Gabler, H-J., and Broman, L., "Overview of European activities in thermal PV", this conference.
7. Hasegawa, H., and Hartnagel, H.L., J. Electrochem. Soc., **123** , 713 (1976).

Modeling of InGaSb Thermophotovoltaic Cells and Materials

M.Zierak*, J.M.Borrego,* I.Bhat*, R.J.Gutmann* and G.Charache+

*Center for Integrated Electronics and Electronic Manufacturing
Department of Electrical, Computer and Systems Engineering
Rensselaer Polytechnic Institute, Troy, New York 12180
+Lockheed Martin, Inc., Schenectady, New York 12301

Abstract. A closed form computer program has been developed for the simulation and optimization of $In_XGa_{1-X}Sb$ thermophotovoltaic cells operating at room temperature. The program includes material parameter models of the energy bandgap, optical absorption constant, electron and hole mobility, intrinsic carrier concentration and index of refraction for any composition of GaInSb alloys.

INTRODUCTION

Recent work on the performance of thermophotovoltaic (TPV) cells operating with black body radiation sources in the temperature range between 1000 °K and 2000 °K shows that the energy bandgap must be in the range between 0.5 eV and 0.6 eV [1]. Since there are no elemental or binary compound semiconductors with energy bandgap in the above range, ternary alloy systems must be considered. The two most important ternary alloy systems being considered at present are InGaAs and InGaSb. In previous publications [2,3]the theoretical and experimental performance of TPV cells using InGaAs alloys have been presented. This paper presents the methodology we have used for calculating the performance of TPV cells using InGaSb alloys.

The methodology we have developed consists of two parts. The first part consists of modeling the material parameters which determine the performance of TPV cells such as energy bandgap, optical absorption constant, mobility of both electrons and holes, intrinsic carrier concentration and index of refraction. The second part of the methodology consists of developing a computer program which calculates the performance of TPV cells such as internal quantum efficiency, J-V characteristics under dark and illuminated conditions using blackbody radiation source with temperature in the range between 1000 °K and 2000 °K.

The paper is arranged in four parts. In the first part the calculation of the important material parameters of the GaInSb alloy system as a function of composition is described. The second part describes briefly the internal quantum efficiency calculation. In the third part some of the results of the simulation are presented, with special emphasis on the maximum efficiency that can be obtained for a given temperature of the blackbody radiation source. The last part deals with conclusions obtained from this modeling of the performance of GaInSb TPV cells.

CP401, *Thermophotovoltaic Generation of Electricity: Third NREL Conference,*
edited by Benner/Coutts
© 1997 The American Institute of Physics 1-56396-734-0/97/$10.00

MODELING OF In$_X$Ga$_{1-X}$Sb ALLOY MATERIAL PARAMETERS

In order to calculate the performance of In$_X$Ga$_{1-X}$Sb alloy TPV cells it is necessary to know some of the material parameters as a function of alloy composition X. The material parameters which determine the performance of a TPV cell are: the energy bandgap, E$_G$; the optical absorption constant, α; the hole, μ_P, and electron, μ_N, carrier mobility; the intrinsic carrier concentration, n$_i$; and the optical index of refraction, N$_R$. The following expression given by Roth et al. [4] for the energy bandgap was used:

$$E_G(X) = 0.714 - 0.945X + 0.397X^2 \tag{1}$$

This relationship, depicted in Figure 1, shows that the alloy composition X for obtaining material with energy bandgaps of 0.6, 0.55 and 0.5 eV are 0.13, 0.19 and 0.25 respectively. Roth et al [4] have also given expressions for the variation of the energy bandgap with temperature as shown in Figure 2 for four different compositions of special interest. The curves in the figures indicate that the variation of E$_G$ with temperature is very linear and of the order of 0.5 mV / ^{0}C . If the dark saturation current of GaInSb TPV cells is diffusion limited, then the variation of the open circuit voltage V$_{OC}$ with temperature is the same as the variation of E$_G$ and of the order of 0.5 mV / ^{0}C. If the dark saturation current is limited by generation-recombination in the depletion layer, then the variation of V$_{OC}$ with temperature is half of the variation of E$_G$ with temperature i.e. 0.25 mV / ^{0}C.

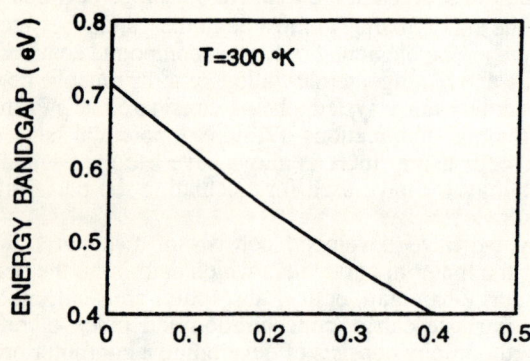

FRACTIONAL COMPOSITION X

FIGURE 1. Energy Bandgap of InGaSb Alloys as Function of Composition

56

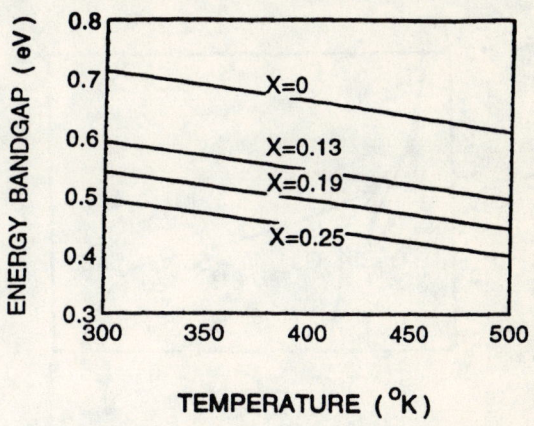

FIGURE 2. Variation of the Energy Bandgap with Temperature for InGaSb Alloys

The exact calculation of the absorption constant for a $In_XGa_{1-X}Sb$ alloy is rather complicated, requiring knowledge of the variation of the critical points of the band structure with alloy composition. A simpler approach, which has shown to be valid for AlGaAs alloys, is to consider only the most important critical point of the band structure, namely the energy bandgap. Using this procedure the absorption constant of the alloy can be calculated from the experimentally determined absorption constant of GaSb and InSb. The absorption constant $\alpha(E)$ of an $In_XGa_{1-X}Sb$ alloy with energy bandgap E_G is obtained from the absorption constant, $\alpha_1(E)$, of InSb with energy bandgap E_1; and from the absorption constant, $\alpha_2(E)$, of GaSb with energy gap E_2 using the equation:

$$\alpha(E) = X\alpha_1(E - E_G + E_1) + (1 - X)\alpha_2(E - E_G + E_2) \tag{2}$$

Figure 3 shows the experimental absorption constants of GaSb and InSb given by Seraphin and Bennett [5] and the calculated absorption constant of a $In_XGa_{1-X}Sb$ alloy with energy bandgap of 0.55 eV (X=0.19). Notice that the total distance for photon absorption in the alloy is longer than in GaSb and shorter than in InSb. The GaInSb cells require between 5 to 10 microns of semiconductor material for total photon absorption; therefore, the minority carrier diffusion in that active region should be at least comparable for good collection efficiency.

The carrier mobility of ternary alloys is difficult to model since disordered alloy scattering, a rather complex theoretical problem must be taken into account. Since the simulation of the TPV cells is not strongly dependent upon the exact value of the carrier mobility, Matthiessen's rule was used to calculate the mobility of the alloy:

$$\frac{1}{\mu_{ALLOY}} = \frac{X}{\mu_{InSb}} + \frac{(1-X)}{\mu_{GaSb}} \tag{3}$$

57

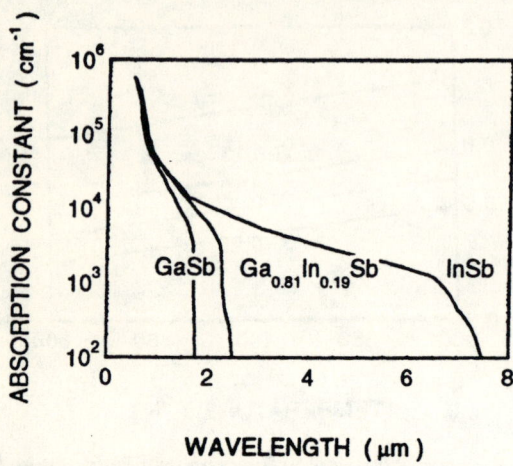

FIGURE 3. Optical Absorption Constant of GaSb, InSb and of an InGaSb Alloy with Energy Bandgap of 0.55 eV

The carrier mobilities of InSb and GaSb were taken from experimental data and fitted to the following empirical equation [6,7,8]:

$$\mu = \mu_{MIN} + \frac{\mu_{MAX} - \mu_{MIN}}{\left(1 + \dfrac{N}{N_{REF}}\right)^A} \tag{4}$$

where N is the impurity concentration and the other constants are listed in Table I for InSb and GaSb materials. Plots of the electron and hole mobilities for $In_X Ga_{1-X} Sb$ alloys of several compositions are given in Figures 4 and 5. Note that alloy scattering is not included.

The intrinsic carrier concentration, n_i, is another material parameter which is needed to properly simulate a TPV cell. The intrinsic carrier concentration of a direct bandgap semiconductor with parabolic energy bands is given by:

$$n_i^2 = N_C(X) N_v(X) \exp - \frac{E_G(X)}{kT} \tag{5}$$

TABLE I. Mobility Parameters used for Modeling the Mobility of InGaSb Alloys

Material/Carrier	μ_{MIN} (cm²/ V-s)	μ_{MAX} (cm²/ V-s)	N_{REF} (cm⁻³)	A -
InSb/h	10^2	7.5×10^2	6×10^{17}	0.6
InSb/e	5×10^3	7.8×10^4	7×10^{16}	0.7
GaSb/h	70	1.4×10^3	2×10^{17}	0.5
GaSb/e	10^2	4.5×10^3	8×10^{17}	0.8

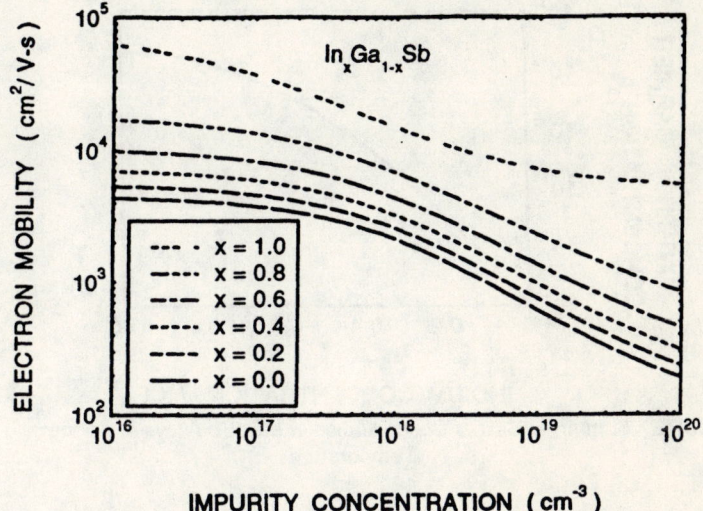

FIGURE 4. Electron Mobility of InGaSb Alloys as Function of Doping Concentration

where $N_C(X)$ and $N_V(X)$ are the density of states of the conduction and valence band, respectively and X is the alloy composition. Since the density of states have a 3/2 power dependence upon the effective masses, the variation of the electron and hole effective masses with alloy composition X is required to calculate n_i . A possible way to calculate this variation is to use the equation [9]:

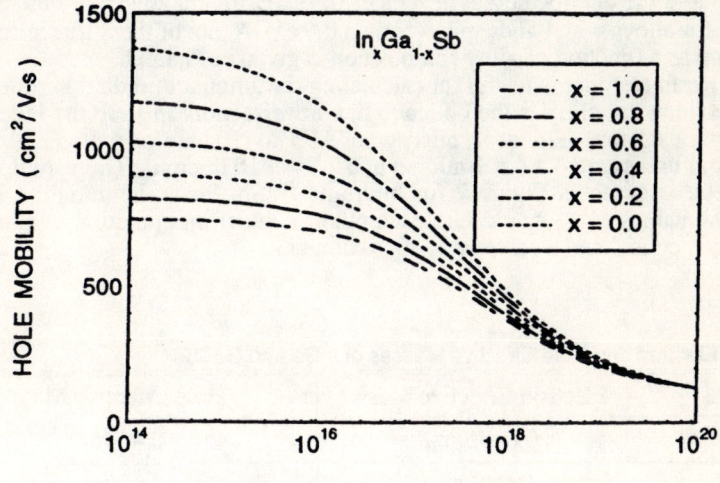

FIGURE 5. Hole Mobility of InGaSb Alloys as Function of Doping Concentration

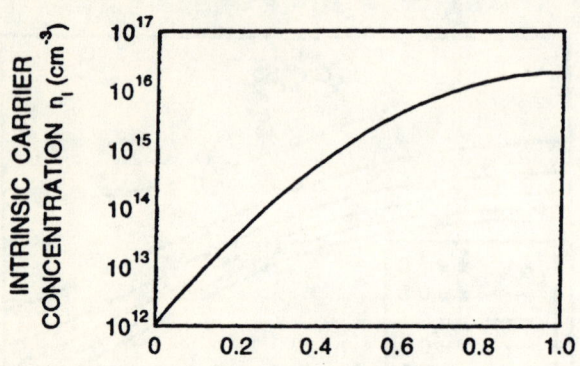

FIGURE 6. Intrinsic Carrier Concentration of InGaSb Alloys at Room Temperature

$$\frac{1}{m^*_{ALLOY}} = \frac{X}{m^*_{InSb}} + \frac{(1-X)}{m^*_{GaSb}} \tag{6}$$

which is based on the *kp* method of calculating energy bands. The effective masses used for electrons and holes in InSb and GaSb are given in Table II. In the calculation of the intrinsic carrier concentration, the contribution from the conduction energy band minima located at the L symmetry point of the Brillouin zone was neglected. This assumption is reasonable since the separation between this minima and the one located at the center of the Brillouin zone (Γ) increases appreciable for alloys with bandgap lower than 0.6 eV. A plot of the intrinsic carrier concentration as a function of alloy composition is given in Figure 6.

The last parameter that is useful for calculating the amount of radiation reflected by the semiconductor alloy is the optical index of refraction. In InSb the index of refraction varies from 4.24 at 1 micron to 3.95 at 10 microns. In GaSb the corresponding values are 4.12 at 1 micron and 3.84 at 10 microns. The variation of this parameter is given in Figure 7 for InSb and GaSb. Since the variation is so small and the values are very close to each other, a linear interpolation with alloy composition is considered a reasonable approximation.

TABLE 2. Electron and Hole Effective Masses of InSb and GaSb

Material	Electron Effective Mass	Hole Effective Mass
InSb	$0.0145\ m_0$	$0.40\ m_0$
GaSb	$0.042\ m_0$	$0.40\ m_0$

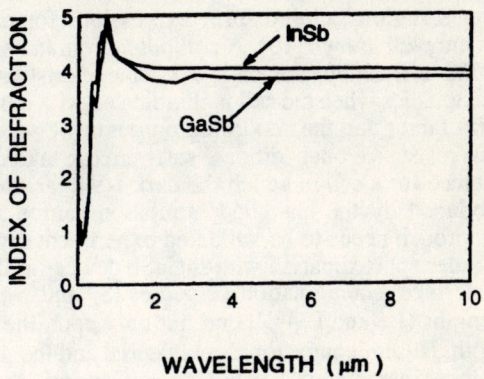

FIGURE 7. Optical Index of Refraction as Function of Wavelength for InSb and GaSb

SIMULATION OF $In_XGa_{1-X}Sb$ TPV CELLS

The performance of TPV cells illuminated by blackbody thermal sources of known temperature can be calculated from the material parameters of the $In_XGa_{1-X}Sb$ as function of alloy composition. Figure 8 shows the structure assumed for the TPV cell which consists of four uniformly doped regions; that is, a neutral emitter region of length W_E, a depletion layer of width W_D, a neutral base region of length W_B and a heavily doped substrate. Additional parameters which are needed for calculating the performance of the TPV cell are the recombination velocity, S_E, at the emitter surface; the recombination velocity, S_B, at the base substrate interface; the diffusion length of minority carriers at the emitter, L_E; and the diffusion length, L_B, of minority carriers in the base. The total internal quantum efficiency $Q(\lambda)$ as function of wavelength λ is the sum of the quantum efficiencies of the emitter $Q_E(\lambda)$, the depletion layer $Q_D(\lambda)$ and the base $Q_B(\lambda)$:

$$Q(\lambda) = Q_E(\lambda) + Q_D(\lambda) + Q_B(\lambda) \qquad (7)$$

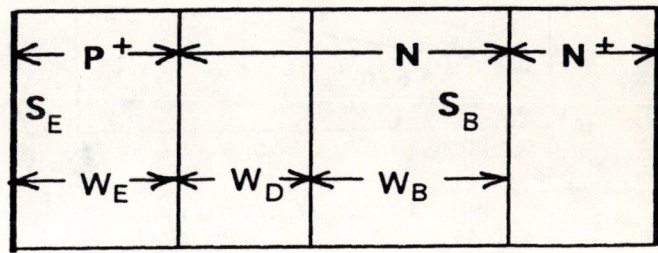

FIGURE 8. Schematic Diagram of a TPV Cell

For uniformly doped regions, closed form expressions for each of the above quantum efficiencies are well known [10]. A computer program was developed that calculates the quantum efficiency, the dark J-V characteristics of the cell, the illuminated J-V characteristics when the cell is illuminated with a blackbody of temperature T_S, the fill factor and the maximum output power density. This model does not include series resistance but additional software can take into account any series or shunt resistance for a cell in which the dark J-V characteristics as well as the photocurrent produced by the blackbody radiation source are known. The validity of such an approach needs to be validated experimentally and can only be considered as a first-order approximation with InGaSb devices at this time.

The surface and interface recombination velocities (S_E and S_B) and the emitter and base diffusion lengths (L_E and L_B) depend not only upon the doping, but also upon the quality of both the semiconductor alloy material and the device processing. Although values of these parameters are not known apriori, their range can be bounded. For a good surface or interface, the value of the recombination velocity is of the order of 10^3 cm/sec and for a bad surface or interface the value is of the order of 10^6 to 10^7 cm/sec. The minority carrier diffusion length is more difficult to bound, but an upper limit to the minority carrier lifetime is given by the radiative recombination limit [11]. The excess carrier lifetime in the radiative recombination limit was calculated as a function of the alloy composition for carrier densities of 10^{15}, 10^{16} and 10^{17} cm^{-3} as shown in Figure 9. For a given alloy composition, the radiative recombination lifetime is inversely proportional to the doping concentration and increases as the bandgap decreases. This figure and the carrier mobilities given in Figures 4 and 5 can be used for determining upper bounds to the diffusion length of minority carriers given a doping concentration and an alloy composition.

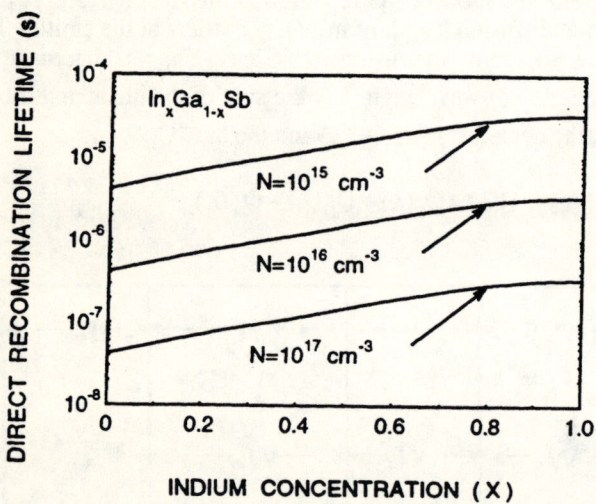

FIGURE 9. Radiative Recombination Lifetime of InGaSb as Function of Alloy Composition for Several Doping Concentrations

SIMULATION RESULTS

Using the computer program the efficiency of N+P and P+N cells with an energy band gap of 0.55 eV was optimized when illuminated by a 1500 °K blackbody radiation source. In order to reduce the parameters to be optimized, the surface emitter and base interface recombination velocities have been assumed to have a low value of 10^3 cm/sec. The minority lifetime has been set to the upper-bound radiative recombination limit and is determined by the doping in that region. Also the depletion layer width has been set to be 0.1 micron, independent of the emitter and base doping. With these assumptions, the efficiency depends only upon four parameters: the widths of the emitter and base regions (W_E, and W_B) and the doping concentrations (N_E, and N_B) in the same regions. The optimum values obtained for these parameters for achieving maximum efficiency are given in Table III and depend upon the type of cell. Note that the diffusion length in the base, L_B, is significantly greater than the diffusion length in the emitter, L_E, since the emitter doping is at least a order of magnitude greater than the base doping.

The efficiency achieved without any below-bandgap filter and with an ideal below-bandgap filter as well as the associated output power density are given in Table IV. An ideal below-bandgap filter reflects all photons with energy below the bandgap to the thermal radiator where these photons are assumed to be recycled without energy loss. In both cases an ideal matching filter for above-bandgap radiation has been assumed. The open circuit voltage is 0.365 V and the fill factor is 0.75 for these conditions. The results given in Table IV show that N+P and P+N cell are capable of achieving the same upper bound to efficiency and output power density if properly optimized. Furthermore, the results of Table III indicate that the

TABLE III. Parameter Values of N+P and P+N TPV Cells for MaximumEfficiency

Cell Type	N_E (cm-3)	N_B (cm-3)	W_E (μm)	W_B (μm)	L_E (μm)	L_B (μm)
N+P	5×10^{18}	10^{17}	0.1	7	1	27.2
P+N	10^{18}	5×10^{16}	0.7	6	6.4	16.8

TABLE IV. Maximum Efficiency of a 0.55 eV InGaSb TPV Cell with /without Ideal Filters

Cell Type	Efficiency without Filter Below-Bandgap (%)	Efficiency with Ideal Filter Below-Bandgap (%)	Output Power Density (W / cm2)
N+P	12.5	34.9	3.58
P+N	12.4	34.6	3.55

total active region of the cell should be of the order of 7 microns, with the doping in the emitter of the order of 10^{18} cm^{-3} and in the base of the order of 10^{17} cm^{-3}. These design values assume that radiative recombination is limiting.Devices fabricated in GaSb and InGaSb to date have lower diffusion lengths.

CONCLUSIONS

In this paper we have presented the methodology used for obtaining the material parameters of $In_xGa_{1-x}Sb$ alloys which are important for calculating the performance and efficiency of TPV cells. Using these material parameters the influence of many of the cell parameters on cell efficiency and performance can be evaluated. Assuming low values of surface recombination velocity and assuming the lifetime to be limited by radiative recombination, a closed form computer model is used to optimize a cell of 0.55 eV bandgap illuminated by a blackbody radiation source at 1500 ^{0}K. The maximum possible efficiency is independent of the cell type, N+P or P+N, and equal to 12.5% without any below-bandgap filter and 35% with an ideal below-bandgap filter which reflects all the below-bandgap radiation to the thermal radiator in a reusable form. Factors such as series resistance and non-diffusive junction current (eg. tunneling current) will reduce these upper-bound values in actual devices, as well as the mobility and lifetime assumptions explicity indicated previously.

REFERENCES

1. P.F. Baldasaro et al., *1st NREL Conference on Thermophotovoltaic Generation of Electricity*, AIP Proceedings **321**, 29, (1994).
2. M. Zierak, et al., *1st NREL Conference on Thermophotovoltaic Generation of Electricity*, AIP Proceedings **321**, 457, (1994).
3. S. Wojtczuk, *1st NREL Conference on Thermophotovoltaic Generation of Electricity*, AIP Proceedings **358**, 387, (1995).
4. A.P. Roth et al., *Canadian Journal of Physics*, **58**, 560, (1980).
5. B.O. Seraphin and H.E. Bennett in *Semiconductors and Semimetals* Vol. 3, R.K. Willardson and A.C. Beer, eds., Academic Press, New York, 1967, p. 469.
6. J.D. Wiley, *Semiconductors and Semimetals* Vol.10, R.K. Willardson and A.C. Beer, eds., Academic Press, New York 1975, p. 154,157
7. C. Hilsum and A.C. Rose-Innes, *Semiconducting III-V Compounds,* Pergamon Press, New York, 1961, p. 128-140.
8. A.G.Milnes and A.Y.Polyakov, *Solid State Electronics Journal,* **36**, 803, (1993).
9. S.Adachi, *Physical Properties of III-V Semiconductor Compounds,* J.Wiley and Sons, New York, 1992, p.91.
10. S.M. Sze, *Physics of Semiconductor Devices*, 2nd. Edition, J.Wiley and Sons, New York, 1981, p. 800.
11. W. Van Roosbroech and W. Shockley, *Phys. Rev.* **94**, p. 1558, (1954).

Growth and Characterization of $In_{0.2}Ga_{0.8}Sb$ Device Structures Using Metalorganic Vapor Phase Epitaxy

H. Ehsani, I. Bhat, C. Hitchcock and R. Gutmann

Department of Electrical, Computer and Systems Engineering
Center for Integrated Electronics and Electronics Manufacturing
Rensselaer Polytechnic Institute, Troy, NY 12180-3590

G. Charache and M. Freeman

Lockheed Martin Inc., Schenectady, New York 12301-1072

Abstract: $In_{0.2}Ga_{0.8}Sb$ epitaxial layers and thermophotovoltaic (TPV) device structures have been grown on GaSb and GaAs substrates by metalorganic vapor phase epitaxy (MOVPE). Control of the n-type doping up to $1x10^{18}cm^{-3}$ was achieved using diethyltellurium (DETe) as the dopant source. A Hall mobility of greater than 8000 cm^2/Vs at 77K was obtained for a $3x10^{17}cm^{-3}$ doped $In_{0.2}Ga_{0.8}Sb$ layer grown on high-resistivity GaSb substrate. The $In_{0.2}Ga_{0.8}Sb$ epilayers directly grown on GaSb substrates were tilted with respect to the substrates, with the amount of tilt increasing with the layer thickness. Transmission electron microscopy (TEM) studies of the layers showed the presence of dislocation networks across the epilayers parallel to the interface at different distances from the interface, but the layers above this dislocation network were virtually free of dislocations. A strong correlation between epilayer tilt and TPV device properties was found, with layers having more tilt providing better devices. The results suggest that the dislocations moving parallel to the interface cause lattice tilt, and control of this layer tilt may enable the fabrication of better quality device structures.

INTRODUCTION

$In_xGa_{1-x}Sb$ material system is attractive for application in TPV cells since the energy bandgap can be varied from 0.17 eV to 0.72 eV by varying the composition(1).By adding arsenic into the system, lattice-matched $In_xGa_{1-x}As_ySb_{1-y}$ structures can be grown on GaSb substrates. Even though antimonides may have some advantages compared to an $In_xGa_{1-x}As$ system for TPV applications, the

CP401, *Thermophotovoltaic Generation of Electricity: Third NREL Conference,*
edited by Benner/Coutts
© 1997 The American Institute of Physics 1-56396-734-0/97/$10.00

growth technology for $In_xGa_{1-x}Sb$ is not as advanced. MOVPE growth of antimonides has emphasized lattice-matched $In_xGa_{1-x}As_ySb_{1-y}$ or the binary compounds GaSb and InSb. Doping studies and dislocation reduction techniques for lattice-mismatched epitaxy have not been investigated in detail for $In_xGa_{1-x}Sb$ layers grown on GaSb substrates (2). Studies on doping characteristics and on methods to reduce dislocation densities in the active layers are required for better antimonide devices.

Preliminary results on the growth and doping of $In_{0.2}Ga_{0.8}Sb$ by MOVPE (3) and p-type doping using Si have been presented before (4). In this paper, recent results on the n-type doping of $In_{0.2}Ga_{0.8}Sb$ by Te and structural characterization of $In_{0.2}Ga_{0.8}Sb$ layers are presented. These epilayers are tilted with respect to the substrates, with the amount of tilt increasing with layer thickness. This result has important implications for the design of step grading techniques for epitaxial growth of device quality $In_{0.2}Ga_{0.8}Sb$.

EXPERIMENTAL PROCEDURES

The ternary epilayers were grown on (100) oriented GaSb and semi-insulating (100) GaAs substrates in a low pressure, rf heated, horizontal MOVPE reactor. Trimethylgallium, trimethylindium, trimethylantimony and diethyltellurium (DETe) were used as the Ga, In, Sb and Te sources, respectively. After degreasing the wafers in organic solvents, the GaAs substrates were etched in Caro's etch (solution containing $H_2SO_4:H_2O_2:H_2O$, 5:1:1 by volume) for two minutes and the GaSb wafers were etched in 1% bromine-methanol solution for 30 seconds. The GaSb substrates were of low resistivity p-type at room temperature, but were of high resistivity at 77K (sheet resistance of 600   per square) so that Hall measurements could be made on these at 77K.

Double crystal x-ray diffraction was used to determine layer tilt with respect to the substrate, crystalline quality of the layers, composition of the layers, and lattice relaxation. Variation of the peak separation between the epilayers and the substrates as a function of rotational angle (azimuth) was used to determine the lattice mismatch and the epilayer tilt. The residual strain in some films was measured by x-ray diffraction spectra of non-symmetric planes such as (115) and symmetric plans such as (004). Details of the measurement technique can be found elsewhere (5).

RESULTS AND DISCUSSIONS

Earlier work (3) reported n-type doping of $In_{0.2}Ga_{0.8}Sb$ with Te for DETe mole fractions higher than $1x10^{-6}$, where the carrier concentration actually decreased as the DETe mole fraction was increased. This decrease was attributed to the formation of Te precipitates or Ga-Te and Sb-Te reactions at the growth surface. These reactions probably resulted in inclusions of undesired precipitates in the layer, causing degradation of crystal quality. Secondary ion mass spectrometry (SIMS) measurements showed that only 2-3% of the incorporated Te was active when the total Te concentration was about 2×10^{19} cm^{-3}. The mobility of the layers also decreased as the DETe mole fraction was increased.

In this work, this study was extended to lower concentrations of Te, by modifying the reactor with the addition of a "double dilution" scheme to deliver DETe. In this method, hydrogen through an additional line is used to dilute the DETe/H$_2$ gas mixtures. Only a small portion of the DETe/H$_2$ is delivered to the reactor, and the rest is passed through the exhaust lines. The mole fraction of DETe in the reactor can be controlled over a wide range (10-8 to 10-5) using this method.

Figure 1 shows the measured carrier concentration at 77K as a function of the DETe mole fraction for layers grown on GaAs and GaSb substrates. The carrier concentration increases linearly with the DETe flow initially, but beyond 10^{18} cm^{-3} the carrier concentration actually decreases with the DETe flow. The continuous line in the graph corresponds to n   (DETe)1.0 suggesting a high degree of Te activation at doping concentrations as high as 1×10^{18} cm^{-3}. TEM of lightly Te-doped In$_{0.2}$Ga$_{0.8}$Sb ($<1 \times 10^{18}$ cm^{-3}) does not show presence of any precipitates.

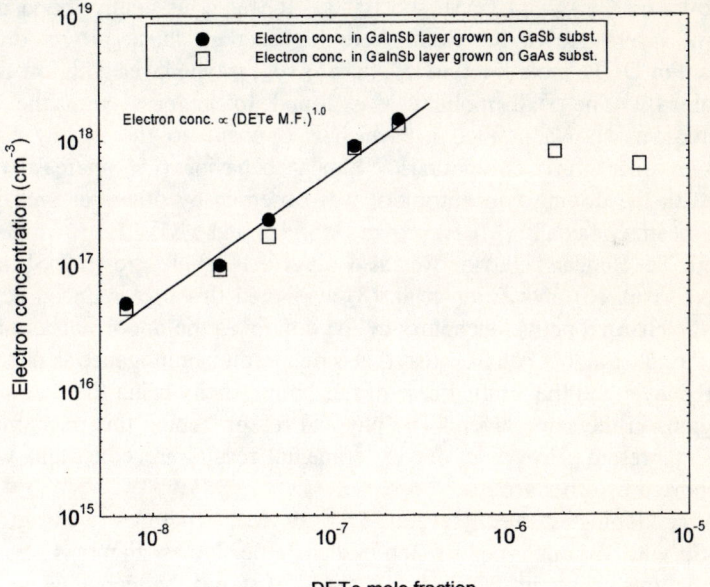

Figure 1. Carrier concentration at 77K versus DETe mole fraction in In$_{0.2}$Ga$_{0.8}$Sb grown on GaAs and GaSb substrates.

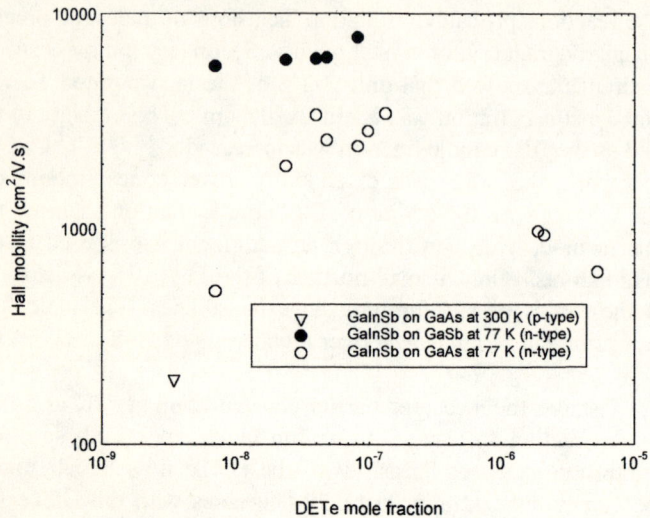

Figure 2. Hall mobility versus DETe mole fraction for $In_{0.2}Ga_{0.8}Sb$ layers grown on GaAs and GaSb substrates.

Figure 2 shows the Hall mobility at 77K versus DETe mole fraction for layers grown on GaAs and GaSb substrates. It was consistently found that the mobility of layers grown on GaSb were higher than those grown on GaAs substrates. For DETe mole fraction less than 5×10^{-9}, p-type layer with low mobility was obtained. The Hall mobility was found to increase with the doping concentration as shown in Fig. 3, which is opposite to the usually observed behavior in other III-V compounds. Similar behavior (i.e., increase of Hall mobility with the doping concentration) was observed by others as well in both molecular beam epitaxially (MBE) grown GaSb (6) and MOVPE grown GaSb (7) doped with Te. Similar behavior was also observed in bulk grown InSb material (8). Turner et al. (6) and Zitter et al. (8) attributed this to a reduced screening effect of the charged native acceptors by the donors as the donor concentration is reduced. Pascal et al. (7) believed that this is due to an inhomogeneous distribution of Te in the layer and the limiting case of this homogeneity being the presence of p and n regions in the same layers. The physical reason behind this phenomenon is not clear at present. However, our experimental results are consistent with the results reported by other groups.

After doping studies, several TPV device structures were grown in $In_{0.2}Ga_{0.8}Sb$ with various types of step-graded buffer layers. Figure 4 shows two device structures grown in this study. Figure 4(a) shows the technique commonly used, in which several step layers are grown with equal distribution of thickness and composition between the active layer and the substrate. This technique has been found to be successful in GaInAs/InP TPV structures (9). This method is particularly useful when the lattice mismatch between the active layer and the

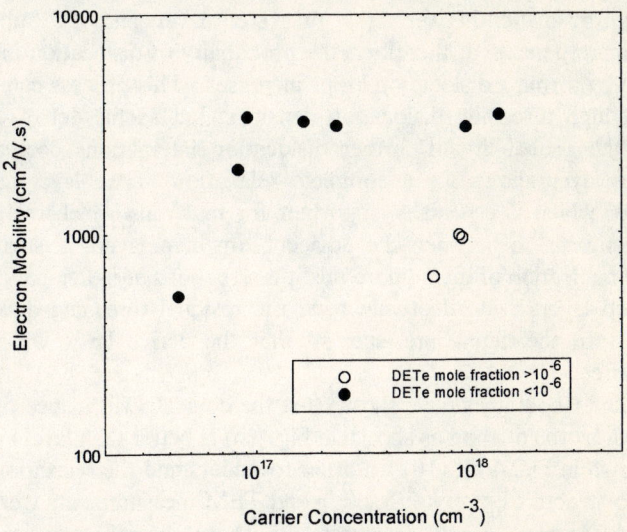

Figure 3. Hall mobility at 77K versus carrier concentration for $In_{0.2}Ga_{0.8}Sb$ layers grown on GaAs substrate. Data for high flow of DETe (mole fraction > 10^{-6}) are also plotted.

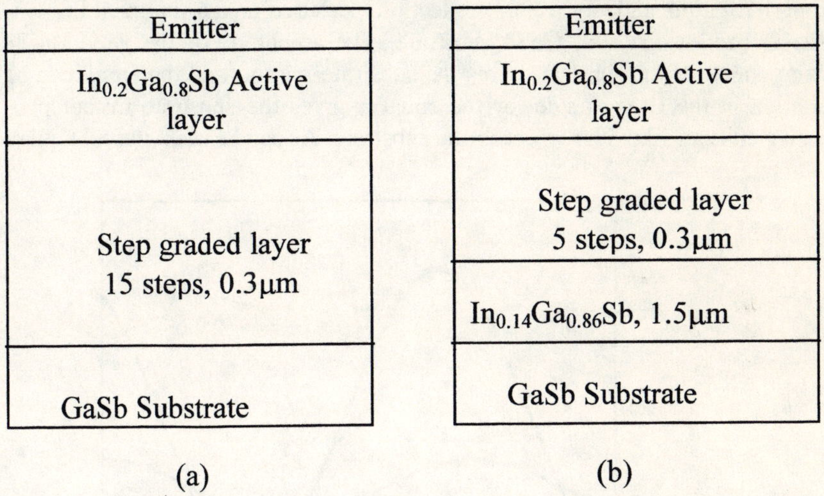

Emitter
$In_{0.2}Ga_{0.8}Sb$ Active layer
Step graded layer 15 steps, $0.3\mu m$
GaSb Substrate

(a)

Emitter
$In_{0.2}Ga_{0.8}Sb$ Active layer
Step graded layer 5 steps, $0.3\mu m$
$In_{0.14}Ga_{0.86}Sb$, $1.5\mu m$
GaSb Substrate

(b)

Figure 4 Two types of step grading methods used in this study

substrate is small, and for materials that relax very quickly beyond the critical layer thickness. Figure 4(b) shows another step-grading technique in which a thick layer with a large lattice mismatch is grown as the first step layer, followed by the growth of a few thin step layers that are closely lattice-matched.

The ideas behind the second method are as follows: (1) Since there is a large lattice mismatch between the first step layer and the substrate, many dislocations are generated; therefore, the probability of dislocation interaction and annihilation by forming dislocation loops increases. This process can be extremely effective at high threading dislocation densities, but as the defect concentration decreases, the probability of further dislocation interactions decreases. (2) A thicker step layer allows for a complete relaxation of the layer at the growth temperature, which is especially important for materials which relax slowly. (3) The lattice mismatch between the adjacent top step layers is small enough to prevent the nucleation of many more misfit dislocations at individual interfaces. (4) The top step layers will effectively bend the residual threading dislocations that propagate from the initial interface, so that the active layer will be relatively dislocation-free.

Preliminary studies have shown that the device performance of $In_{0.2}Ga_{0.8}Sb$ layers grown by the method as shown in Fig.4(b) is better than layers grown by the method shown in Fig. 4(a). (10) In order to understand the relaxation phenomena in these two structures, x-ray diffraction and TEM measurements were carried out on selected samples. X-ray measurements show that the epitaxial layers are actually tilted with respect to the substrates.

Figure 5 shows the variation of the relative Bragg angle as a function of the azimuth rotation angle for a (004) reflection measured on a 6 µm thick $In_{0.2}Ga_{0.8}Sb$ epilayer grown on (100) GaSb substrate. The amplitude of the variation in the Bragg angle for the substrate gives the substrate miscut, and the amplitude of the variation in the Bragg angle for the epilayer gives the substrate miscut plus the relative epilayer tilt with respect to the substrate. As can be seen, the tilt angle

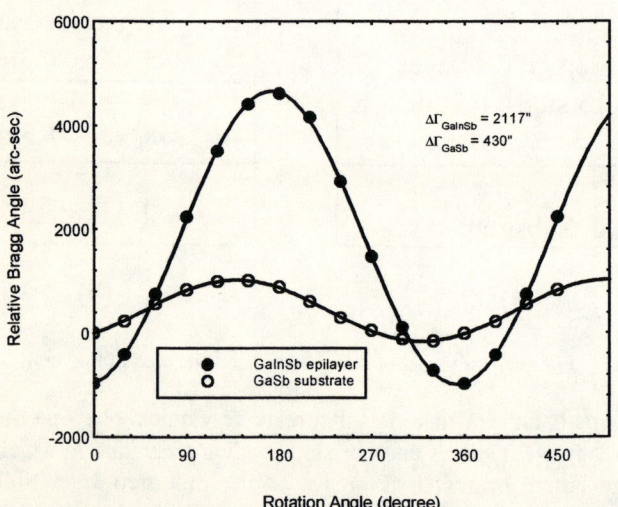

Figure 5. Relative Bragg angle versus azimuth rotation angle. The amplitude difference between the epi and the substrate indicates the layer tilt with respect to the substrate.

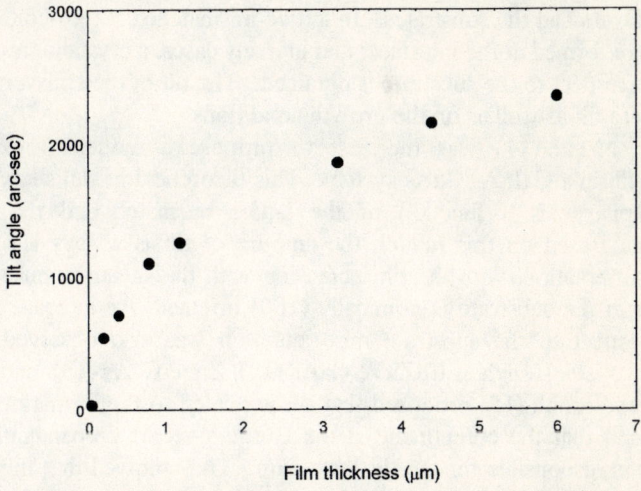

Figure 6. Epilayer tilt with respect to the substrate for different values of layer thickness.

between the $In_{0.2}Ga_{0.8}Sb$ epilayer and the GaSb substrate is much higher than the substrate misorientation, which indicates that formation of the epilayer tilt in $In_xGa_{1-x}Sb$/GaSb system is not initiated by the substrate misorientation alone. To determine whether the total tilt angle is formed in the vicinity of the interface (where the density of misfit dislocation is large) or whether it gradually increases, a systematic study was conducted. In this set of experiments, epilayers of different thickness were grown keeping all other parameters constant. Figure 6 shows that the tilt angle increases as the layer thickness increases, indicating that the layer tilt is formed continuously throughout the epilayer.

The full width at half maximum (FWHM) of the layers which is related to the dislocation density was measured and the data are shown in Fig. 7. Several points should be noted in this figure: (1) The FWHM of the 0.2 μm $In_{0.2}Ga_{0.8}Sb$ layer is lower than that of thicker layers, which could be attributed to the layer not being fully relaxed even though the layer is thicker than the critical layer thickness for lattice relaxation. (2) The FWHM increases as the thickness increases until the thickness reaches 0.6 μm, since a large population of misfit and threading dislocations form due to the 1.3% lattice mismatch. (3) The FWHM decreases between 0.6 μm and 2.5 μm, owing to the annihilation of the dislocations. (4) The FWHM of the layers do not decrease significantly beyond 2.5 μm. Also, note that the tilt of the epilayer continuously varies with the thickness (see Fig. 6). This variation of tilt with thickness contributes to the broadening of the x-ray peaks. Hence, it was concluded that the actual FWHM of the layers should be lower than that shown in Fig. 7.

A long-standing problem in the area of epitaxial growth is the lack of understanding of the relaxation rate of the epitaxial layers and the tilt formation between the layers and the substrates. In lattice-mismatched hetero-epitaxy, misfit dislocations are formed at the interface; and in many cases, a crystallographic tilt of the layer with respect to the substrate is obtained. The tilt of the epilayers depends on the substrate tilt as well as on the growth conditions.

Nagai (11) was the first to propose a model to explain the formation of tilt in a GaInAs/GaAs system. This theoretical model shows that the tilt of the epilayer is a function of the lattice mismatch and the substrate misorientation. Based on this model, the amount of tilt is always less than the substrate misorientation, with the tilt increasing with the substrate misorientation and vanishing if the substrate is nominally (100) oriented. An increase in the tilt angle with respect to the substrate misorientation was also observed in many hetero-epitaxy systems such as CdZnTe/GaAs (12), ZnSe/GaAs (13), and ZnSe/Ge (14). Olsen and Smith (15) proposed that tilt is related to the formation of misfit dislocations and that the component of the Burgers vector perpendicular to the growth plane is responsible for the tilt formation. They showed that the tilt angle of epitaxial layers is directly proportional to the misfit strain, except when misfit is relieved by pure edge dislocations. Their model can only predict an upper limit for the magnitude of tilt.

To date, the experimental results showed that the amount of tilt angle between the epitaxial layers and the substrates with exact (100) orientation (0.1-0.2° off) is less than a few hundred arc-sec. To the authors knolwledge, this is first time that a large amount of tilt was observed for growth on nominally (100) oriented substrates, and also that the epilayer tilt increases with the thickness.

To investigate the propagation behavior of dislocations, TEM microscopy was carri0 ed out on several samples. Figures 8(a) and 8(b) show the TEM bright field images for two different layers of nominally 2 μm thick $In_xGa_{1-x}Sb$ epilayers grown on GaSb substrate with a 0.3 μm thick GaSb buffer layer. The figures clearly show a zone of dislocations buried at a depth of 1.5-1.7 μm below the surface. On these and other samples that were studied, long dislocations parallel to the misfit networks were observed just above the misfit zone; and in several part of the samples these dislocations seem to bend parallel to the surface at different depths from the surface. Most of the threading dislocations which originate at the misfit networks are deflected parallel to the interface (converted to misfit dislocations) and then terminated. Beyond approximately 1μm thick layers, few dislocations were observed in the area studied by TEM. These dislocations which are parallel to the interface may be responsible for the lattice tilt. Since these dislocations are found to move parallel to the interface at different distances from the interface, the tilt is also a function of the layer thickness.

The inclination angle of the layers grown with the two step grading schemes described before was compared. It was found that when the step graded layer consists of 15 steps with equal thickness and 0.09% lattice mismatch between adjacent layers, the value of tilt angle was very low (about 35 arc-sec). On the other hand, a large tilt angle was observed (about 2100 arc-sec) when a thick

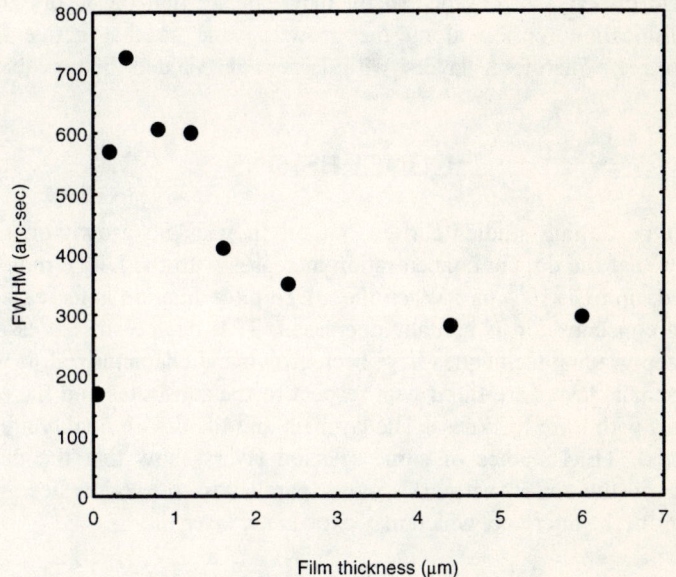

Figure 7. The FWHM of InGaSb layers grown on GaSb substrates as a function of layer thickness.

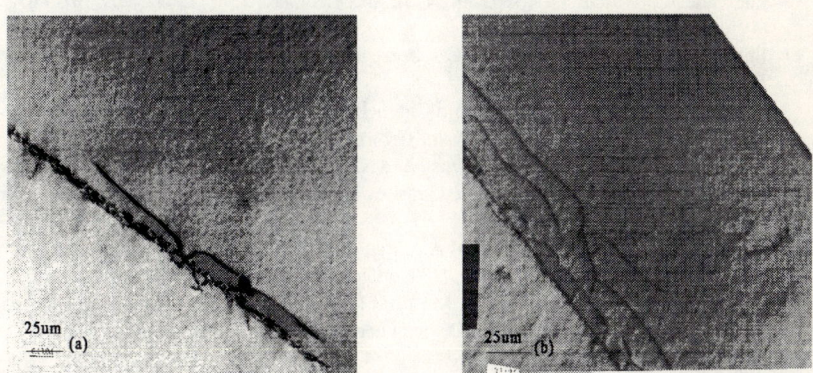

Figure 8. Bright field TEM micrographs of a Si doped (a) and Te doped (b) $In_{0.2}Ga_{1-x}Sb$ layer grown on GaSb with a 0.3μm thick buffer layer. Note the deflection of dislocations parallel to the interface, and at different distances from the interface.

initial layer of larger composition and lattice mismatch was grown followed by a few thinner steps. Since the lattice tilt does not intrinsically have any effect on the device performance, the presence of tilt may indicate that the layers are relaxed through dislocation motion along the growth plane so that active layers are dislocation-free. Therefore, layers with larger tilt yielded better quality TPV devices.

CONCLUSIONS

N-type doping studies carried out on $In_{0.2}Ga_{0.8}Sb$ grown on GaAs and GaSb show that the doping concentration increases with the DETe mole fractions as expected, up to $1x10^{18}cm^{-3}$. When the DETe mole fraction is increased further, the doping concentration is actually decreased. TPV device structures using two different step-grading techniques have been grown and characterized. It was found that the epitaxial layers are tilted with respect to the substrates and the amount of tilt increases with layer thickness. The layer tilt and the device quality are found to be correlated. TEM studies of some selected layers show that the dislocations originating at the misfit networks move parallel to the interface at various distances from the interface, which may explain the layer tilt.

REFERENCES

1. Milnes, A. G. and Polyakov A. Y., *Solid State Electronics*, **36**, 803 (1993).
2. Bougnot J., Delannoy F., Pascal F., Grosse P., Giani A., Kaoukab J., Bougnot G., Fourcade R., Walker P. J., Mason N. J., and Lambert B., *J. Cryst. Growth*, **107**, 502 (1991).
3. Ehsani H., Bhat I., Hitchcock C., Borrego J., and Gutmann R., *AIP Conference Proceedings*, **358**, 423 (1996).
4. Ehsani H., Bhat I., Gutmann R., and Charache G., *Appl. Phys. Lett*, **69**, 3867 (1996).
5. Leiberich A. and Levkoff J., *J. Crystal Growth*, **100**, 330 (1990).
6. Turner G. W., Eglas S. J., and Strauss A. J., *J. Vac. Sci. Technol.*, **B11**, 864 (1993).
7. Pascal F., Delannoy F., Bougnot J., Gouskov L., Bougnot G., Grosse P., and Kaoukab J., *J. Electron. Mater.*, **19**, 187 (1990).
8. Zitter R. N., Strauss A., and Attard A., *Phys. Rev.* **115**, 266 (1959).
9. Sharps P. R. and Timmons M. L., *AIP Conference Proceedings*, **358**, 458 (1996).
10. Hitchcock C., Ehsani H., Freeman M., Bhat I., Charache G., Borrego J., and Gutmann R., "GaInSb and GaInAsSb TPV Device Fabrication and Characterization", this conference
11. Nagai H., *J. Appl. Phys.*, **45**, 3789 (1974).
12. Ahlgren W. L., Johnson S. M., Smith E. J., Ruth R. P., Johnson B. C., Kalisher M. H., Cockrum C. A., James T. W., Arney D. L., Zieger C. K., and Lick W., *J. Vac. Sci. Tech.*, **A7**, 331 (1989).
13. Ohki A., Shibata N., and Zembutsa S., *J. Appl. Phys*, **64**, 694 (1988).
14. Kleiman J., Park R. M., and Mar H. A., *J. Appl. Phys*, **64**, 1201 (1988).
15. Olsen G. H., and Smith R. T., *Phys. Stat. Sol*, **A31**, 739 (1975).

Lattice-Matched Epitaxial GaInAsSb/GaSb Thermophotovoltaic Devices

C.A. Wang, H.K. Choi, G.W. Turner, D.L. Spears, and M.J. Manfra,

Lincoln Laboratory, Massachusetts Institute of Technology, Lexington, MA 02173-9108

G.W. Charache

Lockheed Martin, Inc., Schenectady, NY 12301

Abstract. The materials development of $Ga_{1-x}In_xAs_ySb_{1-y}$ alloys for lattice-matched thermophotovoltaic (TPV) devices is reported. Epilayers with cutoff wavelength 2 - 2.4 μm at room temperature and lattice-matched to GaSb substrates were grown by both low-pressure organometallic vapor phase epitaxy and molecular beam epitaxy. These layers exhibit high optical and structural quality. For demonstrating lattice-matched TPV devices, p- and n-type doping studies were performed. Several TPV device structures were investigated, with variations in the base/emitter thicknesses, and some with the incorporation of a high-bandgap GaSb or AlGaAsSb window layer. Significant improvement in the external quantum efficiency and open circuit voltage is observed for devices with an AlGaAsSb window layer compared to those without one.

INTRODUCTION

Recent developments of thermophotovoltaic (TPV) systems are based on thermal sources which operate in the temperature range 1100 - 1500K [1]. For high conversion efficiency, the cutoff wavelength of the photovoltaic cell should closely match the peak in emissive power of the thermal source, which for this temperature range corresponds to 1.9 - 2.6 μm. Consequently, optimized cells will

CP401, *Thermophotovoltaic Generation of Electricity: Third NREL Conference,*
edited by Benner/Coutts

be based on low-bandgap semiconductor materials. For example, InGaAs grown on InP substrates has been pursued [2,3]. However, the alloy composition that satisfies this wavelength range is lattice mismatched to the InP substrate, and defect filtering schemes must be incorporated to reduce crystalline defects. In spite of this limitation, TPV devices have exhibited external quantum efficiency (QE) as high as 50% at 2 μm [3].

An alternative low-bandgap materials system is the $Ga_{1-x}In_xAs_ySb_{1-y}$ quaternary alloy, which has the advantage of being lattice matched to either GaSb or InAs substrates. The energy gap is dependent primarily on the In content, while As determines the lattice matching. Growth on GaSb substrates is preferred over InAs substrates due to thermodyamic considerations [4], electronic band structure [5], and mechanical stability [6]. Thermodynamically stable alloys with a cutoff wavelength of 2.39 μm have been grown on GaSb by liquid phase epitaxy (LPE) [7]. Therefore, the $Ga_{1-x}In_xAs_{1-y}Sb_y$ alloys are of particular interest for TPV systems. Recently, GaInAsSb TPV devices grown by LPE and molecular beam epitaxy (MBE) have been demonstrated, and external QE exceeding 40% at 2 μm has been obtained [6,8-9].

In this paper, we report the growth of $Ga_{1-x}In_xAs_ySb_{1-y}$ alloys lattice matched to GaSb substrates by both organometallic vapor phase epitaxy (OMVPE) and MBE. Doping studies were performed, and the electrical, optical, and structural properties of these alloys grown using the different techniques are presented and compared. P-on-n $Ga_{1-x}In_xAs_ySb_{1-y}$ devices were grown on GaSb substrates and evaluated. The effects of base/emitter thickness, surface passivation layer, and higher-bandgap AlGaAsSb window layers on external QE and open circuit voltage V_{oc} are presented.

EPITAXIAL GROWTH AND CHARACTERIZATION

For OMVPE growth, $Ga_{1-x}In_xAs_ySb_{1-y}$ epilayers were grown on (100) Te-doped GaSb or semi-insulating (SI) GaAs substrates misoriented 2° toward (110) or 6° toward (111)B. A vertical rotating-disk reactor with H_2 carrier gas at a flow rate of 10 slpm and reactor pressure of 150 Torr was used [10]. All organometallic sources including solution trimethylindium (TMIn), triethylgallium (TEGa), tertiarybutylarsine (TBAs), and trimethylantimony (TMSb) were used with diethyltellurium (DETe) (50 ppm in H_2) and dimethylzinc (DMZn) (1000 ppm in H_2) as n- and p-type doping sources, respectively [11]. The total group III mole fraction was typically $3.5 - 4 \times 10^{-4}$ which resulted in a growth rate of ~2.7 μm/h. The V/III ratio was typically 1.1 - 1.3. The growth temperature ranged from 525 - 575°C. AlGaAsSb lattice matched to GaSb substrates was grown with tritertiarybutylaluminum (TTBAl), TEGa, TBAs, and TMSb as previously described [12].

For MBE growth, epilayers were grown on (100) Te-doped GaSb or SI GaAs substrates in a solid-source EPI Gen II system. Conventional effusion cells were used to provide Ga, In, and Sb_4 fluxes, and a valved As cracker to provide As_2 as described previously [13]. The growth temperature was 500 - 510°C, and the growth rate was ~1 μm/h. Be was used as the p-type dopant, and GaTe as the n-type dopant.

The surface morphology was examined using Nomarski contrast microscopy. Double-crystal x-ray diffraction (DCXD) was used to measure the degree of lattice mismatch to GaSb substrates. Photoluminescence (PL) was measured at 4 and 300K using a cooled PbS detector. Electrical properties were obtained from Hall measurements based on the van der Pauw method. The composition of epilayers was determined from DCXD splitting, the peak emission in PL spectra, and the energy gap dependence on composition based on the binary bandgaps [14]:

$$E(x,y) = 0.726 - 0.961x - 0.501y + 0.08xy + 0.415x^2 + 1.2y^2$$
$$+ 0.021x^2y - 0.62xy^2,$$

where $y = 0.867x/(1 - 0.048x)$, the condition for lattice matching to GaSb.

GROWTH RESULTS

For OMVPE growth, the sensitivity of As incorporation (which controls the lattice matching on GaSb substrates) in $Ga_{1-x}In_xAs_ySb_{1-y}$ (x ~ 0.13), was established by growing epilayers with various TBAs vapor phase concentration ratios, y_v = [TBAs]/([TBAs]+[TMSb]). The results, Figure 1, show that the lattice mismatch varies linearly with little deviation as a function of y_v,

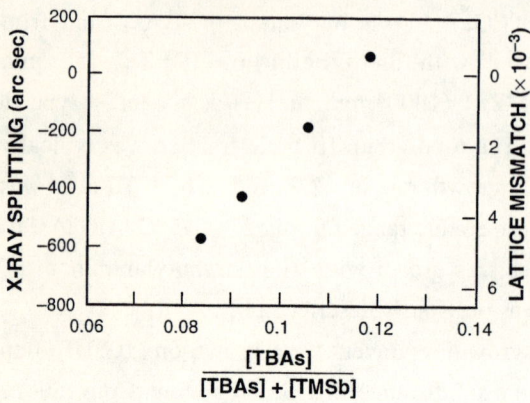

Figure 1. Dependence of lattice mismatch of GaInAsSb epilayers grown at 575°C by OMVPE on GaSb substrates as function of TBAs gas phase concentration.

indicating that TBAs provides excellent controllability of lattice-matching conditions. Similar control of lattice matching was obtained for epilayers grown by MBE.

The surface morphology of lattice-matched GaInAsSb epilayers grown by OMVPE on (100) substrates with a 2° toward (110) misorientation is mirror-like to the eye, but for $x > 0.1$, exhibits a slight texture under Nomarski contrast. For epilayers grown on substrates with a 6° toward (111)B misorientation, the surface is mirror-smooth. The morphology is mirror-smooth for MBE-grown epilayers. Cross-hatching was observed for all layers with a lattice mismatch $> 5 \times 10^{-3}$. Figure 2 shows the DCXD scan for a 2-μm-thick $Ga_{0.9}In_{0.1}As_{0.08}Sb_{0.92}$ layer. A narrow full width at half-maximum (FWHM) of 21 arc sec, which is comparable to 22 arc sec for the GaSb substrate, indicates the excellent structural quality of this layer. The x-ray splitting of 39 arc sec corresponds to a lattice mismatch of 3×10^{-4}. For lattice-matched epilayers, the DCXD scans are similar whether the layers are grown by OMVPE or MBE.

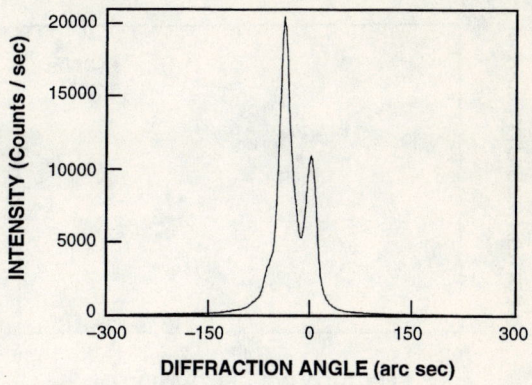

Figure 2. Double-crystal x-ray difffraction scan of $Ga_{0.9}In_{0.1}As_{0.08}Sb_{0.92}$ grown at 575°C on GaSb by OMVPE.

The optical quality of $Ga_{1-x}In_xAs_ySb_{1-y}$ epilayers was evaluated by comparing the FWHM of PL spectra measured at 4K. Figure 3 summarizes the results for epilayers grown by OMVPE and MBE. The composition of epilayers was varied to cover the 300K energy range 0.55 - 0.72 eV, corresponding to 2.4 - 1.9 μm. In general, the FWHM values are comparable for layers grown by OMVPE and MBE for 4K PL peak energy $E_{pk} > 0.58$ eV. For lower E_{pk}, FWHM values are slightly larger for layers grown by OMVPE. Also shown for comparison are data for layers grown by LPE [15] and OMVPE [16].

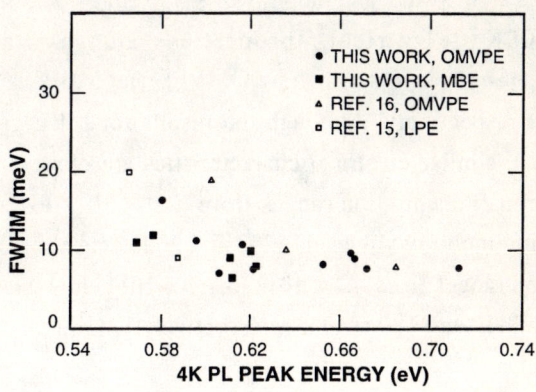

Figure 3. Full width at half-maximum of PL spectra measured at 4K of GaInAsSb layers grown on GASb substrates by OMVPE and MBE.

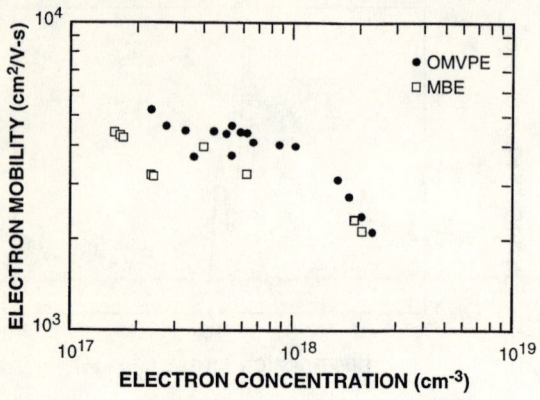

Figure 4. Electrical properties measured at 300K on n-$Ga_{0.87}In_{0.13}As_{0.12}Sb_{0.88}$ grown by OMVPE and MBE.

The electrical properties were measured at 300K for nominally undoped $Ga_{0.87}In_{0.13}As_{0.12}Sb_{0.88}$ layers grown on SI GaAs substrates. This composition corresponds to a cutoff wavelength of 2.2 µm at 300K. Since the lattice mismatch between $Ga_{1-x}In_xAs_{1-y}Sb_y$ (lattice matched to GaSb) and GaAs is 8%, growth was first initiated with a GaSb buffer layer. For layers grown by OMVPE at 550°C, nominally undoped epilayers are p type with a typical hole concentration of 5 - 8 x 10^{15} cm^{-3} and hole mobility 450 - 580 cm^2/V-s. Nominally undoped GaInAsSb layers grown by MBE are p type with a hole concentration of 2 x 10^{16} cm^{-3} and mobility of ~ 300 cm^2/V-s.

The 300K electrical properties of n- and p-doped $Ga_{0.87}In_{0.13}As_{0.12}Sb_{0.88}$ layers grown by OMVPE and MBE are summarized in Figures 4 and 5, respectively. Although the results for MBE-grown layers are somewhat limited, similar electrical characteristics are observed. For OMVPE layers, the electron concentration ranges from 2.3 x 10^{17} to 2.3 x 10^{18} cm^{-3}, with corresponding mobility values of 5208 and 2084 cm^2/V-s, respectively. The hole concentration ranges from 4.4 x 10^{15} to 1.7 x 10^{18} cm^{-3} with corresponding mobility values of 419 and 180 cm^2/V-s, respectively.

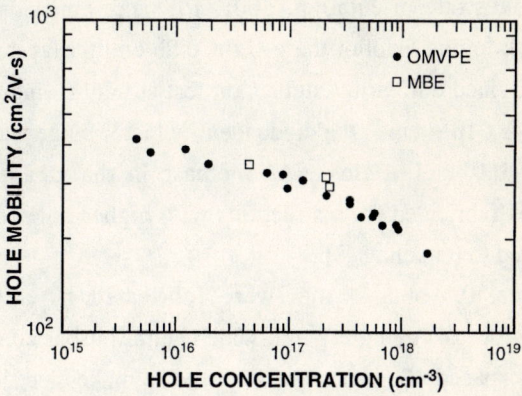

Figure 5. Electrical properties measured at 300K of p-Ga$_{0.87}$In$_{0.13}$As$_{0.12}$Sb$_{0.88}$ grown by OMVPE and MBE.

DEVICE STRUCTURES AND FABRICATION

Several different TPV structures were grown for comparison. The basic structure consists of an n-GaInAsSb base layer and p-GaInAsSb emitter layer grown on a GaSb substrate. Variations to the structure included a variation in base/emitter layer thicknesses and incorporation of an AlGaAsSb/GaSb window layer. Device structures grown by OMVPE were on (100) GaSb substrates with either 2° toward (110) or 6° toward (111)B misorientation, while structures grown by MBE were on exactly (100) GaSb substrates. Table 1 summarizes the

TABLE 1. GaInAsSb/GaSb TPV structures

Wafer	Base (µm)	Emitter (µm)	AlGaAsSb (µm)	GaSb (µm)	Misorientation	300K PL (µm)	Δa/a (x10-3)
379*	3	0.2	0	0.05	2-(110)	2.15	0
459*	3	1	0	0	2-(110)	2.24	2
462*	3	1	0	0	6-(111)B	2.24	1
463*	1	3	0	0	6-(111)B	2.24	1.5
542*	1	3	0.1	0.025	6-(111)B	2.26	5
543*	1	3	0.1	0.025	6-(111)B	2.26	2.5
548*	1	3	0.1	0.025	6-(111)B	2.26	-1.2
041+	1	3	0	0	0	2.14	1
068+	1	3	0.1	0.025	0	2.2	0.4

* OMVPE
+ MBE

device structure, substrate orientation, 300K PL peak emission, and lattice mismatch $\Delta a/a$. The doping level of the p-GaInAsSb emitter layer was designed at ~2 x 10^{17} cm^{-3}, since our earlier studies on test structures indicated that for structures with p ≤ 2 x 10^{17} cm^{-3}, the diode ideality factor range was 1.1 - 1.3 for current density of 0.01 - 1 A/cm^2 . An increase in the ideality factor was observed for diodes fabricated from structures with higher hole concentrations, which may be related to tunneling [8].

Mesa diodes, 0.5 and 1 cm^2, were fabricated by a conventional photolithographic process. A single 1-mm-wide central busbar connected to 10-μm-wide grid lines spaced 100 μm apart was used to make electrical contact to the front surface. Ohmic contacts to p- and n-GaSb were formed by depositing Ti/Pt/Au and Au/Sn/Ti/Pt/Au, respectively, and alloying at 300°C. Mesas were formed by wet chemical etching to a depth of 5 μm. No antireflection coating were deposited on these test devices.

DEVICE RESULTS

The external QE as a function of wavelength for devices OMVPE-379, -459, -462, and -463 are shown in Figure 6. The value of $\Delta a/a$ of these structures

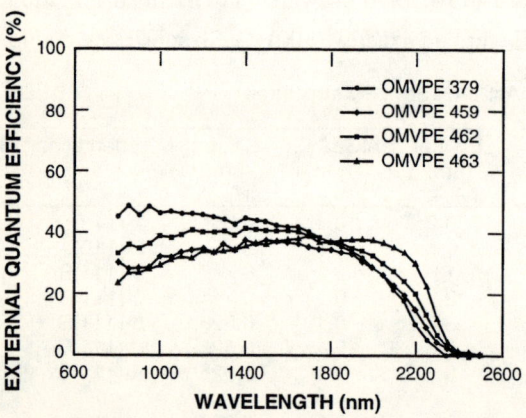

Figure 6. External QE of OMVPE-grown TPV devices described in Table 1.

is less than 2×10^{-3}. The highest QE near the bandedge is observed for OMVPE-463 with a 3-μm-thick emitter layer/1-μm-thick base layer, which results because of the higher minority carrier diffusion length in p-type GaInAsSb compared to n-type GaInAsSb. However, at shorter wavelengths, the QE of OMVPE-463 is lower than OMVPE-462, which consists of 1-μm-thick emitter layer/3-μm-thick base layer. Since carriers are predominantly generated in the base layer for OMVPE-462, this result suggests that these GaInAsSb devices are highly susceptible to surface recombination. The highest QE at wavelengths below 1.6 μm is measured for OMVPE-379, which has a GaSb window layer. In general, the performance of devices grown on (100) 2° toward (110) substrates (OMVPE-459) are inferior to those grown on (100) 6° toward (111)B substrates (OMVPE-462). The QE of TPV MBE-grown devices (MBE-041 with 3-μm-thick emitter layer/1-μm-thick base layer) is similar to the results measured for OMVPE-grown devices.

Figure 7 shows the QE as a function of wavelength for OMVPE-548, which consists of a 1-μm-thick base layer, 3-μm-thick emitter layer, and 0.1-μm-thick lattice-matched $Al_{0.25}Ga_{0.75}As_{0.02}Sb_{0.98}$/GaSb window layer. Higher-bandgap window layers are often incorporated to improve the performance of GaAs and InP solar cells [17]. For OMVPE-548 with $\Delta a/a = -1.2 \times 10^{-3}$, the QE is as high as 55% over the whole wavelength range 1.4 - 2.0 μm and nearly 50% at 2.2 μm, which is about 1.5 times higher than OMVPE-463 (also shown in

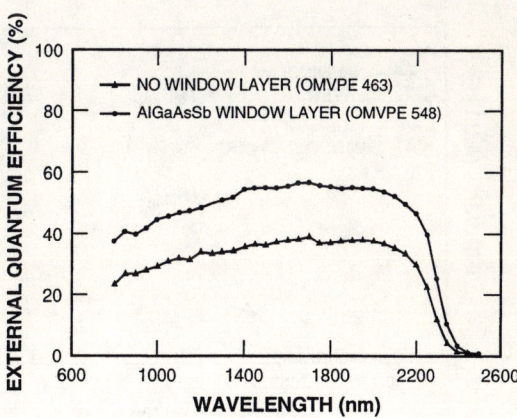

Figure 7. External QE of OMVPE-grown TPV devices with (OMVPE-548) and without (OMVPE-463) AlGaAsSb window layer.

83

Figure 6) without the AlGaAsSb window layer, higher than has been previously reported for GaInAsSb/GaSb TPV devices [6,8-9], and approaching the ~70% limit for uncoated devices. Compared to OMVPE-548, the QE is slightly lower by about 5% for OMVPE-543 with $\Delta a/a = 2.5 \times 10^{-3}$ and lowest by about 10% for OMVPE-542 with $\Delta a/a = 5 \times 10^{-3}$, indicating that precise lattice matching is important in determining the performance of these devices. High values of QE were also measured for devices fabricated from MBE-grown structures. For MBE-068, the QE is about 58% for the wavelength range 1.4 - 2.0 μm. The QE is comparable to lattice-mismatched InGaAs/InP devices, which had a maximum QE of nearly 60% at 1.65 μm and dropped off to about 20% at 2.2 μm [8]. For both OMVPE- and MBE-grown devices with the AlGaAsSb window layer, measured values of V_{oc} for short circuit current density I_{sc} range 0.1 - 1 A/cm^2 are comparable, with $V_{oc} \sim 300$ mV at $I_{sc} = 1$ A/cm^2. This value is also similar to that measured for lattice-mismatched InGaAs/InP devices [8]. Without the window layer, V_{oc} values are 30 - 40% lower.

The 300K PL spectra of TPV structures with (OMVPE-548) and without (OMVPE-463) the AlGaAsSb window layer are shown in Figure 8. The PL efficiency is more than 5 times higher for the structure with the window layer. Since carriers are generated near the surface in these PL experiments (excitation

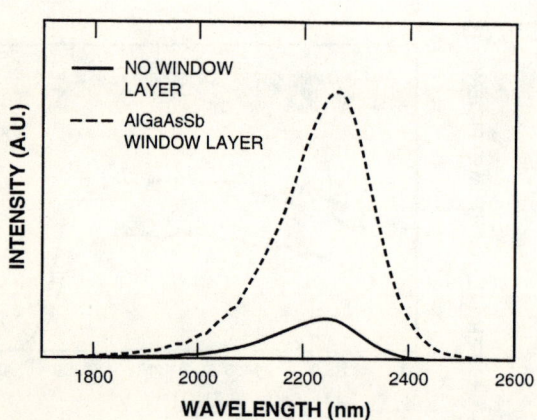

Figure 8. PL spectra of TPV devices with (OMVPE-548) and without (OMVPE-463) AlGaAsSb window layer.

source is 647 nm), these results indicate that the AlGaAsSb is epecially effective in passivating the surface of the underlying GaInAsSb and effectively reduces the surface recombination velocity. Furthermore, standard calculations [18] of external QE suggest that the surface recombination velocity may be reduced by over an order of magnitude with the AlGaAsSb window layer and that the minority electron diffusion length in our lattice-matched GaInAsSb is about 5 μm. Further characterization of GaInAsSb/AlGaAsSb/GaSb devices should be performed to assess the potential of this materials system for TPV systems.

CONCLUSIONS

High-quality GaInAsSb epilayers were grown lattice matched to GaSb substrates by OMVPE and MBE. The use of a higher-bandgap AlGaAsSb window layer is particularly effective in increasing the external QE by reducing surface recombination velocity, and results in overall improved performance. External QE ranging 55 - 58% between 1.4 and 2 μm and V_{oc} ~ 300 mV have been measured for both OMVPE- and MBE-grown devices. The present results suggest that the GaInAsSb materials system is promising for high-performance TPV systems with source temperatures operating 1100 - 1500K.

ACKNOWLEDGMENTS

The authors gratefully acknowledge D.R. Calawa, C.K. Chen, J.W. Chludzinski, M.K. Connors, F.P. Hermann, C.D. Hoyt, P.M. Nitishin, D.C. Oakley, S.A. Paul, R.J. Poillucci, D.D. Santiago, and V. Todman-Bams for technical assistance, K.J. Challberg for manuscript editing, and B-Y. Tsaur for continued support and encouragement.

REFERENCES

1. Benner, J.P., Coutts, T.J., and Ginley, D.S., 2nd NREL Conference on the Thermophotovoltaic Generation of Electricity, AIP Conf. Proc. Vol. 358, Woodbury, NY, 1995.

2. Wanlass, M.W., Ward, J.S., Emery, K.A., Al-Jassim, M.M., Jones, K.M. and Coutts, T.J., "$Ga_xIn_{1-x}As$ thermophotovoltaic converters," Solar Energy Mater. Solar Cells **41/42**, 405–417 (1996).

3. Wojtczuk, S., Colter, P., Charache, G., and Campbell, B., "Production data on 0.55 eV InGaAs thermophotovoltaic cells," Proc. 25th IEEE Photovoltaic Specialist Conference, pp. 77–80 (1996).

4. Cherng, M.J., Jen, H.R., Larsen, C.A., Stringfellow, G.B., Lundt, H., and Taylor, P.C., "MOVPE growth of GaInAsSb," J. Cryst. Growth **77**, 408–417 (1986).

5. Milnes, A.G. and Polyakov, A.Y., "Indium arsenide: a semiconductor for high speed and electro-optical devices," Mater. Sci. Eng., **B18**, 237-259 (1993).

6. Uppal, P.N., Charache, G., Baldasaro, P.F., Campbell, B.C., Loughin, S., Svensson, S., and Gill, D., "MBE growth of GaInAsSb p/n junction diodes for thermophotovoltaic applications," to appear in J. Cryst. Growth.

7. Tournie, E., Pitard, F., and Joullie, A., "High temperature liquid phase epitaxy of (100) oriented GaInAsSb near the miscibility gap boundary," J. Cryst. Growth **104**, 683–694 (1990).

8. Charache, G.W., Egley, J.L., Danielson, L.R., DePoy, D.M., Baldasaro, P.F., Campbell, B.C., Hui, S., Frass, L.M., and Wojtczuk, S.J., "Current status of low-temperature radiator thermophotovoltaic devices," Proc. 25th IEEE Photovoltaic Specialist Conference, pp. 137–140 (1996).

9. Shellenbarger, Z.A., Mauk, M.G., DiNetta, L.C., and Charache, G.W., "Recent progress in InGaAsSb/GaSb TPV devices," Proc. 25th IEEE Photovoltaic Specialist Conference, pp. 81–84 (1996).

10. Wang, C.A., Patnaik, S., Caunt, J.W., and Brown, R.A., "Growth characteristics of a vertical rotating-disk OMVPE reactor," J. Cryst. Growth **93**, 228–234 (1988).

11. Wang, C.A. and Choi, H.K., "OMVPE growth of GaInAsSb/AlGaAsSb for quantum-well diode lasers," to appear in J. Electron. Mater.

12. Wang, C.A., "Organometallic vapor phase epitaxial growth of AlSb-based alloys," J. Cryst. Growth **170**, 725–731 (1997).

13. Turner, G.W., Choi, H.K., Calawa, D.R., Pantano, J.V., and Chludzinski, J.W., "Molecular-beam growth of high-performance midinfrared diode lasers," J. Vac. Sci. Technol. B **12**, 1266–1268 (1994).

14. DeWinter, J.C., Pollock, M.A., Srivastava, A.K., and Zyskind, J.L., "Liquid phase epitaxial $Ga_{1-x}In_xAs_ySb_{1-y}$ lattice-matched to (100) GaSb over the 1.71 to 2.33 μm wavelength range," J. Electron. Mater. **14**, 729–747 (1985).

15. Tournie, E., Lazzari, J.-L., Pitard, F., Alibert, C., Joullie, A., and Lambert, B, "2.5 μm GaInAsSb lattice-matched to GaSb by liquid phase epitaxy," J. Appl. Phys. **68**, 5936–5938 (1990).

16. Sopanen, M., Koljonen, T., Lipsanen, H., and Tuomi, T., "Growth of GaInAsSb using tertiarybutylarsine as arsenic source," J. Cryst. Growth **145**, 492–497 (1994).

17. Hovel, H.J., *Solar Cells*, Vol. 11 of *Semiconductors and Semimetals*, Academic Press, NY, 1975.

18. Moller, H.J., *Semiconductors for Solar Cells*, Artech House, Inc., Boston, 1993.

GaInSb and GaInAsSb Thermophotovoltaic Device Fabrication and Characterization

C. Hitchcock*, R. Gutmann*, J. Borrego*, H. Ehsani*, I. Bhat*, M. Freeman†, and G. Charache†

*Center for Integrated Electronics and Electronics Manufacturing
Department of Electrical, Computer, and Systems Engineering
Rensselaer Polytechnic Institute, Troy, New York 12180

†Lockheed Martin, Inc., Schenectady, New York 12301

Abstract. Thermophotovoltaic (TPV) devices have been fabricated using epitaxial ternary and quaternary layers grown on GaSb substrates. The GaInSb layers were grown by organometallic vapor phase epitaxy (OMVPE) and the InGaAsSb lattice-matched layers were grown by liquid phase epitaxy (LPE). Device fabrication steps include unannealed p-type ohmic contacts, annealed Sn/Au n-type ohmic contacts, and a thick Ag top-surface contact using a lift-off process. Devices are characterized primarily by dark I-V, photo I-V, and quantum efficiency measurements, which are correlated to microscopic and macroscopic material properties. Particular emphasis has been on material enhancements to increase quantum efficiency and decrease dark saturation current density. TPV device performance is presently limited by the base diffusion length, typically 1 to 2 microns.

I INTRODUCTION

This paper presents the fabrication and characterization of ternary (GaInSb) and quaternary (GaInAsSb) device layers grown on GaSb substrates, extending earlier results [1]. For current devices, all active layers have bandgaps in the 0.52eV to 0.56eV range. The lattice constants of the ternary structures are, of necessity, not matched to that of the binary (GaSb). However, the added complexity of the fourth element (As) in quaternaries allows matching the lattice constant of the epilayers to the substrate while varying the bandgap. The epitaxial growth conditions and characterization of the epitaxial material are presented elsewhere [2]. Here the fabrication and

CP401, *Thermophotovoltaic Generation of Electricity: Third NREL Conference*,
edited by Benner/Coutts

Technology	Generation 1	Generation 2	Generation 3
Ternary OMVPE	X	X	X
Quaternary LPE	X	X	

TABLE 1. Device Fabrication and Characterization

characterization of thermophotovoltaic devices using the epitaxial layers as starting material are presented.

II EPITAXIAL STRUCTURES AND JUNCTION PROCESSING

Both ternary (GaInSb) and quarternary (GaInAsSb) epitaxial layers grown on GaSb substrates have been used in fabricating and characterizing p-n junctions for TPV applications. The ternary device structures were produced by OMVPE, with the quaternary device structures produced by LPE. The junctions that have been fabricated and characterized are grouped into three "generations," corresponding to the general time frame of fabrication. An overview of fabrication and characterization to date is shown in table 1, where the 'X' indicates one or more device process lots to be presented. Most complete results have been obtained with generation 1 and 2 devices.

The ternary device structures are shown in figure 1. These generation 1 and generation 2 structures were initial structures with parameters chosen primarily based upon preliminary material studies and first-order device design. The generation 3 structure, however, contains layers with variable thicknesses and dopings in order to study the effects of individual parameter variations on device performance. In all three cases, the grown epi-layers were not lattice matched to the underlying GaSb. A series of step layers varying from $Ga_{1.0}In_{0.0}Sb$ at the substrate to the approximately $Ga_{0.8}In_{0.2}Sb$ of the device region, separate the substrate from the base region.

The quaternary device structures shown in figure 2 were fabricated by LPE. Generation 1 and 2 were identical except that generation 1 relied upon parasitic gallium anti-site defects for emitter doping, while the generation 2 emitter was doped with germanium. For both structures, the quaternary epi-layers were approximately $Ga_{0.8}In_{0.2}As_{0.2}Sb_{0.8}$, producing the desired bandgap and a lattice match to the GaSb substrate.

At the start of the fabrication process, the samples were prepared by cleaning in solvents. Successive immersions in trichloroethane, acetone, methanol,

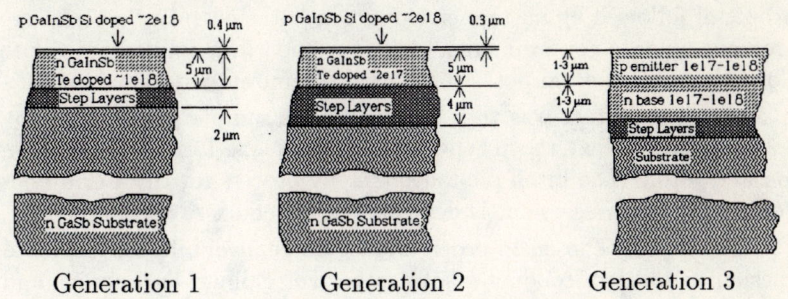

FIGURE 1. Ternary Structures

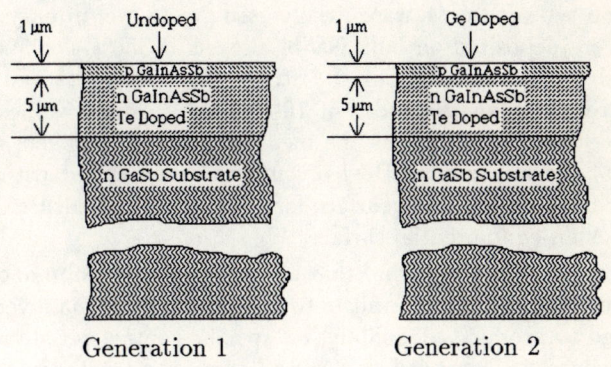

FIGURE 2. Quaternary Structures

1. Sample Cleaning and Oxide Removal

2. Back Contact Deposition and Anneal

3. Front Contact Deposition

4. Mesa Etch

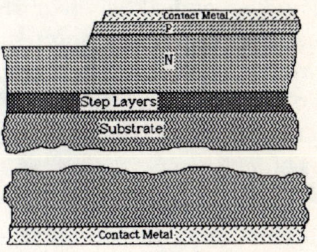

FIGURE 3. Baseline Device Processing and Finished Device Structure

and isopropanol followed by an immersion in concentrated sulfuric acid were used to remove organic contaminants. An immersion in a solution of dilute hydrochloric acid and sodium potassium tartrate removed native oxide [3].

A blanket n-type contact was then applied to the sample back surface by e-beam evaporation. Next the p-type front contact was formed by e-beam deposition and liftoff. The liftoff photolithography process for thick grid lines necessary for low series resistance is described in appendix A.

Contact metals were chosen in order to provide good ohmic contacts and good adhesion to the semiconductor substrates. For gallium antimonide and related compounds containing up to 20% indium and arsenic, the Fermi level is pinned near the valence band of the semiconductor, regardless of the work function of the contact metal [4]. As a result, contacts to p-type materials are generally ohmic as deposited, while those to n-type material must be alloyed.

For p-type contacts, 2000Å of an alloy of gold with 8% zinc, popular for gallium arsenide p-type contacts, was initially used [5]. In preliminary studies, these contacts were deposited on bulk GaSb, alloyed at 350°C for 5 seconds, and measured using the method of Cox and Strack [6,7]. The measured specific contact resistances were on the order of 10^{-5} Ωcm^2. Actual values of the contact resistance were uncertain, as the measurements were taken near the detection limit of the apparatus. Device contacts were capped with 1 to 2 microns of silver to provide a good surface for probing and to ensure that the contact metal was an equipotential surface.

During device fabrication, we found that alloyed gold-based ohmic contacts to thin p-type emitters shorted the emitter to the base. Since nonalloyed ohmic contacts appeared to produce adequately low specific contact resistances, the fabrication procedure was changed to avoid alloying the front contact. For some devices, the metallization was changed to 300Å of titanium capped with 1 to 2 microns of silver.

For n-type contacts, an evaporation of 100Å tin followed by 2000Å of gold was employed [8]. In this case, the unalloyed contact was definitely non-ohmic, and an anneal was required. Like the gold-based p-type contact, the gold-based n-type contact shorted the device emitter to the base when alloyed on a thin emitter of these n-on-p devices. Because of the lack of shallow n-type ohmic contacts, all of the devices described have p-emitters. The n-type backside device contacts of these p-on-n devices were capped with 1 to 2 microns of silver.

After the front contact deposition, individual devices on the samples were isolated by etching the device layers in the non-active areas. A hydrochloric acid, sodium potassium tartrate, and hydrogen peroxide solution was used [9], with photoresist protecting the optically sensitive areas of the devices. In some cases, the contacts were also protected, while in others the contacts were left exposed to self align the mesas to the edges of the contacts and reduce the uncertainty in the optical areas of the devices.

III JUNCTION CHARACTERIZATION TECHNIQUES

In order to determine the underlying physics of the semiconductor junctions, the current flowing through each junction was measured as a function of voltage applied across the device to obtain the junction I-V characteristics. A curve tracing instrument (oscilloscope and associated drive electronics) was used to measure junction behavior qualitatively, while a microcomputer with an analog-to-digital/ digital-to-analog adapter was used to take more quantitative measurements.

Due to the high current, low voltage nature of thermophotovoltaic devices, Kelvin contacting methods were used for both front and rear contacts. Contact was made to the front of the device using both a current and a voltage microprobe. Contact to the back of the device was achieved using a metal clad insulating board which had been patterned to produce independent voltage and current contacts to the device backside metallization. The deposition of a thick layer of metal on both front and rear device contacts during processing was critical to the assumption that the metal contacts could be considered as equipotential surfaces, a necessary condition for Kelvin contacting.

Room temperature dark I-V measurements were taken in both the forward and reverse directions using the apparatus described above. Comparison was made between the experimental results and theoretical models for expected device behavior using the following relationship:

$$I = I_s \cdot \left[exp \left(\frac{q(V - I \cdot R_e)}{nkT} \right) - 1 \right] + \frac{V - I \cdot R_e}{R_u} \tag{1}$$

where I represents device current, I_s represents saturation current, V represents the applied voltage, n is an ideality factor that should according to theory range between 1 and 2 for various regions of device operation, R_e is the series resistance, R_u is the shunt resistance, and kT is the thermal energy. This equation fit the room temperature I-V data quite well in most cases. The circuit diagram corresponding to equation (1) is shown in figure 4.

In addition to measuring I-V curves at room temperature, dark I-V curves were also measured at temperatures down to 10K. Observing the temperature dependences of the ideality factor, saturation current, and the resistances can be used to determine their physical origins. These measurements were less conclusive, but clearly can be used to assist in delineating fundamental transport mechanisms.

In addition to measuring I-V characteristics of non-illuminated devices, optically sensitive devices were also measured under high illumination conditions. This measurement is of importance not only because the test device is under conditions close to the working environment of a device in a TPV power generation system, but also because these measurements provide information about

93

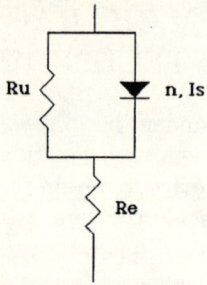

FIGURE 4. Circuit Model for Dark I-V Measurements

the physical origins of junction parameters. In the simplest case of a TPV device described accurately by the model of equation (1) with $R_e = 0$ and R_u infinite, the modification is to put a current source corresponding to carrier generation in series with the entire device, leading to the device model:

$$I = I_s \cdot \left[exp \left(\frac{qV}{nkT} \right) - 1 \right] - I_L \qquad (2)$$

where I_L is the magnitude of the light generation current. For the more intricate model with parasitic resistances, the problem is complicated by the fact that some of the circuit elements will be in series with the generation current source, while others will be in parallel. Establishing a circuit model which fits the data will determine which of the circuit elements are associated with the device junction itself and which are associated with parasitic elements.

Several possibilities for simple circuit models are shown in figure 5. The first model is the simplest, formed from the dark model by putting a current source in series with some of the circuit elements. A second, more general model replaces the bypassed circuit elements by a general passive two terminal device. The transfer function of the unknown device can be determined by measuring the dark I-V characteristic of the device. The third model allows for nonlinear circuit elements not associated with the active junction by replacing the non-bypassed series resistor with a general nonlinearity. This could be necessary for non-optimized structures with parasitic rectifying junctions at the substrate-epilayer boundary, or poor ohmic contacts. These models are discussed further in Section IV.

A junction characterization technique which measures device performance under illumination is the quantum efficiency measurement. Quantum efficiency measurements differ from illuminated measurements under TPV conditions in three principal respects. First, quantum efficiency measurements occur under monochromatic illumination, with the photoresponse of the test junction measured at one particular wavelength. Second, standard quantum efficiency measurements occur at much lower levels of illumination than TPV

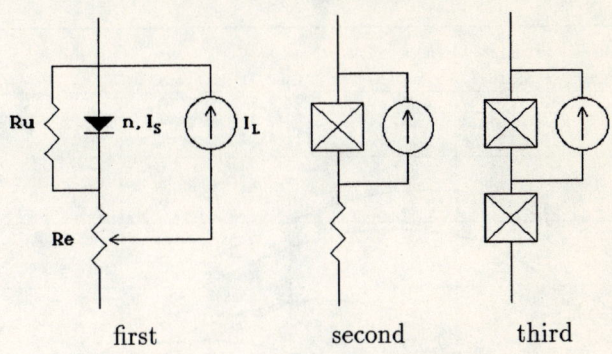

FIGURE 5. Light I-V Device Models

operating conditions. Third, quantum efficiency measurements are not taken with the device biased at the maximum power point of its I-V characteristic or along a range of bias points, but only at the zero voltage (short circuit) bias point. The result of the zero bias condition is that the exponential characteristic present in the models of equations (1) and (2) evaluates to zero. The result of the lower illumination is that the voltage drop across the parasitic series resistance is insignificant. The remaining circuit model is simply the current source associated with the junction illumination:

$$I = -I_L \tag{3}$$

This current can be normalized to the illumination intensity to obtain the fraction of incident photons which produce collected electron-hole pairs as a function of wavelength. If the material absorption coefficient is know as a function of wavelength, the measured quantum efficiency can be fitted to a theoretical quantum efficiency to obtain carrier diffusion lengths in the device regions.

While the standard quantum efficiency measurement and dark current measurement provide information that can be combined to produce a theoretical operational curve for the device under any bias and illumination condition, these theoretical extrapolations depend on the linear nature of the device. A test of this assumption is to illuminate the device with a DC high intensity broad spectrum light source as well as an AC low intensity monochromatic light source and measure the AC intensity variations. If the device response to light is linear for illuminations below the high intensity DC level, the standard and high intensity quantum efficiency curves should be identical. Any variation indicates a nonlinear optical response, which can be caused by high level injection, tunneling currents, or other physical effects.

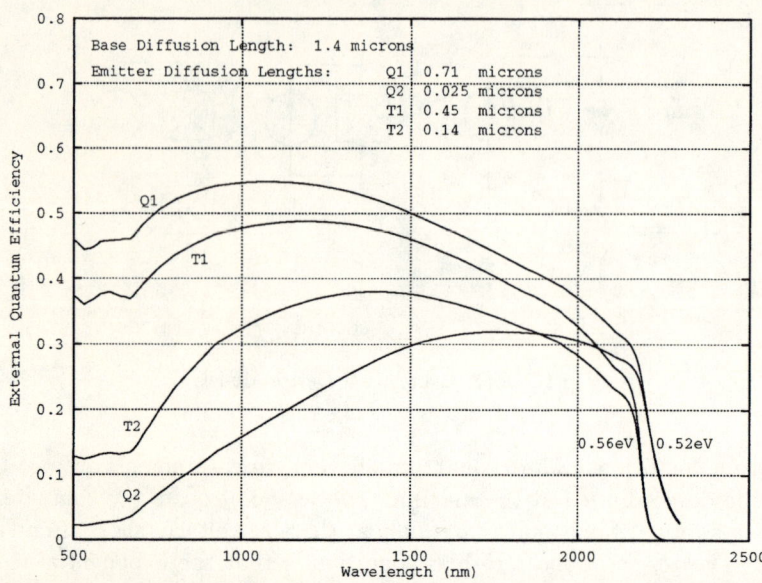

FIGURE 6. Fitted quantum efficiency curves for ternary (OMVPE) and quaternary (LPE) junctions, generations 1 and 2

IV JUNCTION EXPERIMENTAL RESULTS AND DISCUSSION

Fitted quantum efficiency curves for the generation 1 and generation 2 devices are shown in figure 6. The curves were obtained by graphing, for each individual device, all of the experimental quantum efficiency curves on a single scale. Quantum efficiency curves were then simulated using the known parameters of the device and varying the diffusion lengths in the emitter and base to obtain the best fit. In this model a front surface recombination velocity of zero was assumed, as these thin emitter regions did not allow experimental separation of surface and bulk recombination. The base-to-substrate or base-to-graded layer surface recombination velocity was also assumed to be zero, but this approximation is not significant since the base thickness is more than 3 diffusion lengths for all generation 1 and 2 devices.

Using this approach, the emitter diffusion lengths were found to vary considerably, in a way that does not correlate well with any controlled or measured device parameters (see insert in figure 6). We postulate that this behavior is due primarily to effective surface recombination velocity variations at the front surface. The base diffusion lengths, however, were approximately 1.4 microns for the four devices, which is quite surprising due to the different

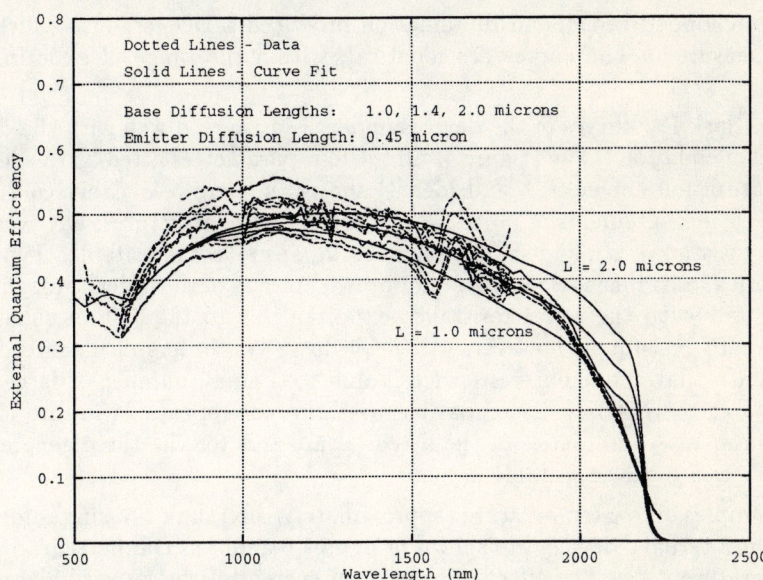

FIGURE 7. Quantum efficiency data and fitted curves for the generation 1 ternary device

growth techniques and process variables.

Figure 7 illustrates the curve fitting technique using the data from the generation 1 ternary device. A base diffusion length of 1.4 microns produced the best fit, while using 1.0 and 2.0 microns produced quantum efficiencies which were too low and too high, respectively. Varying the effective emitter diffusion length has a greater impact in the lower wavelength region. With such a model, the base diffusion length in the four device types shown in figure 6 was determined to be between 1.0 and 2.0 microns.

In TPV device operation, the majority of incident usable light will be in the wavelength region just below the bandgap of the device. As the absorption coefficient in this region will be lower than that in the shorter wavelength regions, the quantum efficiency in this region is determined primarily by the efficiency of the base region. Improving the 1.0 to 2.0 micron diffusion length in the base is therefore a priority in the material, processing, and device research program.

For all quantum efficiency measurements, the intensity of the monochromatic light source was approximately $50\mu W/cm^2$. In order to investigate device performance under higher intensity illumination, the monochromatic light source was supplemented, for certain measurements, with a white light source having an intensity of approximately $5mW/cm^2$. Measurements were taken by chopping the monochromatic source but not the white source, and measuring the resulting AC response synchronously with a lock-in amplifier.

The unchopped broadband illumination produced a DC response which was not measured. The curves are identical, within the range of experimental error.

The dark I-V curves of the devices under study agreed well with the device model of equation (1) and figure 4. Of the four parameters fitted by this model, the saturation current, I_s, and ideality factor, n, were reasonably consistent between measurements of nominally identical device structures; however, the series resistance, R_e, and shunt resistance, R_u, were less repeatable. Typically the series resistance varied by a factor of two but was of the order of magnitude consistent with the series resistance expected due to the $400\mu m$ substrate. The shunt resistance, however, varied by up to two orders of magnitude. A postulate that the shunt resistance is due to a small number of large area defects at the junction is compatible with the non-repeatable nature of this measurement. Summaries of the fitted parameters for the three generations of devices are shown in table 2.

Examples of low temperature (approximately 10K) dark I-V curves for generation 2 ternary devices are shown in figures 8 and 9 in comparison to room temperature data. The effective saturation current of the forward biased device decreases by almost two orders of magnitude although the reverse current decreases only by a factor of two. These relationships have not been quantified and the mechanisms for junction transport have not been delineated, particularly with the very soft reverse characteristics.

An example of a high intensity light I-V curve, taken using a photographic flash lamp to avoid heating the sample, is shown in figure 10. If this result were amenable to the analysis of either of the first two models of figure 5, it would be possible to make the light I-V curve and the dark I-V curve coincide by translation, which is not the case. This non-agreement is likely due to either parasitic nonlinearities, or a two-dimensional effect which is not accurately simulated by these one-dimensional models.

Reverse bias C-V measurements of these device junctions have, in general, not been amenable to quantitative analysis. The reverse breakdown of gallium antimonide based devices is sufficiently soft that significant reverse leakage currents obscure the results at even modest applied biases. Figure 11 shows a reverse bias C-V measurement of a generation 2 ternary device. The measurement was taken at low temperature to reduce the leakage current. Ignoring the nonlinear nature of the curve and fitting a tangent line at zero voltage obtains a $0.4V$ junction built in voltage and a doping concentration of approximately 5×10^{16}, which is quite close to the expected 2×10^{17} considering the crude nature of the calculation. Figure 9 indicates that the reverse current rises to the milliampere level at voltages less than half a volt, and this will account for the poor C-V behavior. The effect of leakage current on measured capacitance must be quantitatively determined to fully utilize C-V data as done for larger bandgap materials.

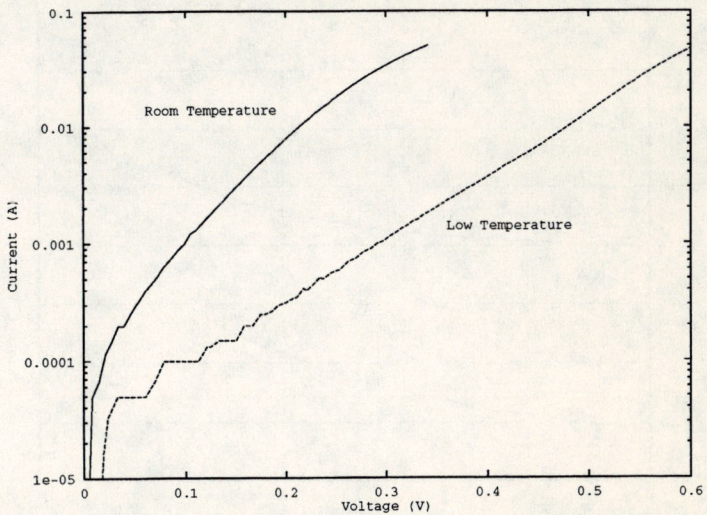

FIGURE 8. Low temperature (approximately 10K) forward bias dark I-V for a generation 2 ternary device (device area: $1.3 \times 10^{-2} cm^2$)

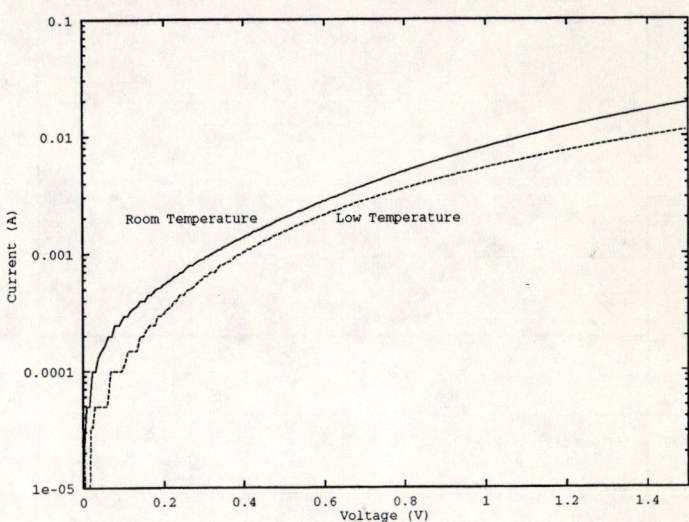

FIGURE 9. Low temperature (approximately 10K) reverse bias dark I-V for a generation 2 ternary device (device area: $1.3 \times 10^{-2} cm^2$)

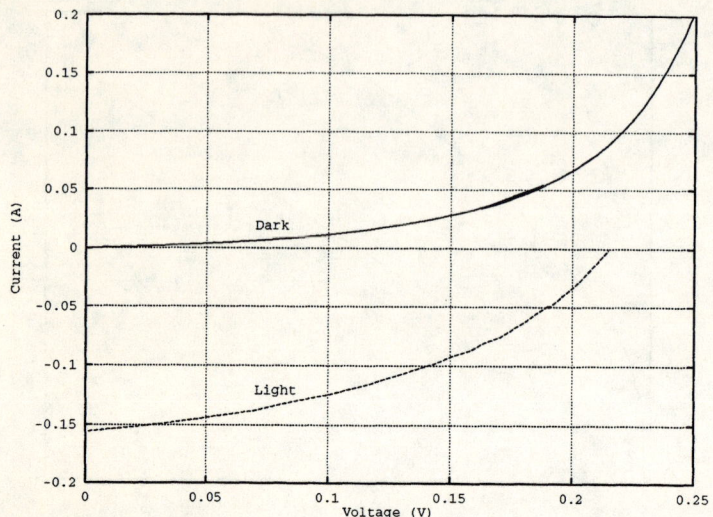

FIGURE 10. Generation 1 Ternary — Dark and Light I-V (device area: $6.6 \times 10^{-2} cm^2$, grid coverage: 58%)

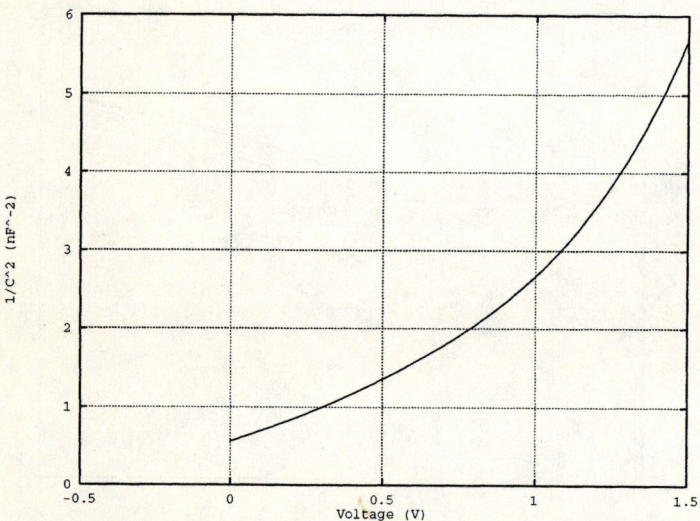

FIGURE 11. Generation 2 ternary — reverse bias C-V measurement taken at 10K (device area: $1.3 \times 10^{-2} cm^2$)

Technology	Generation 1	Generation 2	Generation 3
Ternary OMVPE	$n = 1.11$ $J_s = 0.53\ mA/cm^2$	$n = 1.30$ $J_s = 1.6\ mA/cm^2$	$n = 1.06$ $J_s = 0.12\ mA/cm^2$
Quaternary LPE	$n = 1.04$ $J_s = 30\ mA/cm^2$	$n = 1.19$ $J_s = 0.98\ mA/cm^2$	

TABLE 2. Average Fitted Parameters for Dark I-V Measurements

V SUMMARY AND CONCLUSIONS

A series of working TPV test devices were formed from both ternary and quaternary epilayers on GaSb substrates. The devices exhibited properties consistent with the Schockley theory of semiconductor p-n junctions, exhibited photoresponse over the the wavelengths of interest, and generated electrical power when operated under field conditions. The photoresponse of the devices was poor at energies just above the bandgap, which is attributed to a short diffusion length in the n-type device base region (from one to two microns). The high intensity light I-V was not predicted accurately by a simple device model, which is attributed to either two-dimensional device effects, tunneling current, and/or a parasitic device nonlinearity.

REFERENCES

1. Ehsani, H., *et. al.*, *Second NREL Conference on Thermophotovoltaic Generation of Electricity*, 423–433 (1995).
2. Ehsani, H., *et. al.*, *Third NREL Conference on Thermophotovoltaic Generation of Electricity*, (1997), to be published.
3. Buglass, J. G., *et. al.*, *J. Electrochem. Soc.* **133**, 2565–2567 (1986).
4. McCaldin, J., *et. al.*, *J. Vac. Sci. Tech.*, **13**, 802–806 (1976).
5. Shur, M., *GaAs Devices and Circuits*, New York: Plenum Press, 1987, ch. 3, pp. 147–156.
6. Cox, R. H., and Strack, H., *Solid-State Elec.* **10**, 1213–1218 (1967).
7. Carver, G. P., *et. al.*, *IEEE Trans. Elec. Dev.* **35**, 489–496 (1988).
8. Heinz, C. H., *Int. J. Elec.* **54**, 247–254 (1983).
9. Gomez Zazo, L. J., *et. al.*, *J. Electrochem. Soc.* **136**, 1480–1484 (1989).

A LIFTOFF PROCESSING

A complex photoresist process was used to form the device front contacts in order to fabricate a thick contact with the poor step coverage required for

1. Spin on AZ4620 at 7000 RPM for 40s.

2. Hot plate bake 15 minutes at 90°C.

3. Flood expose 1350mJ/cm².

4. Spin on AZ4620 at 7000 RPM 40s.

5. Hot plate bake 1 minute at 120°C.

6. Mask expose 450mJ/cm².

7. Develop in AZ401K 1 minute.

TABLE 3. Photoresist Processing

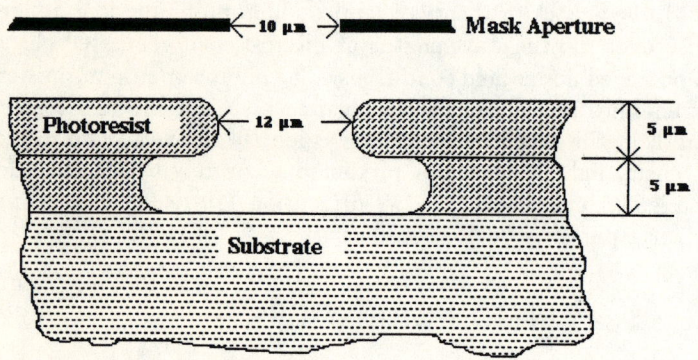

FIGURE 12. Photoresist Structure

liftoff. The photoresist processing steps are shown in table 3. Two layers
of Hoescht-Celanese AZ4620 photoresist were deposited. The first layer was
weakened by a flood UV exposure, and the second layer was added and exposed in the desired grid pattern. When the photoresist was developed, the
exposed sections of the upper photoresist layer dissolved, allowing the devel-
oper to attack the lower layer photoresist and undercut the upper layer. A
schematic of the photoresist structure after develop is shown in figure 12, with
a micrograph of the actual photoresist structure shown in figure 13. When
metal was evaporated onto this structure, the metal condensing on the sample
surface was not connected to that on the photoresist, as shown in figure 14.
Therefore, immersion in a photoresist solvent lifted off the unwanted metal
successfully without damaging the contact metal.

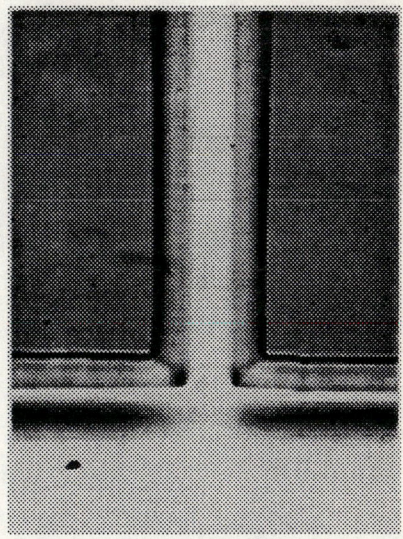

FIGURE 13. Thick Photoresist at the Intersection of a Grid Line and a Bus

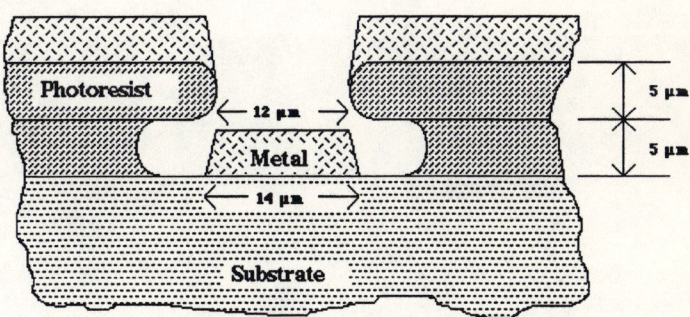

FIGURE 14. Structure after metal deposition

GaSb Based Ternary and Quaternary Diffused Junction Devices for TPV Applications

V.S.Sundaram, S.B.Saban, M.D. Morgan, W.E. Horne, B.D. Evans, J.R. Ketterl, M.B.Z. Morosini*, N.B. Patel* and H. Field**

EDTEK Inc. 7082 S. 220th Street Kent, WA 98032
* Instituto de Fisica, UNICAMP, Campinas, Brasil.
** NREL, Golden, Colorado.

Abstract. In this work we report the characteristics of ternary, GaInSb (Eg=0.70eV) and quaternary, GaInAsSb (Eg=0.58eV) diffused junction photovoltaic devices. The unique feature of the quarternary device is the extended long-wavelength response to 2.1 microns enabling the efficient use of the blackbody-like thermal sources operating at 1373 K in thermophotovoltaic energy conversion systems. The ternary device was fabricated by diffusing zinc into a n-type (100) oriented GaInSb substrate. For the quarternary, a four micron thick Te doped GaInAsSb layer grown by LPE on a n-type GaSb(100) wafer was used as the starting substrate for zinc diffusion. The ternary device exhibits an open circuit voltage of 0.38V, Fill Factor of 0.63 and a short circuit current of 0.8 A/cm^2 , while the corresponding values for the quarternary device are 0.25V, 0.58 and 0.8 A/cm^2, respectively. The peak internal quantum efficiency for the ternary is over 90% and that of the quarternary is above 75%. Process optimization should improve the performance charcateristics of the quarternary.

INTRODUCTION

Thermoelectric devices that convert radiant heat energy into electric energy are currently used as power generators in deep space missions (1). Limitations on the overall system efficiencies attainable with the presently available thermoelectric devices have generated an interest in developing new power generating schemes employing diversified technologies that provide efficiency improvements. The most promising of these is the radioisotope thermophotovoltaic (TPV) generator which uses photovoltaic (PV) conversion of photons emitted from a radioisotope-heated source to produce electrical energy(2). TPV devices fabricated from III-V compound semiconductors can increase system efficiencies by a factor of four to five over that attainable with the thermoelectric devices. TPV devices are currently being evaluated for deep space missions and terrestrial applications. Schock, et al.(3) have reported a design of a radioisotope-powered TPV power generator in which the heat source mass, cost, and fuel loading are reduced 60%. Overall generator mass is reduced 50% and the system specific power is tripled over that achieved by conventional thermoelectric generators . The thermal generator designated to be used with this TPV power source is the Department of Energy's space-qualified General Purpose Heat Source (GPHS) (4). As a result of materials constraints and safety concerns, this heat source was designed to operate with a surface temperature not exceeding 1373 K; the equivalent blackbody emission peak

CP401, *Thermophotovoltaic Generation of Electricity: Third NREL Conference,*
edited by Benner/Coutts
© 1997 The American Institute of Physics 1-56396-734-0/97/$10.00

occurs at 2.1 μm. The best commercially available photovoltaic converter material to be used in this system, whose photoresponse spectrum is compatible with the ≈1373 K blackbody-like source spectrum, is GaSb. This material has a bandgap of 0.72 eV, absorbing radiation of wavelength shorter than 1.72 μm within a micron thickness. Furthermore, it has been established (2) that system efficiencies are improved dramatically by using TPV devices in conjunction with IR bandpass filters that pass near bandgap radiation to the TPV device, and reflect back to the source all other short and long wavelength components in the black body spectrum. By recycling these "unused" photons, over-all system efficiency is increased.

GaSb photovoltaic devices, originally developed for GaAs/GaSb tandem solar cells (5), have found wide application in the TPV area. GaSb, a direct bandgap III-V compound semiconductor with a bandgap of 0.72 eV, is well suited when the source temperature is about 1500°C. GaSb TPV devices have been fabricated by a simple diffusion process, similar to that used in the silicon solar cell industry, and have exhibited excellent diode like behavior (5). Large scale manufacturability with excellent batch-to-batch yield has been established.

Even higher TPV conversion efficiency could be achieved with lower source temperatures if the photovoltaic (PV) device responded to longer wavelength radiation. In other words, photovoltaic device efficiency is optimal if the device bandgap is matched with the peak wavelength of the black body emission. For example, Gruenbaum et al (6) using an ideal band pass filter have shown that GaSb PV devices exhibit a maximum efficiency when the black body heat source is maintained near 1500°C, while PV cells with a bandgap of 0.6 eV achieved a maximum efficiency with a source temperature near 1100°C. The availability of such lower bandgap TPV devices is a key enabling technology for a viable terrestrial TPV system by reducing materials constraints and manufacturing costs associated with long-term operation of higher temperature sources.

Significant progress has been achieved in recent years in the fabrication of low bandgap photonic devices. Most efforts have involved the epitaxial growth of ternary or quaternary compound semiconductors on binary III-V compound semiconductor substrates. For example, Wojtczuk, et al. (7) and Wilt, et al. (8) have described PV cells based on Metal Organic Chemical Vapor Deposition (MOCVD)-grown, n-on-p homojunctions in several ternary compositions based on $Ga_{1-x}In_xAs$ formed on InP substrates. When x=0.53, the ternary is lattice matched to the InP substrate. This condition yields efficient PV cells with V_{oc} = .30 V. The bandgap of $Ga_{0.47}In_{0.53}As$ is 0.74 eV (1.65 μm cutoff). Higher In concentrations are required to reach lower bandgaps; $x \geq .65$ for $E_g \leq 0.60$ eV. However, these ternary compositions are not lattice matched to the InP substrate. Wojtczuk, et al. have demonstrated that, in the composition range x >0.6, V_{oc} dropped abruptly. For example, when the bandgap, E_g, was lowered to 0.55 eV (x = 0.72), V_{oc} dropped to 0.087 V. In addition, Wilt reported that high dark currents contributed to poor cell performance. It was concluded that these effects were due to large recombination currents at the ternary-binary interface, a result of the misfit dislocation density due to lattice mismatch between $Ga_{1-x}In_xAs$ ($a_o \geq 5.94$Å) and the InP substrate (a_o = 5.86875Å), i.e., $\Delta a_o/a_o \geq 1.2 \times 10^{-2}$, when $x \geq 0.7$. Layers of graded ternary GaInAs composition have been reported to partially suppress misfit dislocation generation, but these structures add to cell complexity and fabrication cost. Since these structures are not economical for high-volume terrestrial applications, other techniques for realizing low lattice mismatch are sought.

The addition of InSb to GaSb results in a ternary GaInSb compound whose bandgap can be adjusted by control of InSb molar fraction in ternary. For example, $Ga_{0.85}In_{0.15}Sb$ has a bandgap near 0.50 eV. However, the incorporation of larger In ion (0.81Å) by substitution on a Ga ion site (0.62Å) causes significant local lattice distortion. This dilatation contributes to the lattice energy, resulting in low solubility for InSb in GaSb, and yields large internal strains in these ternary compounds. Therefore, attempts to grow bulk GaInSb crystals with InSb mole fraction above a few percent have been unsuccessful. In the present study we used a (100) oriented Te doped GaInSb substrate supplied by a vendor.

The quaternary GaInAsSb is "lattice matched" to GaSb over a wide compositional range. For example, the lattice spacing a_O for InAs is $a_O = 6.0584$Å, and for GaSb $a_O = 6.094$Å, so that here $\Delta a_O/a_O \leq 6 \times 10^{-3}$. Furthermore, the quaternary $Ga_{1-x}In_xAs_ySb_{1-y}$ has been employed over the past two decades to demonstrate detectors sensitive to wavelengths longer than 1.7 μm (9) and to demonstrate 2-to-5-μm diode lasers, some with cw operation at room temperature (10). Several workers have demonstrated that an approximate 0.6-eV bandgap can be achieved with $x \approx 0.1$ and $y \approx 0.9$ by both Liquid Phase Epitaxy (LPE) and Molecular Beam (MBE) growth techniques (11).

This paper presents the results of experiments with diffused junction photovoltaic cells fabricated from low-bandgap, ternary GaInSb, and with quaternary GaInAsSb. Both ternary and quaternary devices exhibit excellent behavior in the diffusion current limited region. These devices demonstrate that the lower bandgap extends their responsivity to longer wavelengths past the GaSb cutoff (1.75 μm) to 2.2 μm for the quaternary.

EXPERIMENTAL

Cell Fabrication

Zinc was diffused into a (100) oriented , n type ternary GaInSb substrate by a process similar to that used in the fabrication of aerospace silicon cells. For the quaternary system, we used a (100) oriented, n-type GaSb substrate with a 4 micron thick epitaxial, GaInAsSb, lattice matched layer grown on it by LPE technique as the starting substrate for zinc diffusion. The nominal n type carrier concentration in the base region was in the range of 3 to 5 $\times 10^{17}$ cm-3. The bandgap of the LPE grown quaternary and the ternary substrate was discerned from optical absorption measurements. These bandgap results were confirmed from the long wavelength cut-off of the device photo response. In order to compare the extended, long-wavelength response of the ternary and quaternary cells, GaSb cells were also fabricated by similar processing scheme.

Standard photolithographic processes were used to define the cell area first. Then, shallow p/n junctions were created by zinc diffusion in a "pseudo-closed box" technique. The cell geometry included an optimized current collection grid design with a low total power loss. Ti/Pt/Au and Au/Sn/Au were used as the p and n ohmic contacts respectfully. Grid line obscuration loss was estimated to be about 7%. An anti reflecting (AR) layer based on silicon nitride was used to minimize reflection losses incurred at the front surface. The antireflection (AR) coating used in these cells had a minimum at about 1.54 micron, while that used in the binary cells had a minimum at 1.4 microns.

Electrical and Optical Measurements

Cell current-voltage (I-V) characteristics were measured with a Spectralab Model XT-10 solar simulator equipped with a 1-kW xenon arc lamp and additional lenses for concentrated solar simulation. Data was gathered, formatted and plotted with a Textronix 370 Programmable Curve Tracer. The sample was mounted on a temperature-controlled copper substrate within a box supplied with vacuum and a quartz window. The infrared irradiance used was about 2.5 W/cm^2.

External quantum efficiency (QE) measurements were made at both EDTEK and the National Renewable Energy Laboratory (NREL) using the standard "comparison" technique.(12)

RESULTS AND DISCUSSION

The I-V characteristics of typical diffused junction quaternary, ternary, and binary GaSb cells at 290K are shown in Figure 1. For the quaternary device, $V_{OC} = 0.250$ V, $I_{SC} = 0.8$ A/cm^2, and fill factor (FF) was 0.58. This quaternary fill factor was not as high as in bulk single-crystal GaSb; however, they are respectable considering the preliminary nature of this effort. The ternary GaInSb exhibits a V_{OC} of 0.38 V and a fill factor of 0.63 at similar light intensities.

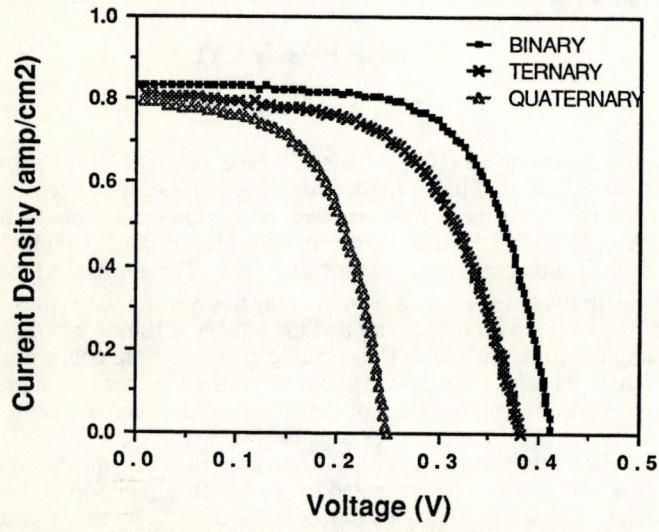

Figure 1. I-V characteristics of binary, ternary and quaternary cells.

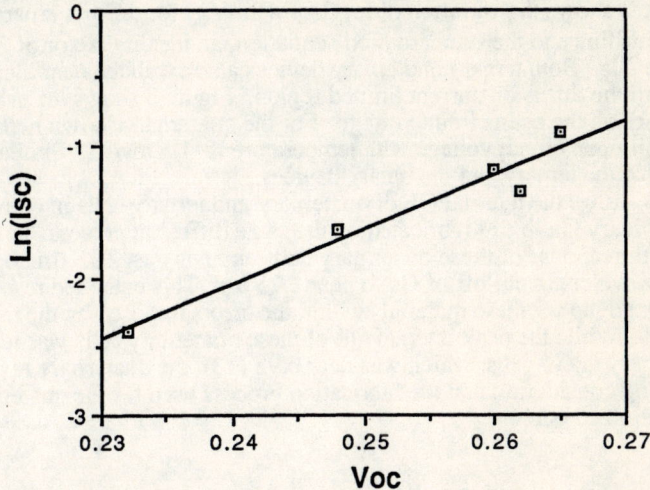

Figure 2. Variation of open circuit voltage with log of short circuit current for the quaternary cell.

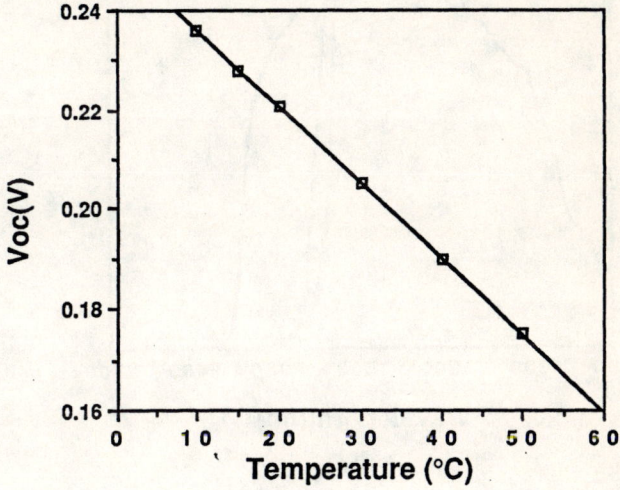

Figure 3. Variation of open circuit voltage with temperature for the quaternary cell.

109

Figure 2 shows the variation of log(Isc) with V_{oc} for the quaternary. From the slope of this line and the standard diode equation, an ideality factor of 2 is derived for this device. Both ternary and binary devices also exhibited near ideal diode behavior in the diffusion current limited region. Figure 3 shows the effect of temperature on the open circuit voltage. For the quaternary shown here, the variation of open circuit voltage with temperature is -1.5 mv/°C. Similar value is obtained for the ternary and the binary devices.

Figure 4 shows the internal QE of quaternary and ternary cells in comparison with a typical binary GaSb cell fabricated by the same diffusion process. The long-wavelength response of these quaternary cells extends past 2.15 µm, 0.40 µm past the long-wavelength cut-off of GaSb near 1.75 µm. This extended response agrees with the bandgap for these materials within the errors imposed by the measurement techniques. While the peak internal QE of these quaternary cells was less than that for the binary GaSb cells, which was near 0.92 (13), the quaternary results are encouraging considering that the fabrication process used for the quaternary was not an optimized one.

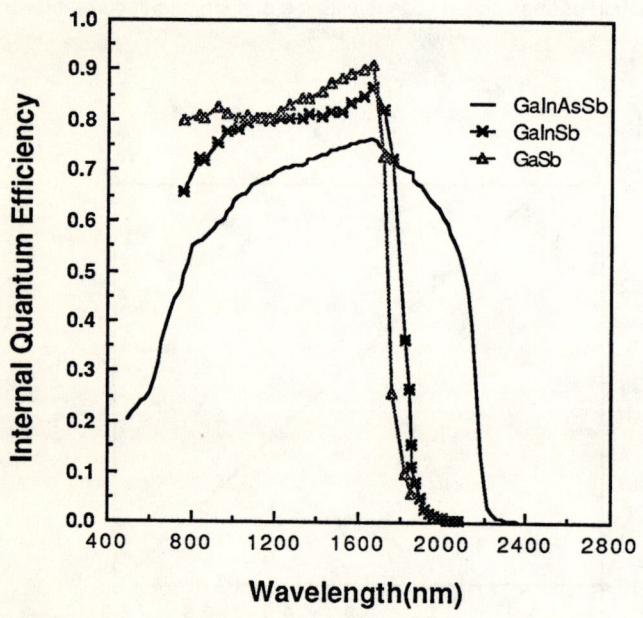

Figure 4. Spectral response of binary, ternary, and quaternary cells

The quaternary spectral response in Figure 4 demonstrates a fall-off in the near band-edge region, 1.8-to-2.1 µm, as compared to the response of the binary GaSb in its sub-gap spectral region of 1.3-to-1.7 µm. The standard solar-cell junction model suggests this was primarily due to electron-hole recombination in the base region (14). In this instance, the base is the quaternary layer lattice matched to a GaSb substrate. This binary-quaternary interface region represents a source of defects: (1) due to a slight lattice misfit between the two compounds, (2) due to remnant impurities such as native sub-oxides not completely removed from the substrate surface prior to epitaxial growth, and (3) polished-derived surface damage also not completely removed prior to eptiaxial growth. Improved epitaxial processes can remove effects (2) and (3). In contrast, the short-wavelength response of the quaternary was similar to the binary response shifted to longer wavelengths. This indicates that the desirable, low surface-recombination velocity demonstrated by binary GaSb cells (in marked contrast to GaAs-based cells), has been carried over to the new quaternary and ternary cells. This also demonstrates that the desirable absence of a surface "dead layer" reported in diffused binary GaSb cells (13), in spite of the very high carrier concentration at the top surface of the p-region, i.e., $\approx 10^{20}$ cm^{-3}, is equally present in the quaternary cell.

TPV Energy Conversion Efficiency: Effect of Device Bandgap and Source Temperature

A theoretical evaluation of TPV conversion efficiency using GaSb-based converters was carried out. The model assumed a blackbody-like radiator corresponding to a source temperature of 1373K with an emissivity of 1.0. This blackbody-like emission was spectrally modified with an actual EDTEK fabricated micro mesh filter. The transmittance of this band pass filter as a function of wavelength is shown in figure 5. Such spectrally modified blackbody emission was used in conjunction with the spectral response data for an EDTEK made GaSb TPV cell. These calculations were extended to GaInAsSb with varying indium arsenide contents. In these calculations, the spectral response of the filter was displaced in wavelength to match the bandgap of GaInAsSb. Open-circuit voltage and fill factor were corrected for bandgap variation following Sze(14) and the results are presented in Figure 6. The model calculations revealed that the efficiency (at the converter level) varied between 18.8% for GaSb and 19.65% for a device with a bandgap of 0.58eV while the maximum power output varied between 0.357W/cm^2 and 0.457W/cm^2. These calculations demonstrate that under these operating conditions, maximum converter efficiency and maximum power are realized with a bandgap of 0.58 to 0.59 eV. The quaternary GaInAsSb device fabricated with an indium arsenide content of about 11% corresponds to this bandgap. The fundamental material advantage of the quaternary system GaInAsSb over the ternary system GaInSb is that the ternary system requires much higher indium antimonide contents, near 20%, to achieve the same low bandgap. This advantage, coupled with the fact that the addition of InAs to GaSb induces an order of magnitude lower internal strain , makes GaInAsSb a more attractive system for TPV applications.

A combination of improved TPV system efficiency together with lower-cost cell fabrication techniques is desirable. As mentioned previously, conventional single-crystal semiconductor growth-from-melt techniques, such as Czocheralski, gradient-freeze, etc. have proven difficult for ternary and quaternary GaSb-based compounds. Recently, EDTEK has explored novel techniques for fabricating quaternary substrates based upon the temperature gradient zone melting method (15). In this method, a moving temperature gradient is used to initiate mass transfer

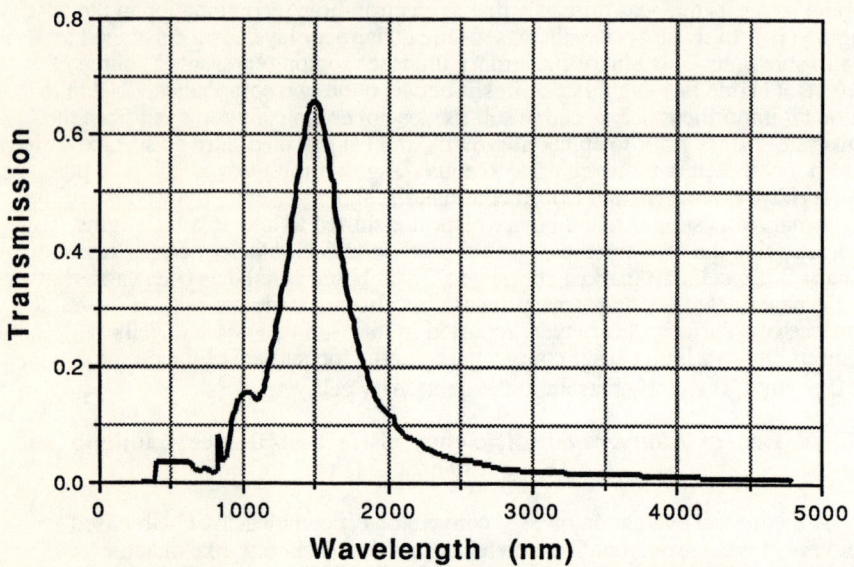

Figure 5. Transmittance of EDTEK Band Pass Filter

from a polycrystalline feed stock to a seed crystal where recrystallization takes place, thereby, growing a monocrystalline material of the same average composition as that of the polycrystalline feed stock. The process is carried out below the melting temperature of the constituent materials minimizing problems associated with contamination and with overpressure requirements when high vapor-pressure constituents, such as arsenic, are present. This technique has recently realized small diameter GaInAsSb substrates demonstrating a bandgap near 0.60 eV.

In summary, photovoltaic cells based upon a diffused, p-on-n homo-junction in modified GaSb have been demonstrated. Prototype cells based on bulk ternary GaInSb exhibited a long-wavelength response extended to 1.9 μm, a Voc of 0.38 V and a fill factor of 0.63, while cells based on LPE-grown GaInAsSb lattice matched to GaSb had a long-wavelength response extending past 2.1 μm, with a Voc of 0.25 and a FF of 0.58.

112

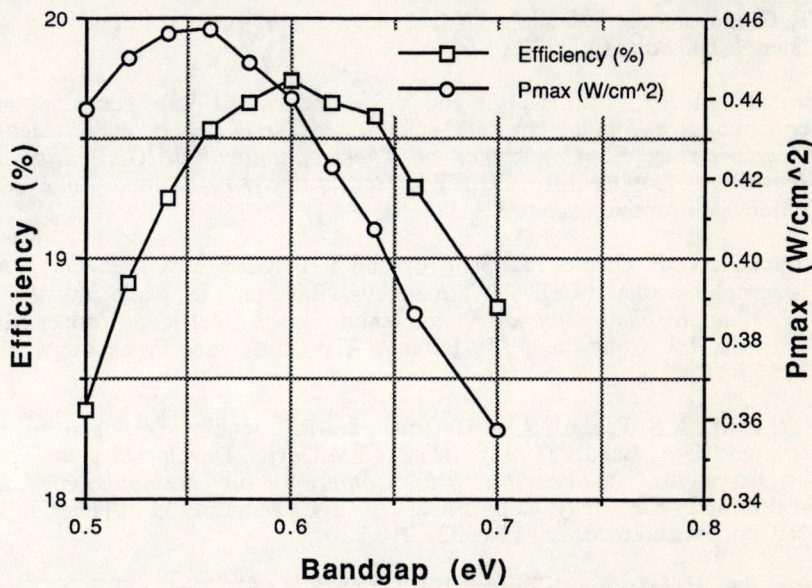

Figure 6. Variation of TPV conversion efficiency(%) and power at maximum power point (W/cm^2) as a function of cell bandgap for a source at 1373K

REFERENCES

1. Schock, A (1993) "RTG Options for Pluto Fast Flyby Mission", in Proc. of the 44th Congress of the International Astronautical Federation, Graz, Austria, IAF-93-R.1.425b.

2. Morgan, M.D., W.E. Horne, and P.R. Brothers. (1993) "Radioisotope Thermophotovoltaic Power System Utilizing the GaSb IR Photovoltaic Cell", in *Proc. of the 10th Symposium on Space Nuclear Power Systems,* CONF-930103, AIP Conference Proc. No. 271:217-225.

3. Schock, A. and V. Kumar (1994) "Radioisotope Thermophotovoltaic System Design and its Application to an Illustrative Space Mission", in *The First NREL Conference on Thermophotovoltaic Generation of Electricity*, Copper Mt., CO., T.J. Coutts and J.P> Benner, Eds., AIP Conference Proceedings No. 321: 139-152

4. Schock, A. (1980) "Design Evolution and Verification of the General-Purpose Heat Source." #809203, in *Proc. of 15th Inersociety Energy Conversion Engineering Conference*, held in Seattle, WA. 18-22 August 1980, 2:1032-1042.

5. P.E. Gruenbaum, V.T. Dinh nad V.S. Sundaram (1994), Solar Energy Materials and Solar Cells, 32:61-69

6. Gruenbaum, P.E., M.S. Kuryla and V.S. Sundaram (1995) "Technical and Economic Issues for Gallium Antimonide Based Thermophonovoltaic Systems, in *The First NREL Conference on Thermophonovoltaic Generation of Electricity*, Copper Mt., CO., T.J. Coutts and J.P. Benner, Eds., AIP Conference Proceedings No. 321:357-367.

7. Wojtczuk, S., E. Gagnon, L. Geoffroy and T. Parodos (1995) "$In_xGa_{1-x}As$ Thermophotovoltaic Cell Performancevs. Bandgap, in *The First NREL Conference on Thermophonovoltaic Generation of Electricity*, Copper Mt., CO., Eds. T.J. Coutts and J.P. Benner, AIP Conference Proceedings No. 321:177-187.

8. Wilt, D.M., N.S. Fatemi, R.W. Hoffman, Jr., P.P. Jenkins, D. Scheiman, R. Lowe and G.A. Landis (1995) "InGaAs PV Device Development for TPV Power Systems," in *The First NREL Conference on Thermophonovoltaic Generation of Electricity*, Copper Mt., CO., T.J. Coutts and J.P. Benner, Eds., AIP Conference Proceedings No. 321:210-220.

9. Astles, M., H. Hill, A.J. Williams, P.J. Wright and M.L. Young (1986) "Studies of the $Ga_{1-x}In_xAs_{1-y}Sb_y$ Quaternary Alloy System I. Liquid-Phase Epitaxial Growth and Assessment," *J. Electron. Materials*, 15(1):41-49.

10. Herrera-Perez, J.L., M.B.Z. Morosini, A.C.F. da Silveira and N.B. Patel (1992) "Low thresholdcurrent density GaInAsSb/GaAlAsSb lasers emitting at 2.2 μm," *Proc. of the 18th Internat. Sym. on Gallium Arsenide and Related Compounds*, Seattle, WA, 9-12 September 1991, IOP Publishing Ltd., Bristol, England, Inst. Phys. Conf. Ser. No 120, 483-486.

11. Tournie, E., J.-L. Lazzari, F. Pitard, C.Alibert and A. Joullie (1990) "2.5 μm GaInAsSb Lattice-Matched to GaSb by Liquid Phase Epitaxy," *J. Appl. Plys.*, 68 (11):5936-5940.

12. Evans, B.D., V.S. Sundaram, M.D. Morgan, W.E. Horne, J.R. Ketterl, S.B. Saban, M.B.Z. Morosini, and N.B. Patel. "A New GaInAsSb-Based Photovoltaic Cell For Use With Sources At ≤ 1073K in Thermophotovoltaic Power Conversion Systems.", *Proc. of the 14th Symposium on Space Power and Propulsion.*, Albuquerque, NM, M.S. El-Genk, Ed., AIP Conference Proceedings 387:1585-1591

13. Sundaram, V.S. and P.E. Gruenbaum (1993) "Zinc Diffusion in GaSb," *J. Appl. Phys.* 73(8):3787-3789.

14. Sze, S.M. (1981) *Physics of Semiconductor Devices*, second edition, John Wiley, N.Y.

15. Wald, F.V. and R.O. Bell, "Natural and Forced Convenction During Solution Growth of CdTe by the Traveling Heater Method (THM)," *J. Crystal Growth*, 30, 29 (1975).)

Improvements in
GaSb-Based Thermophotovoltaic Cells

Z.A. SHELLENBARGER, M.G. MAUK, J.A. COX, M.I. GOTTFRIED,
P.E. SIMS, J.D. LESKO, J.B. MCNEELY, and L.C. DINETTA

AstroPower, Inc., Solar Park, Newark, DE 19716-2000
tel: 302-366-0400   fax: 302-368-6474 e-mail: zane@astropower.com

R.L. MUELLER

Jet Propulsion Laboratory California Institute of Technology
4800 Oak Grove Drive   Pasadena, CA 91109

Abstract.   This work seeks to improve the performance of GaSb-based thermophotovoltaic (TPV) devices. Previously, we demonstrated InGaAsSb (~0.53 bandgap) cells with very high internal quantum efficiencies at wavelengths of 2 microns. Enhanced efficiency should be possible using more sophisticated double heterostructures with wide-bandgap, lattice-matched AlGaAsSb front surface passivating "window" layers and Back Surface Field (BSF) cladding layers. The double heterostructure also provides more design flexibility for improved fill factor—by reducing series resistance—and more effective photon recycling—by avoiding high doping and thereby minimizing parasitic optical absorption. We demonstrate a near ten-fold reduction in reverse-saturation current (to as low as 10^{-5} A/cm^2), higher fill factors (over 60%), and much improved short-wavelength spectral response using a wide-bandgap (~1.1 eV) AlGaAsSb window layer to passivate the front surface.   The effective use of a BSF layer is less straight-forward, as very non-ideal current voltage characteristics can result from inclusion of a BSF cladding layer.   Some preliminary results indicate these problems can possibly be avoided by more detailed optimization of the base and BSF doping.

CP401, *Thermophotovoltaic Generation of Electricity: Third NREL Conference,*
edited by Benner/Coutts
© 1997 The American Institute of Physics 1-56396-734-0/97/$10.00

INTRODUCTION AND BACKGROUND

In the last year we have reported progress in *p-n* homojunction InGaAsSb thermophotovoltaic (TPV) cells [1-2]. This material system has a good spectral match for many practical thermal emitters. These cells have an epitaxial device structure with an absorbing region bandgap of ~0.53 eV and are designed for thermophotovoltaic conversion of ~1200 °C blackbody sources, such as the GPHS (General Purpose Heat Source). TABLE 1 summarizes typical performance of 1-cm x 1-cm InGaAsSb homojunction TPV cells with the design of FIGURE 1a. The most significant results from the first stage of effort were the high internal quantum efficiencies achieved at wavelengths between 1800 and 2100 nm. A later version of these cells (FIGURE 1b) was designed to utilize photon recycling by the addition of a "back mirror" contact to reflect sub-bandgap light back to the thermal source. To effect a mirror, silicon dioxide was deposited on the backside of the substrate, patterned with stripe openings, and then overlaid with a gold coating.

Table I

open-circuit voltage (V)	0.260
short-circuit current (A/cm^2)	2.0
fill factor	0.58
internal QE at λ = 2 μm	95%
internal QE at λ = 1 μm	55%

In this case, a low-doped (10^{16} cm^{-3}) GaSb substrate was used to reduce free-carrier absorption of sub-bandgap energy photons. Sub-bandgap reflection (measured without a front contact grid) was over 90%, but this value is significantly reduced when the device layers and/or substrate are more heavily doped. Although this design achieved the desired optical performance when the doping of the substrate and epilayers was kept low, the cell suffered from reduced fill factors (FF = 40 to 50%) due to high series resistance and possibly non-linear behavior of the InGaAsSb base / GaSb substrate junction. To help reduce series resistance and free-carrier optical absorption, we thin the GaSb substrate to less than 100 microns thickness in a post-growth etching using a KI-I$_2$ aqueous solution or a Br-methanol "polishing" etch. Substrate thicknesses as low as 60 microns on 1-cm x 1-cm devices have been achieved.

InGaAsSb quaternary alloy are good candidates for TPV applications that require high spectral response in the 1500 to 2500 nm wavelength range [3-5]. Depending on its alloy composition (*x,y*), the direct bandgap of the In$_{1-x}$Ga$_x$As$_{1-y}$Sb$_y$ alloy varies from 0.18 eV (InSb) to 1.43 eV (GaAs). The quaternary alloy can be closely lattice-matched to the GaSb substrate provided the composition is restrained to values such that $y \approx 0.1 + 0.9\,x$. With this lattice matching condition, the bandgap of the quaternary alloy ranges from approximately 0.3 to 0.7 eV. However, there is a further limitation due to a wide solid-phase miscibility gap in this quaternary at typical growth temperatures. The miscibility gap evidently precludes bandgaps in the range of 0.35 to 0.5 eV [6]. Therefore, for the spectral range of interest, we assume the lowest attainable bandgap is 0.50 to 0.52 eV. This bandgap range corresponds to an optical absorption edge of 2380 to 2480 nm.

It is worth emphasizing that the use of the quaternary alloy, as opposed to a ternary alloy such as InGaAs, provides the required bandgap with near-exact lattice matching to the GaSb substrate. Lattice-matching is important since even a small degree of lattice mismatch degrades device performance and reliability. Although there are epitaxy techniques to partially ameliorate effects associated with lattice mismatch of ternary alloy layers on binary substrates (e.g. defect-filtering superlattices, interrupted growth regimens, etc.), a simpler and more efficient approach is to use the quaternary alloy to avoid lattice mismatch altogether.

High-quality, lattice-matched heterostructures using alloys of AlGaAsSb and InGaAsSb are also feasible. We believe lattice-matched structures are important in avoiding degradation at the high current densities (1 to 10 A/cm^2) typical in TPV applications. One objective is to extend the high quantum efficiencies to shorter wavelengths by passivating the front surface with a wide-bandgap, lattice-matched AlGaAsSb "window" layer, similar to the designs common in AlGaAs/GaAs heteroface solar cells used for space power. Another objective is to reduce the reverse saturation current for higher open-circuit voltages. This entails the use of a BSF (Back Surface Field) cladding layer. This is the well-established and very effective design approach used in high-efficiency GaAs solar cells. The multilayer structure should also provide more flexibility in achieving reducing series resistance and realizing high fill factors, while still maintaining low parasitic absorption of sub-bandgap photons. For instance, a wide bandgap window layer can help reduce the sheet resistance of the emitter.

We are undertaking a systematic comparison and optimization of the designs shown in FIGURE 1 to determine the best structure for GaSb-based TPV cells, specifically with respect to the utility and optimum implementation of front surface window layers, and back surface field (BSF) cladding layers, as well as comparisons between *p-n* InGaAsSb homojunctions vs. *p-n* AlGaAsSb/InGaAsSb heterojunctions, and *n*-on-*p* vs. *p*-on-*n* cell configurations. The experimental cells are characterized on the basis of internal and external quantum efficiency, sub-bandgap reflection, current-voltage characteristics under pulsed solar simulator illumination, and variation of operating characteristics with cell temperature. This study will be useful in exploiting the full potential of GaSb-based alloys for thermophotovoltaic applications.

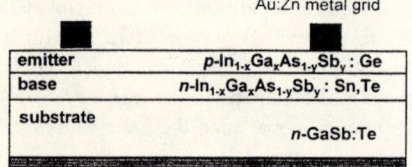

a. *p-on-n* homojunction {$x = 0.85$; $y = 0.83$}

Au:Zn metal grid

emitter	p-In$_{1-x}$Ga$_x$As$_{1-y}$Sb$_y$: Ge
base	n-In$_{1-x}$Ga$_x$As$_{1-y}$Sb$_y$: Sn,Te
substrate	n-GaSb:Te

planar contact

Au:Zn metal grid

emitter	p-In$_{1-x}$Ga$_x$As$_{1-y}$Sb$_y$: Ge
base	n-In$_{1-x}$Ga$_x$As$_{1-y}$Sb$_y$: Sn,Te
substrate	SiO$_2$ n-GaSb:Te

Au metal overlay

b. *p-on-n* homojunction (with back oxide / gold reflector and thinned substrate)

Au:Zn metal grid

emitter	n-In$_{1-x}$Ga$_x$As$_{1-y}$Sb$_y$: Te, Sn
base	p-In$_{1-x}$Ga$_x$As$_{1-y}$Sb$_y$: Ge
substrate	p-GaSb:Zn

planar contact

c. *n-on-p* homojunction {$x = 0.85$; $y = 0.83$}

Au:Zn metal grid

emitter	p-Al$_{1-x'}$Ga$_{x'}$As$_{1-y'}$Sb$_y$: Ge
base	n-In$_{1-x}$Ga$_x$As$_{1-y}$Sb$_y$: Sn,Te
substrate	n-GaSb:Te

d. *p-on-n* AlGaAsSb/InGaAsSb heterojunction

Au:Zn metal grid

emitter	p-In$_{1-x}$Ga$_x$As$_{1-y}$Sb$_y$: Ge
base	n-In$_{1-x}$Ga$_x$As$_{1-y}$Sb$_y$: Sn,Te
BSF/ cladding	n-Al$_{1-x'}$Ga$_{x'}$As$_{1-y'}$Sb$_{y'}$: Sn,Te
substrate	n-GaSb:Te

e. *p-on-n* homojunction with wide bandgap cladding (BSF) layer.

emitter	p-Al$_{1-x'}$Ga$_{x'}$As$_{1-y'}$Sb$_{y'}$: Ge
base	n-In$_{1-x}$Ga$_x$As$_{1-y}$Sb$_y$: Sn,Te
BSF / cladding	n-Al$_{1-x'}$Ga$_{x'}$As$_{1-y'}$Sb$_{y'}$: Sn,Te
substrate	n-GaSb:Te

f. *p-on-n* heterojunction with wide bandgap cladding (BSF) layer.

Au:Zn metal grid

window	p-Al$_{1-x'}$Ga$_{x'}$As$_{1-y'}$Sb$_{y'}$: Ge
emitter	p-In$_{1-x}$Ga$_x$As$_{1-y}$Sb$_y$: Ge
base	n-In$_{1-x}$Ga$_x$As$_{1-y}$Sb$_y$: Sn,Te
BSF / cladding	n-Al$_{1-x'}$Ga$_{x'}$As$_{1-y'}$Sb$_{y'}$: Sn,Te
substrate	n-GaSb:Te

g. *p-on-n* homojunction (with BSF / cladding layer, window layer, back oxide/Au reflector, and thinned substrate)

window	n-Al$_{1-x'}$Ga$_{x'}$As$_{1-y'}$Sb$_{y'}$: Sn,Te
emitter	n-In$_{1-x}$Ga$_x$As$_{1-y}$Sb$_y$: Sn,Te
base	p-In$_{1-x}$Ga$_x$As$_{1-y}$Sb$_y$: Ge
BSF / cladding	p-Al$_{1-x'}$Ga$_{x'}$As$_{1-y'}$Sb$_{y'}$: Ge
substrate	p-GaSb:Zn

h. *n-on-p* homojunction (with BSF / cladding layer, window layer, back oxide/Au reflector, and thinned substrate)

Figure 1:   TPV Cell Designs.

The 'baseline' homojunction device (FIGURE **1a**) is a two-layer epitaxial InGaAsSb structure formed by liquid-phase epitaxy on a GaSb substrate. The (direct) bandgap of the $In_{1-x}Ga_xAs_{1-y}Sb_y$ alloy is 0.50 to 0.55 eV, depending on its exact alloy composition (x,y), and is closely lattice-matched to the GaSb substrate. Internal quantum efficiencies as high as 95% have been measured at a wavelength of 2 microns. At a wavelength of 1 micron, internal quantum efficiencies of 55% have been observed. At a current density of 1.6 A/cm^2, an open-circuit voltage of 0.250 V and a fill factor of 60% have been measured. Our results to date show that the GaSb-based quaternary compounds provide a viable and high performance energy conversion solution for thermophotovoltaic systems operating with 1000 to 1500 °C source temperatures.

EPITAXY AND DEVICE FABRICATION

The TPV devices are epitaxial AlGaAsSb / InGaAsSb structures formed by liquid-phase epitaxy on a GaSb substrates at a growth temperature of 515 °C. Liquid-Phase Epitaxy (LPE) is a well-established technology for III-V compound semiconductor devices. A major advantage of LPE for this application is the high material quality, and more specifically, the long minority carrier diffusion lengths, that can be achieved. This results in devices which are equal or superior in performance to those made by other epitaxy processes such as molecular beam epitaxy (MBE) or metal organic chemical vapor deposition (MOCVD). Another important advantage is that LPE is a simple, inexpensive, and safe method for semiconductor device fabrication. Significantly, the LPE process does not require or produce any highly toxic or dangerous substances—in contrast to MOCVD. Also, the epitaxial growth rate with InGaAsSb LPE is ~2 μm/min., which is 10 to 100 times faster than growth rates for MOCVD or MBE. We have successfully scaled up the LPE process for epitaxial growth in a semi-continuous mode on 3-inch diameter wafers. This, combined with the high growth rates, will dramatically improve the manufacturing throughput compared to traditional and more costly epitaxy processes. The objective is to develop an epitaxial growth technology to produce low-cost, large-area, high efficiency TPV devices.

InGaAsSb photodiodes, light-emitting diodes, and double heterostructure injection lasers made by liquid-phase epitaxy have been previously reported [6-10]. We have adapted this technology for the production of InGaAsSb TPV cells.

A standard horizontal slideboat technique is used for the liquid-phase epitaxial growth of the InGaAsSb. The graphite slideboat is situated in a sealed quartz tube placed in a microprocessor-controlled, programmable, three-zone tube furnace. The growth ambient is palladium-diffused hydrogen at atmospheric pressure with a flow rate of 300 ml/min.

The substrates are 500-micron thick, chemically polished, (100) oriented, *n*-type GaSb wafers obtained from MCP Wafer Technology, Ltd. (Milton Keynes, UK) or Firebird Semiconductor, Ltd. (Trail, BC, Canada). Substrates are doped to 3-5 x 10^{17}

cm^{-3} with tellurium. The substrate resistivity is 9×10^{-3} Ω·cm, and the average etch-pit density is approximately 1000 cm^{-2}.

The growth solutions are indium (X_{In}=0.59), gallium (X_{Ga}=0.19), antimony (X_{Sb}=0.21), and arsenic (X_{As}=0.01). The melts are formulated with 3 to 5 mm shot of high purity (99.9999%) indium, gallium, and antimony metals and arsenic added as undoped polycrystalline InAs material. The total weight of the melt is about 7 g. Prior to growth, the melts are baked out at 700 °C for 15 hours under flowing hydrogen to de-oxidize the metallic melt components and outgas residual impurities. After bake-out, appropriate dopant impurities are added to each melt. The first melt for the growth of the n-type InGaAsSb base layer contains tellurium. The small amount of Te needed to dope the layer (atomic fraction in the melt $\cong 10^{-5}$) is problematic. For reproducible doping, a weighable amount of Te is added as 100 to 200 mg of Te-doped GaSb ($C_{Te}=10^{19}$ cm^{-3}) or as In-Te alloy (1% Te by weight). The second melt for the growth of the p-type emitter contains 1 to 10 mg germanium. We have begun a more detailed and systematic characterization of impurity segregation and doping in the In-Ga-As-Sb quaternary system with the aim of achieving better control and a greater range of doping concentrations.

The melts are equilibrated for 1 hour at 530 °C and then cooled at a rate of 0.7 °C/min. At 515 °C, the substrate is contacted with the first melt for 2 minutes to grow a 5 micron thick n-type InGaAsSb base layer. Next, the substrate is moved to the second melt for 5 seconds to grow a 0.3 micron thick p-type InGaAsSb emitter layer.

For the Al$_{0.3}$Ga$_{0.7}$As$_{0.02}$Sb$_{0.98}$ window and BSF layers, the LPE procedure was similar to that described above except that the melt solution consisted of aluminum (X_{Al} = 0.015), gallium (X_{Ga} = 0.957), antimony (X_{Sb} = 0.028) and arsenic (X_{As} = 0.001). The As is added as undoped polycrystalline GaAs. The bandgap of AlGaAsSb corresponding to this quaternary composition is about 1.2 eV. The layer compositions were verified by electron microprobe analysis.

Front and back ohmic contacts are formed on the epitaxial InGaAsSb/GaSb structure by standard processing techniques. The back of the substrate is metallized by plating with a 200 nm thick electron-beam evaporated Sn:Au layer and alloyed at 300 °C. The front contact grid consists of 10 micron wide metallization lines with 100 micron spacing and a single 1-mm wide center busbar. The grid is formed by a photolithography lift-off process with a 200 nm thick electron-beam evaporated Au:Zn:Au metallization. The front grid is thickened to 5 microns by gold electroplating. The front contact is not sintered. The substrate is masked and patterned to define a 1 cm x 1 cm device and isolation etched with a potassium iodide - iodine etch. Most of our TPV cells are 1 cm x 1 cm in area; although larger cells (2 cm x 2 cm) with comparable performance have also been made. In order to simplify the spectral response analysis, no anti-reflection coating was applied to the cells. FIGURE 2 is a top-view photograph of a 1 cm x 1 cm InGaAsSb TPV cell.

FIGURE 2. Top-view photograph of a 1 cm x 1 cm InGaAsSb TPV cell.

TPV DEVICE DESIGN AND OPTIMIZATION

FIGURE 1 shows various TPV device designs in cross-section. The fabricated cells have a 0.3 to 0.5 micron thick *p*-type emitter with a Ge concentration of approximately 10^{19} cm^{-3}, as indicated by Secondary Ion Mass Spectroscopy (SIMS). A thicker, more heavily doped *p*-layer will reduce the sheet resistance of the emitter and therefore improve the fill-factor, but will tend to reduce spectral response due to higher free-carrier absorption and increased sensitivity to front surface minority carrier recombination.

The base thickness ranges from 3 to 5 microns with a Te concentration of about 10^{17} cm^{-3}, as determined from capacitance-voltage measurements and SIMS. Modeling indicates that base dopings in this range will yield the optimum open-circuit voltages and short-wavelength quantum efficiencies. FIGURE 3 shows the SIMS depth profile indicating the abruptness of the *p-n* junction and the depth uniformity of the doping concentrations. There is apparently very little smearing of the doping profile due to diffusion or segregation of dopants. Discrepancies between the Te dopant concentration measured by SIMS (total impurity concentration) and that implied by capacitance-voltage measurements (net donor concentration) indicate that some of the Te is either not ionized or is compensated. This is a common problem in Te doping of III-V semiconductors, especially in GaSb-based materials, and is probably due to the formation of electrically inactive telluride complexes or compounds in the material. We have also employed tin as an n-type dopant. The optimization of target doping levels for high efficiency, and realization of specified doping in the epitaxy process, are areas where much additional effort is needed.

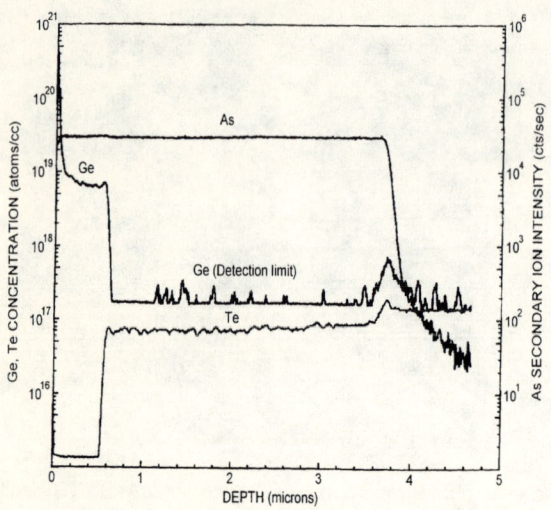

FIGURE 3. SIMS depth profile of doping.

TPV DEVICE EVALUATION

We present external and internal spectral response curves and current-voltage characteristics for 1 cm x 1 cm p-$In_{0.15}Ga_{0.85}As_{0.17}Sb_{0.83}$:Ge / n-$In_{0.15}Ga_{0.85}As_{0.17}Sb_{0.83}$:Te epitaxial cells on n-GaSb:Te substrates produced as described above. The *external* spectral response and *internal* spectral response of typical InGaAsSb TPV cells are shown in FIGURE 4. The lower external spectral response is due to grid shading and reflection of incident light from the uncoated InGaAsSb emitter surface. The grid shading is 18.2%. The absorption edge implied by the spectral response measurements of a number of samples ranged from approximately 2250 to 2300 nm. At a wavelength of 2000 nm, internal quantum efficiencies as high as 95% have been measured, and at a wavelength of 1 micron, internal quantum efficiencies of almost 55% have been observed. The internal quantum efficiency averaged over the spectral region from 1 to 2 microns wavelength is 60%. (It should be noted that for the intended TPV applications, the response of the cell for wavelengths less than 1.5 microns is not important.) Sub-bandgap external reflection was measured for cells with the back mirror structures shown in FIGURE 1b, but before formation of the front surface contact grid, the presence of which complicates interpretation of reflection measurements. With a thinned substrate (< 100 microns) we can achieve sub-bandgap external reflectivities over 90% over a bandwidth from 2 to 6 microns wavelength.

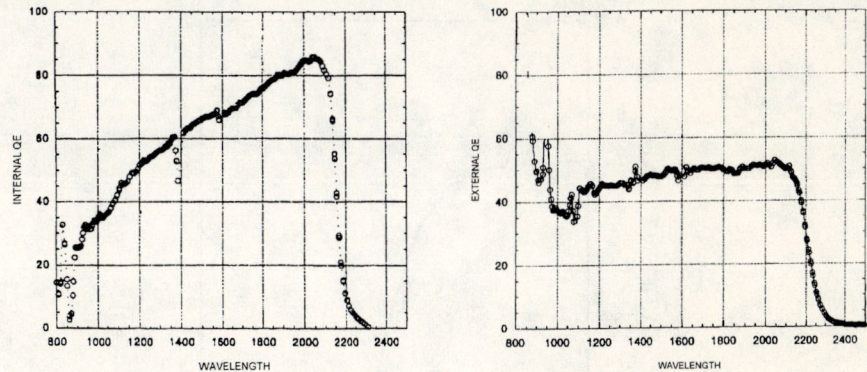

Figure 4. Internal and external quantum efficiencies of InGaAsSb/GaSb *p-n* junction thermophotovoltaic cells.

The 1 cm x 1 cm InGaAsSb TPV cells were tested under simulated infrared light using a ZnSe-filtered tungsten source (Carley Lamps, Inc., Torrance, CA) with a spectral emission in the 800 to 3000 nm wavelength range. Under an illumination intensity corresponding to a short-circuit current density of 2 A/cm^2, open-circuit voltages as high as 0.260 volts have been measured. Cells were also tested at the Jet Propulsion Laboratory using a pulsed solar simulator with an infrared bandpass filter. FIGURE 5 shows the current-voltage characteristics of a 1 cm x 1 cm InGaAsSb TPV cell tested in this manner which gives an output current density of 0.348 A/cm^2. This intensity yields an open-circuit voltage of 0.235 V and a fill-factor of 0.58. FIGURE 6 shows short-circuit current vs. open-circuit voltage for varying light intensity. The diode ideality factor changes from m = 1 to m = 2 at an open-circuit voltage of ~0.08 V. This graph implies a reverse saturation current of $J_0 \approx 1.5 \times 10^{-4}$ A/cm^2.

The addition of a 1.3-eV bandgap AlGaAsSb window layer (as indicated in FIGURE 1g) significantly improves cell performance. The reverse-saturation current is lowered by more than a factor of ten to $J_0 \approx 1.4 \times 10^{-5}$ A/cm^2. Also, the quantum efficiency between 1200 and 1600 nm is improved 30 to 40%. Also, a fill factor of 0.61 has been measured at a current density of 0.7 A/cm^2. The improvements in open-circuit voltage are seen for cells with a high base doping (>10^{17} cm^{-3}). Cells with low base dopings (10^{16} cm^{-3}) do not benefit as much from the presence of a front surface passivation layer. This is consistent with the reasonable assumption that cells with high base dopings have significant emitter components of reverse saturation current that are relatively sensitive to front surface recombination. We have yet to demonstrate any significant advantage of heterojunction cells (FIGURE 1f). To date, such heterojunctions have resulted in cells with very non ideal I-V characteristics.

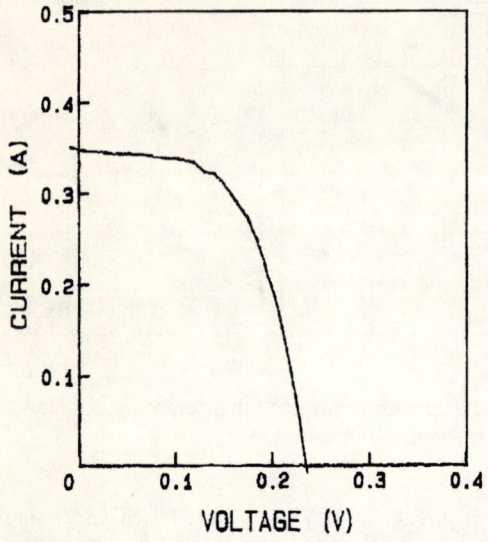

Figure. 5. Current-voltage characteristic of 1 cm x 1 cm InGaAsSb TPV cell.

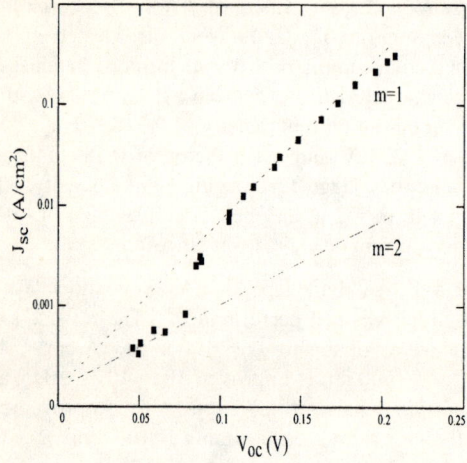

Figure 6. Short-circuit current vs. open-circuit voltage for InGaAsSb TPV cell under varying illumination intensity.

CONCLUSION AND DISCUSSION

Our results to date have demonstrated the potential of InGaAsSb TPV devices made by liquid-phase epitaxy. There is still room for substantial efficiency enhancements in these devices by optimization of the doping levels and layer thicknesses. The effectiveness of a wide-bandgap lattice-matched AlGaAsSb window layers for front surface passivation is clearly demonstrated. The use of AlGaAsSb back-surface field cladding layers to reduce the reverse saturation current and thereby increase the open-circuit voltage is still tentative. As yet, we see no advantage to heterojunctions. Highly doped contact layers will provide lower series resistance, as will substrate thinning. Lower series resistance will lead to higher fill factors. Thinning the substrate will also improve the heat sinking capability of the device. Based on the results reported here, we believe a 0.5-eV bandgap GaSb-based TPV cell with an open-circuit voltage well over 300 mV at current densities of 2 A/cm^2, and with a fill factor of 70%, is a realistic near-term goal.

The required performance of a TPV device is dependent on its system application. Spectral control of thermal emitters, the use of selective filters and reflectors, heat transfer, and photon recycling effects need to be included in the device design and system optimization. These considerations are not usually relevant for conventional photovoltaic devices and therefore the design and optimization rules for TPVs will be significantly different than those for solar cells. For example, grid obscuration and reflection are not necessarily losses in TPV systems if photons reflected from the front surface are re-absorbed by the emitter. At this stage, a better computer-based theoretical model to simulate AlGaAsSb/InGaAsSb TPV cells is needed to guide the optimization and interpret experimental results.

REFERENCES

1. Z.A.Shellenbarger, M.G. Mauk, and L.C. DiNetta, "Recent Progress in InGaAsSb/GaSb TPV Devices" *Conf. Rec. 25th IEEE Photovoltaics Specialists Conf.* (IEEE Press, Piscataway, NJ, 1996) 81-84.

2. Z.A. Shellenbarger, M.G. Mauk, L.C. DiNetta, and G.W. Charache, "InGaAsSb/GaSb Thermophotovoltaic Cells" *Proc. 14th Space Photovoltaic Research and Technology Conf. 1995 (SPRAT XIV)* NASA Conf. Publication 10180 (1996) 215-222.

3. A. Schock, C. Or, and V. Kumar, "Small Radioisotope Thermophotovoltaic (RTPV) Generators", *Proc. 2nd NREL Conf. on Thermophotovoltaic Generation of Electricity,* (New York: American Institute Physics, 1996), pp. 81-97.

4. E. Kittl, "Unique Correlations Between Blackbody Radiation and Optimum Energy Gap for a Photovoltaic Conversion Device", *Proc. 10th IEEE PV Specialists Conf.,* 1973, pp. 103-106.

5. L.D.Woolf, "Optimum Efficiency of Single and Multiple Bandgap Cells in Thermophotovoltaic Energy Conversion", *Solar Cells* **19**, 1986, pp. 19-38.

6. M. Astles, H. Hill, A.J. Williams, P.J. Wright, and M.L. Young, "Studies of the $Ga_{1-x}In_xAs_{1-y}Sb_y$ Quaternary Alloy System I. Liquid-Phase Epitaxial Growth and Assessment", *J. Electronic Materials* **15**, 1986, pp. 41-49.

7. A. Andaspaeva, A.N. Baranov, A. Guseinov, A.N. Imenkov, L.M. Litvak, G.M. Filaretova, and Y.P. Yakovlev, "Highly Efficient GaInAsSb Light-Emitting Diodes ($\lambda = 2.2$ μm, $\eta=4\%$, $T = 300K$)", *Soviet Technical Physics Letters* **14**, 1988, pp. 377-378.

8. C. Caneau, A.K. Srivastava, A.G. Dentai, J.L. Zyskind, and M.A. Pollack, "Room-Temperature GaInAsSb/AlGaAsSb DH Injection Lasers at 2.2 μm", *Electronics Letters* **21**, 1985, pp. 815-817.

9. A.E. Drakin, P.G. Eliseev, B.N. Sverdlov, A.E. Bochkarev, L.M. Dolginov, and L.V. Druzhinina, "InGaSbAs Injection Lasers", *IEEE J. Quantum Electronics* **QE-23**, 1987, pp. 1089-1094.

10. N. Kobayashi, Y. Horikoshi, and C. Uemura, "Liquid-Phase Epitaxial Growth of InGaAsSb/GaSb and InGaAsSb/AlGaAsSb DH Wafers", *Japanese J. Applied Physics* **18**, 1979, pp. 2169-2170.

New Concepts for
III-V Antimonide Thermophotovoltaics

MICHAEL G. MAUK, ZANE A. SHELLENBARGER, MARK I. GOTTFRIED,
JEFF A. COX, B.W. FEYOCK, JAMES B. MCNEELY, and LOUIS C. DINETTA

AstroPower, Inc., Solar Park, Newark, DE 19716-2000
tel: 302-366-0400, fax: 302-368-6474, e-mail: mauk@astropower.com

ROBERT L. MUELLER

**Jet Propulsion Laboratory, California Inst. of Technology
4800 Oak Grove Drive   Pasadena, CA   91109**

Abstract: We survey and assess new concepts for "next-generation" GaSb-based thermophotovoltaic (TPV) devices. The objectives are new device structures with novel back mirror designs to better utilize photon recycling effects, isolation schemes to realize monolithic, series-interconnected TPV arrays, and reduced cost by the use of surrogate substrates in place of GaSb or InAs wafers. The processes considered include 1. liquid-phase epitaxial lateral overgrowth on patterned, masked substrates, 2. epitaxial film transfer or wafer fusion techniques, 3. epitaxial growth of GaSb alloys on high resistivity (lattice matched) ZnTe, 4. realization of semi-insulating III-V antimonides, 5. selective wet oxidation of Al-containing antimonides (e.g., AlAsSb) wherein a the semiconducting epitaxial layer is converted to Al_2O_3 to form a buried insulating layer, and 6. GaSb- and InAs-on-silicon heteroepitaxy.

CP401, *Thermophotovoltaic Generation of Electricity: Third NREL Conference,*
edited by Benner/Coutts
© 1997 The American Institute of Physics 1-56396-734-0/97/$10.00

1. INTRODUCTION

We introduce and assess some new concepts for a next-generation of III-V antimonide thermophotovoltaic (TPV) cells. Our broad objectives are:

- to develop novel device designs and fabrication technologies to better exploit optical effects such as photon recycling,

- to circumvent the disadvantages of the presently-used GaSb and InAs substrates (high cost, limited size, lack of semi-insulating substrates), and

- to explore approaches for monolithic, series-interconnection of TPV devices for high-voltage arrays or mini-modules on a single large-area substrate.

In pursuit of these objectives, we are developing several technologies for their application to GaSb and/or InAs-based TPV devices:

- epitaxial lateral overgrowth on patterned, masked substrates,

- *epitaxial film transfer* or *wafer fusion* wherein an epitaxial device structure is bonded to a surrogate superstrate and the original substrate is removed by selective etching or separated by "lift-off,"

- heteroepitaxy of GaSb and InAs on silicon,

- realization of high resistivity or semi-insulating III-V antimonides, and

- selective "steam oxidation" of Al-containing antimonides

These techniques, alone or in combination, show reasonable prospects for realizing the above-listed aims. While all of these technologies are well established for other III-V semiconductors such as GaAs, there is little or no precedent for their application to III-V antimonides or TPV devices. While some of the above processes can be directly adapted to III-V antimonides in a straightforward way, others are more speculative and long-range; their inclusion at this stage of the work is to assess their feasibility and difficulty of implementation for TPV devices.

2. BACKGROUND

GaSb- and InAs-based alloys are good choices for thermophotovoltaic cells intended for conversion of blackbody radiation sources operating in the 1000 to 1500 °C temperature range since:

1. the requisite epitaxial growth technology is adequately developed,

2. lattice-matched, thermal-expansion matched, high-quality double heterostructures can be grown with bandgaps in the range of 0.5 to 0.8 eV,

3. there is a substantial technology base for III-V compound antimonides owing to their use in mid-infrared detectors, LEDs, and lasers, and

4. previous work in related areas, although not extensive, indicates that Sb-based optoelectronic devices are stable and robust.

In the last year we have reported[1] InGaAsSb/GaSb *p-n* homojunction TPV cells with internal quantum efficiencies as high as 90%. At an illumination level yielding a short-circuit current density of 1.6 A/cm^2, open-circuit voltages of 0.250 V and fill-factors of 60% have been measured. These devices showed no sign of degradation after many hours of operation. Further optimization of AlGaAsSb / InGaAsSb / AlGaAsSb hetero-structure TPV cells is on-going[2]. Preliminary results indicate significant improvements in open-circuit voltage, fill-factor, and short-wavelength quantum efficiency. These results encouraged us to pursue more sophisticated devices with the features cited above.

3. TPV CELLS WITH "BURIED" MIRRORS

In the context of TPV devices, photon recycling refers to the reflection of sub-bandgap energy photons back to the heated source. This can be accomplished by depositing a reflecting layer on the backside of the substrate. For GaSb-based TPVs, the GaSb substrate (with a bandgap of 0.7 eV) is nominally transparent to light not absorbed in the AlGaAsSb/InGaAsSb active layers (~0.5 eV). Nevertheless, free carrier absorption in the substrate can significantly diminish the effectiveness of the backside mirror. An alternative approach is the "buried" mirror design shown in FIGURE 1. Here, a reflective layer is situated between the substrate and the epitaxial device structure. Openings in the mask (called *vias*) provide electrical contact between the substrate and the active layers. This design avoids parasitic substrate absorption in effecting photon recycling.

Several types of films have been employed as "buried" mirror layers including refractory metals such as tungsten, dielectrics such as SiO$_2$ and Si$_3$N$_4$, metal nitrides such as WN, multilayer electron-beam deposited Si/SiO$_2$ quarter wavelength Bragg stacks, and various multilayer combinations of metals and dielectrics. The reflectivities for some of these films deposited on III-V substrates is shown in FIGURE 2. Broad-band, wide-angle reflectors with reflectivities well over 90% can be achieved with many of these film combinations by appropriately "tuning" the thicknesses of the component layers to form interference reflectors.

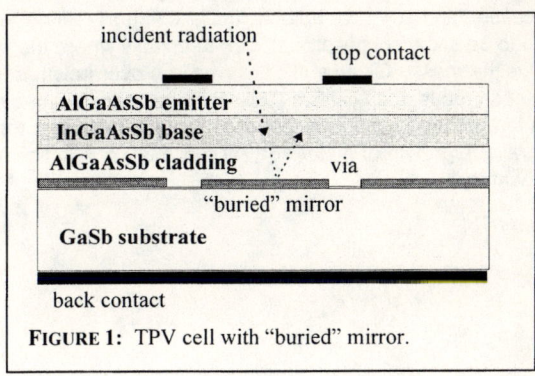

FIGURE 1: TPV cell with "buried" mirror.

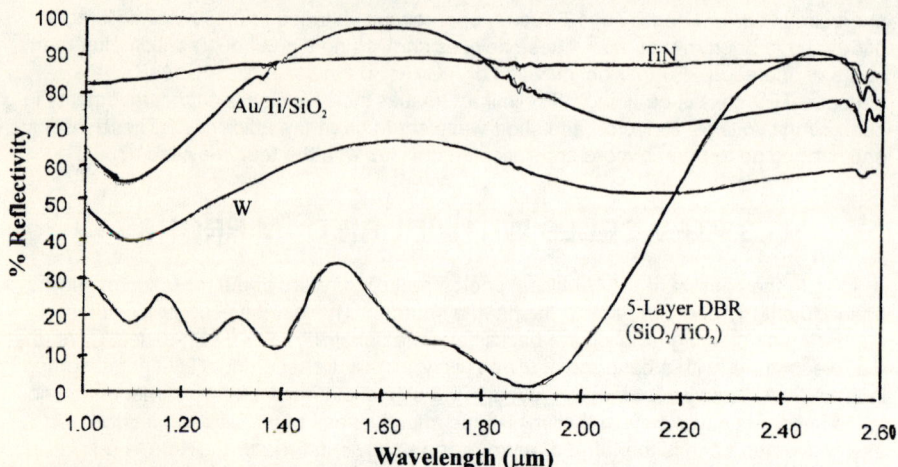

FIGURE 2: Reflection spectra for various electron-beam deposited or sputtered "buried" mirror structures on III-V substrates.

The "buried" mirror structure device structure shown in FIGURE 1 can be realized by epitaxial lateral overgrowth or a wafer fusion technique.

4. EPITAXIAL LATERAL OVERGROWTH

The epitaxial lateral overgrowth process is shown in FIGURE 3. A GaSb substrate is masked with a dielectric, metal, or composite multilayer coating of metals, dielectrics, or amorphous or polycrystalline semiconductors and patterned with stripe openings (**3a**). The patterned, masked substrate is used for epitaxial growth of AlGaAsSb and InGaAsSb films. The stripe openings where the underlying GaSb substrate is exposed serve as sites of preferential nucleation and selective epitaxy; there is virtually no nucleation on the mask itself. FIGURES **3b** to **3e** show successive stages of epitaxy where the selectively grown epilayer overgrows the mask. Given sufficient time, the overgrowth from adjacent vias impinges to form a continuous epitaxial film (**3e**). By this technique, a single crystal layer of a GaSb-alloy is formed on a masked substrate. FIGURE 4 shows a top-view photomicrograph of lateral overgrowth of AlGaAsSb (at a composition lattice matched to GaSb) on a (111)GaSb wafer.

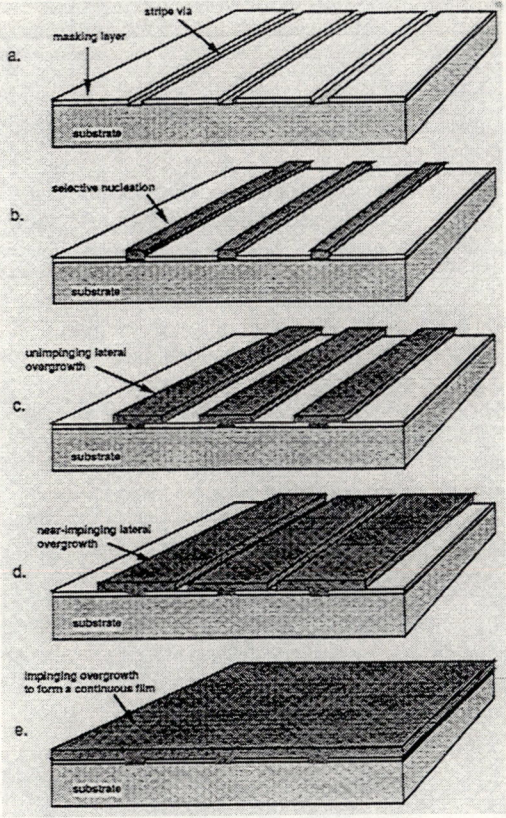

We are able to achieve impinging lateral overgrowth of AlGaAsSb epilayers on oxide-masked substrates with stripe openings spaced as much as 1000 microns apart. We are presently developing the double heterostructure AlGaAsSb / InGaAsSb / AlGaAsSb TPV device structure using epitaxial lateral overgrowth.

Actually, since the masking layer can be a conductor, insulator, or combination thereof, the lateral overgrowth and underlying mask may have several functions: **1.** as a "buried" mirror, **2.** as a buried contact, or **3.** as an insulating layer to dielectrically isolate the epilayers from the substrate and thus facilitate monolithic series interconnection of TPV device elements.

FIGURE 3: Epitaxial lateral overgrowth process in successive stages: **a.** substrate masked and patterned with stripe openings (vias) prior to growth; **b.** preferential nucleation and selective epitaxy; **c.** and **d.** near-impinging growth, **e.** complete lateral overgrowth to form a continuous epitaxial layer.

5. SEMI-INSULATING III-V ANTIMONIDES

The availability of a semi-insulating GaSb or InAs substrate would greatly simplify approaches for a monolithic, series-interconnected GaSb-based TPV array. For example, in related work WILT et al.[3] have described a monolithic, interconnected epitaxial InGaAs TPV module made on a semi-insulating InP substrate.   The low bandgap and resulting thermal carrier generation, and the intrinsic electrically active point defects in the III-V antimonides seems to preclude semi-insulating GaSb material.

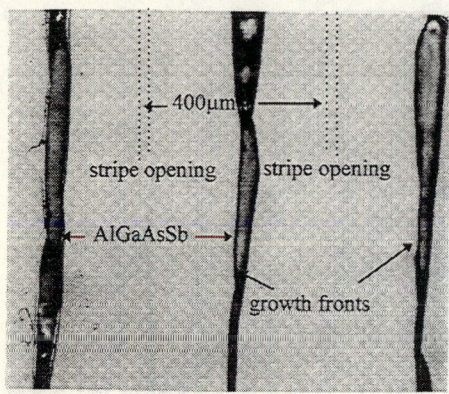

FIGURE 4: Top-view photomicrograph of near-impinging growth of AlGaAsSb epilayer on an SiO$_2$-masked (111)GaSb substrate patterned with 10-micron wide stripe vias, spaced 400 microns apart. (This sample corresponds to epitaxial lateral overgrowth at stage **d** in FIGURE 3.)

The resistivity requirements for substrate isolation in TPV monolithic interconnected arrays—which specifically entails an estimation of the minimum resistivity necessary for a 5- to 10-element series interconnected array of InGaAsSb TPV cells grown on the same substrate—needs to be assessed. It is speculated that high resistivity material in combination with junction isolation may be sufficient for some designs. Based on an extensive literature review, and a worldwide survey of GaSb and InAs substrate suppliers, the prospects for high resistivity or semi-insulating III-V antimonides are not good. The feasibility of lattice-matched, wider bandgap epitaxial layers of AlGaAsSb rendered semi-insulating by doping with transition metals (Cr, Fe, V) and/or rare earths or by using non-stoichiometric compositions to suppress carriers should be explored..4

Such a semi-insulating epitaxial layer sandwiched between the device and substrate wafer could provide a means for electrical isolation of TPV device elements. A similar approach could perhaps be taken by situating a high resistivity ZnTe layer (closely lattice matched) between the GaSb substrate and InGaAsSb epilayers. The compatibility of ZnTe with III-V antimonides is not known. This approach appears relatively less promising than the other isolation methods discussed here.

6. WAFER BONDING TECHNIQUES

Another alternative method of forming TPV devices with backside reflectors (or on an insulating substrates) uses *wafer bonding* or *wafer fusion* techniques, and related epitaxial film transfer processes. This process is straightforward but to our knowledge has not been previously applied to GaSb-based devices. FIGURE 5 shows a general process sequence for transferring a GaSb-based epitaxial device structure to a surrogate superstrate. The important issues are:

- supersubstrate: ceramic, glass, metal, silicon, thermal expansion matching
- type of bonding: adhesive, van der Waals, electrostatic
- contacts: both base and emitter from one side, application
- substrate removal: selective etchants, "stop etch" layers"
- scribing, dicing
- yield

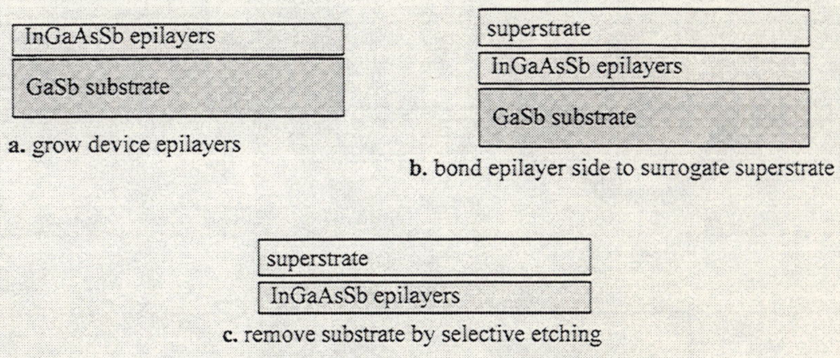

Figure 5: Epitaxial film transfer and surrogate substrate bonding.

7. MONOLITHIC INTERCONNECTS
USING EPITAXIAL LATERAL OVERGROWTH

The lateral overgrowth process using an insulating mask layer can also be adapted for monolithic series-interconnected arrays. FIGURE 6 shows a specific implementation for isolating adjacent TPV elements using an isolation etch on an epitaxial TPV device formed over an insulating layer.

8. INAS AND GASB ON SILICON

All III-V substrates are at least a factor of ten to twenty times more expensive than silicon. In the interest of reducing costs, we are developing simple heteroepitaxy techniques for growing single-crystal films of InAs and GaSb on silicon substrates. FIGURE 7 is a top-view photomicrograph of InAs on silicon grown by a simple atmospheric pressure close-spaced vapor transport process. This InAs-on-silicon structure can be used as a surrogate substrate in place of an InAs bulk wafer. A similar approach is possible with GaSb on silicon.

Although the large lattice mismatch in these systems looks tractable, other important factors in this type of heteroepitaxy include thermal expansion mismatch, cross-doping, anti-phase domain formation, and various defect generation phenomena. We realize the major problems of III-V heteroepitaxy on silicon (e.g., GaAs-on-silicon and InAs-on-silicon) remain unsolved. However, with regard to thermal stress and cross doping, one advantage InAs-on-silicon and GaSb-on-silicon have over GaAs-on-silicon heteroepitaxy is the relatively low growth tempera-tures. Epitaxy of InAs and GaSb alloys is normally at temperatures of 500 to 550 °C, whereas GaAs epitaxy is typically in the range of 800 °C. This ~300 °C lower temperature is an important advantage for heteroepitaxy of these TPV materials on silicon. Also, epitaxial lateral overgrowth has demonstrated good potential for defect reduction. Therefore, a GaSb-on-silicon or InAs-on-silicon heteroepitaxial lateral overgrowth process could yield very low defect, low stress surrogate substrates for III-V antimonide TPV applications.

135

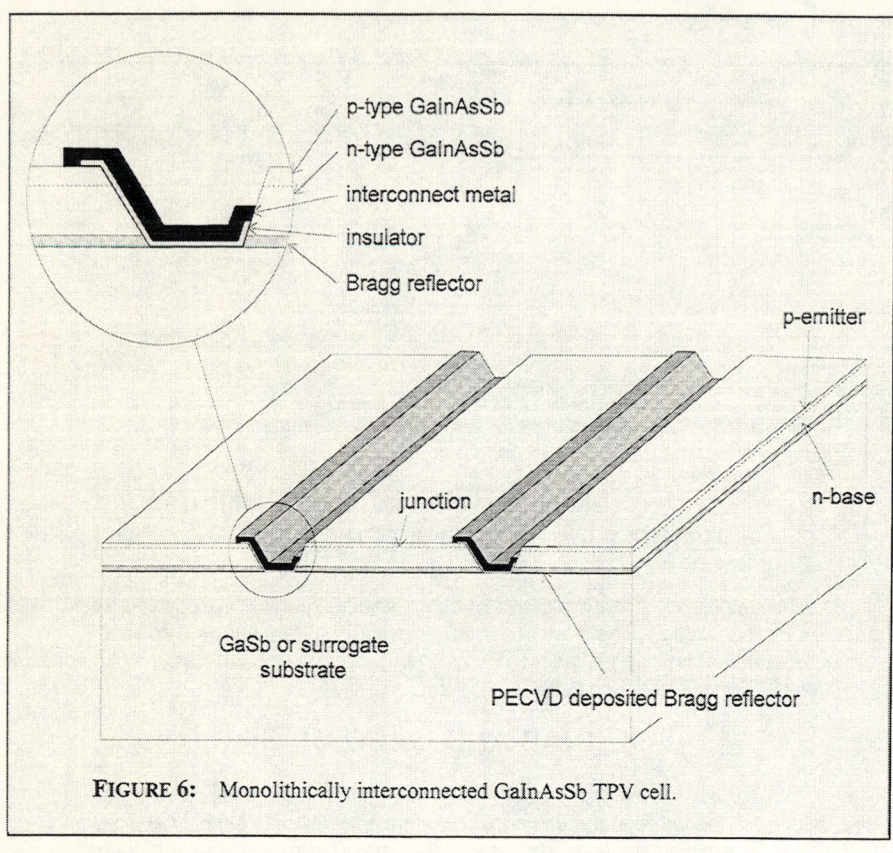

FIGURE 6: Monolithically interconnected GaInAsSb TPV cell.

9. SELECTIVE OXIDATION OF AL-CONTAINING ANTIMONIDES

Finally, we mention a selective wet thermal oxidation process, as described by BLUM *et al.*[4], which converts a "buried" epitaxial AlAsSb layer to an insulating Al oxide layer. The original application of this technique was for current confinement structures in vertical cavity surface-emitting lasers. By etching AlGaAsSb/InGaAsSb/AlAsSb epitaxial heterostructure TPV devices as mesas, the underlying AlAsSb layer could be converted to an insulating layer to isolate the mesas from each other and the substrate. In this case, the oxidation proceeds laterally from the exposed edge of the AlAsSb layer at the sidewall of the mesa. The oxidation occurs at ~350 °C in an N_2 ambient saturated with water vapor at 85 °C. The InGaAsSb/AlGaAsSb layers above the selectively oxidized AlAsSb remain single crystal, but are now "floating" on top of an insulating layer, electrically isolated from the substrate and adjacent mesas. This appears to be a simple process to realize monolithic integration of epitaxial TPV elements with lateral dimensions on the order of 100 microns.

FIGURE 7: InAs on silicon heteoepitaxial layer grown by Close-Spaced Vapor Transport (CSVT).

References

1. SHELLENBARGER, M.G. MAUK, L.C. DiNETTA, and C.W. CHARACHE, "Recent Progress in InGaAsSb/GaSb TPV Devices" *Conf. Rec. 25th IEEE Photovoltaic Specialists Conference* (1996) 81-84.

2. Z.A. SHELLENBARGER *et al.*, "Improvements in GaSb-Based Thermophotovoltaic Cells" submitted to this conference.

3. WILT *et al.*, "Monolithically Interconnected InGaAs TPV Module Development" *Conf. Rec. 25th IEEE Photovoltaic Specialists Conference* (1996) 43-48.

4. O. BLUM, K.M. GEIB, M.J. HAFICH, J.F. KLEM, and C.I.H. ASHBY, "Wet Thermal Oxidation of AlAsSb Lattice Matched to InP for Optoelectronics Applications" *Applied Physics Letters* **68, 22** (1996) 3129-3131.

RF/Microwave Non-Destructive Measurements of Electrical Properties of Semiconductor Wafers for Thermophotovoltaic Applications

S. Saroop[†], J. M. Borrego[†], R. J. Gutmann[†], H. Ehsani[†], I. Bhat[†],
S. Dakshina Murthy[†], A. Ostrogorsky[†], P. Dutta[†], M. Freeman[*],
G. Charache[*]

[†]Center for Integrated Electronics and Electronics Manufacturing,
Department of Electrical, Computer, and Systems Engineering
Rensselaer Polytechnic Institute, Troy, New York 12180

[*]Lockheed Martin Inc., Schenectady, New York 12301

Abstract — A radio-frequency/microwave measurement system has been designed for non-contacting determination of sheet resistance and excess carrier lifetime of low-bandgap materials and junctions, specifically GaSb-based alloys for thermophotovoltaic (TPV) applications. The design incorporates RF circuitry in the 100-500 MHz frequency range and utilizes a Q-switched YAG laser at 1.32 microns to photo-generate electron-hole pairs and conductivity modulate the material and/or junction under test. Supplementary measurements with a GaAs pulsed diode laser at 904 nm provides a faster transient response with near-surface photogeneration. Initial measurements on GaSb substrates, Zn-diffused materials and epitaxially grown layers are presented and discussed.

INTRODUCTION

Conductivity, mobility and excess carrier lifetime are the three basic material parameters which influence the efficiency of thermophotovoltaic (TPV) cells. Conductivity and mobility are determined traditionally from Hall effect measurements, and excess carrier lifetime by several steady state and transient measurements on specially fabricated test structures. These techniques require the formation of ohmic contacts or the fabrication of a complete device and are destructive by nature.

A waveguide microwave reflection technique operating at Ka-band (36 GHz) was previously developed in our laboratory to measure the resistivity of wafers in

CP401, *Thermophotovoltaic Generation of Electricity: Third NREL Conference,*
edited by Benner/Coutts
© 1997 The American Institute of Physics 1-56396-734-0/97/$10.00

the range of 10 Ω-cm to 1000 Ω-cm and, with the use of GaAs and AlGaAs pulsed lasers, to measure excess carrier lifetimes longer than 10 nanoseconds[1-5]. In this work the use of the photoconductivity decay to measure recombination lifetime in silicon wafers[1] and to characterize of the defect free zone in precipitated CMOS wafers was demonstrated[2]. The techniques were also applied to GaAs using both above-bandgap and below-bandgap excitation to characterize both starting semi-insulating wafers[3] and ion-implanted channels for field-effect transistor (FET)-based integrated circuits[4]. The technique has even been successfully applied to semi-insulating InP wafers where both lasers were above-bandgap, but the effective absorption depth was significantly different[5]. With key information in the compound semiconductors available from the pulse amplitude response, mapping of the wafer surface can be used to plot defect structural information [6,7].

In this paper, the modifications performed in the microwave reflection system to allow measurement of semiconductor wafers with resistivity as low as 0.01 Ω-cm and excess carrier lifetimes larger than 100 ns (present limitation 0.7 μs) are presented. Initially, the measurement concept is described, followed by a first-order calculation of laser power required for conductivity modulation. Then the measurement technique implementation is presented, followed by the GaSb-based semiconductor materials and junctions evaluated. While a full understanding of these results has not been realized to date, the measurement system is clearly able to discriminate similar

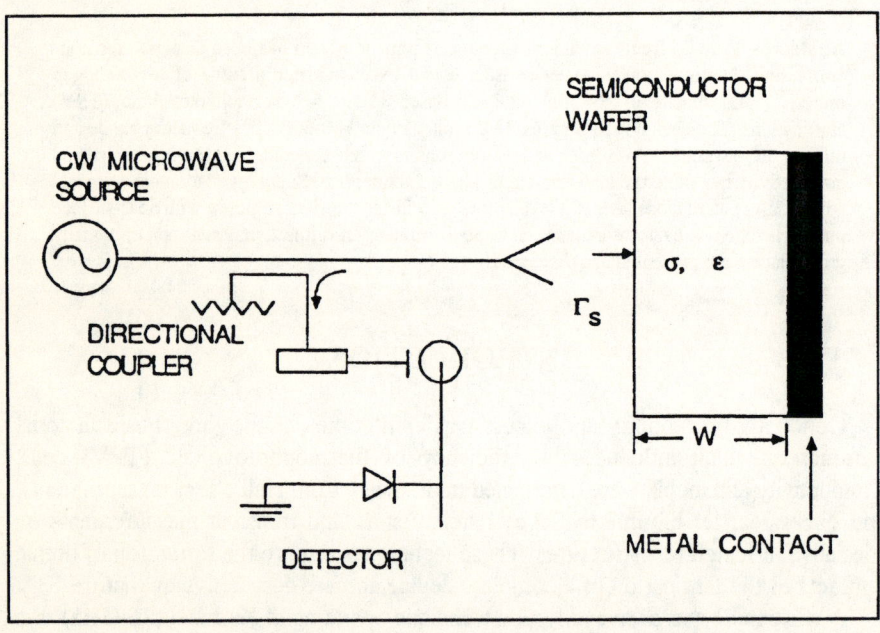

FIGURE 1. Microwave Reflectance System

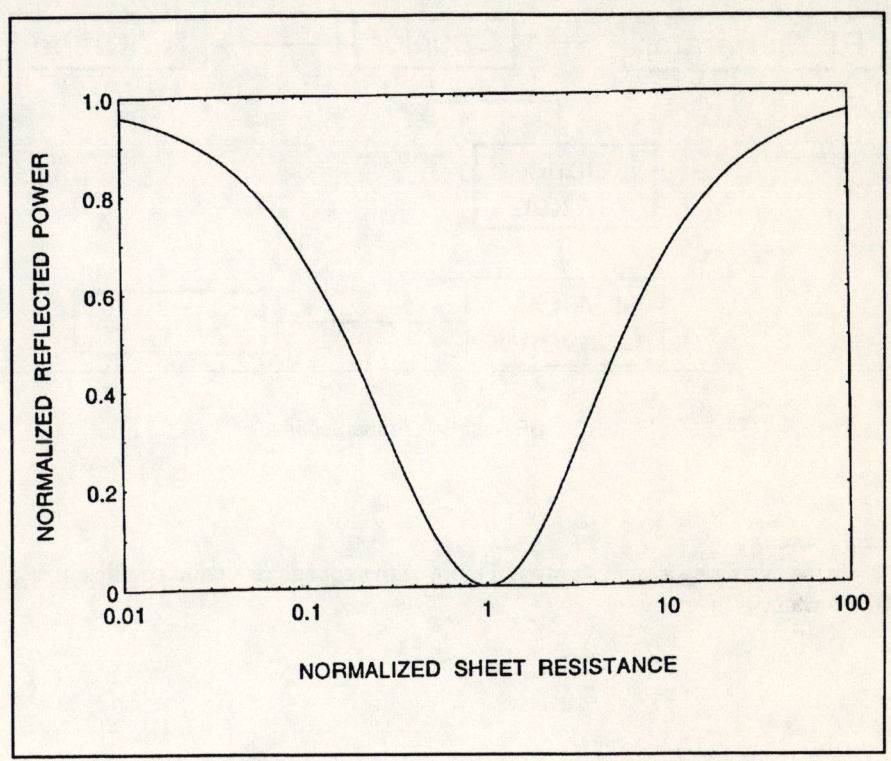

FIGURE 2. Normalized Reflected Power vs. Normalized Sheet Resistance

materials demonstrating that such an approach is useful for characterization and screening of TPV materials and junctions.

RF/MICROWAVE REFLECTION CONCEPT

A schematic diagram of a microwave reflection system is shown in Figure 1. A microwave source provides continuous microwave power which is incident on the sample, and the reflected power is measured using a directional coupler and a microwave detector. The reflected power depends upon the conductivity, dielectric constant and thickness of the sample and the frequency of the continuous-wave (CW) microwave source. Thus, the conductivity of the sample can be determined from the measurement of the reflection coefficient. In addition, the excess carrier lifetime can be measured by recording the decay of excess carriers by monitoring the change in the reflected power after the generation of hole-electron pairs by a pulsed monochromatic

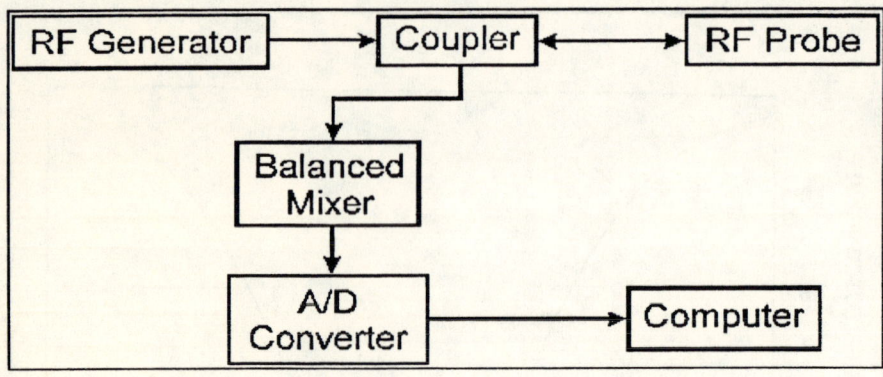

FIGURE 3. RF Resistivity Probe Schematic

light source.

In the measuring system shown in Figure 1, the voltage reflection coefficient, Γ_S, of the wafer is given by:

$$\Gamma_s = \frac{Z_s - Z_o}{Z_s + Z_o} \qquad (1)$$

where Z_S is the wave impedance of the wafer at the top surface of the wafer and Z_0 is the wave impedance of the microwave incident upon the wafer. Assuming that the wave impedance of the wafer is purely resistive and given by R_S, the reflection coefficient is determined by the normalized sheet resistance of the wafer $R_{SN} = R_S / Z_0$. The normalized reflected power $|\Gamma_S|^2$ as a function of the normalized sheet resistance of the wafer is shown in Figure 2. The figure shows that measurement of wafer sheet resistances in the range of $0.1\ Z_0 < R_S < 10\ Z_0$ is possible. Assuming that the microwave incident upon the wafer has a wave impedance like in a waveguide, i.e., 500 Ω, the sheet resistances which can be measured in a waveguide system are in the range between 50 and 5000 $\Omega/\square$. If the wafers have a thickness of approximately 0.020 inches, the wafer resistivities which can be measured are in the range between 2.5 Ω-cm and 250 Ω-cm.

This first-order analysis gives the approximate range of wafer resistivity which can be measured with a waveguide resistivity probe. In order to measure lower resistivity wafers a measurement system with a lower characteristic impedance is required and/or a transformer is needed which changes the sheet resistance of the wafer measured by the measurement system. Both approaches have been used in an RF measurement system which is more suitable for high-conductivity low-bandgap materials of interest for TPV applications.

In order to lower the range of resistivity which can be measured by a microwave reflection system, a coaxial measuring system with a wave impedance of 50 Ω and a transformer to couple the wafer being measured to the measurement system were utilized. While this 50 Ω coaxial system would allow one to measure wafers with resistivity in the range of 0.25 Ω-cm to 25 Ω-cm, this range is not low enough to measure semiconductor wafers for TPV cells. An impedance transformer is used to transform the sheet resistance of lower resistivity wafers to the appropriate range of the RF coaxial system. In order to be able to build impedance transformers without significant parasitics, the frequency of operation of the measurement system was lowered to be in the RF frequency range. However, the frequency has to be large enough to measure transient decays larger than 50 or 100 ns. As a compromise, the system was designed to operate in the frequency range of 100 to 500 MHz using instrumentation similar to that introduced by Yablonovitch[8].

Figure 3 is a schematic diagram of the RF resistivity probe measurement circuit, consisting of an RF generator with two outputs. The variable output provides the RF signal which is sent to the wafer through a directional coupler. The reflected signal

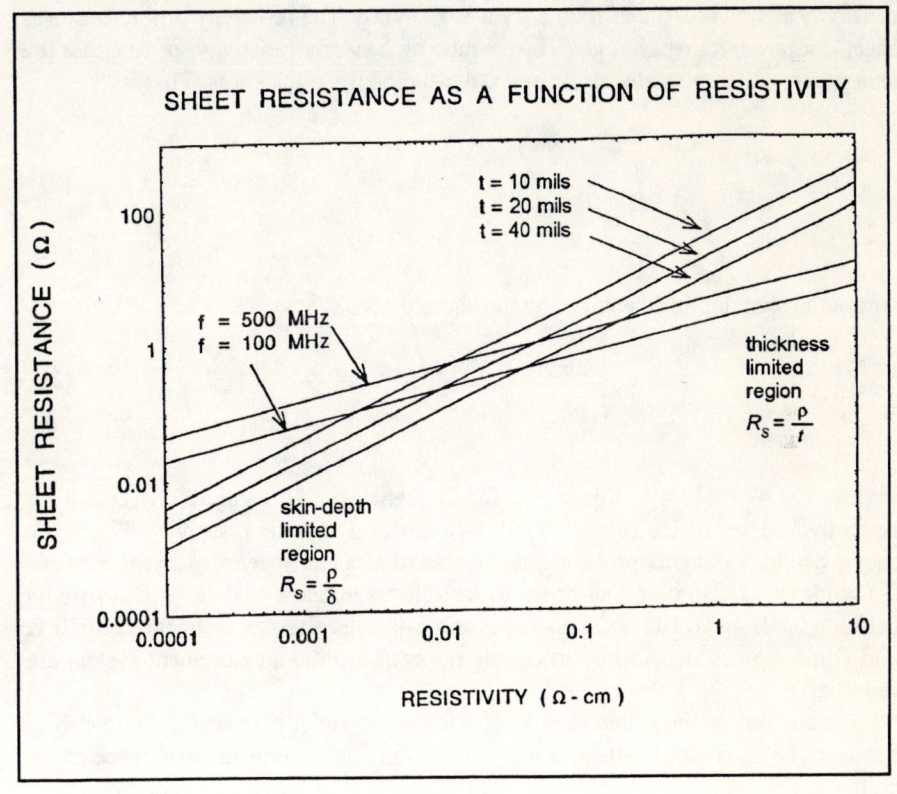

FIGURE 4

143

is directed by the directional coupler to the signal input of a balanced mixer. The fixed output of the RF generator is sent to the local oscillator (LO) input of the balanced mixer. The intermediate frequency (IF) output of the balanced mixer is then digitized and then sent to an IBM PC.

The same setup can be used for measuring the lifetime of excess carriers. By the addition of a pulsed laser, the transient decay of the photoconductivity induced by a pulsed light source can be monitored with the digital oscilloscope. Figure 2 shows that a change in the sheet resistance, as caused by a light source, translates into a change in the power reflected for the microwave direct detection system of Figure 1. Note that if the normalized dark sheet resistance is exactly at the minimum of the curve (i.e. the normalized sheet resistance is equal to unity), the power reflected does not change with optical excitation. However, in the homodyne detection system implemented here, the detected signal is proportional to Γ_S (rather than $|\Gamma_S|^2$), so that high sensitivity can be achieved around a DC operating point where $\Gamma_S=0$.

The effect of the frequency of operation of the RF resistivity probe upon the range of wafer resistivity which can be measured using the skin depth as a function of wafer resistivity and measurement frequency was analyzed. The resistivity probe measures sheet resistance, R_S, which is given by the ratio of the wafer resistivity, ρ, to either the skin depth, δ, or the wafer thickness, t, depending upon thickness. That is:

$$R_s = \frac{\rho}{\delta}, \qquad \delta \ll t$$
$$= \frac{\rho}{t}, \qquad \delta \gg t \tag{2}$$

where the skin depth, δ, is given by the equation:

$$\delta = \sqrt{\frac{2\rho}{\omega\mu}} \tag{3}$$

Figure 4 shows a plot of Equation 2 for frequencies of 100 and 500 MHz and for wafer thicknesses of 10, 20 and 40 mils as a function of wafer resistivity. The figure shows that the resistivity probe should be able to measure sheet resistances between 0.1 and 10 Ω. However, in order to be able to measure wafers with resistivity between 0.001 and 0.1 Ω-cm, a coaxial system with characteristic impedance of 50 Ω and some type of transformer to couple the wafer to the measurement system are required.

An analysis of the impedance, Z, of a flat coil coupled to a semi-infinite slab of conductivity, ρ, resulted in the following expression for the terminal impedance of the coil:

$$Z_s - \frac{\rho}{\delta} N_T^2 A (1-j) \qquad \qquad (4)$$

where N_T is the number of turns per unit length and A is the area of the coil. The above equation shows that a flat coil of an appropriate area and turns per unit length may be needed to measure the desired range of wafer resistivity.

In order to connect the flat coil or transformer to the coaxial measuring system shown in Figure 2, an appropriate microstrip circuit was designed. A coaxial launcher makes the transition to 50 Ω microstrip without introducing appreciable discontinuities. The plane coil or impedance transformer is placed at the end of the microstrip circuit, which has provisions for placing capacitors in shunt and in series to tune out any inductive reactance of the coil or of the microstrip circuit itself.

LASER POWER FOR CONDUCTIVITY MODULATION

The measurement of excess carrier lifetime requires a pulsed increase in the conductivity of the semiconductor material using a pulsed laser with the photoconductivity decay transient monitored after termination of the laser light pulse. The power needed to produce a detectable conductivity change depends upon the dark conductivity of the semiconductor. A lower resistivity semiconductor requires a higher power laser to produce a detectable increase in conductivity. An approximate method for estimating the power of the laser is as follows.

The equation which governs the time behavior of the increase, ΔN, of the carrier concentration, N, is given by:

$$\frac{d}{dt}(AT\Delta N) - AN_{PH}(1-R) - \frac{AT\Delta N}{\tau} \qquad \qquad (5)$$

where T is the thickness of the wafer, A is the area probed in the material, N_{PH} is the photon flux density, R is the reflectivity of the semiconductor wafer, and τ is the effective excess carrier lifetime taking into account bulk and surface recombination effects. For a laser pulse longer than τ, steady state solution is given by:

$$\Delta N - \frac{N_{PH}(1-R)\tau}{T} \qquad \qquad (6)$$

The ratio of photon flux density to carrier concentration becomes:

$$\frac{N_{PH}}{N} \cdot \frac{\Delta N}{N} \frac{T}{(1-R)\tau} \tag{7}$$

Assuming:

$$\frac{\Delta N}{N} \approx 0.2, \ T \approx 0.05\,cm, \ R \approx 0, \ \tau \approx 10^{-7}s \tag{8}$$

the required N_{PH} is approximately 10^5 times larger than the carrier concentration N. Assuming that the carrier concentration is of the order of 10^{17} carriers/cm^3 and that the laser wavelength is approximately 1.3 µm (i.e. the photon energy is approximately 1 eV), the laser intensity is of the order of 1000 W/cm^2. Usually the area probed is of the order of 0.25 to 0.1 cm^2, so the laser power needed is of the order of 100 to 250 W. The new setup was designed for a Q-switched, 1.32 µm Nd:YAG laser with pulses of around 500 W .

RF MEASUREMENT SYSTEM DESCRIPTION AND CALIBRATION OF CAPABILITIES

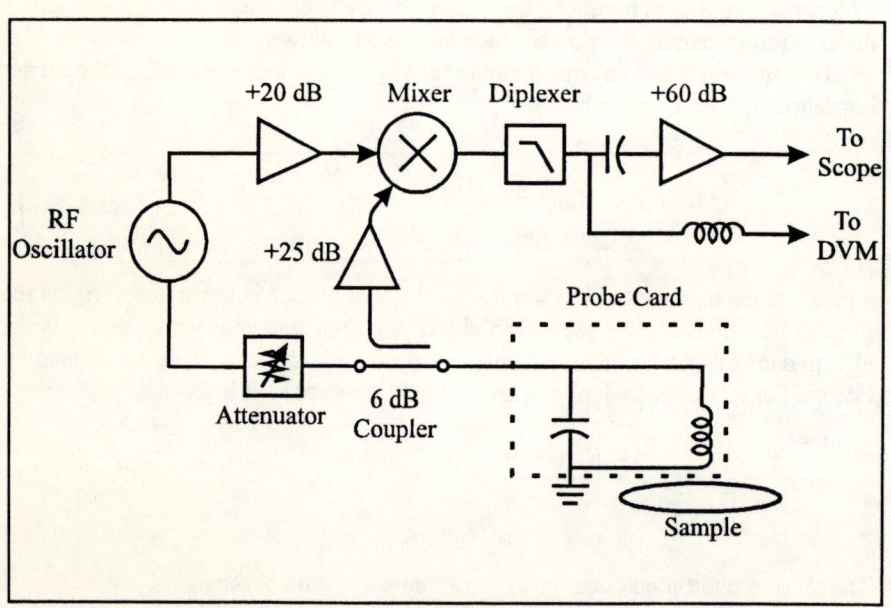

FIGURE 5. RF system layout

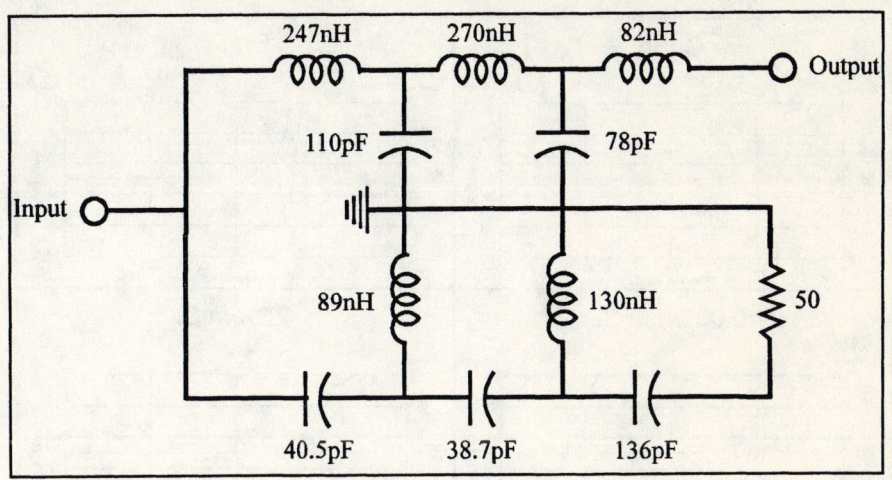

FIGURE 6. 50 Mhz 5th-order Chebyshev diplexer

The photoinduced RF reflectance decay system is a homodyne system implemented around a variable frequency RF oscillator (10~500 MHz) delivering up to +20 dBm (100 mW) into a 50 Ω load. A portion of this signal is fed to a probe card and any signal reflected back from the sample is mixed with the oscillator signal. Since the reflected signal has the same frequency as the oscillator signal, the mixer output will ideally contain two components -- one at twice the oscillator frequency and one at DC. By filtering out the high frequency components, the resultant signal will be proportional to the product of the magnitude of the reflected signal and the cosine of its phase relative to the oscillator output. This signal is then sampled on a digital scope for processing on a computer connected via the general-purpose interface bus (GPIB) interface. By using homodyne detection of the reflected signal, greater sensitivity is achieved over the diode detector of the microwave reflectance system.

A more detailed schematic of the RF system is depicted in Figure 5. The variable attenuator serves both to limit the power being fed to the probe card and to isolate the signal at the local oscillator (reference) port of the mixer from changes in loading. Following the attenuator is a 6 dB directional coupler which samples the signal reflected back from the probe card with a total of 12 dB loss. The reflected signal is then amplified by 25 dB and fed into the RF port of the mixer to be mixed with the 20 dB amplified signal from the oscillator. The latter amplifier is employed to provide a +23 dBm local oscillator signal necessary for good mixer performance, i.e., minimum insertion loss ~ 8 dB. The mixer output is then filtered by a diplexer consisting of a 50 MHz 5th order low-pass section in parallel with a 50 MHz 5th order high-pass section to shunt the higher frequencies into a matched load, as

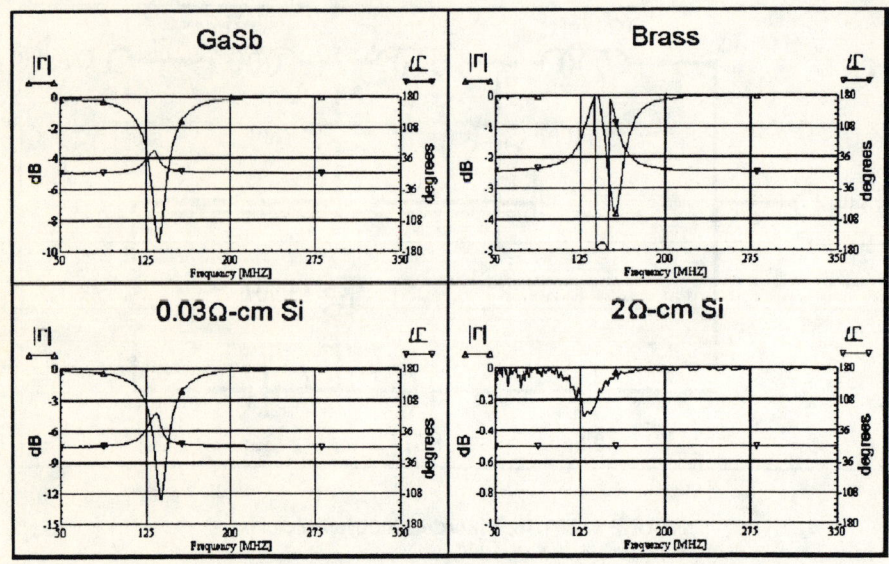

FIGURE 7. Network Analyzer Output with 4 mm 1-turn coil

depicted in Figure 6.

A low-pass filter alone would allow the rejected higher frequencies to be reflected back into the mixer which would distort the DC signal. The capacitor following the diplexer passes all but very low frequencies into a 60 dB amplifier to be sampled by the digital scope. The very low frequencies are passed by the RF choke into a digital DC voltmeter. The DC voltage is a measurement of the sheet resistance of the sample down to a depth on the order of the skin depth of the RF signal in the material. The signal measured on the scope is sensitive to changes in the sample conductivity as brought about, for example, by photogenerated carriers excited during a laser pulse.

The probe card couples RF energy to the sample surface and detects the reflected signal with a minimum of loss. Since the rest of the system uses components with a 50 Ω characteristic impedance, the probe card utilizes 50 Ω microstrip transmission lines on a Duroid® (ϵ_r=2.2) microwave substrate. The probe card circuit is essentially a tuned RF transformer with the physical (variable) capacitor and coil in the primary, and the eddy currents induced in the sample as the secondary when the coil is in close proximity. The sheet resistance of the sample impeding the eddy currents is transformed into the primary circuit on the probe card. Therefore, the impedance is changed, which results in a change in the amount of signal reflected back to the mixer.

The probe card and transformer coils were initially characterized against materials of various conductivities on a network analyzer between 50 and 500 MHz to tune the variable capacitor and optimize the coil design for maximum sensitivity to the change in conductivity. With the circuit unloaded resonance set to 130 MHz using a 4 mm

148

diameter one-turn coil, the probe card data from the network analyzer is shown in Figure 7 for 4 samples (2 Ω-cm Si, 0.03 Ω-cm Si, in-house grown GaSb boule, and a brass block (6.4 μΩ-cm)).

The left axes (△) show the magnitude of the reflected signals, in decibels (dB), and the right axes (▽) show the relative phase shift, in degrees, as functions of frequency (note: different scales are used on the left axes). From the amount of attenuation near resonance, i.e. minimum in |Γ|, this coil appears to have a useful range from 2 Ω-cm to 6.4 μΩ-cm, with a peak sensitivity between 0.03 Ω-cm and the resistivity of GaSb because of their approximation to a matched load. However, other factors, such as the minority carrier lifetime, diffusion length, optical absorption depth and the power of the optical pulses, decrease the actual useful range of the coil in the RF decay system. For the highly conducting brass sample, the phase shift near resonance is close to 180° so that the reflected signal is actually inverted.

The laser sources currently used in the RF decay system include a 10 ns, 80 W/pulse GaAs diode laser at 904 nm and a 1.5 μs, 800 W/pulse Q-switched Nd:YAG laser at 1.32 μm wavelength. The differences between these lasers have been characterized and are presented in Figures 8 and 9.

Figure 8 depicts the normalized open-circuit voltage (V_{oc}) measured with an InGaSb junction grown on a GaAs substrate with photoexcitation by YAG and GaAs lasers, respectively. The YAG response has a characteristic risetime of 0.26 μs and a decay time of 1.22 μs whereas the GaAs response risetime is 20 ns and the falltime is 0.25 μs. The true decay of the GaAs laser pulse is actually much faster than 0.25 μs, with this decay time being more representative of the actual recombination lifetime of the cell. However, since the V_{oc} decay from the YAG laser is much slower, the measured decay rate is an accurate profile of the optical decay of this laser.

Figure 9 shows the RF system response for the same high-resistivity GaSb substrate, demonstrating the effect of the optical pulse shape on the measured decay time. The response from the YAG laser has a characteristic risetime of 0.22 μs and

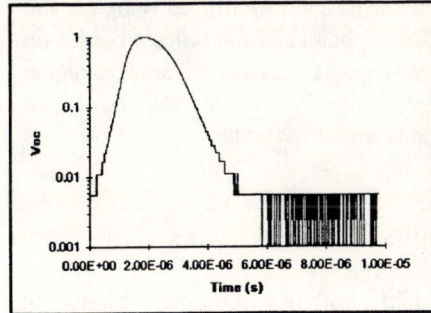

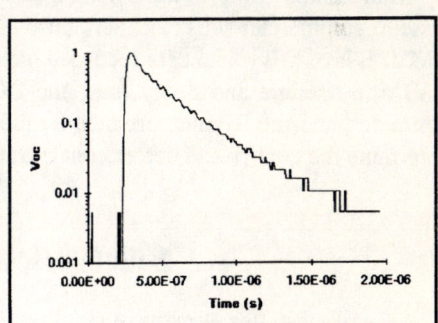

FIGURE 8A. Open circuit response to YAG laser

FIGURE 8B. Open circuit response to GaAs laser

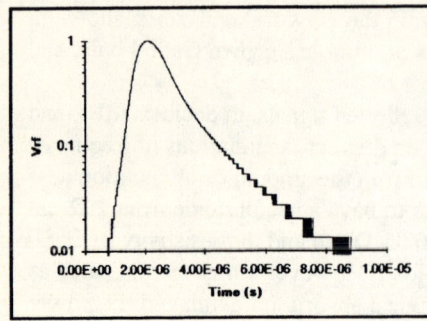
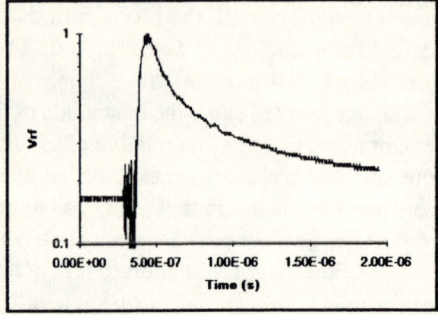

FIGURE 9A. RF Decay with YAG laser   **FIGURE 9B.** RF Decay with GaAs laser

a decay time of 0.78 µs, and the GaAs laser response has a risetime of 25 ns and a decay time of 0.26 µs. The fact that the measured RF risetime is 25 ns implies that the electronics of the RF system are capable of measuring such decay times. Thus, the GaAs laser provides a faster pulse which allows better resolution of the sample carrier lifetime, but has lower power and shorter wavelength which limits absorption depth to approximately 0.1 µm in GaSb. However, the YAG laser is capable of modulating the conductivity of relatively highly doped samples and has an absorption coefficient (α) of 1.4×10^4 cm^{-1} (corresponding to an effective absorption depth of 0.7 µm). In addition, a slow rise and fall time effectively places a lower limit on the measurement of decay to approximately 0.7 µs.

SEMICONDUCTOR MATERIAL MEASUREMENTS

Many GaSb samples have been measured using the RF non-contacting probe system: commercial substrates, in-house bulk substrates, Zn-diffused samples, and OMVPE layers. Typical data recorded includes the height of the reflected pulse (in mV), the risetime and decay time, the DC voltage as measured from the digital voltmeter, and the RF measurement frequency. The goal is to eventually be able to determine the quality and defect concentration in a given sample.

Commercial Substrates

Several substrates were measured from Firebird®, both p- and n-type, with doping concentrations from 5×10^{16} cm^{-3} to 8×10^{17} cm^{-3}. However, the samples doped higher than 10^{17} cm^{-3} did not yield measurable pulse signals (resolution presently 1 to 3 mV). The data for the 5×10^{16} cm^{-3} p-type substrates is summarized in Table 1.

TABLE 1. Measurement on p-type Commercial Substrates

Sample	Pulse Height (mV)		Decay Time* (µs)		DC Voltage (V)
	YAG	GaAs	YAG	GaAs	
FB1 edge	1200	900	0.78	0.25	0.18
FB1 center	32	<1	0.93	N/A	0.28
FB202 edge	1350	832	2.73	0.44	0.25
FB202 center	6	<1	5.31	N/A	0.28

* Some decay times are not available (N/A) due to the low pulse response

The results indicate that the pulse height is very sensitive to the non-uniformities in the samples as the probe is moved from the edge of the sample to the center. The decay time increased by 19 % in FB1 and appears to double in FB202. However, the noise level of the system is around 1 mV making measurement of the decay time on small signals much less precise. In both samples, the use of the GaAs laser gives a pulse height ~30 % less than does the YAG laser (which can be attributed to differences in laser power), but much smaller decay times due to the difference in speed of the lasers. As mentioned above, the GaAs decay times are much closer to the sample carrier lifetime, but due to the much shorter absorption depth, is more sensitive to surface phenomena.

In-House Crystals

A quarter-inch thick, 32 mm diameter, undoped GaSb wafer from an in-house grown boule was mapped with the RF measurement system. This sample was of particular interest because of a twin boundary approximately 6 mm from the edge. Measurements depicted in Figure 10 were taken in 4 mm steps from the center along four perpendicular radii in the 0°, 90°, 180°, and 270° directions. The twin was located along the 270° path at the 12 mm data point.

The DC voltage is azimuthally symmetric and does not vary significantly along the radials. Once again, the pulse height displays the greatest sensitivity to nonuniformities in the wafer. The pulse heights along the 270° scan shows a large increase just before the defect is reached, but the decay times are all similar. These results indicate a sensitivity to material quality, and suggests that the defect is not electrically active, although an interpretation of the result is not clear.

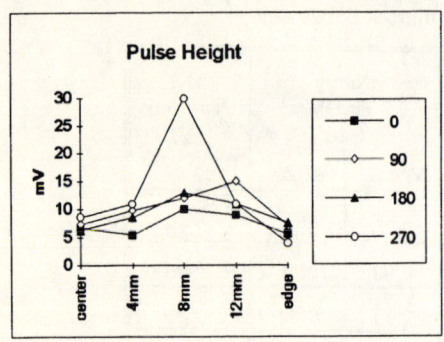
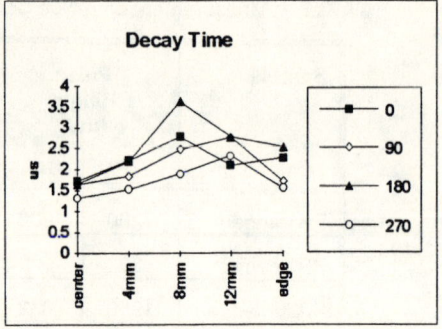

FIGURE 10. Pulse Height and Decay Time Spatial Distribution for Undoped GaSb Wafer

Zn-Diffused Samples

Measurements were taken on commercial substrates from Firebird® (5×10^{16} cm^{-3} Te) and MCP® (10^{17} cm^{-3} Te) which underwent heavy (~10^{19} cm^{-3}) constant source Zn diffusions for various durations at 600°C. With the higher doped MCP® substrates, pulse data could not be obtained and the DC voltages were less than 0.070 V. The Firebird® data is shown in Table 2, ordered by the diffusion time.

The most prominent trend is the decreasing pulse heights (with either laser) as the sample is exposed to longer diffusion times. Since the DC voltage changes very little, the decrease in pulse height is attributed to the presence of higher concentrations of traps; this agrees with the presumption that the sample becomes more defected with

TABLE 2. Measurements on Zn-diffused Samples (FB202 substrate)

Sample	Pulse Height (mV)		Decay Time* (µs)		DC Voltage (V)	Diffusion Time
	YAG	GaAs	YAG	GaAs		
Unprocessed	1350	832	2.73	0.44	0.25	No diffusion
GP37	327	260	1.22	0.72	0.24	1 hr 20 min
GP33	46	40	0.90	0.46	0.28	4 hr
GP38	11	<1	0.95	N/A	0.28	5 hr 20 min
GP30	26	34	0.92	0.74	0.29	10 hr

* Some decay times are not available (N/A) due to the low pulse response

152

deeper Zn diffusions. Also, a slight increase in the DC voltage is observed at deeper diffusions, indicating a higher sheet resistance. This trend suggests that the hole (majority carrier) mobility may be decreasing as well.

Epitaxial Samples

Measurements were taken on OMVPE samples consisting of p/n layers grown on an n-type graded layer on an n-type GaSb substrate, with and without metallization. The metallization is evaporated at low temperatures so that the grown layers are not significantly altered. The results are presented in Table 3 along with the thicknesses and dopings of the p/n layers.

The metallization was deposited in either large area metal windows and dots (#141, 142, and 146) for electrical characterization of the junction, or in a low density metal grid (#146) typical of TPV cells. In all three samples, the highly doped layers result in very low pulse heights, which increase over 30 times for both lasers with the addition of the metal. Also, the DC voltages decrease as a result of the decreased average sheet resistance due to the presence of the metal circles. The exception is the

TABLE 3. Measurements on OMVPE samples

Sample	Structure (all p on n with thicknesses and doping concentrations as indicated)	Pulse Height (mV)		Decay Time* (µs)		DC Voltage (V)
		YAG	GaAs	YAG	GaAs	
141 (bare)	$3\mu m\ 8\times10^{17}$ /$1\mu m\ 5\times10^{17}$	3	<1	2.50	N/A	0.084
141 (metal circ)	same as above	90	20	2.77	2.13	-0.032
142 (bare)	$3\mu m\ 5\times10^{17}$ /$1\mu m\ 5\times10^{17}$	2.8	<1	2.51	N/A	0.137
142 (metal circ)	same as above	200	53	1.26	1.43	-0.019
146 (bare)	$0.3\mu m\ 5\times10^{17}$ /$4.5\mu m\ 7\times10^{17}$	3.3	<1	3.24	N/A	0.028
146 (metal circ)	same as above	110	48	2.21	1.74	0.009
146 (metal grid)	same as above	238	135	2.57	2.08	0.121

* Some decay times are not available (N/A) due to the low pulse response

153

case of the low density metal grid where the pulse height increases by almost two orders of magnitude and the DC voltage increases. The presence of the grid concentrates the eddy currents into loops between the gridlines; therefore, the path length of the currents is increased, presenting a larger effective sheet resistance. As a result, the sensitivity is increased for the highly doped layers. In comparison, signals could not be detected with the 36 GHz microwave system for any of the GaSb samples evaluated.

SUMMARY AND CONCLUSIONS

A system which is capable of non-destructively discriminating defect concentrations in low bandgap materials has been demonstrated based on the pulse height and decay time of photogenerated carriers in the sample. With the current 4 mm diameter probe coil, the system can provide maps of defect levels across the surface of the sample with a 4 mm resolution. The system also has the capability of discerning the differences in the background doping of materials and can be used to map the variations of this quantity across the surface of a sample. Since it was designed around a characteristic impedance of 50 Ω, the RF system is better suited to make measurements on low-bandgap or heavier doped samples than is the microwave system. Evidence of this capability was found with the 2 Ω-cm Si, the 0.03 Ω-cm Si, and the FB1 GaSb substrate using the GaAs laser. The low-doped Si sample yielded a pulse on the RF system over twice as large as on the microwave system with decay times differing by only 3.5 %. A detectable pulse could not be obtained on either the heavily doped Si or the FB1 substrate with the microwave system.

The pulse height is the most sensitive parameter to variations in a given sample. High quality commercial substrates have almost two orders of magnitude variation in pulse height from center to edge. The mapping of the wafer containing a twin boundary, and the measurements on the Zn diffused samples are correlated to the presence of known defects.

The decay time shows less sensitivity to material quality, but has the advantage of being relatively unaffected by the background doping level as compared to the pulse height. The laser sources provide a measurement capability on samples with up to 10^{17} cm^{-3} n-type doping (limited only because of the low pulse heights). A decay time as low as 0.78 μs can be measured with the YAG laser and 20 ns with the GaAs laser. The latter corresponds to the theoretical limit of the system with a 50 MHz low-pass filter. However, decay times on this order can not be measured with the GaAs laser presently due to the low pulse power. That is, the number of excess carriers generated is proportional to the carrier lifetime; in poorer material with a low lifetime a sufficient density of carriers for detection will not be generated. However, decay times as low as 0.25 μs have been demonstrated with the GaAs laser (see Table 1).

The system is sensitive to the positioning of the height of the probe card with respect to the sample. Also, since the laser power is fed to the sample via a fiberoptic cable, the finite divergence angle at the output of the cable means that the intensity of the laser illumination will vary with distance from the sample. Thus, the pulse height will be directly affected by the positioning of the fiber. Therefore, a mechanically stable probe card and fiber optic holder are required to achieve repeatable results.

REFERENCES

1. J. M. Borrego, R. J. Gutmann, N. Jensen, "Non-Destructive Lifetime Measurement in Silicon Wafers by Microwave Reflection", *Solid-State Electronics* **30** No. 2, 195 (1987).

2. J. M. Borrego, R. J. Gutmann, N. Jensen, C. S. Lo, O. Paz, "Characterization of the DFZ in Silicon Wafers by a Non-Destructive Technique", *Mat. Res. Soc. Symp. Proc.* **69**, 373 (1986).

3. R. J. Gutmann, J. M. Borrego, C. S. Lo, M. C. Heimlich, O. Paz, "Microwave-Detected Photoconductivity-Transient Spectroscopy for Non-Destructive Evaluation of GaAs Wafers", *SPIE* **794**, 128 (1987)

4. M.C. Heimlich, R. J. Gutmann, D. Seielstad, D. Hou, "Damage-Induced Effects in Co-Implanted LEC GaAs", *6th Conf. on Semi-ins. III-V Mat.*, 367 (1990)

5. R. J. Gutmann, M. C. Heimlich, M. Tait, T. B. Bylsma, E. Monberg, "Photoinduced Microwave Reflectometry as a Non-Invasive Technique for S. I. InP Wafer Char-acterization", *6th Conf. on Semi-ins III-V Mat.*, 323 (1990)

6. M. C. Heimlich, E. R. Atwood, R. J. Gutmann, "Two-Dimensional Mapping of Implantation and Annealing Phenomena in GaAs by Photoinduced Microwave Reflectometry", *Semicond. Sci. Tech.* **7**, A275 (1992)

7. M. C. Heimlich, R. J. Gutmann, L. Kerber, S. Moreau, J. Vaughan, "Characterization of Active Channel Processing by PIMR", *7th Conf. on Semi-ins Mat.*, 291 (1992)

8. E. Yablonovitch, T. J. Gmitter, "A Contactless Minority Lifetime Probe of Hetero-structures, Surfaces, Interfaces, and Bulk Wafers", *Solid-State Electronics* **35** No. 3, 261 (1992).

Bulk growth of GaSb and $Ga_{1-x}In_xSb$

P.S. Dutta[*], A.G. Ostrogorsky[*] and R.J. Gutmann[†]

Center for Integrated Electronics and Electronics Manufacturing
[*]*Department of Mechanical Engineering, Aeronautical Engineering and Mechanics*
[†]*Department of Electrical, Computer and Systems Engineering*
Rensselaer Polytechnic Institute, Troy, New York 12180

Abstract. Synthesis and growth of bulk GaSb single crystals and GaInSb polycrystals have been carried out by the vertical Bridgman technique, with a baffle immersed in the melt and by complete encapsulation of the melt by low melting temperature alkali halides or oxides. The critical roles of the baffle and the encapsulation are discussed. Efforts in obtaining device grade GaSb with superior structural and electrical properties and compositionally homogeneous GaInSb are described, emphasizing the key steps in the growth cycle developed to obtain good crystalline quality.

INTRODUCTION

GaSb and InGaSb have been demonstrated to be suitable choices for high efficiency thermophotovoltaic (TPV) cells [1]. At present, GaSb technology is in its infancy, and considerable research is needed prior to large scale device manufacturing. However, advancement in crystal growth of GaSb and InGaSb offers the possibility of improved performance and yield of TPV devices, as well as other low band gap devices.

Compared to other III-V compounds such as GaAs, InP, GaN, GaP and InAs, the synthesis and growth of GaSb is much simpler [2]. In particular, Sb has a low vapor pressure and is less toxic than As or P, thus eliminating the complication of using furnaces enclosed in high pressure chambers. Moreover, the synthesis and growth temperature of GaSb ($712^{\circ}C$) is much lower than that of the other compounds. The thermo-physical properties of molten GaSb, the high stacking fault energy (highest amongst III-V compounds), and the high critical resolved shear stress (CRSS) indicate that the growth of high structural quality bulk material should be achievable. All of these factors indicate that low cost growth of such material should be quite feasible.

CP401, *Thermophotovoltaic Generation of Electricity: Third NREL Conference,*
edited by Benner/Coutts

However, GaSb and related alloys have high native defect concentrations which are electrically and optically active, and pose problems in device fabrication. A review of the vast literature on other III-V compounds indicates that the lack of a commercial market for GaSb based devices has resulted in insufficient research for a consistent understanding of these defects, although recent work is very promising [3]. The present investigation is a part of the on-going development program for the growth of antimonide based boules for TPV cells [4].

A combination of several advanced concepts of bulk crystal growth is reported. Each concept has been separately proven to produce high quality materials. The three principal features are summarized below:

(1) The experimental growth setup is a state-of-art multi-zone, computer controlled Mellen furnace, custom built for crystal growth in an industrial environment. This setup allows precise control of the temperature profile during boule growth.

(2) The vertical Bridgman technique, which has been found to result in lower dislocation and defect density when compared to the traditional Czochralski (CZ) method, has been adopted. A few technical problems have been encountered during the growth of GaSb with the CZ technique, especially during seeding. The formation of Ga_2O_3 scum on the melt during synthesis impedes seeding of the charge [5]. The reduction kinetics of the scum is very slow at the low growth temperature of GaSb (712°C); therefore, the scum is difficult to remove from the top of the melt [6]. Similarly, due to the high viscosity of B_2O_3 at the melting temperature of GaSb (conventionally used for encapsulation during CZ growth of III-V materials), seeding becomes problematic [7].

(3) Liquid encapsulation with low melting point salts of alkali halides is used to relieve the stress from the walls of the crucible during growth, and at the same time remains inert with respect to the melt and crystal [8]. The use of a baffle in the melt, which is novel for compound synthesis and subsequent growth of the crystal, is one of the highlights of previous experiments [9]. The submerged baffle has several important roles. Compound synthesis is done in open ampoules (under inert atmosphere) in less than one hour compared to tens of hours using conventional techniques; therefore, impurity incorporation in the melt is reduced. The incorporation of a baffle during the growth encompasses the advantages of the vertical gradient freeze technique, double crucible Czochralski (CZ) method, and traveling heater method (or float zone technique). Ultimately, a low radial temperature gradient, introduction of compositional uniform melt at the S-L interface (as in the case of the double crucible CZ technique), and a small melt depth over the solid-liquid (S-L) interface with low natural convection (diffusive controlled growth) are anticipated. The resultant boules should have lower dislocation (or defect) density and more uniform impurity distribution (or solute incorporation in the crystal). These results have been demonstrated in

the growth of InSb and Ge [10,11], and are demonstrated in our present work on undoped and Te- doped GaSb and InGaSb.

From the above mentioned summary, it can be ascertained that the preparation techniques adopted should result in higher quality material and should be easier to adopt for production. The key results obtained using these growth concepts are reported here.

BOULE GROWTH PROCEDURES

Charge and crucible preparation

Growth of GaSb (undoped and Te-doped) has been carried out mainly in silica crucibles with oriented seeds along <111>B and <100>, with a few experiments in pyrolytic boron nitride (pBN) crucibles. The liquid encapsulants used during the experiments were: LiCl:KCl eutectic (58% : 42%) salt [8], B_2O_3 and Sb_2O_3. These encapsulants were of high purity commercial grade with the LiCl:KCl eutectic mixture prepared in-house. The growth of the InGaSb crystals was carried out without a seed in flat-bottom silica crucibles. The starting materials (In, Ga, and Sb) were of 6N purity which were not chemically treated before synthesis to reduce the chances of enhanced oxidation on the freshly etched surfaces of the elements. The oxidized surfaces form a layer of scum on the melt and lead to sticking of the charge to the ampoule walls. By using untreated starting materials, the formation of scum can be reduced. The silica ampoules and silica baffles or the pBN crucible were cleaned in organic solvents followed by acid treatments and de-ionized (DI) water wash. The ampoules and baffle were thoroughly dried with nitrogen and then with a hot plate.

The starting materials were placed in the crucible which was then loaded in the furnace. For synthesis, flat bottom crucibles were used, while for seeded growth narrow seed regions were used. Since the diameter of the seed region is usually non-uniform along its length, this region was drilled with a diamond coated steel rod dipped in methanol. The seeds were drilled from single crystal boules using a diamond coated core drill of steel. The prepared seed was then etched in CP4 etchant to fit the seed cross-section of the ampoule. In experiments where a pre-synthesized charge was used, the charge was degreased thoroughly in organic solvents and etched in CP4 etchant to obtain a mirror- like surface. Care was taken to avoid any unreacted or oxidized portions of the charge after etching by using DI water to stop the vigorous reaction. The charge was then thoroughly washed in DI water and methanol with ultrasonic vibrations. The charge was dried using nitrogen followed by heating.

Synthesis and growth sequences

After loading the charge in the furnace, a few grams of the encapsulant were inserted in the ampoule and the growth chamber was immediately evacuated to less than 50 mTorr. The encapsulants are highly hygroscopic in nature and after absorption of moisture they stick non-uniformly to the quartz at high temperature. This process results in undesirable sticking of the melt to the ampoule. The salts were preserved under vacuum in dessicators and the exposure times during insertion in the ampoule were kept extremely short. To insure a moisture-free encapsulant, a high temperature baking in vacuum was carried out prior to growth. The synthesis and growth cycle is depicted in Figure 1.

After evacuation to below 50 milli-Torr, the growth chamber was flushed with argon several times to remove any residual oxygen. The furnace was then heated at a rate of 100°C/hr to around 300°C, which was significantly below the melting temperature of the encapsulants. This is important in order to avoid any evaporation of the encapsulant due to melting. The pre-growth baking was carried out for a period of 10 - 12 hours under vacuum. After the baking, the furnace was filled with 1 atm of argon and heated to around 20-30°C above the melting temperature of GaSb (712°C). The synthesis was carried out by vertically moving the baffle (up and down) in the ampoule for a period of 30-40 minutes. The mixing was also done while using a pre-synthesized charge to achieve a uniformly mixed melt.

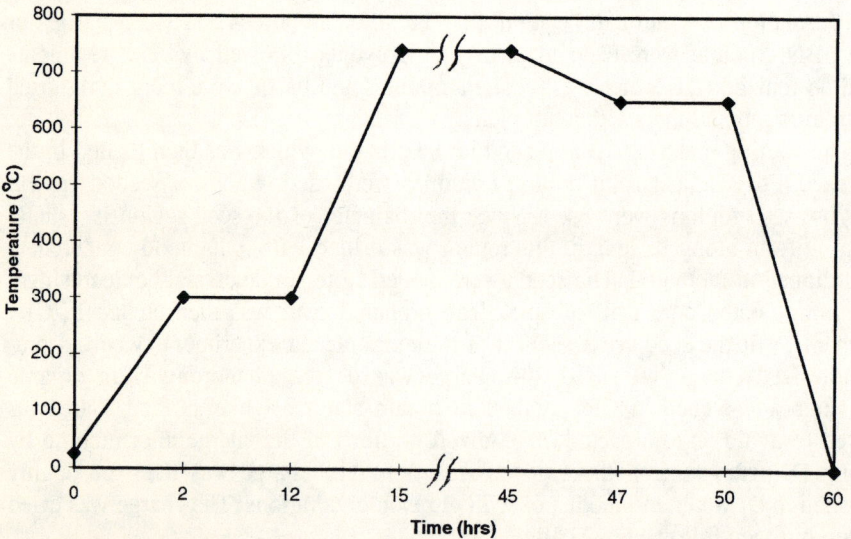

FIGURE 1. Temperature versus time plot for a typical synthesis and growth cycle.

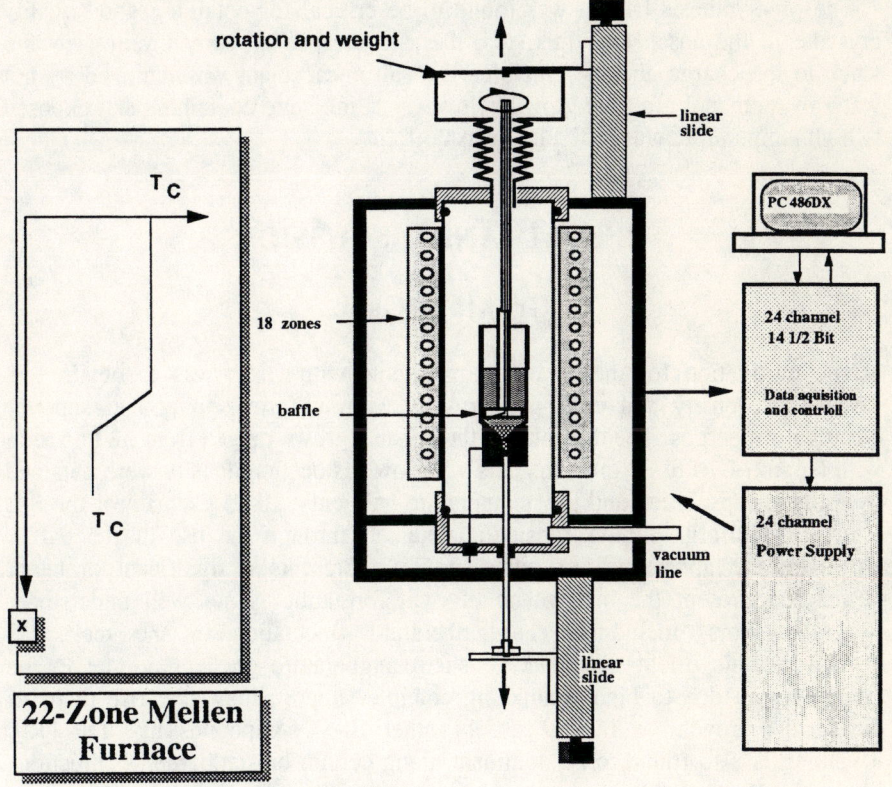

FIGURE 2. Experimental set up for synthesis and crystal growth along with the temperature profile of the furnace.

At the end of the synthesis cycle, the temperature profile of the furnace was adjusted to a typical Bridgman profile (see Figure 2). The baffle was placed 1 cm away from the solid-liquid interface and the ampoule was lowered at a constant rate. For the seeded experiments, a small portion of the seed is re-melted before solidification. The seeding procedure was optimized with respect to the furnace temperature setting and position, and seeding could be repeatedly carried out without the danger of melting the complete seed. During the solidification of the charge, the baffle remained stationary while the ampoule translated along the temperature gradient of the furnace. The distance between the baffle and the S-L interface remained constant during the entire growth cycle, as experimentally verified. The translational rate of the crucible was 3.3 mm/hr in all of the experiments. After the completion of solidification, the furnace was cooled down slowly to room temperature.

The pre-synthesis baking was found to be critical for obtaining good quality crystals. In the absence of this step, the salt encapsulant turned yellowish and stuck to the charge and the crucible. The salt encapsulant was removed by hot water or methanol after the growth. However, a moisture containing salt exposed to high temperature was difficult to wash off.

RESULTS AND DISCUSSION

Growth of GaSb

The motivation for the growth experiments with GaSb was to obtain low dislocation density and twin free crystals with uniform doping and superior electrical properties. The diameter of the crystals grown ranged from 32 to 51mm with length of 70 to 80 mm. Crystals with low dislocation density were obtained by using encapsulation and low temperature gradients (10-15°C/cm) near the S-L interface. Uniform doping density is obtained through the use of the baffle. However, the appearance of twins in GaSb presents a significant problem, because at present the mechanism of twin formation is not well understood. Several factors, including: (1) temperature fluctuation in the melt, (2) crystallographic orientation, and (3) sharp temperature gradient normal to the ampoule wall due to high conducting crucible support, may give rise to twins during the growth, as in the case of other III-V compounds like InP [12]. Twinning is also found to be dominant along certain crystallographic directions like <111>A or <100> as compared to <111>B [13,14]. The twins drastically reduce the yield of the crystals which otherwise show single grain nature.

Through extensive study, the shape of the ampoule and the crucible support were optimized to reduce the probability of twinning. By increasing the conical angle of the ampoule and by preventing the shoulder of the ampoule to touch the graphite crucible support, as shown in Figure 3b, the formation of twins could be avoided in the cone region. The configuration shown in Figure 3a was found to form twins at the junction between the seed and shoulder regions.

The best crystals grown with proper encapsulation exhibited uniformly distributed dislocation densities between 500 and 1000 cm^{-2}. Crystals grown with encapsulation sticking to the quartz ampoule exhibited dislocation density about one order of magnitude higher, i.e., approximately 5000 cm^{-2}. The pBN grown crystals with B_2O_3 encapsulation had a higher dislocation density than those grown with alkali halide salt, but lower than those with sticking problems [15]. The crystal grown with Sb_2O_3 encapsulation was found to stick to the quartz in a glassy-like matrix, and could not be extracted without breaking. The crystal had several areas of Sb inclusions, probably diffusing from the decomposed encapsulation.

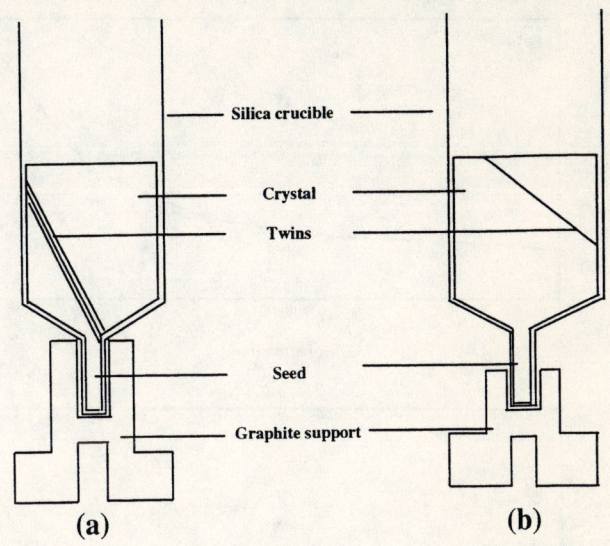

FIGURE 3. Crucible support designs employed in our work. (a) twin formed at the seed-shoulder junction, (b) no twinning occurred.

The Hall mobility, carrier concentration and resistivity of undoped p-type and Te doped n-type (which are indicative of crystalline quality) are listed in Table 1. The electrical properties of the grown wafers are comparable, or in certain cases superior, to those available from commercial suppliers of GaSb.

TABLE I. Hall mobility, carrier concentration and resistivity for p- and n- GaSb

Sample specification	T (K)	type	resistivity (Ω-cm)	carrier concentration (cm⁻³)	mobility (cm²/V.s)
undoped #17	300	p	0.02	1.90×10^{17}	505
undoped #17	77	p	0.05	1.05×10^{17}	1275
undoped #12	300	p	0.05	2.38×10^{17}	500
undoped #12	77	p	0.09	4.10×10^{16}	1575
Te-doped #19	300	n	0.0005	6.28×10^{18}	1876
Te-doped #19	77	n	0.0001	6.81×10^{18}	6281

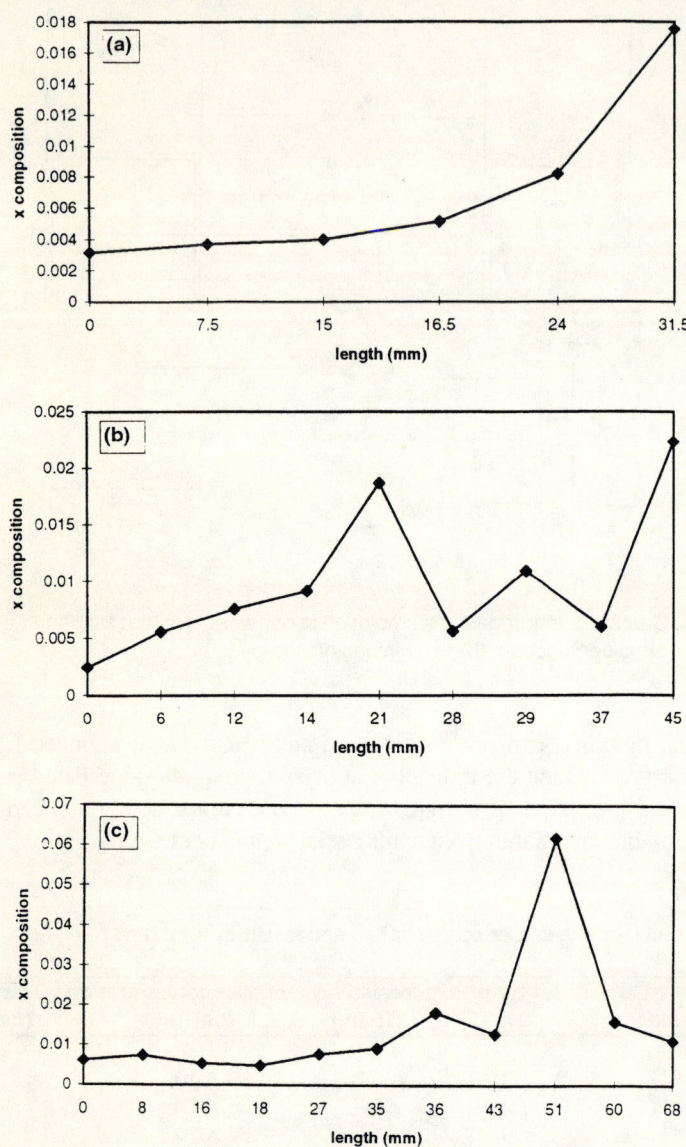

FIGURE 4. Axial profiles of indium in $In_xGa_{1-x}Sb$. (a) without baffle, (b) with stationary baffle and (c) with rotating baffle. The concentration of indium (x) in the melt is 0.03. Note that the scales of x- and y- axes are different in the three plots.

Growth of InGaSb

Growth of InGaSb was carried out in flat bottom crucibles, in order to obtain a compositionally homogeneous material along the radial and axial directions. Once this objective is achieved, the single crystals are expected to be grown by using GaSb seeds. The effect of the baffle in the melt and its rotation, or oscillation, during growth has been found to improve the homogeneity of Ga- doped Ge [10,11]. The same is expected to be true in InGaSb where InSb is rejected at the S-L interface during growth of InGaSb, resulting in a non-uniform axial profile of indium in the grown crystal.

The role of the baffle on the spatial composition of $In_xGa_{1-x}Sb$ crystals can be seen in Figure 4. The initial homogenized concentration of indium (x) in the melt was 0.03. The axial indium distribution in a crystal grown without the baffle is shown in Figure 4(a). The indium profile in the grown crystal clearly shows the dominance of convective conditions during growth. The presence of a baffle (see Figure 4b) gives rise to an initial transient due to build-up of the solute boundary layer at the bottom of the baffle. However, as the concentration of indium in the boundary layer rises, constitutional super-cooling occurs and a sharp increase in the indium concentration is observed. This super-cooling is repeated several times during the whole growth cycle, which leads to the observed wavy shape of the indium profile. The oscillatory rotation of the baffle (see Figure 4c) gives the most homogeneous crystal. Moreover, the indium incorporation in the crystal is higher than in the previous two cases.

The role of encapsulation during synthesis has been found to be significant. Sticking of In, or its oxide, to the silica and subsequent cracking of ampoule has been a major problem reported by other workers [16]. An alkali salt avoids the sticking of the In to the quartz; hence, a homogeneous mixture can be prepared. Imperfections in bulk InGaSb crystals caused by cracks generated due to chemical misfit in the ternary materials [17] must be solved before highest quality crystals can be anticipated.

CONCLUSIONS

The critical issues in the bulk growth of GaSb and InGaSb were discussed describing a particularly attractive approach adopted in this program. GaSb boules with electrical properties comparable to or better than commercial crystals have been obtained. Growth of twin-free <100> GaSb single crystals is essential for device-quality material. Cracks encountered in ternary crystals are a key drawback for InGaSb boules, although enhanced mixing and heavy impurity doping are promising approaches.

ACKNOWLEDGMENTS

The authors wish to acknowledge Dr. Thierry Duffar and Dr. Eric Monberg for providing invaluable scientific information and discussions. Thanks are also due to Dr. David Wark for the EPMA measurements. The growth equipment used in the present work was provided by the A T & T Bell Laboratories at Murray Hill, New Jersey.

REFERENCES

1. Benner, J.P., Coutts, T.J. and Ginley, D.S., Editors, *The Second NREL Conference on Thermophotovoltaic Generation of Electricity*, AIP Conference Proceedings 358, AIP Press, 1996.
2. Sunder, W.A., Barns, R.L., Kometani, T.Y., Parsey, Jr., J.M., and Laudise, R.A., *J. Cryst. Growth* **78**, 9 (1986).
3. Dutta, P.S., Bhat, H.L. and Kumar, Vikram., *J. Appl. Phys.* **81**, 5821 (1997).
4. Charache, G.W., DePoy, D.M., Baldasaro, P.F., and Campbell, B.C., "Thermovoltaic devices utilizing a back surface reflector for spectral control" *The Second NREL Conference on Thermophotovoltaic Generation of Electricity*, AIP Conference Proceedings 358, AIP Press, 1996, pp. 339.
5. McAfee, Jr., K.B., Gay, D.M., Hozack, R.S., Laudise, R.A., Schwartz, G. and Sunder, W.A., *J. Cryst. Growth* **76**, 263 (1986).
6. Cockayne, B., Steward, V.M., Brown, G.T., MacEwan, V.R. and Young, M.I., *J. Cryst. Growth* **58**, 267 (1982).
7. Uemura, C. and Katsui, A., *Jpn. J. Appl. Phys.* **19**, L318 (1980).
8. T. Duffar, Personal Communication, 1995.
9. Ostrogorsky, A.G., *J. Cryst. Growth* **104**, 233 (1990).
10. Ostrogorsky, A.G. and Muller, G., *J. Cryst. Growth* **137**, 64 (1994).
11. Meyer, S. and Ostrogorsky, A.G., *J. Cryst. Growth* **171**, 566 (1997).
12. Monberg, E., "Bridgman and related techniques", Chapter 2 in the *Handbook of Crystal Growth, Vol. 2a, Editor, Hurle, D.T.J.*, North Holland/Elsevier, 1994, pp. 51.
13. Katsui, A. and Uemura, C., *Jpn. J. Appl. Phys.* **21**, 1106 (1982).
14. Moravec, F., Sestakova, V., Stepanek, B. and Charvat, V., *Cryst. Res. Technol.*, **24**, 275 (1989).
15. Koh, H.J. and Ostrogorsky, A.G., (unpublished work).
16. Marin, C., Dutta, P.S., Dieguez, E., Dusserre, P. and Duffar, T., *J. Cryst. Growth* **174**, (1997).
17. Garandet, J.P., Duffar, T. and Favier, J.J., *J. Cryst. Growth* **106**, 426 (1990).

SESSION 3:
SELECTIVE RADIATORS I

Matched Infrared Emitters for Use with GaSb TPV Cells

Luke Ferguson and Lewis Fraas

JX Crystals, Inc., Issaquah, WA

Abstract. Most TPV systems described previously have relied on either filters with near-blackbody emitters or rare earth selective emitters to achieve spectral control. There are serious practical problems with both of these approaches. A new, highly efficient yet practical emitter is presented in this work and will be termed the "matched" emitter because its emissive power spectrum is well matched with the infrared response of GaSb photovoltaic cells. The matched emitter is capable of radiating power that is comparable to a blackbody at the same temperature in the useful portion of the energy spectrum, while at the same time greatly suppressing radiation at non-convertible wavelengths. This behavior is explained in terms of the novel chemistry and materials used in making the matched emitter. Relative emissive power dependence on doping concentration and matrix material is presented. A method for measuring absolute emissive power for matched emitters is presented. Some problems with temperature measurements are discussed and an estimated emittance is presented.

INTRODUCTION AND BACKGROUND

Previously described TPV systems typically used either silicon cells in combination with rare earth oxide "selective emitters", or low bandgap cells in combination with broadband near blackbody emitters. There are problems with both of these approaches.

The major problem with the rare earth oxide selective emitters is that the very narrow emission band widths and mediocre peak emittance (<0.65) results in low power densities unless very high emitter temperatures are used. High emitter temperatures lead to materials lifetime problems and low chemical to radiation coupling efficiencies.

Broadband greybody emitters can work well with low bandgap TPV cells provided that infrared filters are used to reflect the non-convertible infrared energy back to the emitter. Unfortunately, high efficiency systems require near perfect infrared filters with near unity view factors from the emitter to the cell. Many researchers have found that any parasitic losses present in the optical cavity

CP401, *Thermophotovoltaic Generation of Electricity: Third NREL Conference,*
edited by Benner/Coutts
© 1997 The American Institute of Physics 1-56396-734-0/97/$10.00

will quickly defeat the attempt to recycle non-convertible, longer wavelength infrared energy with filters.

A new highly efficient, yet practical, infrared emitter that is nearly ideal for energy conversion using GaSb photovoltaic cells will be described in this paper. The new bandgap "matched" infrared emitters have broader and significantly higher emittance in the useful energy band just above the cutoff wavelength of the cell when compared with "selective" emitters, yet have very low emission at non-convertible wavelengths. The broader emission allows for higher emitter power densities at a moderate emitter temperature of 1400°C, and comparatively little radiation falls in the wavelength band that is non-convertible by the cell.

The matched emitters described here consist of ceramic matrix composites with a Co-doped refractory oxide such as alumina (Al_2O_3), magnesia oxide (MgO), or spinel (Al_2MgO_4). The chemistry and preparation details are propriety and patents are pending. The "matched" emission band results from the use of a d-series transition element (Co) rather than rare earth f-series transition elements. When the appropriate d-series transition element is added as a dopant to a refractory oxide matrix with suitably low infrared emission characteristics, a broadened infrared emission band is observed in the 1.0 to 1.7 micron range. Rare earth f-series transition elements tend to radiate as if they were isolated atoms because the partially filled inner orbitals are shielded from the crystal field by filled outer orbitals. Thus rare earth based emitters such as Er_2O_3 and Yb_2O_3 have narrow line-type emission spectra resulting from the very limited available electronic transitions that can take place. In contrast, the d-series transition elements have partially filled outer electron orbitals that can interact strongly with the crystal field. When the d-series dopant element has anti-symmetric coordination with its neighbor atoms the available discrete energy levels for electronic transitions will tend to split into many closely spaced levels, creating a nearly continuous band of available energy levels. We believe that when the type and concentration of doping elements are optimized, a band of very closely spaced energy levels allows intermediate transitions within the forbidden bandgap of the refractory oxide matrix material. Pure refractory oxides typically have bandgap energies of 3 to 5 eV, restricting electronic absorption and emission to UV or higher energy. Addition of a d-level dopant causes strong radiant emission in a wide range of energies above some cutoff energy that is matched to the bandgap of the photovoltaic cell.

Life testing of matched emitter materials is still in the early stages. Initial testing, however, has already shown that at least two different refractory oxide structures are capable of tightly bonding the dopant elements. After many hours of operation at temperatures above 1523 K, matched emitters have remained quite stable as evidenced by optical testing and undiminished cell currents in working TPV generators produced at JX Crystals, Inc.

EXPERIMENTAL RESULTS

Radiant emitters are typically characterized by emissive power spectra that display power density as a function of wavelength. The total radiated power is equal to the area under the curve. The ideal emitter for TPV would absorb all the available energy and then radiate at energies just greater than the cutoff wavelength of the cell. Matched emitters suppress unwanted radiation at lower energies and enhance radiation useful to the photovoltaic cell.

Figure 1 shows a relative emissive power spectrum for a matched emitter compared to a SiC based near-blackbody emitter. Both of these were dense disks with diffuse flat surfaces, and both were "optically thick". The disks were heated from the front side with a propane torch (the back of the disk was open to the air), and the emissive power spectrum was obtained using a scanning monochrometer. Care was taken to position each disk in an insulating holder such that the view factor and instrument parameters were very nearly constant. The torch flame was adjusted to obtain maximum heating of the SiC disk and a spectrum was recorded. The SiC disk was then replaced with a matched emitter disk and the torch flame (heat input) was reduced until the leading and trailing edge of the matched emitter spectrum closely approximated the spectrum of the SiC disk. The matrix oxide of this particular matched emitter disk was alumina with some impurities. Because both alumina and SiC become highly emissive (e > 0.8) at long wavelengths (greater that about 5 microns) due to lattice vibrations, it is safe to conclude that the SiC disk and the matched emitter are at approximately the same temperature.

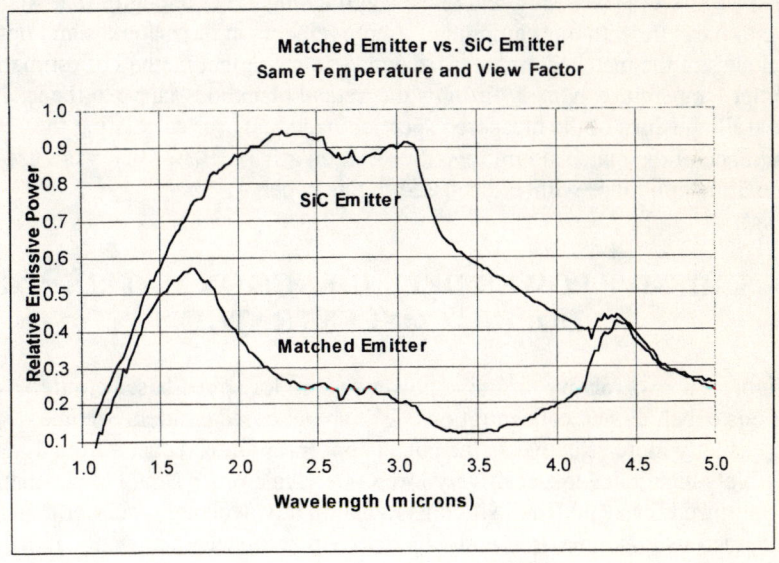

Figure 1. Relative Emissive Power of SiC and Matched Emitters.

171

Several important features of the spectra should be noted. First, undesirable non-convertible radiation of the matched emitter is highly suppressed compared with the near blackbody emitter at the same temperature. At the same time, the matched emitter curve remains relatively close to the near blackbody emitter curve at wavelengths less than about 1.7 microns. Much more useful power is being emitted than could be expected from a rare earth selective emitter such as Er_2O_3, which has a narrow emission peak near 1.5 microns. The emission peaks at about 2.7 and 4.3 microns correspond to water, CO, and CO_2 molecular bond resonances from the propane flame.

In a system where reasonably non-luminous combustion occurs, it is believed that the dominant energy transfer mechanism from combustion gas to emitter is kinetic. Hence both the matched emitter and a blackbody emitter will be expected to absorb roughly the same amount of energy under that same combustion conditions. Assuming approximately equal conduction and convection losses, both the matched emitter and the blackbody emitter must then radiate roughly the same amount of energy, or their temperatures would rise indefinitely. Because the long-wavelength radiation is greatly suppressed by the chemistry of the matched emitter, it is forced to reach higher temperatures, and radiate much more energy at desirable wavelengths. This effect has been qualitatively observed to produce dramatic gains in cell power when blackbody and matched emitters are operated with equivalent fuel input in similar generators.

Figure 2 shows the emissive power for a matched emitter that was recently fabricated at JX Crystals, as well as a theoretical curve for a blackbody at approximately the same temperature. For this matched emitter, note that the emissive power is now suppressed even at the longest wavelengths that we are able to measure with our equipment. Improvements in the materials and optical thickness of the matched emitters have made our original method of estimating emitter temperature, namely fitting a theoretical blackbody curve to the leading and trailing edges of the measured spectra, inadequate for calculating the emittance of the matched emitters. Some alternative methods that were used for estimating emittance will be discussed in this paper.

EMISSIVE POWER DEPENDENCE ON MATRIX AND DOPING CONCENTRATIONS

Figure 3 shows relative emissive power spectra for spinel-based emitters with various cobalt dopant concentrations. The spinel based emitters become increasingly more selective as the cobalt concentration is reduced from a relatively high value to a relatively low value. An emitter is said to be "matched" or optimized for a GaSb cell when its radiation at wavelengths shorter than about 1.75 microns is at a maximum and its radiation at longer wavelengths is minimized. Also shown in figure 3 is a spectrum for a cobalt-doped magnesia based emitter. It is seen that cobalt-doped magnesia based emitters are even more

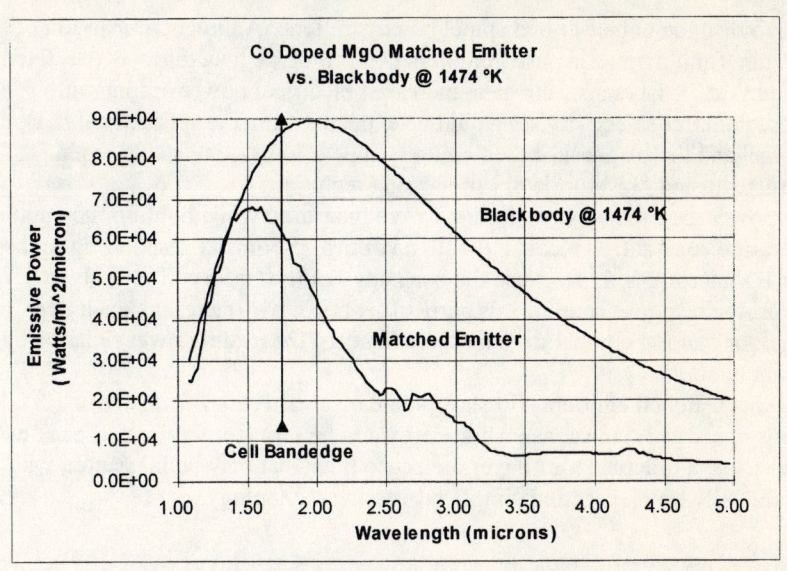

Figure 2. Co-doped MgO Matched Emitter vs. Blackbody at 1474 K.

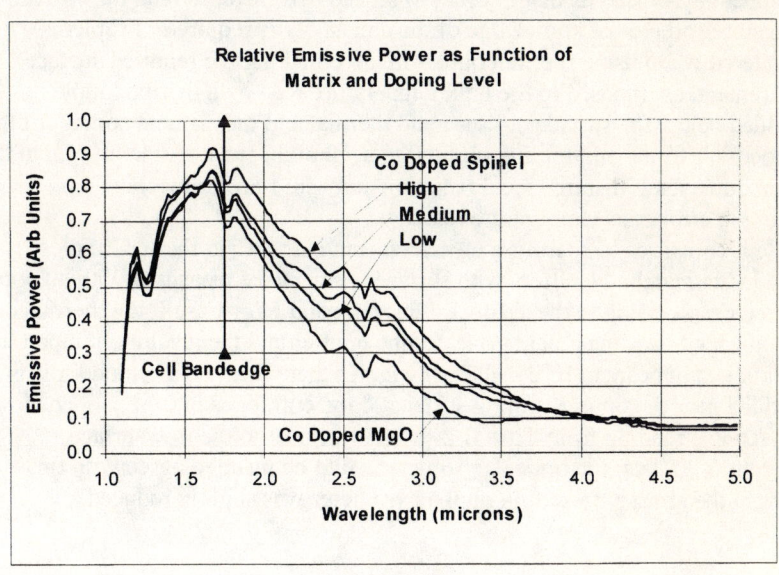

Figure 3 Relative Emissive Power as a Function of Matrix and Doping Level.

selective than the cobalt-doped spinel based emitters. A direct comparison of these spinel and magnesia based emitters in a prototype generator (with a fixed fuel input) at JX Crystals, Inc. gave increased electrical power output with increased emitter selectivity, as would be expected from the spectral data. Both the spinel and the magnesia-based emitters appear to be quite stable when constructed using our proprietary processing technology.

The power spectra shown in figure 3 were measured while holding the emitter temperature constant, as measured with an optical pyrometer responding in the 1.1 to 1.5 micron band. Because the band edge of the GaSb cell is at about 1.75 microns, the response band of this particular optical pyrometer makes it near optimal for making comparative measurements of the useful power radiated by different emitters.

Note that matched emitters still show some residual energy from about 1.8 microns to about 3.6 microns. This energy can be efficiently reflected back to the emitter using a practical multilayer dielectric filter that may be fabricated with comparatively few layers and simple processing technology.

TEMPERATURE MEASUREMENTS

There are problems with using either thermocouples or optical pyrometers when trying to accurately measure the temperature of selective emitters. The problem with optical pyrometry is that in order to set the pyrometer to read the correct temperature some prior knowledge of the emissivity is required. Problems encountered when using thermocouples involve making the required surface measurement, as opposed to the bulk temperature read by a thermocouple embedded below the surface. Also, good thermal and mechanical contact of the thermocouple to the emitter are required to accurately sense temperature, but the thermocouple itself disturbs local radiation and heat transfer mechanisms resulting in erroneous temperature readings.

Table I. compares temperature measurements made using thermocouples embedded in matched emitters with surface temperature measured with an optical pyrometer responding in the 1.1 to 1.5 micron band. For Case I, the thermocouple was embedded near the inner surface of the emitter and the resulting temperature measurement appears to be considerably hotter that when the thermocouple is embedded near the outer radiating surface of the emitter as in Case II. The lower thermocouple reading from Case II is probably closer to the true surface temperature. In fact, the optical pyrometer could be made to agree with this reading if the emissivity setting on the pyrometer was slightly reduced.

174

Case I:	MgO:Co/Alumina Emitter	Temperature (K)
	Thermocouple in substrate	1461
	Pyrometer with ε =0.78, λ = 1.1 to 1.5 μm	<u>1251</u>
	Temperature Difference	210
Case II:	Spinel:Co/Alumina Emitter	
	Thermocouple near surface	1485
	Pyrometer with ε =0.78, λ = 1.1 to 1.5 μm	<u>1400</u>
	Temperature Difference	85

Table I. Comparison of Thermocouple and Pyrometry Measurements

EXPERIMENTAL METHOD FOR DETERMINING ABSOLUTE EMISSIVE POWER

When using the JX Crystals monochrometer fixture alone, only a relative emissive power spectrum may be obtained. The optical pyrometer is used primarily for comparative temperature measurements. It has been possible, however, to measure absolute emissive power spectra for emitters by using a calibrated GaSb cell and calculated view factors. Figure 4 shows the experimental set-up used to determine absolute emissive power. Reflected radiation from insulating cuffs at both ends of the emitter in a real generator would be desirable. But because these significantly complicate the view factor calculation for this measurement, they were eliminated for this experimental set-up. The experimental set-up shown figure 4 is a compromise between closely duplicating a working generator and some simplification necessary for comparative measurements. It will be shown that knowledge of the quantum efficiency (QE) for a particular GaSb cell, along with a measurement of short-circuit current (Isc) and a calculation of the view factor, provides a method for determining absolute emissive power.

VIEW FACTOR CALCULATIONS

Figure 5 shows the results of a radial view factor calculation for a cylindrical emitter about 3 inches in diameter and 2.8 inches long. The horizontal axis represents the axial distance along the emitter with the extreme ends of the scale being the top and base of the emitter. For this experiment, a 1-cm wide cell was located at the midpoint axially along the emitter, as indicated by the two vertical lines in the figure. The figure shows the axial variation in the view factor along

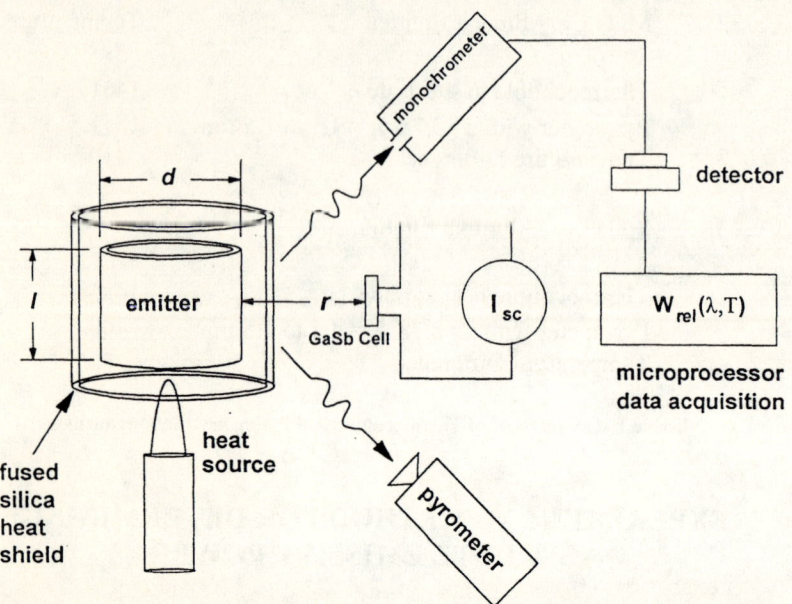

Figure 4. Schematic of the Experimental Set-up to Measure Absolute Emissive Power.

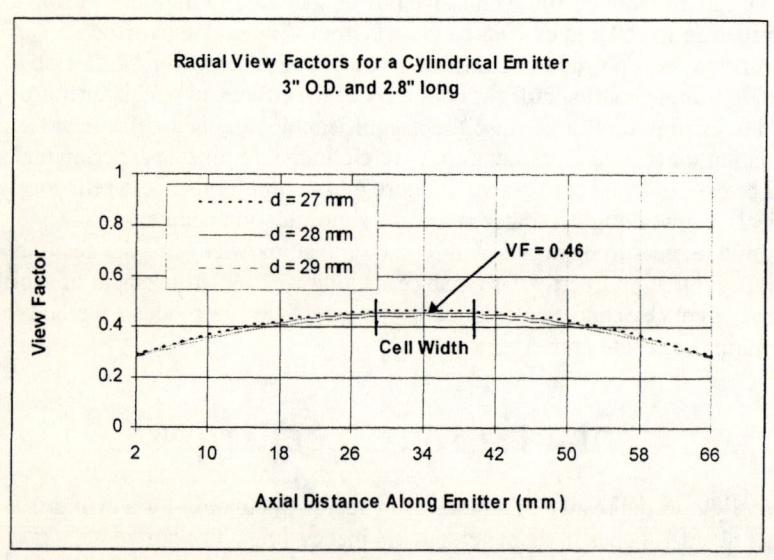

Figure 5. Radial View Factor for a Cylindrical Emitter.

the cell for a cell positioned at radial distance of 28 mm from the emitter surface. Note that for this geometry the view factor is nearly constant over the width of the cell. The radial position of the cell can be measured to within 1 mm, and the dashed lines represent deviations of 1 mm from the measured cell to emitter distance of 28 mm. Again, only small deviations are seen in the view factor.

Currently, emitters with diameters somewhat smaller that 3 inches have been found to reach the highest operating temperatures in JX Crystals generators. Unfortunately, for smaller emitters the view factor varies more rapidly, both axially and radially. Larger emitters are preferred for this experiment because variations in the view factor with cell position are greatly reduced, although they presently do not reach as high an operating temperature.

CALCULATING ABSOLUTE EMISSIVE POWER

The absolute emissive power spectrum of radiant emitters while operating at high temperature may be determined with the following relations.

$$I_{sc} = VF \int W_{abs}(\lambda) QE(\lambda) d\lambda$$
$$S_c W_{rel}(\lambda) = W_{abs}(\lambda)$$

The method for determining the absolute emissive power is as follows. The short circuit current, I_{sc}, is measured using a GaSb photovoltaic cell with known quantum efficiency (QE). The QE of the GaSb cell used here was calibrated at NASA. The source-to-cell view factor, VF, may then be calculated from geometrical considerations, as discussed above. The monochrometer is then used to measure the relative spectrum of emissive power, $W_{rel}(\lambda)$, with the advantage of not having to determine a quantitative view factor for the monochrometer and the source. A trial solution for the absolute emissive power spectrum, $W_{abs}(\lambda)$, is then obtained by multiplying $W_{rel}(\lambda)$ with an estimated scale factor, S_c. Finally, a value for I_{sc} is calculated by integrating the cell QE with a trial solution for the absolute emissive power spectrum, $W_{abs}(\lambda)$, over all wavelengths where the QE is non-zero. When the calculated value for I_{sc} agrees with the measured value, then the scaling factor is correct.

ABSOLUTE EMISSIVE POWER MEASUREMENT AND ESTIMATED EMITTANCE

The absolute emissive power spectrum for the cobalt-doped magnesia matched emitter is shown in Figure 6. This information could be used, along with the radiating surface area of the emitter and the fuel consumption, to calculate an

177

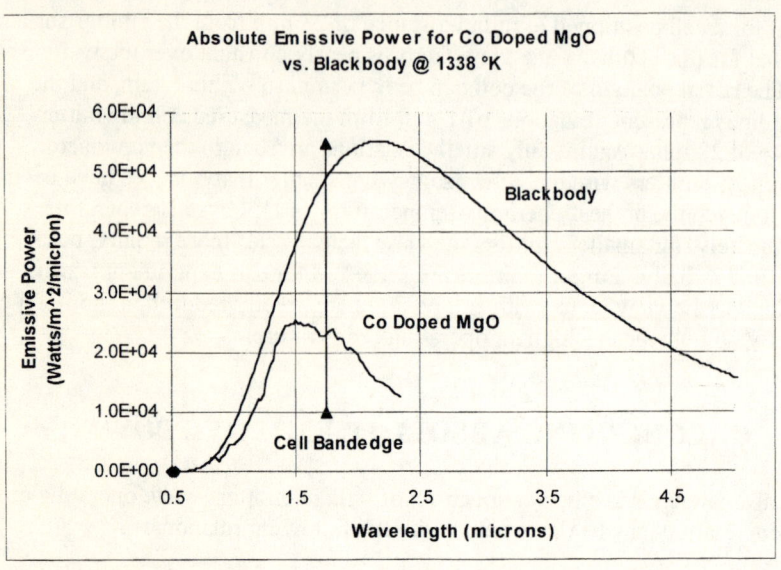

Figure 6. Absolute Emissive Power of a Matched Emitter.

efficiency of the input chemical fuel energy to total radiated power. An efficiency calculation of this type, however, would necessarily involve other system design parameters such as combustion efficiency and heat transfer concerns. These would obscure the performance of the emitter alone.

Based on thermocouple and recalibrated pyrometer readings, the surface temperature of the matched emitter was estimated to be 1338 K. A 1338 K theoretical blackbody curve is also shown in figure 6. By definition, the emittance is taken as the ratio of the absolute emissive power to the blackbody emissive power at the same temperature. Note that the leading edge of the matched emitter spectrum does not quite line up with the blackbody power spectrum.

A conservative estimate of the emittance of a matched emitter is shown in figure 7. The peak emittance is about 0.75, and the emittance decreases rapidly for wavelengths greater than the cutoff wavelength of a GaSb cell. The calculated peak emittance is probably a low estimate because the actual matched emitter temperature is most likely lower than 1338K.

CONCLUSIONS

With matched emitters the thermal load on the TPV cell can be greatly reduced while maintaining high cell current densities. This may be a critical advantage

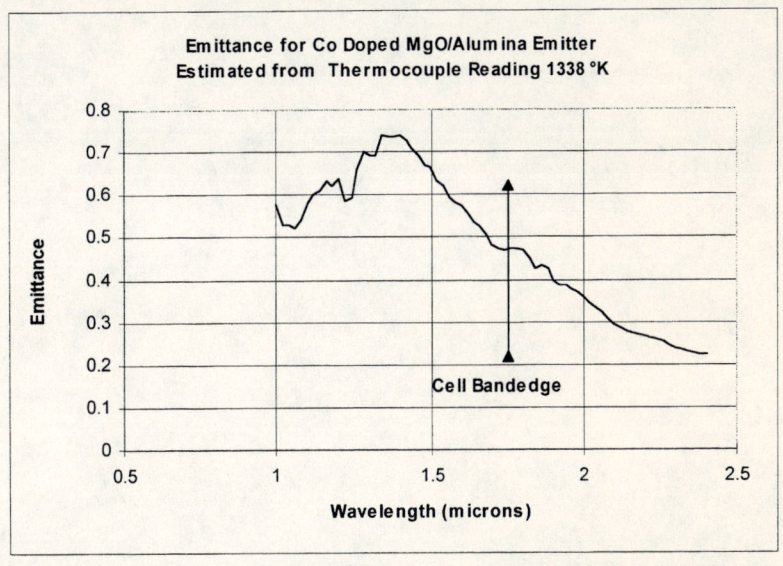

Figure 7. Estimated Emittance for a Matched Emitter.

because photovoltaic cells must be exposed to high radiant power density, and yet must be kept cool, to efficiently convert radiation to electricity. The degree of selectivity of matched emitters is dependent on dopant concentration and the refractory oxide matrix material.

A procedure has been presented for measuring the absolute emissive power of a radiant emitter given a photovoltaic cell with known QE, a measurable view factor, and only a relative emissive power spectrum. Knowledge of the emitter temperature is needed to measure emittance, but accurate temperature measurements of selective emitters are still somewhat challenging and will require additional work. Despite some uncertainty, however, it is believed that measured emittance of emitters for TPV provides the most fundamentally useful and important parameter for comparison and design criteria.

ACKNOWLEDGEMENTS

The authors would like to thank JX Crystals co-workers John Samaras, Russ Ballantyne, She Hui, and Bill Mulligan for their contributions to this work.

Multiple-Dopant Selective Emitter

Zheng Chen, Peter L. Adair, and M. Frank Rose

Space Power Institute, 231 Leach Center
Auburn, Al 36849

Abstract Power efficiency is one of the major concerns in designing and developing a thermophotovoltaic (TPV) generator. Erbium oxide and ytterbium oxide emitter have unique selective line emission, which exhibit a high emittance at a particular wavelength and very low emittance in the rest of the infrared spectrum. The highly selective line emissions are well matched to the response characteristics of some photovoltaic (PV) cells, e.g. erbium oxide emission match well to InGaAs cells and ytterbium oxide emission match well to silicon cells. Obviously, using these emitters can increase emitter efficiency. In addition to efficiency, power density is the other major concern in TPV system design. The disadvantage in using erbium oxide or ytterbium oxide emitters is low photovoltaic convertible power due to the narrow line emissions. One of the ways to increase the photovoltaic convertible power is to broaden the line emissions. In this paper, the authors present a new selective emitter. The emitter contains multiple-doped elements and gives higher efficiency and photovoltaic convertible radiant power than either erbium oxide or ytterbium oxide emitters alone. The emissive spectra, emittance and photovoltaic convertible radiant power of the emitter are also presented. Additionally, the response of a 0.6 eV InGaAs PV cell to the selective band emitter as well as selective line emitters will be discussed.

INTRODUCTION

Research on selective emitters for TPV applications was started in the later 1960's.[1,2] One aspect of this research was focused on the improvement of radiant conversion efficiency or emitter efficiency (η_E). This term has been defined as the ratio of the photovoltaic convertible power (P_E) to the total radiant power (P_R) emitted,

$$\eta_E = \frac{P_E}{P_R} \tag{1}$$

The emitter efficiency is mainly determined by the emitter material type and its structure. Typical emitters currently used in the industries are blackbody emitters. For the 0.6eV InGaAs PV cell, the blackbody emitter efficiency is about 22% at operating temperature (1100 °C). By using rare earth oxides selective emitters, such as erbia, the reported efficiencies reached above 30% at temperatures of 1600 °C.[3] As a result the photovoltaic efficiency, which refers to the ability of the PV cell to convert radiation to electricity, can be improved.

Erbia and holmia emitter have unique selective line emission, both of which exhibit a high emittance at a particular wavelength (1.55 µm for erbia and 2.01 µm for holmia) and very low off band emittance below 4 µm. Both emittance

CP401, *Thermophotovoltaic Generation of Electricity: Third NREL Conference,*
edited by Benner/Coutts

spectra have been measured at Space Power Institute in Auburn University and are shown in Figure 1a. The highly selective line emissions are well matched to the response characteristics of some photovoltaic cells; e.g. erbia line emission match well to 0.75eV InGaAs cells and holmia line emission match well to 0.58eV GaInAsSb cells.[4] Obviously, the cell's photovoltaic efficiency can be improved when the PV cells are illuminated by their matched emitters.

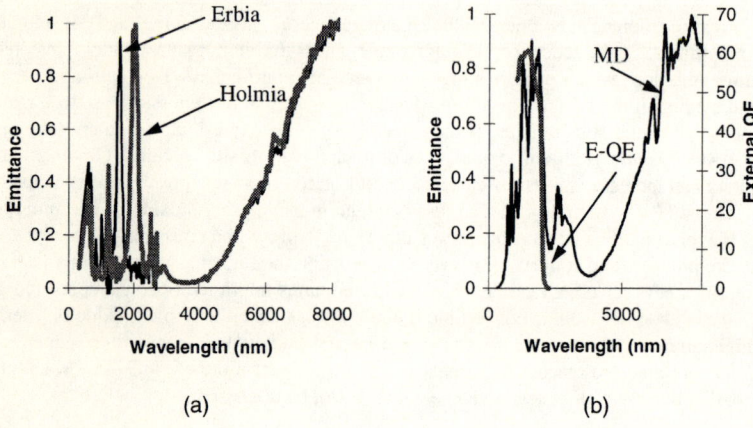

(a)   (b)

Figure 1   (a) Emittances of erbia and holmia emitters. (b) Emittance of multiple-dopant emitter, its spectra matched to the external quantum efficiencies (E-QE) of the 0.6eV PV cell.

Power output density, however, is one of the other major concerns in designing and developing a TPV generator. Based on our calculation, for erbia line emission the emissive power in the wavelengths between 1.45 µm and 1.65 µm is 4.99 W/cm^2, and for holmia line emission the emissive power in the wavelengths between 1.91 µm and 2.11 µm is 4.48 W/cm^2. These calculations were made for the emitters at 1800K. The peak spectral emittance was assumed to be 0.95. If the bandwidth of the selective line emission can be broaden in the wavelengths from 1 µm to 2 µm, even though the average spectral emittance is lower (e.g. 0.5), the emissive power can reach 11.34 W/cm^2. The broader emission doubles the photovoltaic convertible radiant power from the selective line emission. It is clear that the broader selective band emission can increase the power output.

This paper presents a new selective emitter (MD) developed to match a 0.6eV InGaAs PV cell. The emitter contains multiple-doped elements and gives both higher efficiency and higher photovoltaic convertible radiant power than either erbia or holmia emitter alone. The emissive spectra, emittance, and photon convertible power of the various emitters were measured and compared. The energy efficiencies of a 0.6eV InGaAs PV cell were also compared when the cell was illuminated by various emitters.

EXPERIMENTAL PROCEDURE

The emissive power of the emitters were measured using two experimental configurations. One was a flat circular disk in a plane parallel to the flat circular emitter at a distance x. The emitter was 2.5 cm in diameter and was aligned with its normal passing through the center of the disk. The other experimental configuration was a flat circle disk in a plane parallel to the exterior of a cylindrical emitter at a distance x. These schematics for the experimental setups are shown in Figure 2. The disk in Figure 2 represents a thermopile detector (power meter) or a PV cell. For the circular emitters, a H_2/O_2 torch was used as the heat source. The reason for this setup is that the H_2/O_2 flame is able to heat the emitter to high temperatures (over 1600°C) and gives little "contamination" in emitter's inferred spectra.

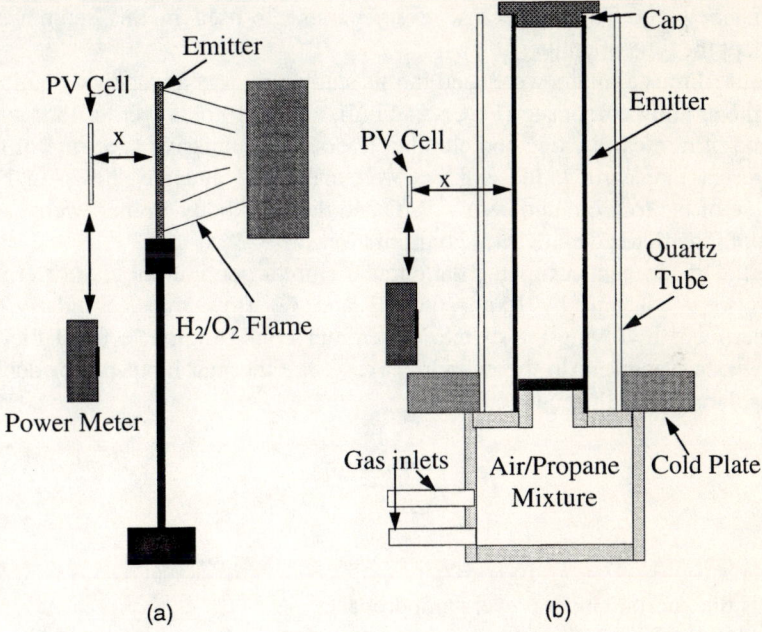

(a)　　　　　　　　　　　(b)

Figure 2　　Experimental setups for energy conversion measurements, (a) flat circular emitter configuration and (b) cylindrical emitter configuration.

For the cylindrical emitter configuration, the emitter was attached to the top of a burner. The burner was made by modifying a Fisher™ Blast Burner. Two gas inlets, one for fuel (propane) and one for air, are located near the bottom of the burner. The flow rates for both propane and air was accurately controlled by Matheson™ flow meters. The burner was cooled by an attached water cooled aluminum plate. This eliminated any potential for flashback of the flame into the chamber. The flame is anchored to the burner by an alloy screen which is fixed to the top of a brass cylinder 2.54 cm in diameter positioned on top of the mixing chamber. A cylindrical emitter 2.74 cm in diameter and 10.16 cm in height is positioned on the brass cylinder. A quartz tube is placed between the emitter and the PV cells to prevent the combustion products from contaminating the PV cell. The quartz tube is 38 cm in diameter and 9.52 cm in height. The thickness of the tube is 1.5 mm. The PV cell was located 3 cm away from the emitter surface. The combustion process is forced to takes place inside the cylindrical emitter by a cap made of fibrous rare earth oxide composite. The thermal energy generated by the combustion heats the emitter. A 0.6 eV InGaAs PV cell was used for measuring photovoltaic efficiency. This experimental setup allows us to quantitatively study the effect of using different emitter types on the overall TPV efficiency, η_T. This is because every parameter used in this setup was fixed, except the type of emitters.

Four fibrous emitters were used in this study; one was a blackbody emitter made of metal alloy composite (Fe-Cr-Al-Si-Y), and the others were selective emitters made of rare earth oxide and alumina fibers. The emissive spectrum of the three different rare earth oxide emitters were measured and are shown in Figure 1. Three of the four emitters, erbia , MD and the blackbody emitter, were also tested in both the flat and cylindrical configuration.

In the experiments using the flat circular emitter configuration, the thermal input was kept constant by fixing the H_2 and O_2 flow rates. Similarly, for the experiment involving the cylindrical emitter configuration, a fixed fuel and air flow rate was used. In the cylindrical case, the thermal input power density was calculated using the equation,

$$P_I = \frac{H_c \rho_f m}{A_1} \qquad (2)$$

where
P_I is the thermal input power from propane,
H_c=46348 kJ/Kg is the heat capacity of propane,
ρ_f=1.86 kg/m^3 is the propane density,
m=0.5590 SLPM, is the flow rate of propane used in this study,
A_1=87.0 cm^2, is the surface area of the emitter.

184

The maximum thermal input power density used for this configuration was 9.2 W/cm^2.

The emissive surface power, ε_s, was calculated based on a radiant power measurement obtained using the power meter. The details of this type of calculation are described in Siegel and Howell.[5]

Radiation power in the wavelength range between 1 μm and 2 μm is photovoltaic convertible by the 0.6eV cell, and was measured by placing a bandpass interference filter between the emitter and the power meter. The filter cuts off all the radiation below 1 μm and above 2 μm and transmits 90% in the wavelengths between 1 μm and 2 μm. Most photovoltaic convertible radiant power, $P_{r(1-2)}$, within the band (1-2 μm) is attributed to selective emissions from rare earth oxides in the selective emitters. The emitter efficiency, η_e, obtained in the experiment can be described as

$$\eta_e = \frac{1.1 P_{r(1-2)}}{P_r} \tag{3}$$

where the 1.1 factor is used to compensation for filter absorption.

Additionally, the temperature of the PV cell surface was measured using a of 0.127 mm Platinum-10% Rhodium vs. Platinum Type S thermocouple.

The electrical power was determined by measuring the current and voltage of the PV cell at different load resistances. The maximum electrical power was then determined from the IV plot. The photovoltaic conversion efficiency, η_p, was determined by the ratio of the maximum electrical power output density, P_e, of the PV cell to the incident radiant power density on the cell, P_r,

$$\eta_p = \frac{P_e}{P_r} \tag{4}$$

RESULTS AND DISCUSSION

The emittance spectra of three rare earth oxide emitters are shown in Figure 1. In Figure 1b, the external quantum efficiency curve of the 0.6eV InGaAs PV cell is superimposed on the emittance of the MD emitter, and clearly shows that the MD spectra is well matches to the response of the 0.6eV PV cell.

The surface emissive power, ε_t, of four circular shaped emitters were measured at a fixed thermal input. The total surface emissive power and the surface emissive power in the wavelengths between 1 μm and 2 μm, $\varepsilon_{(1-2)}$, are listed in Table 1. The blackbody emitter gave the highest total emissive power but the second lowest photovoltaic convertible power. The lowest emitter efficiency is

185

18% from the blackbody emitter. The MD emitter shows the highest photovoltaic convertible power, 27% higher than the erbia emitter and 38% higher than the blackbody emitter. Additionally, its emitter efficiency reaches 39%, which is the highest among these emitters. These measurements demonstrate that the amount of photovoltaic convertible radiant power can be increased by multiply doping elements into the emitter.

Photovoltaic efficiencies were measured by placing the 0.6eV PV cell in the plane parallel to the emitter at a distance x=5 cm (replacing the power meter which was used during the radiant power measurement). The PV cell was positioned with its normal passing through the center of emitter. The electrical power was determined by measuring the current and voltage of the PV cell. The maximum electrical power was determined from the IV plot. The incident radiant power was measured and is listed in Table 2. The performance of the cell when illuminated by the different emitters is also shown in Table 2. The results indicate that the PV cell had the best efficiency and electrical power output when irradiated by the multiple dopant selective emitter.

Table 1 The data of emissive power density measured from four flat circular emitters.

Emitter	ε_t (W/cm^2)	$\varepsilon_{(1-2)}$ (W/cm^2)	η_e (%)
Erbia	7.1	2.6	37
Holmia	8.0	2.2	28
MD	8.5	3.3	39
Blackbody	13.4	2.4	18

Table 2 The data collected from energy conversion measurements. The 0.6 eV PV cell was illuminated by four flat circular emitters, respectively.

Emitter	P_r (W/cm^2)	P_e (W/cm^2)	η_p (%)	FF
Erbia	0.310	0.011	3.5	0.54
Holmia	0.345	0.011	3.2	0.51
MD	0.366	0.015	4.1	0.56
Blackbody	0.578	0.010	1.7	0.44

* P_r, P_e, η_p, and FF are the incident radiant power, the maximum electrical power, photovoltaic conversion efficiency, and fill factor, respectively.

To examine the overall TPV efficiency, three cylindrical emitter types, erbia, MD and blackbody emitters, were used. In order to evaluate the maximum power output, the PV cell was placed at a distance x=3 cm from the emitter. The thermal input was fixed at 803W. Therefore, the constant thermal input power density, P_t, which is the total thermal input divided by the surface area of the cylindrical emitter, is 9.2 W/cm^2. The energy conversion data measured with respect to the three types of emitters are listed in Table 3. The TPV efficiency, η_T, was determined by evaluating the ratio of the maximum electrical power density output from the PV cell to the thermal input density. The data clearly show that the overall TPV efficiency increases 44% when the MD emitter replaces the erbia emitter or 100 % when the MD emitter replaces blackbody emitter. The surface temperature of the PV cell shows 15 °C increase when the MD emitter replaces the erbia emitter but 10 °C decrease when the MD emitter replaces the blackbody emitter.

Table 3 The data collected from energy conversion measurements. The 0.6 eV PV cell was illuminated by three cylindrical emitters, respectively.

Emitter	P_t (W/cm^2)	P_r (W/cm^2)	$P_{r(1-2)}$ (W/cm^2)	P_e (W/cm^2)	η_T (%)
Erbia	9.2	0.154	0.064	0.017	0.18
MD	9.2	0.248	0.113	0.024	0.26
Blackbody	9.2	0.298	0.056	0.012	0.13

The emitter containing multiple doped elements broadens the line emission of the single oxide emitters to a band emission in the wavelengths between 1 μm to 2 μm. These emissions not only matched the quantum efficiency curve of the 0.6 eV cell well, but also increases the photovoltaic convertible radiant power significantly. As a result, the total electrical power output from the PV cell increases by 41% when compared to selective line emission (i.e. erbia).

The causes for low PV conversion efficiency by the cell were not totally understood. However, one explanation may be the inappropriate technique or materials used in bonding the cell to the circuit board. The resulting poor conductive interface bonding may have raised the cell resistance and consequently reduce the cell efficiency.

CONCLUSIONS

The concept of using multiple dopant selective emitter to improve the performance of 0.6 eV InGaAs PV cell was investigated. The emitter demonstrated broad band emission in the wavelength range from 1 μm to 2 μm, and is well matched to the PV cell quantum efficiency curve. The photovoltaic convertible radiant power from the multiple dopant selective emitter also showed a significant increase when compared to the single rare earth oxide emitters (i.e. erbia or holmia). As a result, the total electrical power output and the overall TPV efficiency is increased by simply replacing the erbia, holmia or the blackbody emitter with the multiple dopant selective emitter. From this investigation, the selective emitters also demonstrated their advantages in improving TPV efficiency as compared to a blackbody emitter. Additionally, the multiple dopant selective emitter has demonstrated its mechanical and thermal robust during this study.

Acknowledgments

Authors would like thank Ken Schroeder for his comments and helpful discussion on the paper. Authors are also grateful to Dr. P.R Sharps from Research Triangle Institute for him providing the PV cell and information. Authors wish to acknowledge the financial support of the DOE Office under contract DEFG05-95ER12156 and Army Research Office under contract ARMYDAAL03-92G-0205 for this study.

References

1. Wedlock, B.D., 1963, Proceeding of IEEE, vol. 51, pp. 694
2. White, D.C. and Chwartz, R.J. S. (eds.), 1967, *PIN Structures for controlled Spectrum Photovoltaic Converters* in AGARD Colloquium, Cannes, France, New York, Gordon Breach Science, 897-922
3. Chen, Z. Adair, P.L. and Rose, M. F., 1996, "Selective Emitters for Thermophotovoltaic Energy Converters-A Study Based on Material Prospects," 31st international Energy Conversion Engineering Conference Proceedings, Washington, DC, vol. 2, 1013-17
4. Evans, B.D., Sundaram, Morgan, M.D., Home, W.E., Ketterl, J.R., and Saban, S.B., "A New GaInAsSb-based Photovoltaic Cell for Use with Source at < 1073K in Thermophotovoltaic Power Conversion Systems," The Proceedings of Space Technology and Applications International Forum, Albuquerque, NM 1997.
5. Siegel , R. and Howell, J.R., 1972, "Thermal Radiation Heat Transfer," McGraw-Hill Book Company, New York.

Temperature Measurement Of
High Performance Radiant Emitters

Robert E. Nelson

Quantum Group, Inc.
11211 Sorrento Valley Road; San Diego, CA 92121

ABSTRACT

A procedure is described to determine the effective temperature of high performance emissive structures. The method is especially suitable for the evaluation of fibrous ceramic emitters, including selective IR emitters for applications in thermophotovoltaic systems. The measuring procedure, furthermore, does not contribute insertion errors nor disturb the operating emitter. Although inexact, this simple measurement technique predicts the maximum error expected which, in most cases, is acceptable.

INTRODUCTION

In the study of gas-powered radiant burners, an accurate determination of the temperature of the effective emissive structure is essential to understand the operation of these systems. Among the high performance radiators under consideration here, we include porous matrix ceramic emitters, ported tile radiant burners, reticulated foam porous ceramic emitters, fibrous ceramic mantle emitters, and related fibrous emitter systems.[1] All of these emitters are fabricated from refractory ceramics and, generally, from oxide ceramics for stability in high temperature flames.

---

[1] Nelson, R.E., "Supported Continuous Fiber Burner," *Proceedings of the American Flame Research Committee, 1992 Fall International Symposium*, Coordinated by Arthur D. Little, Inc. and Massachusetts Institute of Technology, Cambridge, MA; Session 7, October 19–21, 1992.

CP401, *Thermophotovoltaic Generation of Electricity: Third NREL Conference*,
edited by Benner/Coutts
© 1997 The American Institute of Physics 1-56396-734-0/97/$10.00

FIBROUS EMITTERS

Fibrous oxide emitters are especially interesting in that small diameter filaments have several attributes essential for high radiant performance. First, an array of small diameter (10μm) filaments couples well to the combustion products of a flame because of the high thermal transfer coefficient associated with these small diameter elements. The fibers reach, therefore, temperatures approaching combustion by-product temperatures. Next, the filamentary structures are thermally stress tolerant, because no significant thermal stress can be built up across the diameter of these very small filaments, and a stress build-up along the axis of a fiber is relieved by flexing. Finally, the combination of a small filament mass and a high thermal transfer coefficient yields a thermal response time on the order of 20 ms which allows these emitters to reach operating temperature quickly or to load-follow where fuel control systems are available.

A special class of oxide ceramic filament emitters has been studied for specific applications to thermophotovoltaic energy conversion.[2] In particular, certain rare earth oxides (i.e., ytterbia, erbia, holmia, and neodymia) in fibrous form, exhibit very selective emissions in the near IR which, in some cases, are very efficient illuminators of available photoconverters. For example, ytterbia has a single emission centered at a wavelength of 0.98 μm with a full width at half maximum of 150 nm. Off-band emission from the fibrous emitter is very low from about 1.2 μm to the onset of highly emissive lattice vibrations at about 5–8 μm. These emitters can provide up to $6 \times 10^4 \text{ W/m}^2$ of silicon-convertible exitance. The temperature dependence of this exitance can be unusually high. Unlike a blackbody, which exhibits a T^4 dependence of total radiant exitance according to the Stefan-Boltzmann Law, the silicon-convertible exitance of an ytterbia emitter exhibits a T^7 dependence at typical operating conditions.[3] Therefore, an accurate determination of emitter temperature is especially important in characterizing these selective emission structures.

---

[2] Nelson, R.E. and Iles, P.A., "Possible Applications of Selective Emitters for Space Power," presented at the ASME International Solar Energy Conference, Washington DC, April 4–9, 1993. "Solar Engineering 1993", Published by the American Society of Mechanical Engineers, New York City, 1993, p. 529.

[3] Nelson, R.E., *Fibrous Emissive Burners: Selective and Broad Band,"* GRI Report No. GRI–93/0384, Prepared for Gas Research Institute, Contract No. GRI-5091–260–2180, Appendix A, October, 1993.

CONVENTIONAL TEMPERATURE DETERMINATION

The most common method of high temperature measurement is thermocouple thermometry. For temperatures up to 2000 K, one may employ a platinum/platinum alloy thermocouple. With fibrous emitters, however, we should be concerned about errors that may arise from this approach. To minimize a distortion in the temperature distribution introduced by the high thermal conductivity of the thermocouple wire, fine thermocouple wire (about 0.001 inch in diameter) is provided by the manufacturers. The thermal conductivity (at 2000 K) of platinum is well documented[4] and is about 80 W/mK. Data on the thermal conductivity of rare earth oxides are meager. The thermal conductivities[5] of ceria, europia, and thoria are (at 2000 K) about 2.5 W/mK, and we shall regard this number as representative of the thermal conductivity of the rare earth oxides in light of the similarity of the physical properties of most of the rare earth oxides.

Let us compare the thermal conductance G of a unit length of a fine platinum wire and a 10 μm diameter rare earth oxide filament.

$$G = kA/l \qquad (1)$$

where k is the thermal conductivity, A is the cross sectional area of the filament, and l is the length of the filament. We find that the thermal conductance of a unit length of the platinum wire exceeds that of a unit length of the rare earth oxide filament by a factor of 200. Therefore, the presence of a platinum thermocouple wire in an array of rare earth oxide filaments acts as a significant thermal shunt, and temperature readings of the thermocouple are suspect if we are interested in the temperature of the ceramic oxide filaments.

OPTICAL PYROMETRY

An alternate practical method of temperature measurement of radiant structures employs a measurement of spectral exitance or brightness. If we neglect transmission losses between our source and our detector, which is a

---

[4] Touloukian, Y.S., Powell, R.W., Ho, C.Y., Klemens, P.G., *Thermophysical Properties of Matter, 1, Thermal Conductivity-Metallic Elements and Alloys*, IFI/Plenum, NY-Washington, p. 262, 1970.

[5] *Engineering Property Data on Selected Ceramics*, III, Single Oxides, Metals and Ceramics Info. Center, Battelle, Columbus Laboratories, Columbus OH, pp. 5.4.9–6 and 5.4.10–5, July, 1981.

realistic condition in our case, then the determination of the exitance at any wavelength will permit the calculation of the source temperature if the source is a blackbody. The problem is that all practical sources are not blackbodies, and the emittance of the source is generally unknown or known only approximately. The emittance of a real body is defined here as the ratio of real body radiation to blackbody radiation at the same true temperature. Generally, the emittance of most oxide ceramics is low at high temperatures, and large uncertainties in temperature determinations by pyrometry are expected if we employ this general method of temperature measurement.

There may be, however, portions of the optical spectrum where the oxide ceramic emittance may be high enough (say, ≥ 0.5) to permit some approximate assessment of emitter temperature. For example, Figure 1 depicts the spectral emittance of a classic fibrous radiative converter – the Welsbach gas lighting mantle, which has highly emissive spectral regions. At long wavelengths (greater than 10 μm) this structure is highly emissive and extends beyond 40 μm. Because of this characteristic, Welsbach mantles have been used as long–wavelength IR sources.[6] As mentioned earlier, this long wavelength emittance arises from lattice vibrations and is common among all oxide ceramics. At the short wavelength end of the spectrum, oxide ceramics become highly emissive because of an absorption edge effect, which arises from electronic transitions across a wide energy gap. Since all materials of interest to us have a high melting temperature, the effective energy gap in oxide ceramics is high — about 3 to 5 ev. Typically, this absorption edge mechanism occurs at ultraviolet wavelengths. Room temperature reflectivity measurements on rare earth oxides by White[7] clearly demonstrate this UV absorption edge phenomenon at wavelengths about 250 nm. At high combustion temperatures, it is expected that this absorption edge will move to longer wavelengths. Figure 1, however, illustrates an absorption edge phenomenon that occurs in the visible rather than at UV wavelengths. It is the significant discovery of Carl Auer[8], the Austrian chemist who invented the Welsbach mantle, that the addition of a small amount (0.7% by weight)

---

[6] LaRocca, A.J., "Artificial Sources." The Infrared Handbook, Edited by W.L. Wolfe and G.J. Zissis, *Prepared by the Infrared Information and Analysis Center, Environmental Research Institute of Michigan, for the Office of Naval Research, Department of the Navy*, Washington DC, pp 2–28, 1978.

[7] White, W.B., "Diffuse-Reflectance Spectra of Rare-Earth Oxides," *Applied Spectroscopy*, Vol. 21, p. 167, 1967.

[8] C. Auer von Welsbach, "Incandescent Lighting Substance," U.S. Patent No. 563,524. Filed September 23, 1885; issued July 7, 1886.

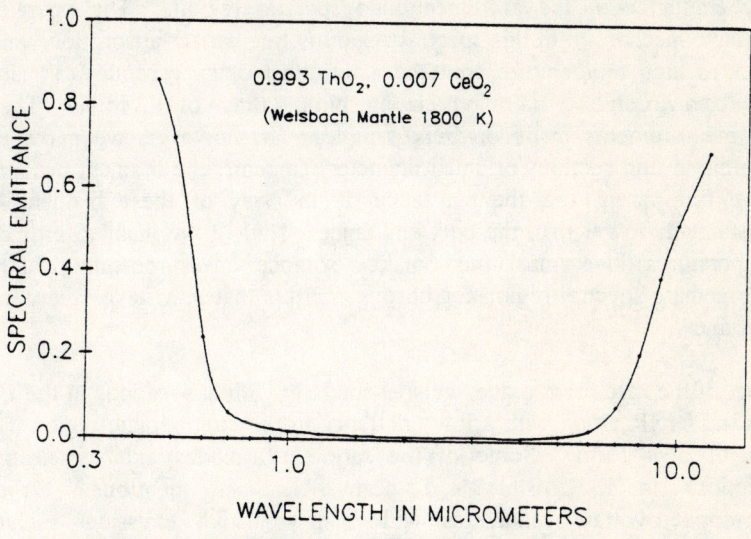

From H. Rubens. Ann. Phys. 18 (1905) 725

FIGURE 1. Spectral Emittance of a Gas Lighting Mantle

of ceria to a host thoria ceramic causes the absorption edge of thoria that normally occurs in the ultraviolet to move to longer wavelengths. The position of this absorption edge depends on the amount of ceria[9] where lesser amounts generate bluer light and greater amounts generate more yellow light. Carl Auer selected 0.7% by weight of ceria to maximize the luminous efficiency of his structure. Although a ceria addition to other host ceramics increases the luminosity of these ceramics, we ordinarily will look in the UV for this absorption edge mechanism when evaluating general ceramic compositions.

Note also the very low spectral emittance in Figure 1 between about 1 μm and the onset of lattice vibrations at about 6 μm. This extraordinarily low emittance accounts, in part, for the luminous efficiency of this structure. The amount of thermal energy wasted in generating unwanted near IR energy is minimized in this engineered component. This low emittance region is a consequence of the fibrous nature of the mantle. Most oxide ceramics, when fabricated into 10 μm or so diameter filaments, exhibit a

---

[9] Ives, H.E., Kingsbury, E.F., Karrer, E., "A Physical Study of the Welsbach Mantle," *Journal of the Franklin Institute*, **186**, p. 401, October, 1918. Continued on p. 585, November, 1918.

low emittance in the aforementioned spectral region. The operative emissive mechanism in this spectral region is free carrier absorption which leads to high temperature emittances for most oxide ceramics (alumina, stabilized zirconia, yttria, magnesia, etc.) in the range of 0.2 to 0.4. These are measurements made on bulk samples. If, however, we make the ceramic in thin sections or small diameter filaments, the filament becomes more transparent and the emittance of an array of these filaments is substantially lower than the bulk emittance. Thus, if we want to estimate temperatures, we must not make exitance measurements in this intermediate spectral region for fibrous emitters that may have a very low emittance.

Most of the rare earth oxides behave similarly with absorptions in the UV and the far IR along with a low emittance in the intermediate region if in fine filament form. Some of the rare earth oxides exhibit selective emissions in the visible and near IR. As mentioned earlier, thermophotovoltaic candidates with their selective emissions include ytterbia (0.98 μm), Erbia (1.55 μm), Holmia (2.0 μm), and Neodymia (2.4 μm). The peak emittances of these selective emitters in the near IR are generally above 0.5, even when in fibrous form. Therefore, we have three spectral regions to consider for optical pyrometry – UV, far IR, and selective emissions from certain rare earth oxides in the near IR. Let us determine which is the best choice for temperature measurement.

ANALYSIS

We begin with Planck's radiation law

$$M_\lambda = C_1\lambda^{-5}[\exp(C_2/\lambda T) - 1]^{-1} \text{ W m}^{-2} \text{ μm}^{-1} \tag{2}$$

where
$$C_1 = 3.741832 \times 10^8 \text{ W μm}^4 \text{ m}^{-2}$$

$$C_2 = 14387.86 \text{ μm K}$$

and where M_λ is the spectral exitance, λ is the wavelength in μm, and T is the temperature in K. We have chosen to investigate structures operating around 1000 K, 2000 K, and 3000 K. We begin by calculating the blackbody spectral exitance from Equation 2 in the spectral range from 0.2 to 32 μm at three reference temperatures (1000 K, 2000 K, and 3000 K). We multiply the blackbody spectral exitance by a graybody emittance ε of

194

0.9, 0.8, 0.7, and 0.6. We then ask what temperature of a blackbody will generate the graybody spectral exitance. We satisfy the condition

$$M_{\lambda(T)} = \varepsilon M_{\lambda\,(Tr)} \tag{3}$$

where T_r is one of the reference temperatures (1000 K, 2000 K, or 3000 K), and T is the temperature of the blackbody that matches the spectral exitance of the 1000 K, 2000 K or 3000 K graybodies. Calculations satisfying Equation (3) are graphed in Figures 2, 3, and 4 for reference temperatures of 1000 K, 2000 K, and 3000 K, respectively. The data for Figures 2, 3, and 4 is tabulated in the Appendix. Note that, in all cases, the deviation between T and T_r increases with wavelength.

END LIMITS

It is worthwhile to evaluate the limits of the aforementioned calculations at very long and at very short wavelengths. Consider first the long wavelength limit. If we substitute the condition

$$C_2/\lambda T \ll 1$$

in Equation (2), then the Planck radiation equation can be approximated by the expression

$$M_\lambda \approx C_1 T/C_2 \lambda^4 \qquad (C_2/\lambda\,T \ll 1) \tag{4}$$

This happens to be the Rayleigh–Jeans radiation law which pre–dated Planck's derivation. If we impose the condition indicated in Equation (3), we obtain the relationship

$$T = \varepsilon T_r \quad \text{when} \qquad \lambda \to \infty \tag{5}$$

which we term the Rayleigh–Jeans limit where the derivation of T from T_r is maximum. These limits are indicated in Figure 2, 3, and 4. At very short wavelengths, we introduce the inequality

$$C_2/\lambda T \gg 1$$

into Equation (2), and the Planck radiation law simplifies to

$$M_\lambda \approx C_1 \lambda^{-5} \exp(-C_2/\lambda T) \qquad (C_2/\lambda T \gg 1) \tag{6}$$

195

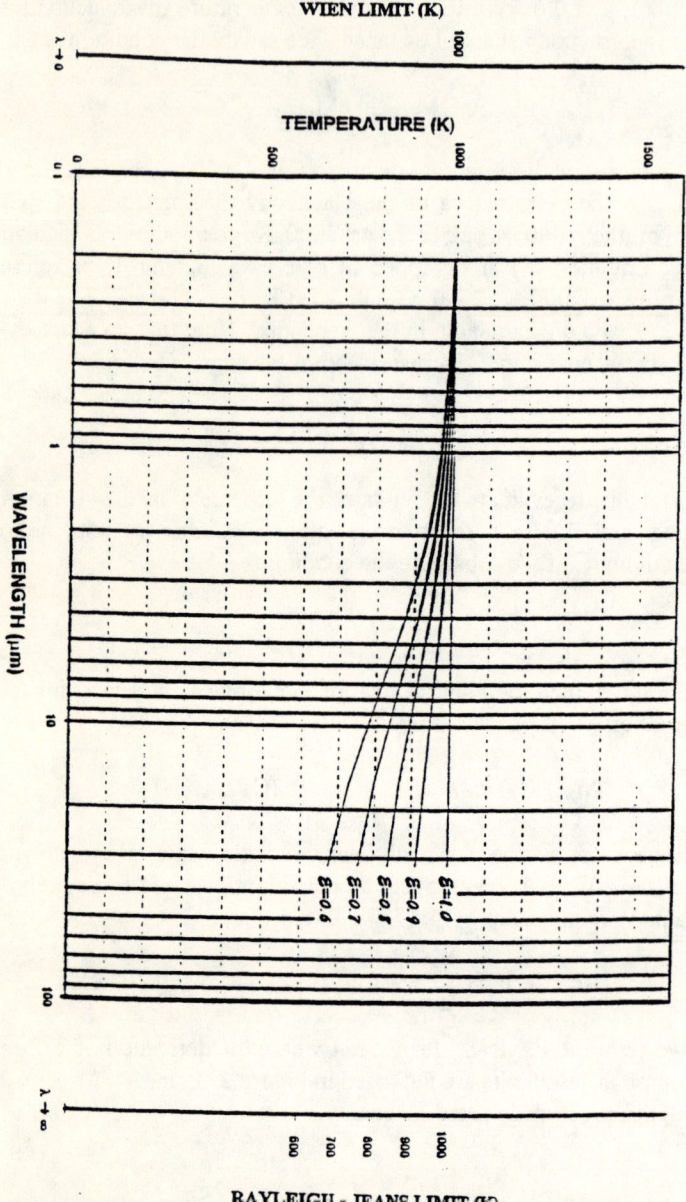

FIGURE 2. Brightness Temperatures of a 1000 K Source

196

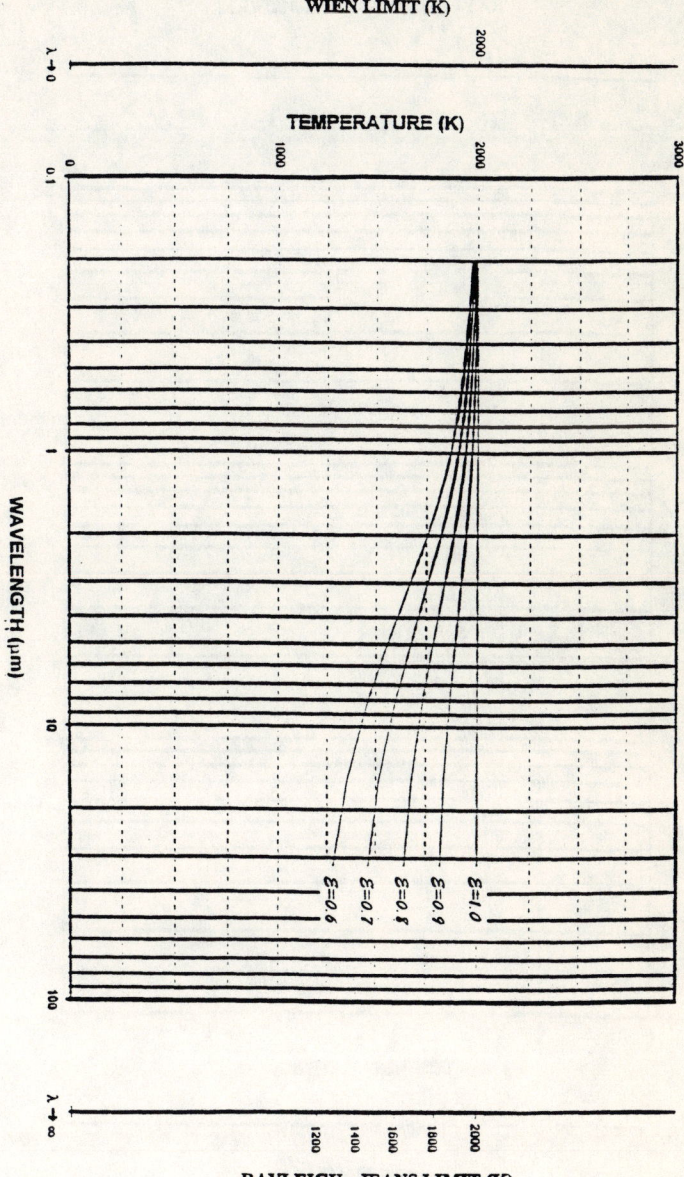

FIGURE 3. Brightness Temperatures of a 2000 K Source

197

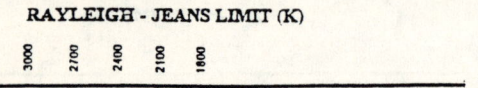

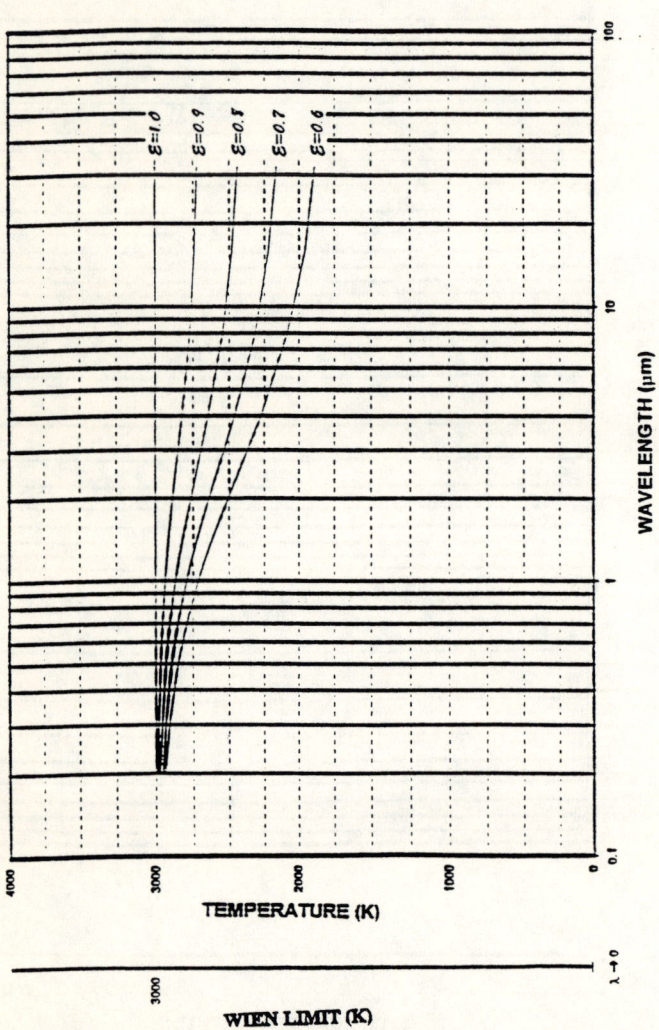

FIGURE 4. Brightness Temperatures of a 3000 K Source

198

which also pre-dates Planck's radiation law as Wien's approximation. If we impose the condition of Equation (3) into Equation (6), we obtain

$$\varepsilon = \exp\,(-C_2/\lambda)(T^{-1} - T_r^{-1}) \tag{7}$$

Equation (7) is a suitable alternative to calculate the data indicated in Figures 2, 3, and 4 for wavelengths less than 0.5 μm. Equation (7) may be rewritten as

$$T = T_r\,(1 - \lambda T_r \ln \varepsilon\,/\,C_2)^{-1} \tag{8}$$

Since $T_r \ln \varepsilon\,/C_2$ is finite, this term when multiplied by λ goes to zero when λ goes to zero. Therefore, we may conclude

$$T = T_r \qquad \text{when} \qquad \lambda \to 0 \tag{9}$$

This Wein limit is indicated on Figures 2, 3, and 4.

CONCLUSIONS

Clearly, the long wavelength portion of the emission spectrum must be avoided in inferring emitter temperatures by optical pyrometry, because this is where the sensitivity to emittance uncertainty is greatest. Conversely, measurements in the UV region are ideal in light of the vanishing effect of emittance on exitance in this wavelength region. UV readings, however, require some compromise in practical measurement techniques. First, very short wavelength UV measurements are complicated by UV absorption in air, and vacuum ultraviolet measurements of flame sources are impractical. Furthermore, the flame temperatures of conventional fuel flames with air oxidants are typically in the vicinity of 2000 K, and the exitance of graybody sources at this temperature is very low and drops rapidly as the UV region is entered. The falling graybody exitance is offset, in part, by a rapidly rising emittance as the absorption edge is approached. There is a UV wavelength region, therefore, that we have found to be practical for emitter temperature determinations. We employ the UV interval between 300 and 400 nm. We seek a maximum exitance in this wavelength interval and convert that exitance into a graybody temperature at that wavelength. With an a priori estimate of 0.5 to unity for the emittance, we obtain a maximum error of about 40 K in the temperature estimate. If, however, we can bracket the emittance, e.g. $0.6 < \varepsilon < 0.8$, which we feel is reasonable, then the total error in temperature may drop to approximately 15 K which we feel is useful. A concluding note is of interest. There are situations where long wavelength emittance determination of fibrous emitters is useful. With the aforementioned procedure, we can determine the operating temperatures of a fibrous emitter at UV wavelengths. With the temperature known along with its uncertainty, then the long wavelength (10μm or so) exitance will yield a direct determination of the emittance. We have used the high sensitivity of emittance in the long wavelength region to estimate the value of emittance.

REFERENCES

1. Nelson, R.E., "Supported Continuous Fiber Burner," *Proceedings of the American Flame Research Committee, 1992 Fall International Symposium*, Coordinated by Arthur D. Little, Inc. and the Massachusetts Institute of Technology, Cambridge, MA; Session 7, October 19–21, 1992.

2. Nelson, R.E. and Iles, P.A., "Possible Applications of Selective Emitters for Space Power," presented at the ASME International Solar Energy Conference, Washington DC, April 4–9, 1993. "Solar Engineering 1993", Published by the American Society of Mechanical Engineers, New York City, 1993, p. 529.

3. Nelson, R.E., *"Fibrous Emissive Burners: Selective and Broad Band,"* GRI Report No. GRI–93/0384, Prepared for Gas Research Institute, Contract No. GRI-5091–260–2180, Appendix A, October, 1993.

4. Touloukian, Y.S., Powell, R.W., Ho, C.Y., Klemens, P.G., *Thermophysical Properties of Matter, 1, Thermal Conductivity-Metallic Elements and Alloys,* IFI/Plenum, NY-Washington, p. 262, 1970.

5. *Engineering Property Data on Selected Ceramics*, III, Single Oxides, Metals and Ceramics Info. Center, Battelle, Columbus Laboratories, Columbus OH, pp. 5.4.9–6 and 5.4.10–5, July, 1981.

6. LaRocca, A.J., "Artificial Sources." The Infrared Handbook, Edited by W.L. Wolfe and G.J. Zissis, *Prepared by the Infrared Information and Analysis Center, Environmental Research Institute of Michigan, for the Office of Naval Research, Department of the Navy*, Washington DC, pp 2–28, 1978.

7. White, W.B., "Diffuse-Reflectance Spectra of Rare-Earth Oxides," *Applied Spectroscopy*, **21**, p. 167, 1967.

8. C. Auer von Welsbach, "Incandescent Lighting Substance," U.S. Patent No. 563,524. Filed September 23, 1885; issued July 7, 1886.

9. Ives, H.E., Kingsbury, and E.F., Karrer, E., "A Physical Study of the Welsbach Mantle." *Journal of the Franklin Institute*, **186**, p. 401, October 1918. Continued on p. 585, November 1918.

APPENDIX

TABLE 1. BRIGHTNESS TEMPERATURES OF A 1000 K SOURCE

ε	Wavelength λ (μm)											
	0.2	0.3	0.5	0.7	1	1.2	1.5	2	4	8	16	32
1.0	1000.0	1000.0	1000.0	1000.0	1000.0	1000.0	1000.0	1000.0	1000.0	1000.0	1000.0	1000.0
0.9	998.5	997.8	996.4	994.9	992.7	991.2	989.1	985.6	972.3	953.0	933.7	919.3
0.8	996.9	995.4	992.3	989.2	984.7	981.7	977.2	969.9	942.9	904.7	866.7	838.4
0.7	995.1	992.6	987.8	982.9	975.8	971.1	964.1	952.8	911.7	854.7	798.8	757.1
0.6	992.9	989.5	982.6	975.7	965.7	959.1	949.4	933.7	878.0	802.6	729.6	675.2

TABLE 2. BRIGHTNESS TEMPERATURES OF A 2000 K SOURCE

ε	Wavelength λ (μm)											
	0.2	0.3	0.5	0.7	1	1.2	1.5	2	4	8	16	32
1.0	2000.0	2000.0	2000.0	2000.0	2000.0	2000.0	2000.0	2000.0	2000.0	2000.0	2000.0	2000.0
0.9	1994.2	1991.3	1985.5	1979.7	1971.1	1965.5	1957.3	1944.5	1906.0	1867.5	1838.7	1820.8
0.8	1987.7	1981.6	1969.5	1957.4	1939.8	1928.3	1911.7	1885.9	1809.4	1733.5	1676.8	1641.4
0.7	1980.4	1970.7	1951.6	1932.9	1905.5	1887.9	1862.4	1823.4	1709.4	1597.6	1514.1	1461.8
0.6	1972.0	1958.3	1931.4	1905.2	1867.4	1843.2	1808.6	1756.0	1605.2	1459.3	1350.5	1281.9

TABLE 3. BRIGHTNESS TEMPERATURES OF A 3000 K SOURCE

ε	Wavelength λ (μm)											
	0.2	0.3	0.5	0.7	1	1.2	1.5	2	4	8	16	32
1.0	3000.0	3000.0	3000.0	3000.0	3000.0	3000.0	3000.0	3000.0	3000.0	3000.0	3000.0	3000.0
0.9	2986.9	2980.4	2967.4	2954.6	2936.0	2924.2	2907.9	2884.3	2824.0	2773.7	2740.6	2721.3
0.8	2972.3	2958.7	2931.8	2905.4	2867.5	2843.8	2811.0	2763.9	2644.9	2546.3	2480.8	2442.6
0.7	2956.0	2934.5	2892.5	2851.6	2793.6	2757.7	2708.3	2638.0	2462.2	2317.1	2220.3	2163.6
0.6	2937.4	2907.1	2848.3	2792.0	2712.9	2664.3	2598.2	2505.1	2274.7	2085.6	1959.0	1884.4

SESSION 4:
TPV DEVICES BASED ON InGaAs

Comparison of 0.55eV InGaAs Single-Junction vs. Multi-junction TPV Technology

Steven Wojtczuk

Spire Corporation
One Patriots Park
Bedford, MA 01730-2396

Abstract: Indium gallium arsenide ($In_xGa_{1-x}As$) single-junction and multi-junction thermophotovoltaic (TPV) cell technology is examined for the lattice-mismatched, extended wavelength response, $In_{0.72}Ga_{0.28}As$ composition (0.55eV bandgap). One of the theoretical advantages of the multijunction technology is that the current is lowered by the number of junctions compared to a single-junction of the same area, potentially leading to a much lower I^2R loss (if R is the same). Using statistical data from 2500 single-junction 0.55eV InGaAs TPV cells, we extract a 300K average dark current (2×10^{-5} A/cm^2) from the average Voc (283mV) and an average series resistance (0.03 Ω-cm^2) for use in comparing the power output of the two cell types including resistance effects. From a 0.74eV $In_{0.53}Ga_{0.47}As$ eight-junction InGaAs cell, we also extract a series resistance (2.2Ω). We present Hall mobility data on both the 72% and 53% InGaAs showing they are essentially the same, and argue that the resistance of a 0.55eV multijunction cell will be similar to that measured for the 0.74eV device. We then calculate the power output including the respective series resistance of each type. For the devices examined, multijunction cells do not hold a significant advantage in lower I^2R ohmic loss; however, other aspects not included in the comparison, such as a potentially better reflector technology to reject unused infrared radiation for MJTPV devices may weigh more heavily than ohmic power loss when the overall system is evaluated.

INTRODUCTION

Multiple-junction lattice-matched 0.74eV $In_{0.53}Ga_{0.47}As$ photovoltaic cells were first demonstrated (1) by Spire in 1994 for laser power converter applications. Recently, Wilt et al. (2) realized that significant optical benefits to a TPV system could accrue from use of such cells.

CP401, *Thermophotovoltaic Generation of Electricity: Third NREL Conference,*
edited by Benner/Coutts
© 1997 The American Institute of Physics 1-56396-734-0/97/$10.00

In this paper, we refer to such cells as multijunction TPV (MJTPV) in keeping with a notation in which SJTPV stands for single-junction TPV cells; however, the MJTPV cells have also been called "monolithically interconnected modules" or MIMs.

Figure 1 shows a simple sketch of a SJTPV and an 8-junction MJTPV cell. In this paper, we are primarily interested in examining whether the I^2R loss of the MJTPV cell is significantly less than that of the SJTPV cell. If the same light intensity is falling on each cell type, the SJTPV cell will have a lower voltage (it is a single junction) and a higher current (it has a larger area) than the MJTPV cell which has a higher voltage (8X higher, for the 8-junction cell depicted) and a smaller current (1/8th the current). The ohmic I^2R power losses for the MJTPV cell could be 1/64th as much as the SJTPV cell if only the series resistance was the same for both cell types.

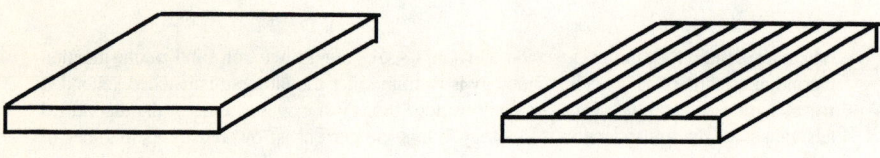

Figure 1. *Sketch of SJTPV (left) and MJTPV (right) cells.*

Unfortunately it is much easier to make a lower resistance vertical-current-flow SJTPV device (in which the currents tend to flow through large cross-sectional areas) than to make a low resistance MJTPV device in which the current must flow laterally along a thin, small-cross-sectional-area buried conduction layer (similar to that used in IC bipolar junction transistor technology). Whether the MJTPV device will have a lower I^2R loss than SJTPV devices depends on whether MJTPV devices can be made with a resistance that is less than "I-squared" greater (e.g. 64X for 8-junctions) than the resistance of the SJTPV devices. Table 1 has a brief comparison of some of the main points of both technologies. SJTPV cells look like standard III-V solar cells, except with more grid metal than customary. However, the MJTPV technology is not as standard, and Figures 2 through 4 are included to help guide the reader.

Table 1   *Comparison of Features of SJTPV and MJTPV Technology*

Single-Junction	Multi-Junction (e.g. 8)
Electrical contact to top and bottom	Both (+) and (-) contacts on top
Either N/P or P/N cell type	P/N due to lower Rsheet in N buried layer (Rsh of P would be ~30X higher)
e.g. for 0.55eV cell, 0.3V and 8A	e.g. for 0.55eV cell, 2.4V and 1A
If 50% FF at 8A then Pout 1.2W/cm2	More Pout if R is not 64X higher than SJTPV
Simple process (higher yield) 2 planar photolithographic steps 2 metal evaporations	More complex process (lower yield) 4 photolithograph. steps (3 nonplanar) 2 metal evaporations
Conducting InP Wafers free-carrier absorption	Semi-insulating InP Wafers little free-carrier absorption
Front surface reflectors (e.g. dielectric and plasma filters)	Can use back surface reflectors (e.g. mirrors)

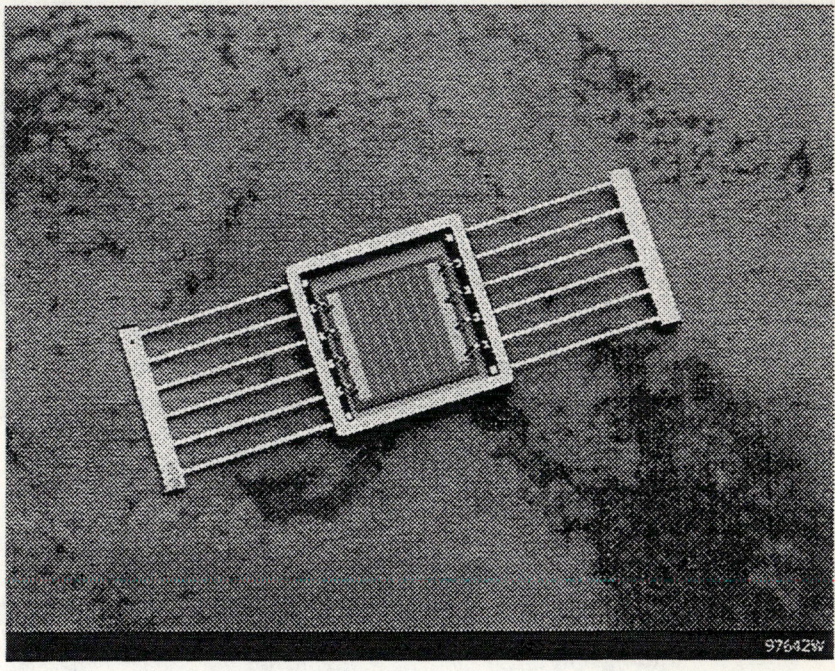

Figure 2   *Prototype of a packaged eight-junction InGaAs power converter.*

207

Figure 3   *Four junctions of an 8-junction device (1). The many gridlines are used to reduce the I^2R loss of the upper P+ InGaAs layer. The long stripes are the interconnect trenches, featured below.*

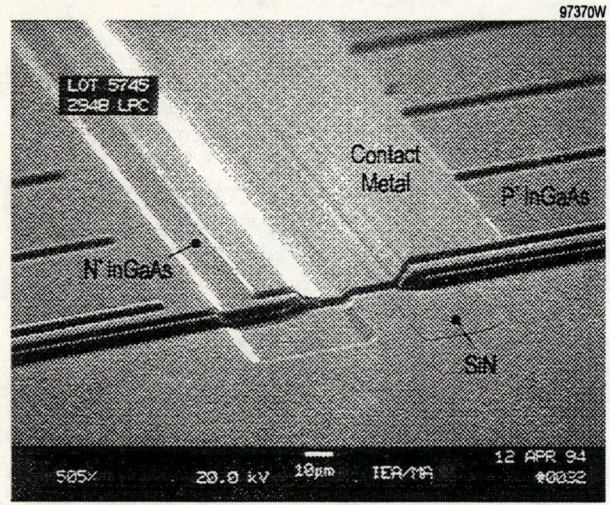

Figure 4   *Details of the interconnect layer between two junctions in a MJTPV device(1). Identical, adjacent epilayer islands have a P/N InGaAs junction. Selective etching is used to reveal the buried N+InGaAs layer at left. Silicon nitride is then patterned over rightside junction edge to prevent interconnect metal from shorting junction. Finally a single layer of grid metal is added as the interconnect.*

208

SIMPLIFIED 0.55eV CELL STRUCTURES

For SJTPV cells, an N-on-P cell is normally used so that the higher mobility majority carrier electrons allow a low sheet resistance in the relatively thin, heavily-doped N emitter layer. The N-on-P design also allows for longer minority-carrier electron diffusion lengths in the more lightly doped P base layer, of great importance since most of the longer-wavelength blackbody irradiation is near the bandgap and is absorbed in the base layer.

$In_xGa_{1-x}As$ is only lattice-matched to the InP wafer as $In_{0.53}Ga_{0.47}As$ (0.74 eV). When grown at compositions away from lattice match (e.g. 0.55 eV $In_{0.72}Ga_{0.28}As$), dislocation defects are created which can thread through the cell junction, increasing the dark current. A compositionally-graded layer is normally used to reduce the threading dislocation density (3).

In the MJTPV case, we assume that this grading layer is heavily-doped and serves as the buried layer on the semi-insulating InP wafer. For the MJTPV, we assume that the cell will be a P-on-N design so that the ~30X higher mobility of the electrons is put to good use in lowering the sheet resistance of the buried layer. In our simple study, we assume that the epilayers are as shown in Figure 5.

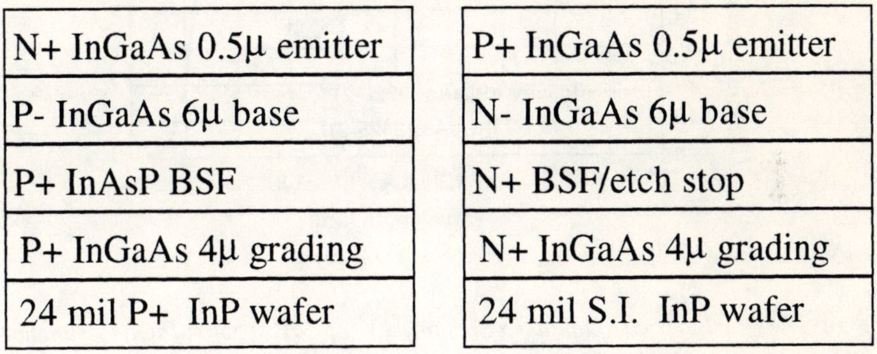

N+ InGaAs 0.5µ emitter
P- InGaAs 6µ base
P+ InAsP BSF
P+ InGaAs 4µ grading
24 mil P+ InP wafer

P+ InGaAs 0.5µ emitter
N- InGaAs 6µ base
N+ BSF/etch stop
N+ InGaAs 4µ grading
24 mil S.I. InP wafer

Figure 5   *Simplified epilayer structures for SJTPV (left) and MJTPV (right).*

QUANTUM EFFICIENCY

For the purposes of this study, we have assumed that the quantum efficiency of both P/N MJTPV and N/P SJTPV devices are roughly the same, since small changes will not affect our rough comparison, and we will later deal just with the photocurrent density. As general information, N/P structures will have better "intrinsic" quantum efficiency due to the ~5X longer diffusion lengths of electrons in the P-type base layer versus holes in an N-type layer (the diffusion lengths scale roughly as the square root of the ratio of the mobilities). However, a mirror on the back of the P/N MJTPV cells (2) could allow an additional pass of long-wavelength light, compensating for the lower hole mobility. A typical quantum efficiency is shown in Figure 6 for a 2.5μm N/P device - a thicker base (such a 6μm shown in Fig. 5) should allow a harder "knee" at the long wavelength cutoff.

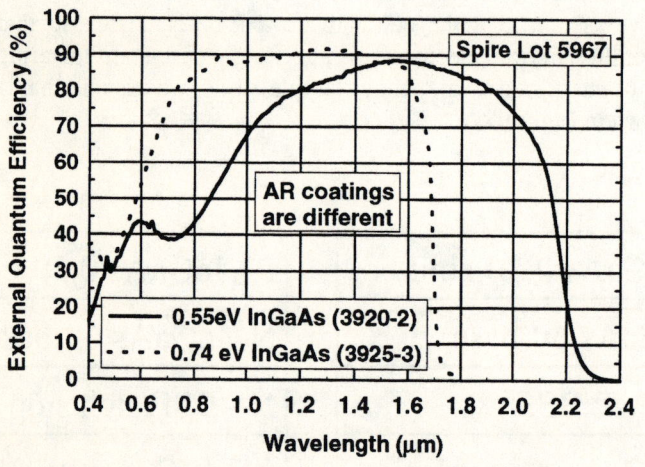

Figure 6   *External quantum efficiency (3) of N/P 0.55eV InGaAs junction with a 2.5μm base layer (0.55eV data courtesy K. Emery of NREL). The dotted curve is a 0.74eV InGaAs cell shown for comparison. The dip at 700nm is due to a reflectance spike in the broadband MgF₂/ZnS antireflection coating applied.*

210

I-V MODELING

In order to compare the power output of the SJTPV and MJTPV cell types, we need an I-V model of the form:

$$J = Jo \; [exp \; (\; q \; (\; V - J \, Rs \;)/kT) - 1] - Jsc$$

We first extracted Jo, an average diffusion dark current density for the 0.55eV cell material, from the average open-circuit voltage (283mV) of 2500 0.55eV N/P InGaAs cells (4) at an illumination sufficient ($\sim$1.2A/cm^2) to drive the cells into the region where the I-V curve is dominated by the diffusion current, arriving at a value of Jo of $\sim$2x10^{-5} A/cm^2. We used this value for both the P/N MJTPV and N/P SJTPV cells since the photovoltage depends only logarithmically on the dark current, and differences of the order of a factor of 2 or 3 are relatively minor in our I^{2}R loss comparison. We also iteratively fit the observed fill-factor for these cells (60% at the 1.2A test current) to obtain an average series resistance Rs for the N/P SJTPV technology (0.03Ω-cm^2, mainly due to contact to P-InP wafer). We then used an I-V (Figure 7) of an available 8-junction 0.74eV InGaAs multijunction cell to model its 51% FF in terms of its Rs and obtained 2.2Ω (roughly equally due to upper P+ and lower N+ InGaAs).

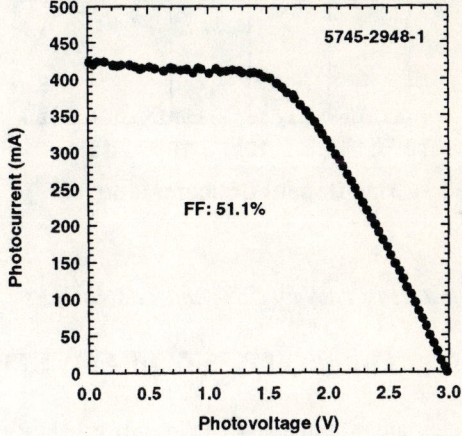

Figure 7   *Illuminated I-V curve of an 8-junction 0.74eV InGaAs power converter. We are using its estimated series resistance 2.2Ω (modeled from the 51% FF) as an estimate of what a similar 0.55eV MJTPV converter would exhibit.*

211

HALL DOPING-MOBILITY DATA for 0.55eV InGaAs

The validity of our assumption in the previous section that the 0.74eV InGaAs MJTPV Rs is similar to the 0.55eV Rs is dependent primarily on whether the electron and hole mobilities and maximum doping of 53% and 72% InGaAs are similar. Figure 8 shows that they are indeed similar. We theorize this is due to the higher electron mobility of InGaAs compositions closer to InAs compensating for the dislocations due to lattice-mismatch at these same InGaAs compositions.

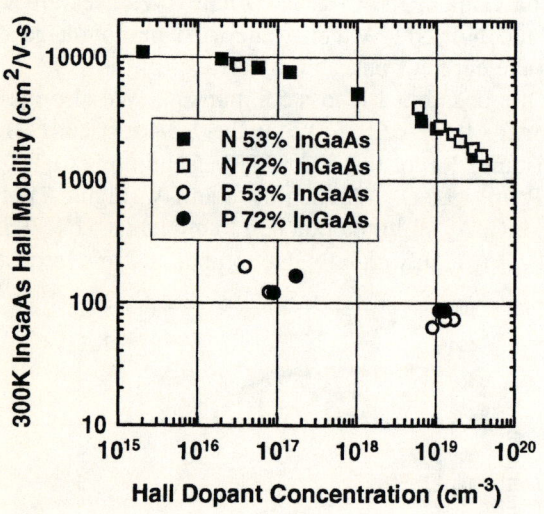

Figure 8   *Measured Hall data for 53% and 72% InGaAs.*

Calculated Power Output of 0.55eV InGaAs SJTPV vs MJTPV

Finally, we calculate whether the I^2R loss of a MJTPV cell is much less than a SJTPV device. We can easily see that for an 8J TPV device, in which the current I is 1/8th that of a SJTPV cell, the series resistance Rs can be no higher than 64X that of a SJTPV device to see any benefit in lower I^2R loss. Our data show that the 2.2Ω Rs of the MJTPV is 73X that of the 0.03Ω of the SJTPV, so that no benefit is expected.

Table 2 shows power outputs based on the I-V models. Eventually, the Rs of both cell types can be decreased. 0.55eV InGaAs SJTPV cells have been made with Rs approaching 0.015Ω-cm^2, while we estimate that with redesign the 8-junction InGaAs MJTPV cell could have a resistance of about 0.7Ω.

Table 2   *Calculated Power Output Including Rs for 0.55eV SJ versus MJ TPV*

Type	Jsc A/cm^2	Isc A	Voc V	FF %	P W
SJ	1	1	0.28	63	0.18
SJ	4	4	0.32	45	0.58
8J	1	1/8	1.8	55	0.12
8J	4	1/2	2.1	36	0.38

SUMMARY

We investigated I^{2}R loss in SJTPV versus MJTPV 0.55eV InGaAs cells to see whether the MJTPV cells would realize a large benefit due to their lower current (reduced by the number of junctions per area) compared to a single junction of the same area. An eight-junction InGaAs cell could have no more than 64X the resistance of a SJTPV cell of the same area to realize such a benefit. MJTPV resistance is higher since current must flow laterally in a buried conducting layer with a narrow cross-section, as well as in an upper thin P-type layer (which is aided by gridline coverage). I-V modeling of the FF of an 8-junction cell estimates Rs is 73X greater than Rs for a SJTPV cell. However, improved back surface mirror technologies for MJTPV compared to front-surface reflectors for SJTPV cells could argue for MJTPV use in many applications when the overall optical system is included in the model.

REFERENCES

1.   Wojtczuk, S., Parodos, T., and Walker, G. "P/N In(Al)GaAs Multijunction Laser Power Converters," in *Proc. of XIII Space Photovoltaic Research and Technology Conf.*, NASA Conf. Pub. 3278, 1994, pp. 363-371.

2.   Wilt, D.M., Fatemi, N.S., Jenkins, P.P., Hoffman, R.W., Landis, G.A. and Jain, R.K., "Monolithically Interconnected InGaAs TPV Module Development," *Proc. of 25th IEEE Photovoltaic Specialists Conf.,* 1996, pp. 43-48.

3.   Wojtczuk, S., Colter, P., and Jhabvala, M. "High-quantum-efficiency 2.2μm InGaAs MOCVD Photodiodes," in *SPIE Proc. Vol. 2999,* 1997, paper 31.

4.   Wojtczuk, S., Colter, P., Charache, G. and Campbell, B. "Production Data on 0.55eV InGaAs Thermophotovoltaic Cells," *Proc. of 25th IEEE PVSC.*, 1996, pp. 77-80.

Electrical and Optical Properties of Degenerately-Doped N-type $In_xGa_{1-x}As$

GW Charache, DM DePoy, JL Egley, RJ Dziendziel,
MJ Freeman, PF Baldasaro, and BC Campbell

Lockheed Martin, Inc., Schenectady, NY 12301-1072

PR Sharps, ML Timmons

Research Triangle Institute, Research Triangle Park, NC 27709-2194

RE Fahey, K Zhang

OFC Corp., Natick, MA 01760

JM Borrego

Rensselaer Polytechnic Institute, Troy, NY 12301-3590

Abstract. Degenerately-doped ($> 10^{19}$ cm^{-3}) n-type $In_xGa_{1-x}As$ ($x > 0.53$) possesses a number of intriguing electrical and optical properties relevant to electro-optic devices and thermophotovoltaic devices in particular. Due to the low electron effective mass of this material ($m_n^* \sim 0.1$) and the demonstrated ability to incorporate n-type dopants into the mid-10^{19} cm^{-3} range, both the Moss-Burnstein bandgap shift and plasma reflection characteristics are particularly dramatic. These properties are investigated for $In_xGa_{1-x}As$ as a function of doping concentration, dopant type, and growth conditions. For undoped InGaAs with a nominal bandgap of 0.6 eV, doping this material to 5×10^{19} cm^{-3} increased the effective optical bandgap to 1.1 eV and has a plasma turn-on wavelength of 5 microns. This filter was coupled to a non-absorbing interference filter, creating a functional tandem filter for thermophotovoltaic applications.

Introduction

$In_xGa_{1-x}As$ ($x > 0.53$) thermophotovoltaic (TPV) devices have demonstrated the highest performance of any infrared material system utilized for energy conversion of blackbody radiators [1]. Unique electrical and optical properties of this material system offer to improve its performance for both TPV devices and mid-infrared optoelectronic devices. Due to the low electron effective mass of this material ($m_n^* \sim 0.1$) and the ability to incorporate n-type

CP401, *Thermophotovoltaic Generation of Electricity: Third NREL Conference,*
edited by Benner/Coutts

dopants into the mid-10^{19} cm^{-3} range, both the Moss-Burnstein bandgap shift and plasma reflection characteristics are particularly dramatic. These properties are investigated for In$_x$Ga$_{1-x}$As as a function of dopant type, dopant concentration, and growth conditions.

The electron-concentration dependence of the near bandgap absorption has been investigated for a variety of III-V semiconductor materials since the mid-1950's [2], and recently in n-InAs and lattice-matched n-InGaAs [3,4]. It has been shown that a theoretical description of the optical bandgap shift is due to both band filling and bandgap narrowing of the gamma valley. However, when electrically active carrier concentrations approach the mid-10^{19} level, a two-valley model may be required. The effective optical bandgap (i.e., the position of the Fermi-level above the valence band) for a parabolic conduction band is given by

$$Eg_{opt} = Eg_0 + \left(\frac{h^2}{2m^*}\right)(2\pi^2 N)^{2/3} - E_{bgn}$$

where, Eg_0 is the nominal semiconductor bandgap, h is Planck's constant, m^* is the conduction band effective mass, q is the electron charge, N is the net electrically active carrier concentration, and E_{bgn} is the effective bandgap narrowing due to both bandtailing and many-body effects. When adding the effects of a non-parabolic conduction band, the effective mass is modified to take into account the filling of the conduction band as carrier concentration increases [5].

The spectral reflection and transmission of a semiconductor plasma filter are determined from the complex index of refraction and film thickness. The complex index of refraction, $\bar{n}$, is given by n-ik, where n is the real part and k is the imaginary part. The parameter k is also referred to as the extinction coefficient and is related to the absorption coefficient, α, at a wavelength, λ, by the following: $\alpha = 4\pi k/\lambda$. The complex index of refraction for a plasma reflector is given by the Drude relation:

$$(n-ik)^2 = \varepsilon_\infty - \frac{Nq^2}{\varepsilon_0 m^* \cdot (\omega^2 - i\gamma\omega)},$$

where ε_∞ is the high frequency limit of the dielectric constant, N is the free carrier concentration, ε_0 is the permittivity of free space, m^* is the free carrier effective mass, ω is the angular frequency, and γ is the damping rate. The angular frequency is related to wavelength, λ, by the following: $\omega = 2\pi c/\lambda$, where c is the speed of light; and the damping rate is related to the free carrier mobility, μ, by the following: $\gamma = q/\mu m^*$. Solution of this equation for the real and imaginary parts of the refractive index as a function of wavelength allows the calculation of the reflection, transmission, and absorption for a semiconductor plasma filter for a given thickness.

Due to the high carrier concentrations that have been obtained in $In_xGa_{1-x}As$ [Table 1] and low electron effective mass, both the Moss-Burstein optical bandgap shift and plasma reflection characteristics are expected to be particularly dramatic. As noted in Table 1, higher electrically active carrier concentrations are observed as the indium fraction increases. Thus, carrier concentrations approaching 10^{20} cm^{-3} should be readily attained with a variety of dopant sources for indium fractions greater than 0.53.

Indium Fraction (x)	Dopant Source	Highest Electrically Active Carrier Concentration (cm^{-3})	Reference
0	SiH_4, Si_2H_6, TESiH	5×10^{18}	6
0	DETe	1×10^{19}	6,7,8,9
0	TESn	$1 - 1.5 \times 10^{19}$	6,8,10
0.53	Sn	1×10^{20}	11
0.53	$SiBr_4$	3×10^{19}	12
0.53	TESn	6×10^{19}	13
0.53	Si	6×10^{19}	14
1	Te	$1 - 2 \times 10^{20}$	15
1	Se	8×10^{20}	15
1	S	1×10^{21}	15

Table 1 - Highest reported electrically active carrier concentrations in $In_xGa_{1-x}As$.

Electrical Properties

For this study single layers of n-$In_xGa_{1-x}As$ (0.53 < x < 0.7) were grown using atmospheric pressure organometallic vapor phase epitaxy (AP-OMVPE) on semi-insulating InP wafers. Electrical properties were optimized by varying the following experimental parameters: dopant type (hydrogen selenide, methylallyltelluride), dopant mole fraction, and V/III ratio. Figures 1 and 2 illustrate the Hall Effect results for both Se- and Te-doped samples as a function of dopant mole fraction. Figure 1 also includes the data of Pizone [13] for Sn-doping of $In_{0.53}Ga_{0.47}As$. These samples had substrate orientations of <100> 2°-off toward <110>, growth temperatures of ~$640°C$, and V/III ratios greater than 50. Higher carrier concentrations were achieved with Te-doping, which is believed to be due to the lower vapor pressure of Te versus Se. For both 0.6 eV (x = 0.67) and 0.73 eV (x = 0.53) material and Te dopant mole fractions greater than 10^{-5}, the carrier concentration remains fixed at a maximum level of 5×10^{19} and 2×10^{19} cm^{-3}, respectively; while the mobility remains fixed or decreases. This suggests a significant amount of inactive tellurium dopant, which is confirmed by secondary ion mass spectroscopy (SIMS) analysis [Fig. 3]. One important feature of this work is the high carrier mobility (> 1000 cm^2/V-s) achieved for lattice mismatched films with electrically active carrier concentrations greater than 10^{19} cm^{-3}. It is these high mobility values that are required to demonstrate high quality (sharp turn-on) plasma filters.

Figure 4 illustrates the effects of varying the V/III ratio on electrically active carrier concentrations for tellurium mole fractions greater than 10^{-5}. Although it was expected that decreasing the V/III ratio would increase the dopant incorporation, the results show no significant trend.

Optical Properties

Reflection and transmission measurements (FTIR) were utilized to extract the effective photonic bandgap as a function of doping level for $In_{0.67}Ga_{0.33}As$ [Fig. 5]. The measured photonic bandgap along with calculated values for both parabolic and non-parabolic band approximations are included. The deviation of the measured and calculated values at doping levels greater than 10^{19} cm^{-3} may be due to two effects: filling of the satellite L-valley and bandgap narrowing. The position of the L-valley in $In_{0.67}Ga_{0.33}As$ was estimated from published values of the L-valley bandgap in GaAs (E_g = 1.70 - 1.73 eV) and InAs (E_g = 1.07 - 1.43 eV) [16-18] and the L-valley bowing parameter (c ~ 0.5 eV) [15,17]. Depending on the position of the L-valley in InAs, the L-valley bandgap in

$In_{0.67}Ga_{0.33}As$ can range from 1.16 - 1.6 eV. The lower end of this range may effect the measured photonic bandgap as shown in Fig. 5. Thus, it is unclear whether filling of the satellite valley or bandgap narrowing is the dominant effect.

Long wavelength (> 2 μm) reflection scans were measured to investigate the plasma reflection characteristics for $In_{0.67}Ga_{0.33}As$ [Fig. 6]. These films demonstrate the characteristic plasma-edge shift toward shorter wavelengths as the free-electron concentration increases. The sharp turn-on of the sample doped 5×10^{19} cm^{-3} is due to the high mobility (low effective mass) of $In_{0.67}Ga_{0.33}As$. Modeling the reflection characteristics of these films with the Drude-theory allows the extraction of the below bandgap optical constants. Using the measured Hall carrier concentration and mobility for films greater than 1 micron in thickness, allow the fitting of the reflection characteristics with the effective mass as the only fitting parameter. Figure 7 illustrates calculated values of the real (n) and imaginary (k) parts of the refractive index as a function of wavelength and mobility for a free carrier concentration of 5×10^{19} cm^{-3}. For this calculation $\varepsilon = 12.5$, $m^* = 0.1$, and $n_i = 1$ (i.e., air). Figure 8 plots the calculated reflection and absorption as a function of wavelength and film thickness for a free carrier concentration of 5×10^{19} cm^{-3}, a mobility of 1400 cm^2/V-s, and an effective mass of 0.1. These properties represent the best InGaAs plasma filter fabricated to-date. This figure illustrates that these InGaAs plasma filters demonstrate high below bandgap reflection and negligible above bandgap absorption for films 1-2 microns in thickness.

These films coupled with an appropriate non-absorbing interference filter yield a highly effective tandem filter concept for TPV spectral control. Figure 9 illustrates the reflection versus wavelength for an initial tandem filter concept. For this concept, an interference filter was directly deposited on the InGaAs plasma and an anti-reflection (AR) coating was deposited on the backside of the InP substrate, in order to maximize above bandgap transmission. This filter demonstrates a spectral utilization factor of ~65% for a 2000 F blackbody radiator [19].

Conclusions

The low effective mass of degenerately-doped n-$In_xGa_{1-x}As$ makes this material an ideal plasma filter for TPV applications. The effective photonic bandgap nearly doubles as the free electron concentration increases from 1×10^{17} to 5×10^{19} cm^{-3}. This has tremendous impact on the design of many mid-infrared optoelectronic devices, including thermophotovoltaics. Recently, the

Moss-Burstein shift was used to design an improved InGaAs resonant cavity photodiode [20]. For TPV devices, this effect may be utilized to create transparent emitter or cap layers for n-on-p devices, eliminating the need for front surface passivation; transparent graded layers for p-on-n devices that utilize a back surface reflector [21]. In addition, it was found necessary to utilize the effective optical bandgap of n^+ InGaAs in order to effectively model the performance of SPIRE-fabricated TPV devices. Finally, the low effective mass ensures high values of mobility even at carrier concentrations exceeding 1×10^{19} cm^{-3}. This sharp plasma resonance due to these high mobilities ensures very low parasitic absorption in the above bandgap region (1.0 - 2.2 μm wavelength) and very high reflection (>95%) just beyond the plasma turn-on. The InGaAs plasma filter was successfully coupled to a non-absorbing interference filter that yielded a spectral utilization factor of ~65% for an 1100 °C blackbody radiator.

References

[1] G.W. Charache, et al, "Current Status of Low-Temperature Radiator Thermophotovoltaic Devices," in *Proceedings of the 25th IEEE PVSC*, 1996, pp. 137-141.

[2] E. Burstein, "Anomalous Optical Absorption Limit in InSb," *Phys. Rev.* **92**, 632 (1954).

[3] Y.B. Li et al, "Infrared Reflection and Transmission of Undoped and Si-doped InAs Grown on GaAs by Molecular Beam Epitaxy," *Semicond. Sci. Technol.*, **8**, 101 (1993).

[4] D. Hahn, et al, "Electron Concentration Dependence of Absorption and Refraction in n-$In_{0.53}Ga_{0.47}As$ Near the Bandedge," *Jl. Electron. Mat.*, **24**, 1357 (1995).

[5] E.O. Kane, "Band Structure of Indium Antimonide", J. Phys. Chem. Solids, **1**, 249 (1957).

[6] M. Weyers, et al, "Gaseous Dopant Sources in MOMBE/CBE", *Jl. Crystal Growth*, **105**, 383 (1990).

[7] S.Z. Sun, et al, "Zinc and Tellurium Doping in GaAs and AlGaAs grown by MOCVD," *Jl. Crystal Growth*, **113**, 103 (1991).

[8] J. Musolf, et al, "Doping of GaAs and InP in MOMBE Using DEZn, TESn, and DETe," *Jl. Crystal Growth*, **107**, 1043 (1991).

[9] Y.M. Houng and T.S. Low, "Te-doping of GaAs and AlGaAs using DETe in LPOMVPE," *Jl. Crystal Growth*, **77**, 272 (1986).

[10] C. R. Abernathy, et al, "Sn Doping of GaAs and AlGaAs by MOMBE Using Tetraethyltin," *Jl. Crystal Growth*, **113**, 412 (1991).

[11] M.B. Panish, et al, "Very High Tin Doping of GaInAs by Molecular Beam Epitaxy," *Appl. Phys. Lett.*, **56**, 1137 (1990).

[12] S.L. Jackson, et al, "Silicon Doping of InP, GaAs, InGaAs and InGaP Grown by Gas Source and Metalorganic Molecular Beam Epitaxy Using a $SiBr_4$ Vapor Source," *in Proceedings of the InP and Related Materials Conference*, 1994, pp. 57-60, "High Efficiency Silicon Doping of InP and InGaAs in Gas Source and Metalorganic Molecular Beam Epitaxy using Silicon Tetrabromide," *Appl. Phys. Lett.*, **64**, 2867 (1994).

[13] C.J. Pizone, et al, "Tin-doped n^+-InP and GaInAs Grown by AP-MOCVD," *Electron. Lett.*, **25**, 1315 (1989). C.J. Pizone et al, "Heavily-doped n-type InP and InGaAs Grown by Metalorganic Chemical Vapor Deposition Using Tetraethyltin," *Jl. Appl. Phys.*, **67**, 6823 (1990).

[14] T. Fujii, et al, "Heavily Si-Doped InGaAs Lattice Matched to InP Grown by MBE," *Electron. Lett.*, **22**, (1986).

[15] V.M. Glazov, et al, "Investigation of the Relationship Between the Electron Density and Solubilities of Sulfur, Selenium and Tellurium in Indium Arsenide," Sov. Phys. Semicond., **10**, 378 (1976).

[16] S. Tiwari and D.J. Frank, "Empirical Fit to Band Discontinuities and Barrier Heights in III-V Alloy System", *Appl. Phys. Lett.*, **60**, 630 (1992).

[17] M. Levenshtein et al, Eds., *Handbook Series on Semiconductor Parameters vol. 1*, , New York: World Scientific, 1996.

[18] S. Adachi, "Bandgaps and Refractive Indices of AlGaAsSb, GaInAsSb, and InPAsSb: Key Properties for a Variety of the 2-4 μm Optoelectronic Device Applications", *Jl. Appl. Phys.*, **61**, 4869 (1987).

[19] P.F. Baldasaro et al, "Experimental Assessment of Low-Temperature Voltaic Energy Conversion," *AIP Conference Proceedings*, **358**, 29 (1995).

[20] S.S. Murtaza et al, "Resonant Cavity Photodiode at 1.55 Microns with Burstein-Shifted $In_{0.53}Ga_{0.47}As$/InP Reflectrors," *Appl. Phys. Lett.*, **69**, 2462 (1996).

[21] G.W. Charache, et al, "Thermophotovoltaic Devices with a Back Surface Reflector for Spectral Control," *AIP Conference Proceedings*, **321**, 371 (1996)

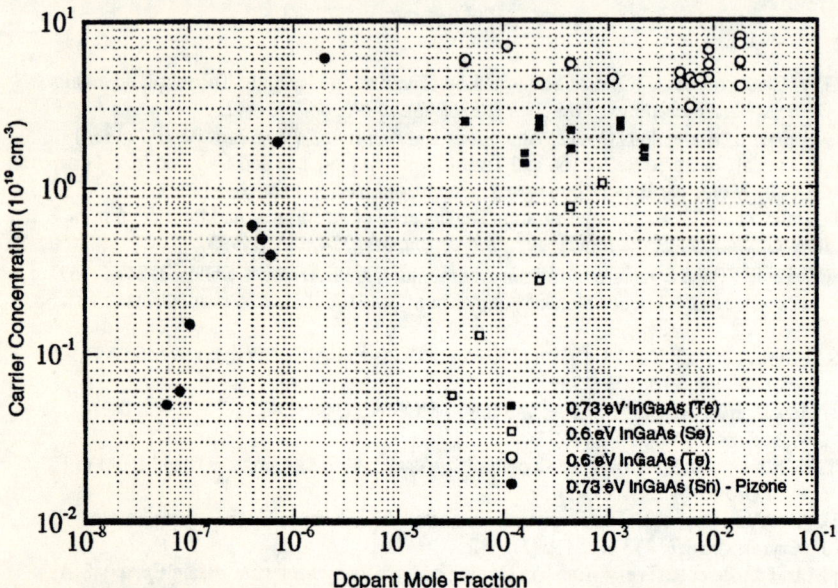

Figure 1 - Measured carrier concentration versus dopant mole fraction for Se- and Te-doped $In_xGa_{1-x}As$.

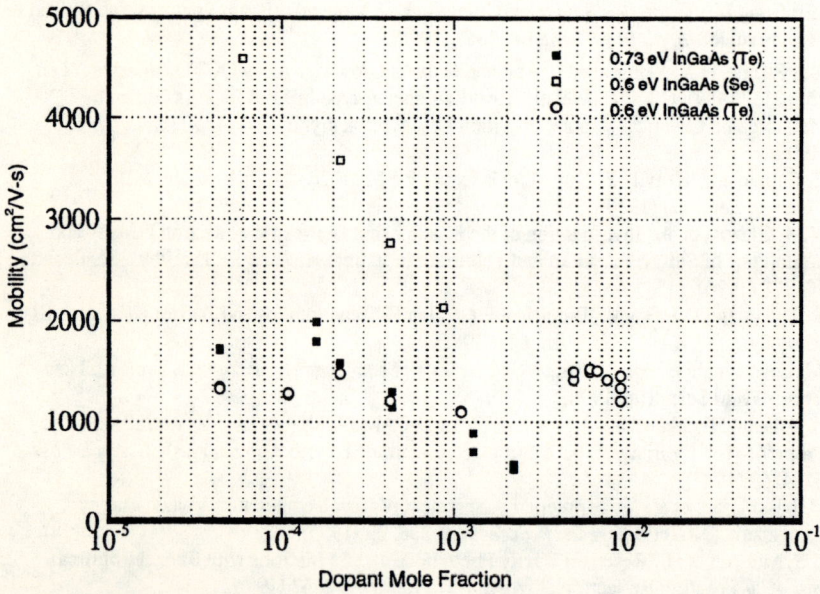

Figure 2 - Measured carrier mobility versus dopant mole fraction for Se- and Te-doped $In_xGa_{1-x}As$.

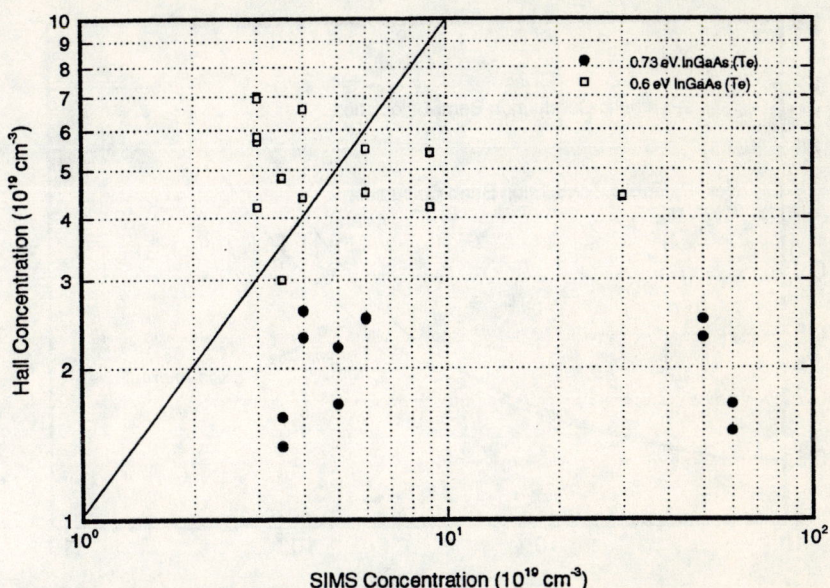

Figure 3 - Electrical activation of tellurium dopant in $In_xGa_{1-x}As$.

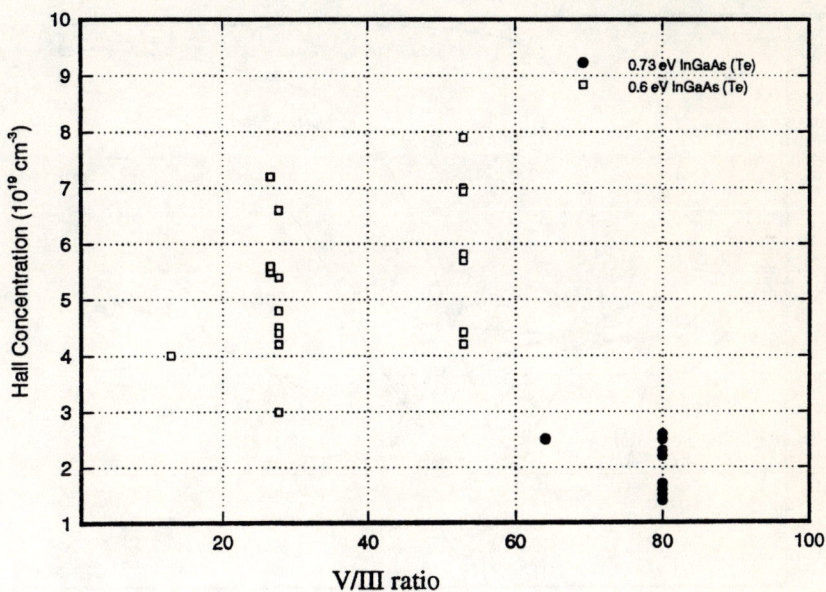

Figure 4 - Dependence of the V/III ratio on the electrically active carrier concentration in $In_xGa_{1-x}As$.

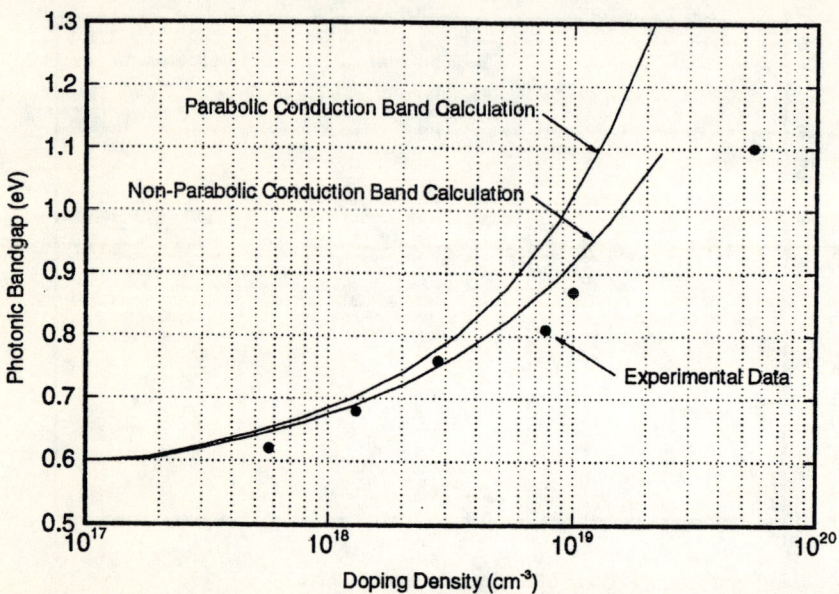

Figure 5 - Theoretical and measured values of the effective photonic bandgap for 0.6 eV InGaAs.

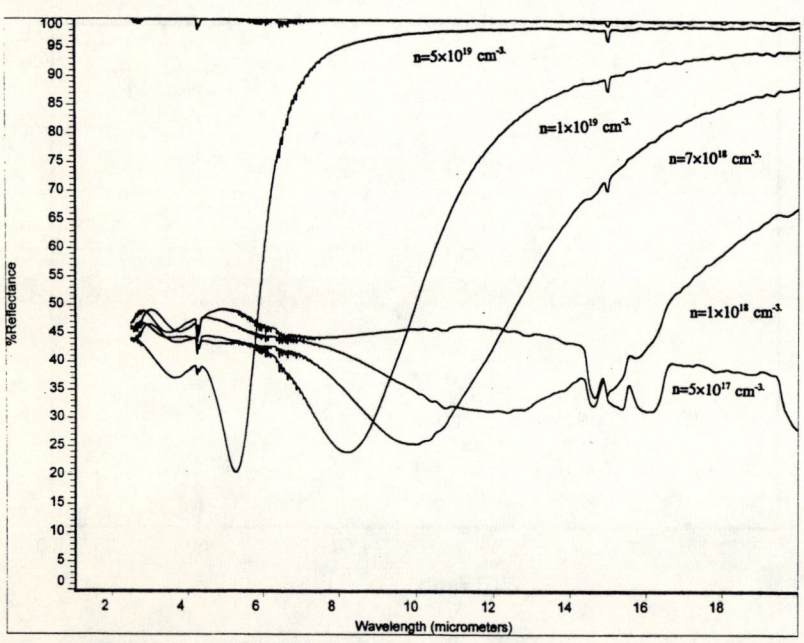

Figure 6 - Measured plasma reflection characteristics of 0.6 eV InGaAs.

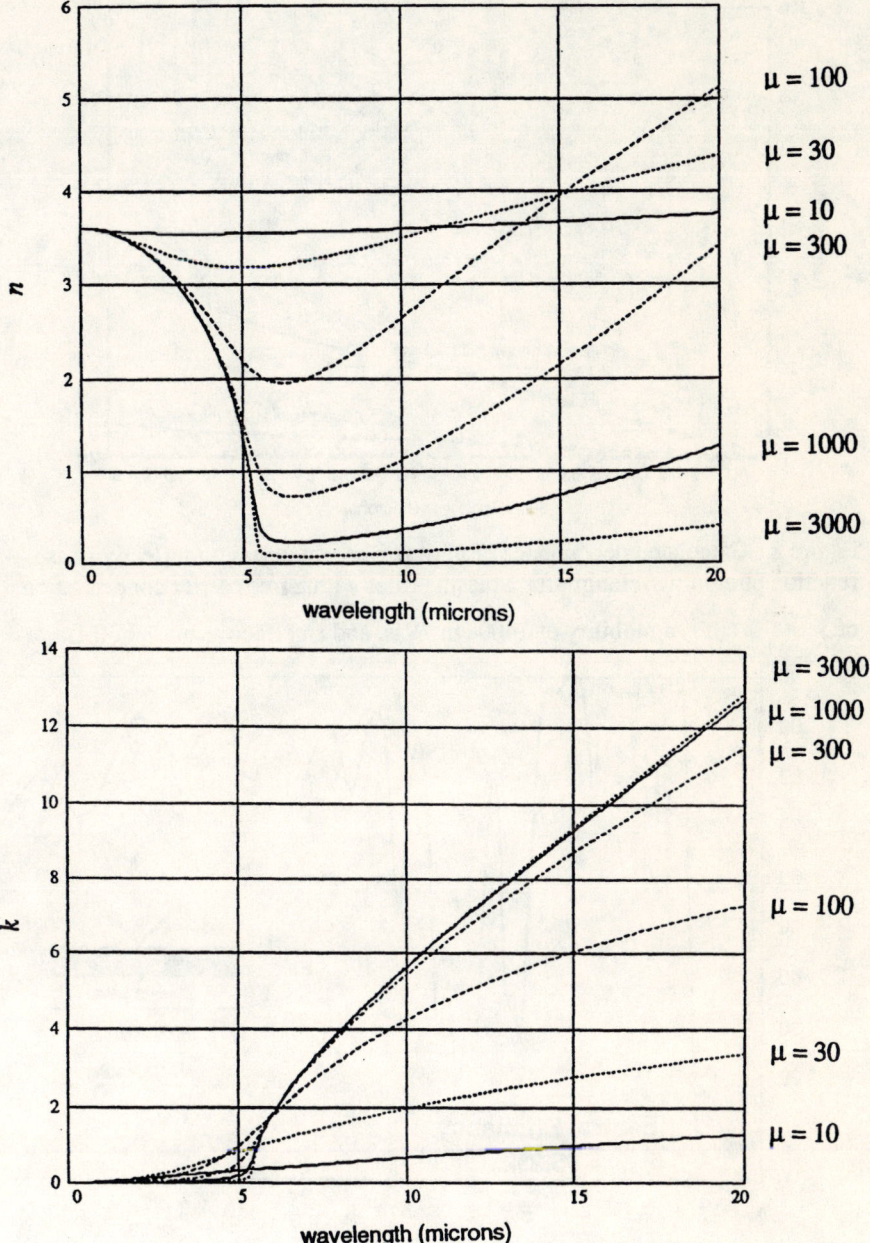

Figure 7 - Calculated values of the real (n) and imaginary (k) components of the refractive index as a function of wavelength and mobility for a free carrier concentration of 5×10^{19} cm^{-3}, $\varepsilon = 12.5$, $m^* = 0.1$, and $n_i = 1$ (i.e., air).

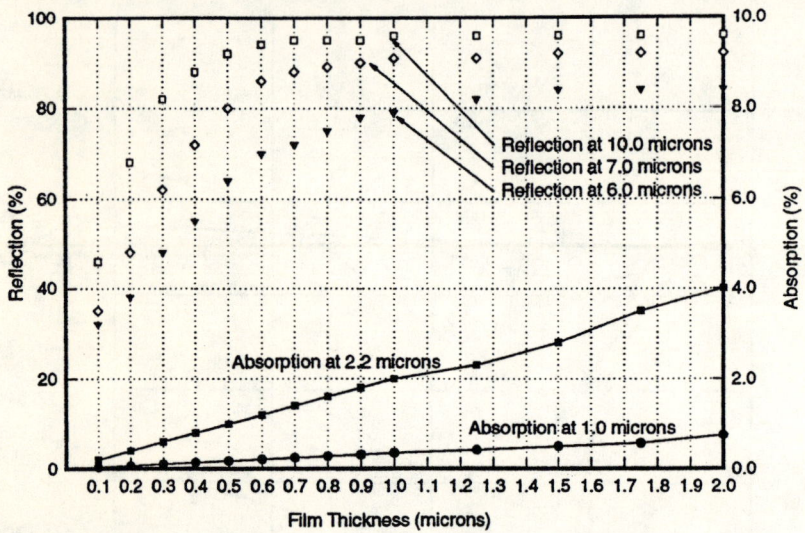

Figure 8 - Calculated dependence of plasma filter thickness on reflectivity as function photon wavelength for a plasma filter with a free carrier concentration of 5×10^{19} cm^{-3}, a mobility of 1400 cm^2/V-s, and an effective mass of 0.1.

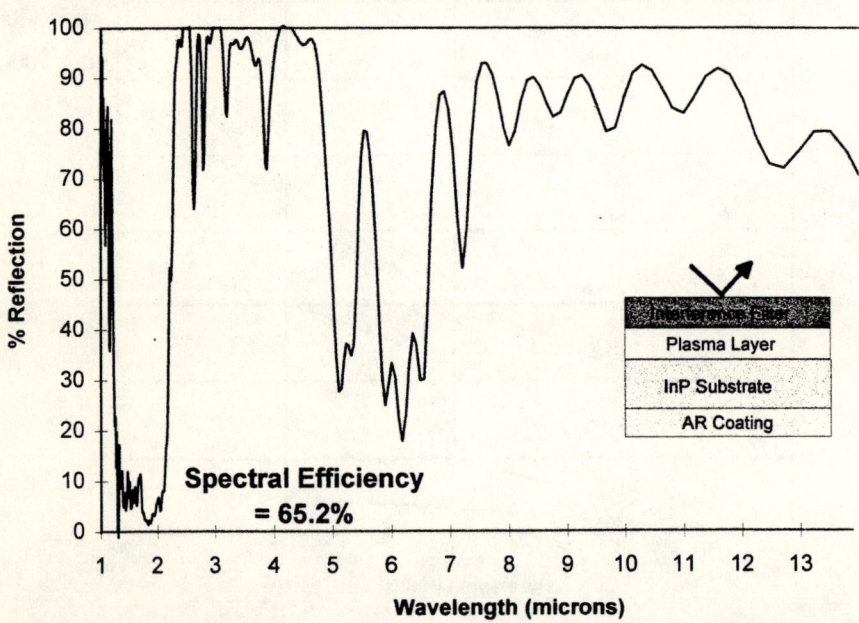

Figure 9 - Measured reflection versus wavelength for a 0.6 eV InGaAs plasma / non-absorbing interference filter tandem filter.

A Novel Design for Monolithically Interconnected Modules (MIMs) for Thermophotovoltaic Power Conversion

J. S. Ward, A. Duda, M. W. Wanlass, J. J. Carapella, X. Wu,
R. J. Matson, T. J. Coutts, and T. Moriarty

National Renewable Energy Laboratory, Golden CO

C. S. Murray and D. R. Riley

Westinghouse, Bettis Atomic Power Laboratory, Pittsburgh, PA

Abstract

The design for the fabrication of monolithically interconnected modules (MIMs) for thermophotovoltaic (TPV) power conversion described in this paper utilizes a novel, interdigitated contacting scheme that increases the flexibility in the size of the component cells, and hence, the output current and voltage of the module. This flexibility is gained at the expense of only minimally increased grid obscuration. Because the design uses the grid fingers of the component cells as the interconnect structure, the area of the device used for this purpose becomes negligible. In this paper, we report on the specifics of the design and on issues related to the fabrication of the modules. Preliminary performance data for representative modules are also offered.

CP401, *Thermophotovoltaic Generation of Electricity: Third NREL Conference*,
edited by Benner/Coutts
© 1997 The American Institute of Physics 1-56396-734-0/97/$10.00

Introduction

Thermophotovoltaic (TPV) systems generate electricity by the direct photovoltaic conversion of photons emitted from a radiant heat source. Monolithically interconnected modules (MIMs) are being developed in the GaInAs/InP material system for TPV applications because of several potential advantages (1). Firstly, small, series-connected devices provide a means of increasing the voltage and decreasing the current generated per unit area of the module. Secondly, both electrical contacts are made on the front surface of the MIM; therefore, free-carrier absorption is decreased with the use of semi-insulating substrates and back-surface reflectors (BSRs). Thirdly, the fabrication of MIMs can simplify the assembly of arrays and offer mechanical and thermal coupling advantages.

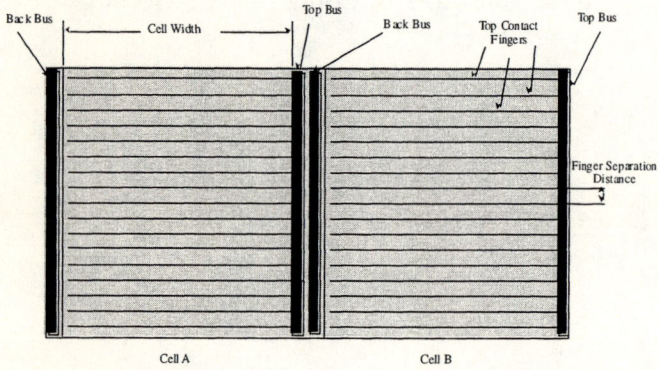

Figure 1:   Plan view of the conventional MIM design (interconnect omitted)

The conventional approach to MIM fabrication is to use a front-surface grid structure (Figure 1) for the component cells, while relying on a low-resistivity back-contact layer for lateral current transport to a single back-contact terminal (Figure 2). For efficient TPV system operation, it is necessary to return sub-bandgap photons to the radiator, which can be achieved with a BSR. Unfortunately, the low-resistivity back-contact layer necessarily absorbs a significant portion of the sub-bandgap photons. The thickness and doping level of the back-contact layer determine the balance between power losses due to free-carrier absorption of sub-bandgap photons and spreading resistance. A model is being developed that will allow one to minimize the sum of these two power-loss terms.

Because there is a great variety of potential TPV system configurations, it is important that the basic converter design be as flexible as possible in terms of geometry and output parameters (i.e., operating voltage and current). In the conventional approach to MIM design, the sheet resistance of the back-contact layer determines the maximum allowable component cell width, and therefore, the output voltage of the array per unit length. If a greater cell width is required, the thickness of the back contact layer must be increased to reduce the sheet resistance. A thicker back-contact layer results in enhanced absorption of sub-bandgap photons by free-carriers. For some requirements, the conventional approach may prove to be satisfactory. Clearly, when more flexibility and higher TPV system efficiency is required, an alternative design must be developed.

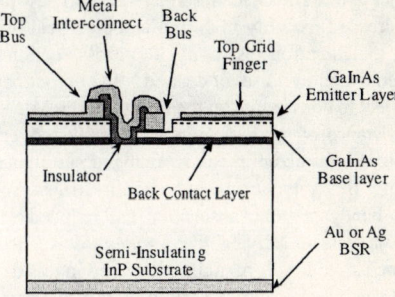

Figure 2:   Cross-section of conventional MIM design for lattice-matched GaInAs TPV converter structures grown on InP substrates

A New Approach to MIM Design

The primary goal of the new MIM design is to gain greater control of the output parameters of the module, while simultaneously increasing the output power density and reducing I^2R and free-carrier absorption losses. In addition, the design needs to be easy to fabricate with a minimum of individual process steps. We realized these goals with a device design that used interdigitated front and back contacts and a novel interconnect scheme that minimized the loss of active area by using the grid fingers of the component cells as the interconnect structure. This technique is referred to as a Grid-Finger Interconnect (GFI). Figure 3 is a simplified plan view of this design.

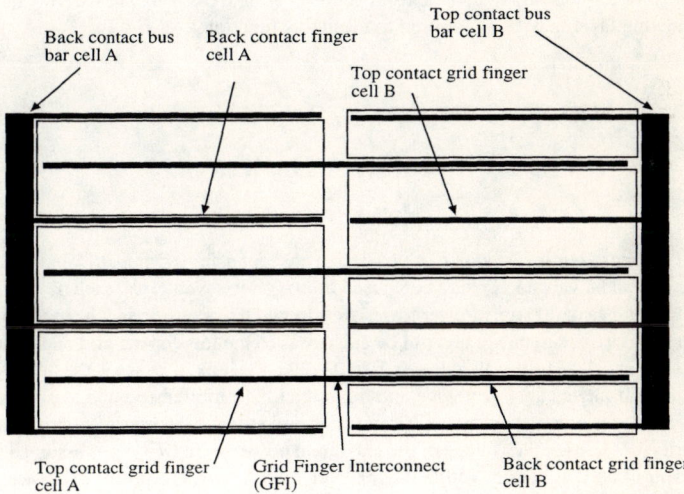

Figure 3:   Simplified plan view of the grid-finger interconnect (GFI) MIM concept.

229

The gridded back contact (GBC) used in the GFI MIM design provides natural solutions to our primary design objectives. The GBC introduces several new parameters to the MIM design that can be exploited to increase flexibility in design and electrical output, while reducing losses. For example, the spacing, metallization resistivity, and cross-sectional area of the grid fingers in the GBC can all be adjusted to achieve a set of desired MIM output parameters under particular operating conditions and subject to constraints on the design of the MIM device structure (e. g., a specific value for the sheet resistance of the back-contact layer to reduce free-carrier absorption). Thus, output flexibility and loss minimization can be realized simultaneously. The above is simply not possible in general using the conventional approach to MIMs. With the GBC, GFI MIM, output flexibility, reduced sub-bandgap optical absorption, and reduced electrical losses are all gained at the expense of a marginal increase in active converter area lost due to the presence of the GBC. However, the abovementioned areal loss is partially recovered in this design because the GFI requires no dedicated interconnect structure because the grid fingers of the component cells are used to form the series connections to complete the module.

The interdigitated GFI MIM device may be fabricated with a simple processing sequence composed of the four basic steps outlined below:

1. <u>Back-contact etch</u>: trenches for the back-contact grid are etched into the device structure, stopping at the surface of the back-contact layer.

2: <u>Cell-isolation etch</u>: a pattern defining the perimeter of the component cells is etched into the device structure to the surface of the insulating substrate.

3: <u>Insulator deposition and patterning</u>: SiO_2 is deposited on the sample via CVD and patterned to allow contact by the metallization, where appropriate.

4: <u>Metallization</u>: a lift-off procedure is used to electron-beam (e-beam) evaporate the top-contact grid structure, the back-contact grid structure, and the interconnects simultaneously.

A final step for the deposition of an anti-reflection coating (ARC) may be used or, alternatively, the SiO_2 insulating layer may be engineered to fulfill this function.

Prototype Devices

We began our investigation of this design concept using a symmetric, interdigitated contact structure. It can be viewed as being composed of two interwoven grids, each of which has a finger separation distance of 200 μm. Power losses are computed by considering the back-contact grid and the low-resistivity back-contact layer to be analogous to the top-contact grid and emitter layer of a traditional solar cell. The metallization is Ti/Pd/Ag/Pd, the bulk of which is the thick Ag conduction layer. The grid fingers are 7 μm wide and 6 μm thick. To limit absorption losses, the emitter and back-contact layer's sheet resistance (R_s) were modeled at 100 ohms/square. For the lattice-matched composition of GaInAs, the voltage of the maximum power point (V_{mpp}) was assumed to be 360 mV, independent of current density within the range of 1 - 5 A/cm². The power losses were found to be dominated by the spreading resistance in the emitter/back-contact layers and joule losses in the metal fingers. Figure 4 is a graphical representation of these two power-loss terms for one of the grid structures as a function of current density. It can be seen that within the current density range of 1 - 3 A/cm^2, the expected power losses for this design are relatively modest.

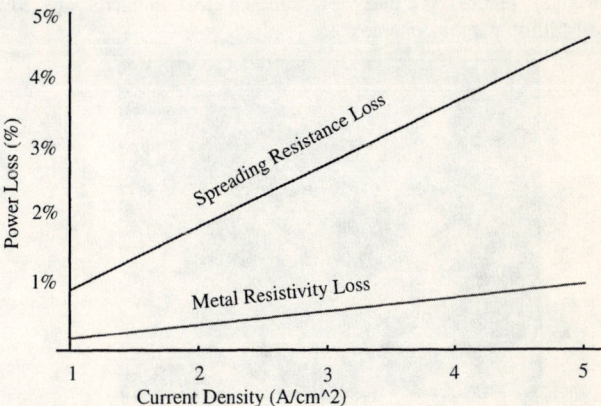

Figure 4:   Modeled power losses for the prototype 1-cm^2 8-cell GFI MIM design as a function of current density for R_s = 100 ohms/square.

We have chosen a 1-cm^2 8-cell module to facilitate comparisons to alternative approaches, but note that this design is fully scaleable without changing the power-loss analysis. Figure 5 is a representation of the prototype 8-cell, 1-cm^2 GFI MIM. In this illustration, every tenth grid finger is shown for clarity. The interconnects are circled and connect back-contacts of the component cell on the left to the top-contacts of the component cell to its right.

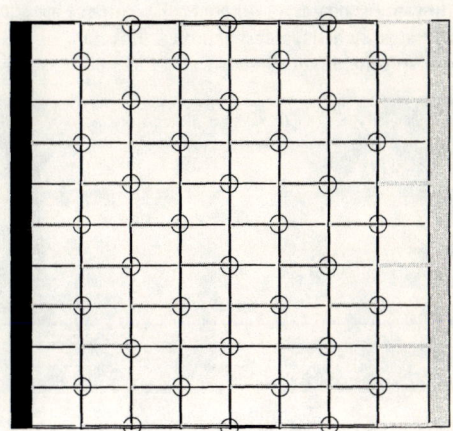

Figure 5:   Plan view of the prototype 1-cm^2, 8-cell GFI MIM design.

To date, we have used this design with n/p and p/n GaInAs device structures lattice-matched to InP ($E_g = 0.74$ eV). All of the device structures were grown by atmospheric-pressure metalorganic vapor phase epitaxy in a vertical reactor. Details of the growth system and device structures have been published previously (2). Figure 6 is a plan view scanning electron microscope (SEM) image of a completed, GFI MIM illustrating the interconnect.

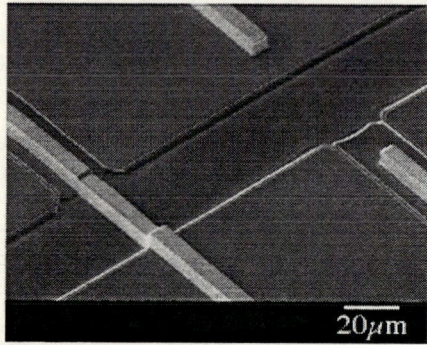

Figure 6:   SEM image of a GFI MIM illustrating the interconnect.

There are two primary ways in which the fabrication of this device differs from that of a simple solar cell. The first is the fact that the metallization forming the interconnect is of a multi-level nature. The second is the use of a deposited dielectric layer as an insulator. The success of this design depends on having the capability to deposit a metal grid finger that can continuously bridge the back-contact region of one cell to the top-contact region of the adjacent cell. This is accomplished by exercising control over the sidewall profile of the etched features.

We routinely employ selective, wet-chemical etchants for the InP/GaInAs material system. HCl is used to etch InP, whereas an etchant composed of $3H_3PO_4:4H_2O_2:1H_2O$ (3:4:1) is used to etch GaInAs. The devices are grown on clockwise-polished, Fe-doped, semi-insulating InP substrates with a (100) orientation 2° toward the nearest (110) plane. Experience has shown that perpendicular to the major flat of these substrates, etched features exhibit a "dovetail" profile. Parallel to the major flat, they exhibit a "V groove" profile. Figure 7 shows an SEM micrograph of the sidewall profiles

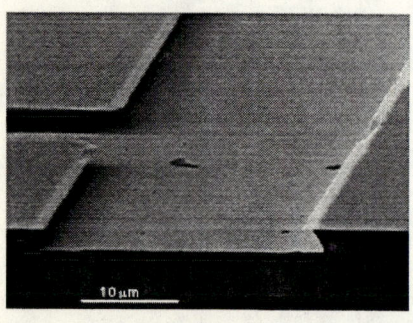
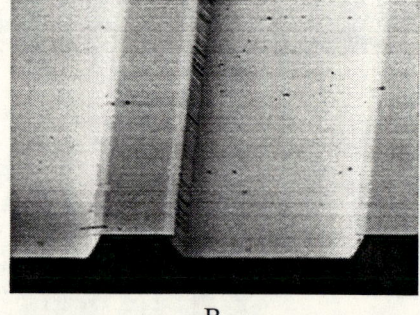

A   B

Figure 7:   Sidewall profiles obtained in GaInAs. A) Dovetail perpendicular to the major
flat of the substrate, and B) V-groove parallel to the major flat.

232

obtained by wet-chemical etching of GaInAs in the 3:4:1 etchant. The "V groove" profile is required to facilitate a continuous metal interconnect.

Figure 8 is an illustration of what can happen to the interconnect if the orientation, and hence, the sidewall profile, is not taken into consideration. In this case, the overhang of the "dovetail" profile causes a discontinuity in the interconnect structure. The desired orientation and result is illustrated in Figure 9. In this case, the etched feature was formed parallel to the substrate's major flat, resulting in a "V groove" etch profile.

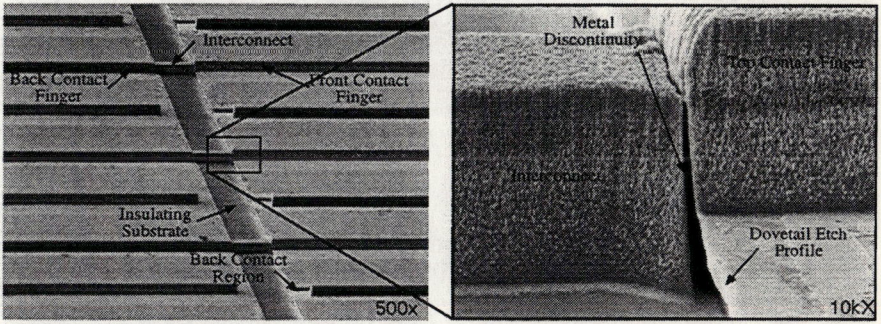

Figure 8:   Discontinuous interconnect caused by improper substrate orientation.

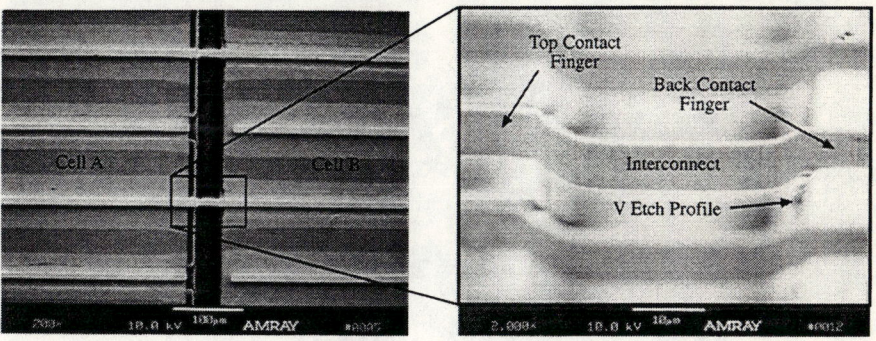

Figure 9:   Continuous interconnect with proper substrate orientation.

The integrity of the SiO_2 insulating layer is critical to successfully fabricating of this device. We are currently depositing 2000 Å of SiO_2 via low pressure chemical vapor deposition (LPCVD). Figure 10 is a cross-sectional SEM image of the SiO_2 converage of the "V-groove" etch profile. This material is of very high quality, with a measured breakdown voltage in excess of 1×10^7 V/cm. Pinhole density was determined to be the primary problematic aspect of the material. Control of pinholes was determined to be a function of proper cleaning of the deposition system and placement of the sample within the system.

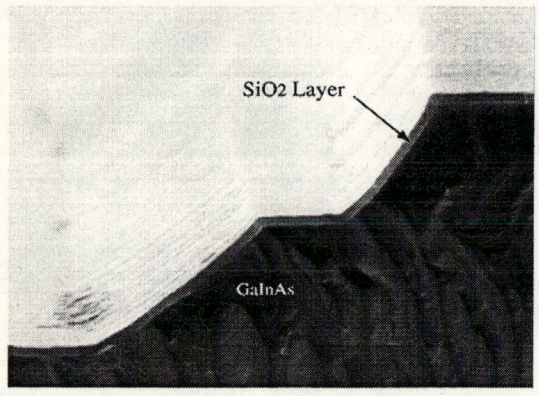

Figure 10:   Cross-sectional SEM image illustrating SiO$_2$ coverage of etched features.

Results and Discussion

Figure 11 is a photo of a completed 3 x 3 array of 1-cm^2 GFI MIMs. This array was fabricated on a 2-inch-diameter InP wafer. The nine modules may be series connected to form an interconnected array with an open-circuit voltage (V_{oc}) in excess of 30 V. The loss of active area associated with the implementation of this design is less then 15%, excluding the bus bars.

Figure 11:   Array of nine 1-cm^2 GFI MIMs fabricated on a 2-inch-diameter InP wafer.

The electrical performance characteristics of representative p/n, 0.74-eV, GaInAs/InP MIMs were measured as a function of operating temperature under spectral conditions simulating an ideal, 1000°C blackbody at full intensity. Results of these measurements for three different temperatures are given in Figure 12. The photovoltaic output parameters for each case are listed in Table 1. The

234

current density refered to in the table is the current density generated within the component cells of the module. The power density output in the table is for the module. The series-connected, multi-cell nature of the MIM are quite apparent from the high measured open-circuit voltages. As the operating temperature is increased, the MIMs show increased short-circuit current densities and lower open-circuit voltages and fill factors. The short-circuit current density increase arises from a negative temperature coefficient of the GaInAs bandagap, which allows the MIM to absorb a larger fraction of the available blackbody photon spectrum as the operating temperature is increased. The reverse-saturation current density of the p/n GaInAs converter diodes also increases with temperature, thus reducing the open-circuit voltages and fill factors of the individual cells in the MIM. The measured maximum output power densities for the MIMs are quite encouraging. Values approaching 0.3 W/cm² were measured for the lower operating temperatures. The MIMs did not have anti-reflection coatings (ARCs) applied, so we can expect power-density increases of ~30% once well-

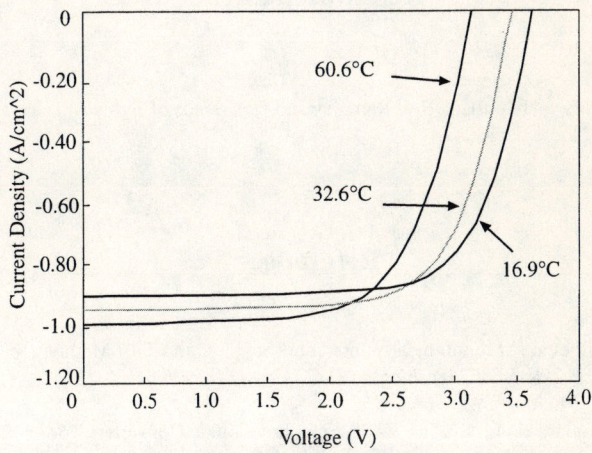

Figure 12:   J-V characteristics of a prototype GFI MIM.

Temp. (°C)	V_{oc} (V)	J_{sc}/cell (A/cm²)	Fill Factor (%)	P_{max}/module (W/cm²)
16.9	3.62	0.907	72.04	0.292
32.6	3.47	0.948	70.40	0.284
60.6	3.14	0.991	67.07	0.257

Table 1:   Photovoltaic output parameters of a prototype GFI MIM.

Conclusions

We have developed a new design for fabricating monolithically interconnected modules (MIMs) that addresses some of the limitations of the conventional approach. This design provides a means of achieving the flexible output (current and voltage) that will be required for many applications. It can be implemented with a relatively simple processing regime composed of four basic steps. The performance of these modules is encouraging, and further research aimed at increasing their output power density is under way.

Acknowledgments

The authors wish to thank Ron Ramaley and Don Seeds of Bettis APL for assistance in characterization of these devices.

References

1.   D. M. Wilt et al., "Monolithically Interconnected InGaAs TPV Module Development," *Proc. IEEE 25th PVSC, Washington D.C., May 13-17 1996.*

2.   M. W. Wanlass et al., "$Ga_xIn_{1-x}As$ Thermophotovoltaic Converters," *Solar Energy Materials and Solar Cells,* 41/42, pp. 405-417, 1996.

Electrical and Optical Performance Characteristics of 0.74 eV p/n InGaAs Monolithic Interconnected Modules

David M. Wilt[*], Navid S. Fatemi[**], Phillip P. Jenkins[**],
Victor G. Weizer[**], Richard W. Hoffman, Jr. [**], Raj K. Jain[†],
Christopher S. Murray[‡] and David R. Riley[‡]

[*]*NASA Lewis Research Center, Cleveland, Ohio*
[**]*Essential Research, Inc., Cleveland, Ohio*
[†]*National Research Council*
[‡]*Westinghouse Electric Corporation, West Mifflin, PA*

Abstract. There has been a traditional trade-off in thermophotovoltaic (TPV) energy conversion development between system efficiency and power density. This trade-off originates from the use of front surface spectral controls such as selective emitters and various types of filters. A monolithic interconnected module (MIM) structure has been developed which allows for both high power densities and high system efficiencies. The MIM device consists of many individual indium gallium arsenide (InGaAs) cells series-connected on a single semi-insulating indium phosphide (InP) substrate. The MIM is exposed to the entire emitter output, thereby maximizing output power density. An infrared (IR) reflector placed on the rear surface of the substrate returns the unused portion of the emitter output spectrum back to the emitter for recycling, thereby providing for high system efficiencies.

Initial MIM development has focused on a 1 cm^2 device consisting of eight series intercon-nected cells. MIM devices, produced from 0.74 eV InGaAs, have demonstrated V_{oc}=3.2 V, J_{sc}=70 mA/cm^2 and a fill factor of 66% under flashlamp testing. Infrared (IR) reflectance measurements (>2 μm) of these devices indicate a reflectivity of >82%. MIM devices produced from 0.55 eV InGaAs have also been demonstrated. In addition, conventional p/n InGaAs devices with record efficiencies (11.7% AM0) have been demonstrated.

INTRODUCTION

In thermophotovoltaic (TPV) energy conversion, an emitter is heated to incandescence and a photovoltaic device is placed in view of the emitter to convert the radiant energy into electrical energy. Research in TPV has been renewed recently due to the development of new emitter, filter and photovoltaic cell technologies (10). Most current efforts in TPV research have concentrated on using front surface spectral control elements such as selective emitters (1) or graybody

CP401, *Thermophotovoltaic Generation of Electricity: Third NREL Conference,*
edited by Benner/Coutts

emitters combined with plasma, dielectric or dipole filters (2, 3) in order to improve system efficiency to the 20–40% range predicted by theory.

The front-surface spectral control approach generally produces systems with low power densities (W/cm^2). Selective emitters, for example, have demonstrated in-band emittances ranging from 0.7 to 0.8 (4), with efficiencies of ~40% (i.e. 40% of the emitted energy is convertible by the photovoltaic device). In order to recuperate the non-convertible energy, filters are used to reflect the long-wavelength photons back to the selective emitter. Unfortunately, there are no filters available which provide both 100% transmission in the usable wavelength region and 100% reflection elsewhere. Thus, a selective emitter emittance of 0.8 coupled with a typical filter transmission of 80% leads to a reduction in the power density of 36%. This is an expensive loss, particularly given the cost of TPV cells. A graybody-emitter based system using the same filter would show a similar, although smaller reduction in power density.

A different approach involves the use of rear-surface spectral controls. Using this technique, the entire radiant output from the emitter is incident upon the photovoltaic (PV) device, thereby providing high output power densities. Photons which the PV device is unable to convert, pass through the cell structure and reflect off of a rear reflector back to the emitter for recycling. Researchers have developed TPV cells which utilize low-doped substrates and reflective rear contacts to provide photon recycling (5, 6). Other researchers have developed series-interconnected, monolithic cells for laser, fiber-optic and TPV applications (7, 8). We are developing a cell which combines the advantages of both of these approaches (11).

The Monolithic Interconnected Module or MIM consists of series-connected indium gallium arsenide (InGaAs) devices on a common, semi-insulating indium phosphide (InP) substrate (Fig. 1). An infrared reflector is deposited on the rear surface of the InP substrate to reflect photons back toward the front surface of the cell. This provides a second pass opportunity for photons capable of being converted by the cell. In addition, long wavelength photons are returned to the emitter for "recycling", improving the system efficiency.

The MIM design offers several advantages. Firstly, small series-connected cells provide high voltages and low currents, reducing I^{2}R losses. In addition, the small size of the cells permits an array to be comprised of series/parallel strings rather than a single series-connected string of larger cells. This should improve the reliability of the TPV module since the failure of a single cell would not debilitate the entire array. In addition, the cell size and distribution may be easily adjusted to minimize the losses associated with emitter non-uniformity (i.e. variation in view factor, temperature, etc.).

Secondly, the MIM design maximizes output power density since losses associated with front-surface spectral controls are eliminated. This represents a significant simplification of TPV system design and thermal management since there are no filters to cool. Thirdly, the rear surface of the device is not electrically active, therefore the cell may be directly bonded to the substrate/heat sink without concern for electrical isolation. This greatly simplifies the array design and improves the thermal control of the cells. Lastly, photons which are weakly absorbed have the possibility of multiple passes through the cell structure. This feature is particularly important for lattice-mismatched devices, where poor minority carrier diffusion length can be partially offset by making the cell thin, forcing the carrier generation to occur closer to the p/n junction.

Although the MIM design has many beneficial attributes, there are limitations. The device is produced on an InP substrate using organo-metallic vapor phase epitaxy (OMVPE) growth techniques and as such may be too expensive for many commercial applications. The simplification of array fabrication may partially offset the higher cost of the MIM's compared to conventional TPV devices.

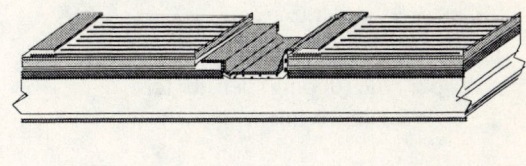

a

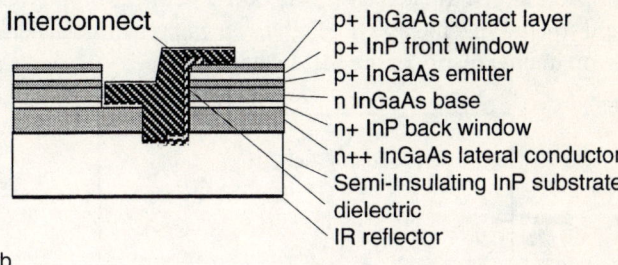

Interconnect
p+ InGaAs contact layer
p+ InP front window
p+ InGaAs emitter
n InGaAs base
n+ InP back window
n++ InGaAs lateral conductor
Semi-Insulating InP substrate
dielectric
IR reflector

b

Figure 1. a) A 3-dimensional view of two cells of a MIM. b) A cross-sectional view of a MIM showing the individual layers.

MIM DEVELOPMENT

Optical Development

Successful development of the MIM device requires balancing trade-offs between optical performance (mid IR reflectivity) and electrical performance. To address the optical performance issues, the free carrier absorption (FCA) for both n and p-type InGaAs as a function of dopant type, level, thickness and wave-length was determined. Calibration samples with doping levels ranging from 5×10^{18} to 3×10^{19} cm^{-3} were fabricated on semi-insulating InP substrates. Absorption measurements were conducted using a spectrophotometer for the near IR (1–3 μm) and a FTIR for the mid IR (3–10 μm). The spectrophotometer data was fitted to determine the actual absorption for a single pass through the material (i.e. the measured data was corrected for reflection at the air/front-surface interface, epi/substrate interface and the rear substrate/air interface) (Fig. 2). The corrected data was fitted to the following equation:

$$\text{absorption } (\lambda) = 1 - \exp(-\alpha(\lambda)t) \tag{1}$$

where:

$$\alpha\,(\lambda) = (C(\lambda)n) \text{ for n-type material} \tag{2}$$

$$\alpha\,(\lambda) = (C(\lambda)p) \text{ for p-type material} \tag{3}$$
$$n = \text{electron (doping) density (cm}^{-3})$$
$$p = \text{hole (doping) density (cm}^{-3})$$
$$t = \text{thickness in cm}$$

The analysis indicates that for an equivalently doped InGaAs layer, the p-type material will have a FCA (averaged from 1.9 to 3 microns) 17× higher than the equivalently doped n-type material ($C=7.97\times10^{-17}$cm^2 for p-type InGaAs, $C=4.48\times10^{-18}$cm^2 for n-type InGaAs). This is an important consideration when determining the optimum polarity of the MIM device.

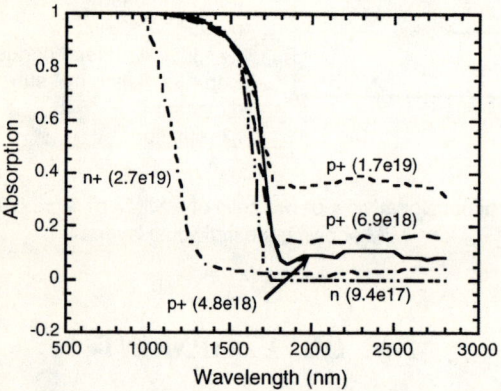

Figure 2. Absorption measurements for 3 μm thick 0.74 eV InGaAs layers with various doping levels.

Another interesting feature of the absorption measurements is the shift in apparent bandgap (0.3 eV) for the heavily doped n-type InGaAs. We have determined that this shift is caused by a Burstein-Moss shift in the degenerately doped material. The use of this material as the lateral conduction layer (LCL) (Fig. 1) allows the base of the cell to be thinned for incomplete absorption. Photons which are not absorbed in the cell on the first pass are able to pass through the LCL, bounce off of the back surface reflector (BSR) and have a second pass through the cell. This approach may not be optimal for 0.74 eV material where the FCA in the n^{++} LCL represents a significant loss (3.6% absorption/pass for 3 μm LCL) with only a minor benefit to the device performance. The technique may be best applied to lattice-mismatched material which suffers from poor minority carrier lifetimes in the base, where reducing the base thickness should increase the current as well as the voltage.

The FTIR data indicated that the n-type material has low absorption up to the plasma frequency which shifts from 7 to 12 microns as the doping density is varied from 3×10^{19} to 8×10^{18} cm^{-3}. The choice of doping density represents another trade-off between FCA and resistive losses. At this point of the MIM development, we are focusing on the electrical development and will address the optimization with respect to optical performance once the fabrication process is well established. The other elements of the optical performance, such as BSR and contact reflectivity and anti-reflection coating development, are covered in a related paper presented at this conference (9).

Electrical Development

The MIM device is currently being developed for use with a low temperature (1200 K) blackbody emitter. Assuming a quantum efficiency of one and a view factor of one, a 0.74 eV device would produce a J_{sc} of 0.87 A/cm^2 and a 0.55 eV device would produce 3.72 A/cm^2. Based on these current densities, cell structures were determined which limit the resistive losses in the LCL and emitter to 1% for each layer. The device structures are shown in Figure 3.

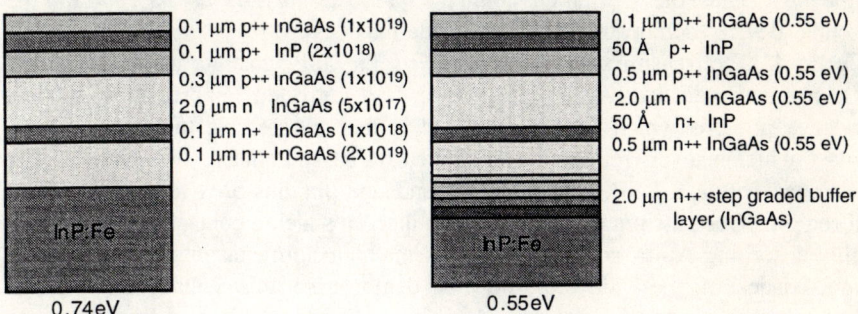

FIGURE 3. 0.74 and 0.55 eV device structures for operation with a 1200 K blackbody.

The thin (0.1 micron) p^{++} InGaAs top layer was added as a contact layer so that non-alloyed ohmic contacts could be used for both the emitter and base contacts.

241

The InGaAs contact layer was removed from between the grid fingers after metallization.

The MIMs reported in this paper were all fabricated as 1 cm^2 devices consisting of eight cells with 300 micron interconnects and 7 micron grid fingers on 100 micron centers. Subsequent processing developments have reduced the interconnect width to 50 microns and the grid finger widths to 5 microns (9). Using these dimensions, a new mask set has been developed to produce 5×5 mm MIMs consisting of eight 500 micron wide cells with the 50 micron interconnects. Test structures have been successfully processed with this design although the results will be presented at a later date.

RESULTS

Conventional planar p/n InGaAs devices were produced using the active cell layers shown in Figure 3 (note: the emitter doping was reduced to 1×10^{18} cm^{-3} for these devices) in order to verify the basic material quality. The I-V curve shown in Figure 4 demonstrates the quality of the baseline devices. The efficiency (11.7% AM0) represents a record for 0.74 eV p/n InGaAs. Calculations indicate that reducing the grid shadowing from the 16% on the test device to the 5% normally used in AM0 devices would increase the efficiency to >13%, a record for any 0.74 eV InGaAs device (p/n or n/p).

The external quantum efficiency for a 0.74 eV baseline device with a dual layer anti-reflective coating is shown in Figure 5. As was stated earlier, the base region was intentionally grown thin so that the effect of the BSR would be demonstrated. It was initially puzzling to observe the high bandedge photoresponse from the conventional cell (with no BSR). Optical modeling indicates that only 62% of the bandedge photons (1600 nm) are absorbed in the thin base region, assuming a single pass. Thus, the internal QE could not be greater than 62%. At 1600 nm the baseline device demonstrated a 74% internal QE (66% external QE, 10% reflection). The transmission characteristic of a n$^+$ InP substrate was measured at 1600 nm and indicated >45% transmission (not corrected for reflection). Thus bandedge photons which are not absorbed in the cell are able to reach the back contact, which is a very reflective non-alloyed Au based contact. It is believed that this contact acts as a BSR, reflecting the bandedge photons back toward the active cell region. Our past p/n devices had all utilized a sintered contact which forms a highly absorbing Au_2P_3 compound at the semiconductor/metal interface. The QE characteristics of these devices did not demonstrate this enhanced bandedge photoresponse. To test this theory, p/n planar cells were fabricated from the same epitaxial InGaAs material with and without sintered back contacts. The QE test data confirmed the reduction in long wavelength response with sintered back contacts compared to non-alloyed contacts.

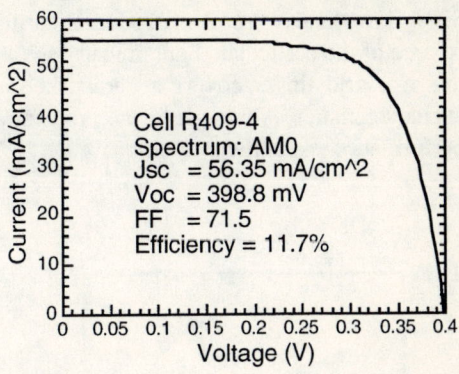

FIGURE 4. AM0 I-V characteristic of 0.74 eV baseline p/n InGaAs device.

A negative aspect of this feature is that the reflection is diffuse in nature. Thus non-convertible photons may be totally internally reflected and add to the thermal load of the cell. A benefit of the diffuse reflection is that convertible photons will generally have a longer path length in the active cell layers, improving the probability for absorption. Given the high absorption coefficient for InGaAs, this is a marginal benefit.

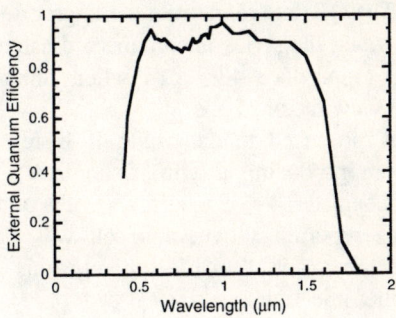

FIGURE 5. External QE of 0.74 eV p/n baseline structure with dual AR.

The I-V curve for a 0.74 eV MIM device is shown in Figure 6 under flashlamp testing. The data indicates an average voltage of 400 mV per cell. This particular device was produced prior to the development of the high quality Si_3N_4 dielectric and therefore is not expected to demonstrate optimum performance. The I-V characteristics of individual cells were examined and found to vary greatly within a single MIM device. The cell characteristics ranged from high quality diodes to heavily shunted or even shorted devices. We believe that this variation is caused by defects in the dielectric isolation layer and/or particulate related defects in the epitaxial material. Again, all of these devices were produced prior to the optimization of the Si_3N_4 layer and were also produced during a major building rehab which was the source of the particulate contaminates in the epitaxial material. We have observed that if a grid finger covers a particulate damaged region, the device shows a shunt characteristic. Although, if the problem area is removed by cleaving, the device performance is significantly improved.

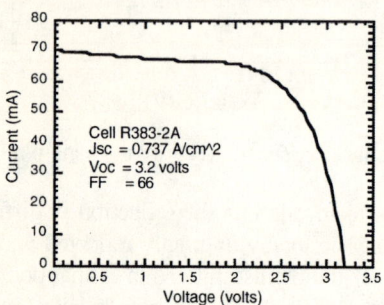

FIGURE 6. I-V characteristic of 0.74 eV MIM under flashlamp testing.

The external QE curve for the 0.74 eV device is shown in Figure 7 (without an anti-reflective coating). The QE data represents the aggregate worst response from across the entire device, given the series interconnected nature of the MIM design. This device is expected to produce 48.5 mA when illuminated by a 1200 K blackbody emitter with a view factor of one.

A 0.55 eV MIM was produced to determine if there were any unforeseen difficulties or problems in producing a MIM from lattice mismatched material. Figure 8 shows the I-V characteristic of a 0.55 eV MIM under AM0 testing. As with the 0.74 eV device reported above, this cell was produced prior to the optimization of the dielectric material. Unfortunately, this device was destroyed prior to I-V testing at higher injection levels.

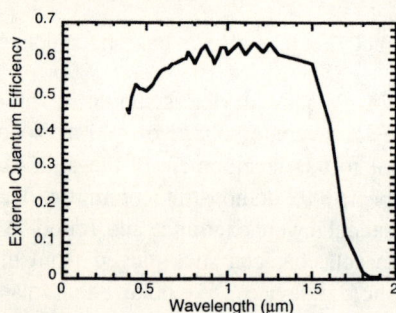

FIGURE 7. External QE of 0.74 eV MIM (no AR).

244

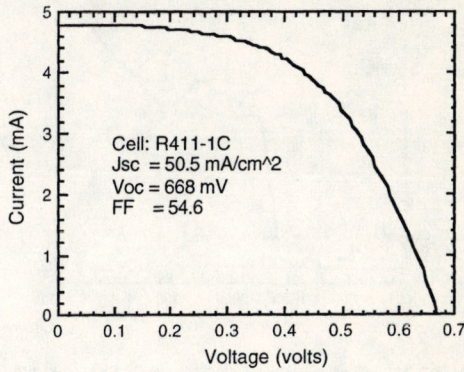

FIGURE 8. AM0 I-V characteristic of 0.55 eV MIM device (no AR).

Figure 9 shows the external QE characteristic for the 0.55 eV MIM (without AR). Given the rudimentary nature of the buffer layer used to produce this device and the limited development of the cell layers, the results were very promising.

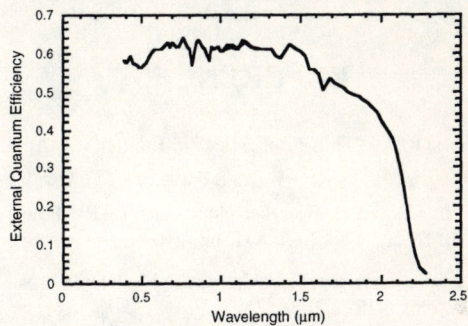

FIGURE 9. External QE of 0.55 eV MIM (no AR).

Figure 10 shows the measured reflectivity for a 0.74 eV MIM device (without an AR coating). This particular device had a 3 μm LCL and a low doped emitter (1×10^{18} cm^{-3}). Optical modeling suggests that IR reflectivity's of >90% are possible with optimized device structures.

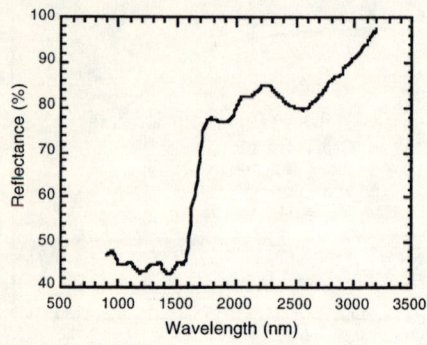

FIGURE 10. Reflectance of 0.74 eV MIM (no AR).

ACKNOWLEDGMENT

The authors wish to acknowledge Dave Scheiman of NASA Lewis Research Center for the I-V and QE measurements.

REFERENCES

1. Chubb, D.L. and Lowe, R.L., "Thin-Film Selective Emitter", J. Appl. Phys. **79** (9), 1993, pp. 5687–5698.

2. Wilt, D.M., et al., "InGaAs PV Device Development for TPV Power Systems", 1st NREL Conf. on TPV Gen. of Elect., 1994, AIP 321, pp. 210.

3. Horne, W.E., et.al., "IR Filters for TPV Convertor Modules", Proc. 2nd NREL Conf. on TPV Gen. of Elect., 1995, AIP 358, pp. 35.

4. Chubb, D.L., et.al., "Review of Recent TPV Research at Lewis Research Center", Proc. 14th SPRAT Conf, 1995, NASA CP–3324, pp. 191.

5. Charache, G.W., et.al., "Thermophotovoltaic Devices Utilizing a Back Surface Reflector for Spectral Control", Proc. 2nd NREL Conf. on TPV Gen. of Elect., 1995, AIP 358, pp. 339.

6. Iles, P.A. and Chu, C.L., "TPV Cells with High BSR", Proc. 2nd NREL Conf. on TPV Gen. of Elect., 1995, AIP 358, pp. 361.

7. Wojtczuk, S., "Multijunction InGaAs Thermophotovoltaic Power Convertor", Proc. 14th SPRAT Conf, 1995, NASA CP–3324, pp. 223.

8. Spitzer, M.B., et.al., "Monolithic Series-Connected Gallium Arsenide Convertor Development", Proc. IEEE 22nd PVSC, (1991), pp. 142–146.

9. Fatemi, N.S., et al., "Materials and Process Development for the Monolithically Interconnected Module (MIM) InGaAs/InP TPV Cells", Proc. 3rd NREL Conf. on TPV Generation of Electricity, 1997.

10. Wilt D.M., et al., "High Efficiency InGaAs Photovoltaic Devices for TPV Power Systems", Appl. Phys. Lett. **64** (18), 1994.

11. Wilt, D.M., et al, "Monolithically Interconnected InGaAs TPV Module Development", Proc. IEEE 25th PVSC, May 1996, pp. 43.

Materials and Process Development for the Monolithic Interconnected Module (MIM) InGaAs/InP TPV Devices

Navid S. Fatemi[1], David M. Wilt[2], Phillip P. Jenkins[1],
Richard W. Hoffman, Jr.[1], Victor G. Weizer[1],
Christopher S. Murray[3] and David Riley[3]

[1]*Essential Research, Inc., Cleveland, OH*
[2]*NASA Lewis Research Center, Cleveland, OH*
[3]*Westinghouse Electric Corp., West Mifflin, PA*

INTRODUCTION

Four major components of a thermophotovoltaic (TPV) energy conversion system are a heat source, a graybody or a selective emitter, spectrum shaping elements such as filters, and photovoltaic (PV) cells. Most TPV systems under consideration envision utilizing planar one-junction low-bandgap (Eg) PV cells. III-V materials such as InGaAs/InP, GaSb, and InGaSbAs/GaSb are examples of the cells currently used. These devices commonly have high output short-circuit current densities (J_{sc}), in the range of 2–10 A/cm^{-2}, and low open-circuit voltages (V_{oc}), in the range of 450–550 mV. In a TPV system, however, these cells are connected in series to create a high-voltage/low-current configuration, to keep resistive power losses (i.e., I^2R) at acceptable levels.

Another approach to achieving a high-voltage/low-current configuration is to fabricate a device, where small area PV cells are monolithically series connected. We have termed this device a monolithic interconnected module (MIM). A MIM has other advantages over conventional one-junction cells, such as simplified array interconnections and heat-sinking, and radiation recycling capability via a back-surface reflector (BSR). These advantages, as well as the electrical performance of MIMs are discussed in detail in the preceding paper in these proceedings (1). We will, therefore, confine the contents of this article to the MIM materials, process development, and some optical results.

To fully take advantage of the BSR, MIM devices require the use of semi-insulating substrates that are transparent to infrared (IR) radiation. Also, to facilitate processing and fabrication, stop-etch layers need to be incorporated into the cell structure. As a result, in practical terms, only the InGaAs/InP cell

CP401, *Thermophotovoltaic Generation of Electricity: Third NREL Conference,*
edited by Benner/Coutts
© 1997 The American Institute of Physics 1-56396-734-0/97/$10.00

materials are suitable candidates for MIM fabrication. The MIM structures we have processed and tested were, therefore, comprised of several p/n InGaAs cells deposited on semi-insulating InP and monolithically interconnected in series. The front, back, and interconnect metallizations were performed at the same time, and access to both positive and negative terminals were made from the top side of the module. The series connections from one cell to the next were made by interconnects that were electrically insulated from active regions of the module by dielectric barrier layers. InGaAs MIM with two different bandgaps were successfully processed: 0.55 and 0.74 eV. Schematic plan and cross-sectional views of a MIM are shown in Figure 1.

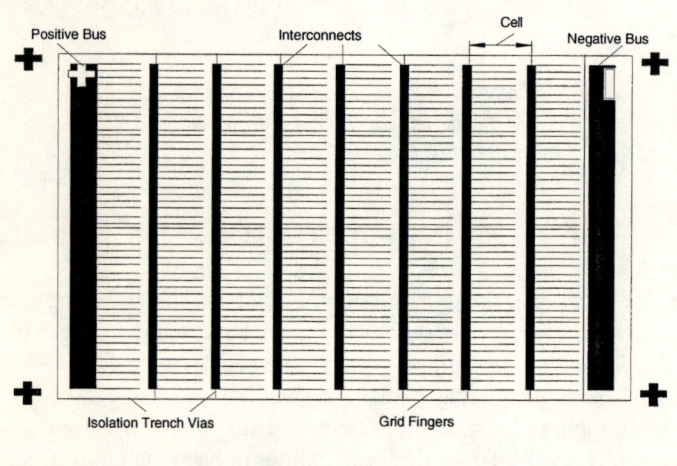

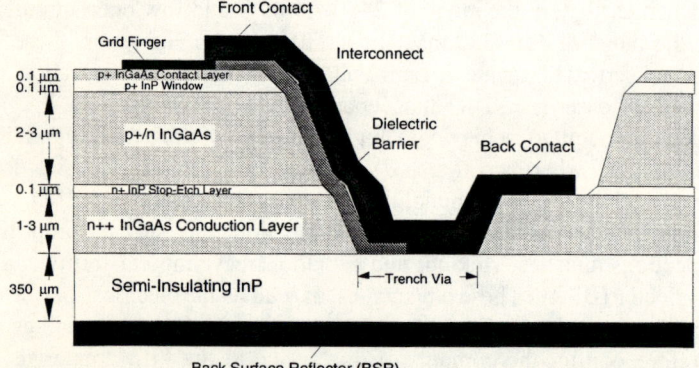

FIGURE 1. Schematic view of a MIM, plan view (top), and cross-sectional view (bottom).

250

The successful fabrication of InGaAs/InP MIM devices entails the development and optimization of several key components and processes. These include: isolation trench via geometry, selective chemical etching, contact and interconnect metallization, dielectric isolation barrier, back surface reflector (BSR), and anti-reflection (AR) coating.

The selection, development, and testing of the materials and processes described above for MIM fabrication will be described in the following sections.

Isolation Trench Via Geometry and Chemical Etching

Minimizing the non-photoactive areas in MIMs will result in maximizing the cell output power density. The non-photoactive areas include the isolation trench vias around the cells, positive and negative busbars, and metal grid fingers and interconnects. The isolation trench vias are responsible for electrically isolating the cells, as well as isolating the MIM from other devices on the InP wafer. They are made by chemically etching the InGaAs and InP layers of the MIM structure down to the semi-insulating InP substrate. (See Fig. 1.)

To investigate the minimum trench via width that could be imaged by ultraviolet (UV) photolithography and etched by wet chemical solutions, we designed a photomask with space widths ranging from 1 to 10 μm. The photolithography process employed in our labs increased the actual space widths in the photoresist by about 2 μm. As a result, the smallest opening in the photoresist increased to about 3 μm and the widest to about 12 μm.

Selective etchant solutions were used to create trench vias of desired depth. Room temperature concentrated HCl was used to etch InP layers. This etchant does not attack InGaAs. $H_3PO_4:H_2O_2:H_2O$ (3:4:1 vol.) was used to etch InGaAs layers with bandgaps (Eg) of 0.55 and 0.74 eV. InP is impervious to this solution. The smallest width via (i.e., 3 μm) was successfully etched down to the semi-insulating InP substrate. In addition to vertical etching, both solutions etched the semiconductor laterally underneath the photoresist. This lateral undercut was about 3–4 μm on each side of the via, for a vertical etch depth of 7–8 μm. We also used an alternative InGaAs selective etchant solution of $H_2SO_4:H_2O_2:H_2O$ (1:2:4 vol.). This solution also laterally etched the semiconductor underneath the photoresist. The undercut produced by this etchant was, however, 1–2 μm greater than what was observed with the $H_3PO_4:H_2O_2:H_2O$ solution.

As a result, using the HCl and the H_3PO_4-based solutions, the narrowest trench via that was produced was about 12 μm wide. A cross-sectional SEM micrograph and a schematic drawing for this via is shown in Figure 2.

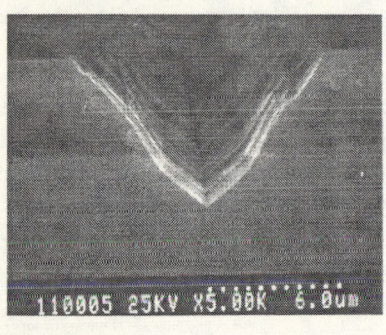

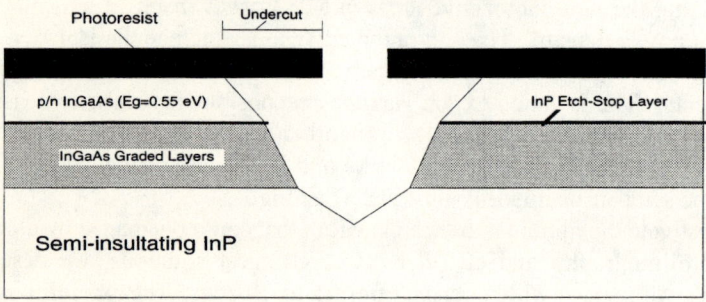

FIGURE 2. Cross-sectional SEM micrograph of a chemically etched trench via (top), and schematic representation (bottom).

As shown in the figure, (111) and (211) semiconductor planes produced by etching the (100) surface, result in v-shaped vias. The outward slope of the via walls are crucial for successful MIM fabrication because they allow for the subsequent deposition of the dielectric barrier layer. V-shaped vias are only attained when the device geometry is properly aligned with the crystallographic orientation of the semi-insulating InP wafer. Using Sumitomo supplied semi-insulating InP wafers, for example, the long dimensions of the busbars and inter-connects in Figure 1 are aligned parallel to the wafer primary flat (PF) to produce v-shaped vias. We also found that alignments that are a few degrees (±5°) off will not change the slope geometry significantly. An alignment perpendicular to the PF, on the other hand, will result in inwardly sloped (111) and (211) planes.

In addition to minimizing the width of the trench vias, we designed a photomask set that would reduce the interconnect width from our original design value of 300 μm to a new value of about 50 μm. The interconnect area (see Fig. 1) includes the trench via, the back contact, and the front contact (excluding grid fingers and busbars).

In our original process sequence, trench vias were etched to a depth of 7–8 μm down to the semi-insulating InP. This was followed by another etch (~3 μm) down to the lateral conduction layer (LDL), for the back contact (see Fig. 1). Our

252

new design reversed this procedure by carrying out the back contact etch first, followed by the trench via etch. In this case, the via is located inside the back contact etch area. This configuration is shown in Figure 3. By performing the via etch after the back contact etch, the via etch depth is reduced from its original value of 7–8 μm to ~5 μm. This, in turn, results in smaller lateral undercutting. As seen in Figures 2 and 3, the total via width was reduced from 12 to 8 μm.

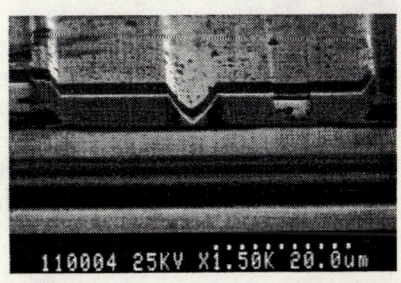

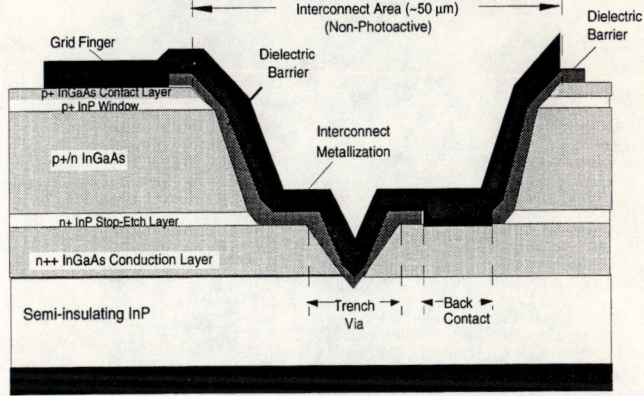

FIGURE 3. Cross-sectional SEM micrograph of an interconnect area with Cr-Au contacts (top), and schematic representation (bottom).

Also shown in Figure 3 are the dielectric barrier layer and the back contact metal area. The interconnect design of Figure 3 allows for the widths of the metal contacts and etch vias to be minimized, and at the same time, provides sufficient tolerance for alignment inaccuracies and dimensional changes that occur in photo-lithography and chemical etching processes.

Contact and Interconnect Metallization

The two most important requirements for a suitable metallization system to our MIMs are: 1) adequate adhesion of the contacts to InP, InGaAs, and dielectric barrier layers, and 2) minimum specific contact resistivity (ρ_c) values in the

253

10^{-6} Ω-cm^2 range to both n- and p-type InGaAs. In the following sections these requirements are addressed.

Metal Adhesion

Most contact metallization systems we tested adhered very well to both InP and InGaAs. They were not, however, very adherent to any dielectric layers. We experimented with several contact systems. These were: Au-Ge-Au, Ag-Au, Cr-Au, and Ti-Au. We had used the Au-Ge-Au system in our previous works with InGaAs planar cells (2). This contact system had exhibited ultra-low ρ_c values on both InGaAs and InP doped layers, but its adhesion to dielectric layers or to angled semiconductor surfaces proved to be poor. This is illustrated in Figure 4, where Au-Ge-Au was deposited on Ta$_2$O$_5$, InGaAs, and InP in an interconnect area.

FIGURE 4. Cross-sectional SEM micrograph of an interconnect area with Au-Ge-Au (0.1/0.1/3 μm) metallization.

To improve adhesion, we replaced the Au layer in contact with the substrate with Ag, Cr, and Ti layers. These layers are known to have better sticking characteristics than Au to most substrate materials. We found that Ag behaved similarly to Au (Fig. 5); however, Cr and Ti, capped with Au, were very adherent to all surfaces (Figs. 6 and 3). Both Cr-Au and Ti-Au, deposited on Ta$_2$O$_5$, passed our tape-pull and scratch tests.

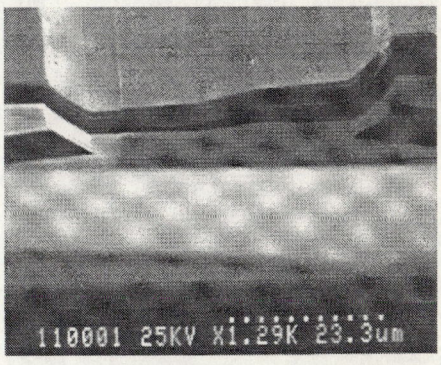

FIGURE 5. Cross-sectional SEM micrograph of an interconnect area with Ag-Au (3/1 μm) metallization.

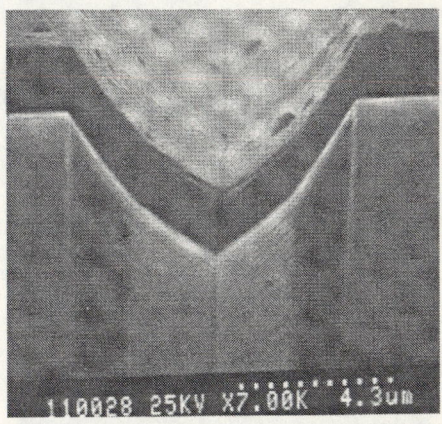

FIGURE 6. Cross-sectional SEM micrograph of a trench via area with Cr-Au (0.05/2 μm) metallization resting on Ta_2O_5.

As seen in Figures 4 and 5, Au-Ge-Au and Ag-Au contacts have lifted from dielectric coated areas, as well as from some of the semiconductor corner areas. This, no doubt, is due to the presence of internal stress forces in the deposited films. The "air bridges" thus formed may pose long term reliability issues with these cells. The use of adherent contacts, such as Cr-Au (Fig. 6) and Ti-Au will alleviate such concerns.

Contact Resistivity

In addition to being very adherent, the metal contacts to MIMs must have low contact resistivities. We measured the ρ_c values for each of the four metal systems tested in the previous section. The metals were deposited onto highly-doped p-type (1×10^{19} cm^{-3}) and n-type (1×10^{19} cm^{-3}) InGaAs epi-layers. The transmission line method (TLM) was used to make the resistivity measurements. None of the contact systems were heat treated prior to TLM measurements. The ρ_c measurement results are shown in Figures 7 and 8.

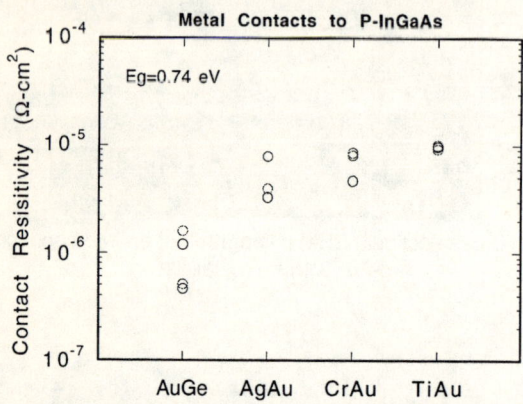

FIGURE 7. Specific contact resistivities for Au-Ge-Au (0.1/0.1/3 μm), Ag-Au (3/1 μm), Cr-Au (0.05/2 μm), and Ti-Au (0.05/2 μm) on p-InGaAs.

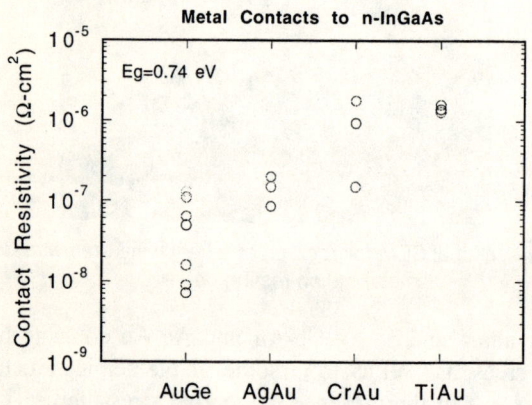

FIGURE 8. Specific contact resistivities for Au-Ge-Au (0.1/0.1/3 μm), Ag-Au (3/1 μm), Cr-Au (0.05/2 μm), and Ti-Au (0.05/2 μm) on n-InGaAs.

As seen in the above figures, ρ_c values for Au-Ge-Au to both n- and p-type InGaAs were extremely low. In fact, some of the values measured on n-InGaAs were near the theoretical minimum (in the 10^{-9} Ω-cm^2 range). Cr-Au and Ti-Au contacts, on the other hand, had ρ_c values in the high 10^{-6} Ω-cm^2 range on p-type, and in the high 10^{-7} Ω-cm^2 range on n-type InGaAs. These values were, on average, about an order of magnitude higher than what is measured with Au-Ge-Au. They are, however, low enough that their contribution to the I^2R resistive losses in the cell remain negligible.

Dielectric Isolation Barrier

The function of the dielectric barrier is to electrically isolate the p- and n-type regions of each cell, thereby, preventing the shorting of the cell by the interconnect metallization (see Fig. 1). The dielectric must, therefore, be highly insulating and be free of pinholes. We examined three dielectric materials. They were Ta_2O_5, spin-on glass (SOG), and Si_3N_4. Ta_2O_5 was electron-beam deposited. Accuglass 211 SOG, purchased from Allied Signal, was spin coated at 3000 RPM, and subsequently heat treated at 250 °C for 15 min. Plasma-enhanced chemical vapor deposition (PECVD) was used to deposit Si_3N_4. All dielectrics had thicknesses in the range of 2000–3000 Å.

We designed metal-semiconductor-dielectric test structures to characterize these dielectrics. These test structures enabled us to measure the current-voltage (C-V) characteristics of the dielectrics (i.e., dielectric constant, resistivity, and breakdown strength), and to measure the presence of pinholes.

Unlike Si_3N_4, Ta_2O_5 and SOG appeared to contain many pinholes and shunting paths. Consequently, their dielectric constant and breakdown strength could not be measured. Si_3N_4, on the other hand, showed far better dielectric properties than the other dielectrics. In Figure 9, the measured resistivity values for all the dielectrics tested are shown.

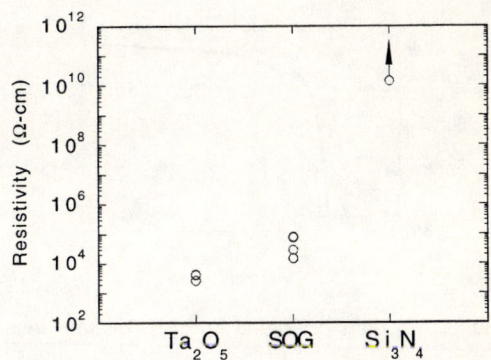

FIGURE 9. Resistivity data for Ta_2O_5, SOG, and Si_3N_4.

As seen in the figure, Si_3N_4 had far greater resistivity than either Ta_2O_5 and SOG. Also, the true resistivity of Si_3N_4 was, in fact, greater than the value of 1.4×10^{10} Ω-cm shown in the figure. We could not determine an exact resistivity value because our C-V measurement set up is limited to a maximum DC resistance of 100 MΩ. It should also be noted that the shunting effects of pinholes are represented in the measured values of resistivities for Ta_2O_5 and SOG. In other words, the actual resistivity values for Ta_2O_5 and SOG may be higher if pinholes were not present in the structure.

In addition to having a very high resistivity, Si_3N_4 had a breakdown strength of greater than 1.3 MV-cm^{-1}, and a dielectric constant of 5.3. These values indicate that the Si_3N_4 fabricated in our facilities is suitable for use in MIM fabrication.

Back Surface Reflector

A back-surface reflector (BSR) is deposited on the non-electrically active (i.e., semi-insulating InP) backside of a MIM (see Fig. 1). Its function is twofold. Since the semi-insulating InP substrate is transparent to infrared radiation, a BSR layer reflects the long-wavelength IR radiation (i.e., out-of-band) back to the heat source. This optical recycling of the non-useful radiation increases the total TPV system efficiency. A BSR also reflects that portion of the in-band radiation that is not absorbed by the active regions of the cell in the first pass through. This, in effect, enables in-band radiation to pass through the cell twice, thus increasing the probability of absorption. This has the effect of increasing the cell power output.

An ideal BSR should exhibit 100% reflectivity in all wavelengths. Practical BSRs, such as pure gold and silver, exhibit reflectivities in the range of 90% for most of the wavelengths of interest. Au and Ag reflectivities, as measured through semi-insulating InP, are shown in Figure 10.

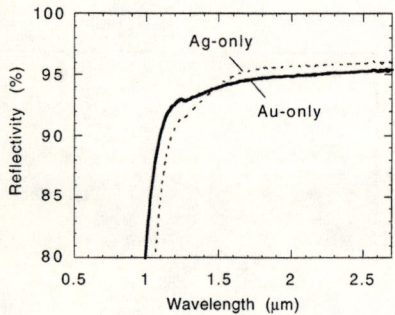

FIGURE 10. Reflectivity data for Au and Ag BSRs on semi-insulating InP.

Our modeling of optimum BSR layers indicated that by interposing a MgF_2 layer (~4300 Å thick) between InP and Au or Ag, longer wavelength (>2 μm) reflectivity can be enhanced. In addition, the MgF_2 layer physically isolates the metallization from the InP substrate, thereby eliminating any long-term room temperature aging effects, which may lead to metal-InP interfacial optical degradation (3). The reflectivity measurements for the MgF$_2$/Au and MgF$_2$/Ag on semi-insulating InP are shown in Figure 11.

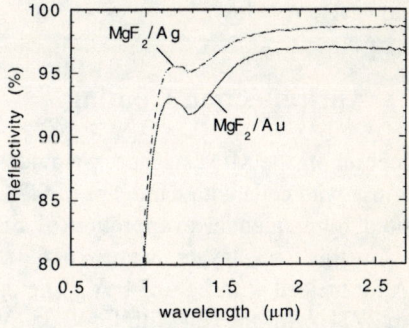

FIGURE 11. Reflectivity data for MgF$_2$/Au and MgF$_2$/Ag BSRs on semi-insulating InP.

A comparison of data in Figures 10 and 11 indicates that the presence of MgF$_2$ enhances the reflectivity for both Au and Ag by a few percent for $\lambda \geq 2$ μm. We found, however, that MgF$_2$/Ag and MgF$_2$/Au are not practical BSRs due to the very poor adhesion of Au and Ag to MgF$_2$. Both Au and Ag were easily scratched off the MgF$_2$ surface by gently rubbing them with tissue paper. Ag or Au films could also be readily lifted off the MgF$_2$ surface by pulling them with tweezers.

To improve adhesion, we introduced very thin (~50 Å) layers of Cr and Ti between MgF$_2$ and Au. We then measured the reflectivity of MgF$_2$/Cr-Au and MgF$_2$/Ti-Au on semi-insulating InP. This is shown in Figure 12.

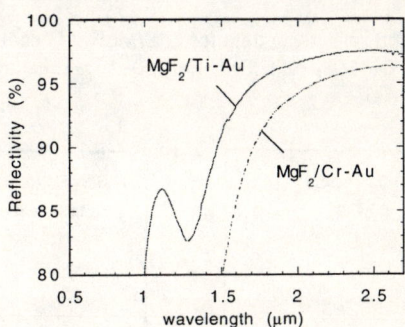

FIGURE 12. Reflectivity data for MgF$_2$/Ti-Au and MgF$_2$/Cr-Au BSRs on semi-insulating InP.

259

As seen in the figure, the reflectivity of MgF$_2$/Cr-Au was unacceptably low in all the wavelengths of interest. MgF$_2$/Ti-Au, on the other hand, showed comparable reflectivity to MgF$_2$/Au in the >2 µm region. The reflectivity in the shorter wavelength range was, however, very low.

Antireflection Coating

To minimize the reflection off the MIM surface, we modeled two antireflective (AR) coating designs for use with cells with bandgaps of 0.74 and 0.55 eV. These AR coatings are deposited by sequential evaporation of ZnS and MgF$_2$. The optimized thicknesses for the ZnS layers were 1686 Å for the cell with Eg=0.55 eV, and 1413 Å for the cell with Eg=0.74 eV. The optimized thicknesses for the MgF$_2$ layers were 2771 Å for the cell with Eg=0.55 eV, and 2343 Å for the cell with Eg=0.74 eV. The calculated reflectivities for these AR coatings on InP (window) on InGaAs are shown in Figures 13 and 14.

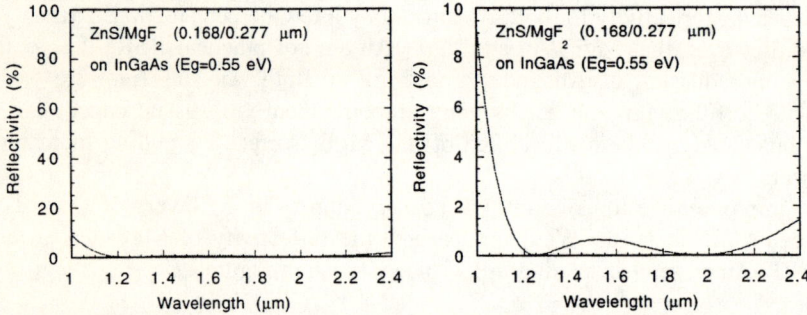

FIGURE 13. Calculated reflectivity data for ZnS/MgF$_2$ AR coating on InGaAs with Eg=0.55 eV, 0–100% reflectivity (left), 0–10% reflectivity (right).

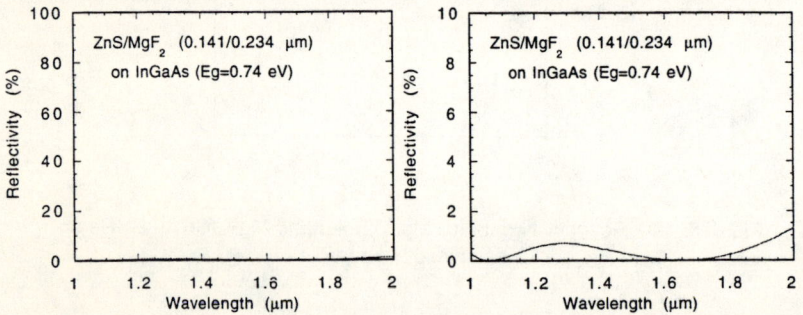

FIGURE 14. Calculated reflectivity data for ZnS/MgF$_2$ AR coating on InGaAs with Eg=0.74 eV, 0–100% reflectivity (left), 0–10% reflectivity (right).

As shown in the figures, the calculated reflectivities for both structures are ≤1%, for the desired range of wavelengths. The performance of the deposited ZnS and MgF$_2$ layers match their calculated values closely. We routinely deposit these films, with a range of thicknesses, for photovoltaic applications, and their measured reflectivity matches our modeling results quite consistently.

SUMMARY

The selection, development, and testing of the materials and processes used for MIM fabrication was described. These included: isolation trench via geometry, selective chemical etching, contact and interconnect metallization, dielectric isolation barrier, back-surface reflector, and antireflection coating. The main specific results obtained in our work is summarized below:

1. Etched trench vias, as small as 8 μm wide were developed. Interconnect areas with widths of only 50 μm were also fabricated.

2. The selective etchants for fabricating these vias were: concentrated HCl to remove InP, and the H$_3$PO$_4$:H$_2$O$_2$:H$_2$O (3:4:1 vol.) solution to remove InGaAs.

3. Cr-Au and Ti-Au metallizations systems were determined to be suitable for use as ohmic contacts to n- and p-InGaAs. They were also found to be very adherent to various dielectric isolation barrier films.

4. The dielectric properties of PECVD-deposited Si$_3$N$_4$ barrier layer was found to be far superior to Ta$_2$O$_5$ or SOG.

5. Ti-Au/MgF$_2$, Ag-only, and Au-only BSRs showed very high reflectivity in the IR region, when deposited on semi-insulating InP. Also, all three BSRs adhered well to the substrate.

6. We designed and modeled two optimized AR coatings for use with MIMs with bandgaps of 0.55 and 0.74 eV. The double-layer AR coatings were comprised of ZnS and MgF$_2$ layers. The calculated reflectivities for both structures were ≤1%, for the desired range of wavelengths.

ACKNOWLEDGMENT

We wish to thank Nick Veraljay and Dr. Sam Altrovitz, of NASA Lewis Research Center, for depositing the Si$_3$N$_4$ films and for the ellipsometry measurements.

REFERENCES

1. Wilt, D.M., Fatemi, N.S., Jenkins, P.P., Weizer, V.G., Hoffman, R.W., Jr., Murray, C.S., and Riley, D., "Electrical and Optical Performance Characteristics of 0.74 eV p/n InGaAs Monolithically Interconnected Modules," Third NREL TPV Conference, May 18–21, 1997, Colorado Springs, CO.

2. Fatemi, N.S,. Weizer, V.G, Wilt D.M., Hoffman R.W., Jr., "Ultra-Low Resistance, Non-destructive Contact System for InP/InGaAs/InP Double Heterostructure TPV Devices," 25th IEEE Photovoltaic Specialists Conference (PVSC), Washington, D.C., May 13–17, 1996, p. 85.

3. Fatemi, N.S and Weizer V.G., "Humidity-Induced Room-Temperature Decomposition of Au Contacted InP," *Appl. Phys. Lett*., **57**, 500 (1990).

SESSION 5:
SELECTIVE RADIATORS II

Influence of Ytterbium Concentration on the Emissive Properties of Yb:YAG and Yb:Y$_2$O$_3$

J.-C. Panitz, M. Schubnell, W. Durisch, F. Geiger

General Energy Research Department, Paul Scherrer Institut, CH-5232 Villigen, Switzerland

Abstract. The Paul Scherrer Institut has initiated a project to develop a thermophotovoltaic (TPV) converter for residential heating applications. By economic reasons we have decided to design a thermophotovoltaic generator based on Si-cells and, accordingly, to make use of the selective emission properties of ytterbium containing emitter materials. In this contribution, we focus on the emitter materials and present results of an experimental study on the influence of ytterbium concentration on the emissive properties of ytterbium doped yttrium aluminum garnet (YAG) and yttria. The emitter materials were prepared by decomposition of nitrates and by co-precipitation methods with subsequent calcination at elevated temperatures. The polycrystalline materials are characterized by X-Ray diffraction (XRD), particle size distribution (PSD) measurements and Raman and fluorescence microscopy. The emissive properties of the Yb-doped materials have been measured at about 1400 K. Additionally, curent-voltage (I-V) curves of the Si-cells used in a prototype TPV generator have been recorded for selected emitter materials.

INTRODUCTION

In TPV residential heating systems, selective emitter materials should ideally guarantee efficient operation during the lifetime of the system. With respect to that application, the principal drawback of Welsbach type selective emitters is their inferior mechanical stability. Selective emitters based on solid ceramics or coatings supported on these ceramics offer better mechanical properties, but one often faces problems with heat conduction losses that deteriorate the radiation output of these emitter materials. In this article, we present results concerning the preparation of ceramic powders suitable for the development of emitter materials with improved emission characteristics. In order to find an optimum concentration of ytterbium centers in the host lattice, we have varied the ytterbium content of the samples prepared. The samples were characterized by XRD and PSD measurements, Raman and fluorescence spectroscopy and thermal emission

CP401, *Thermophotovoltaic Generation of Electricity: Third NREL Conference,*
edited by Benner/Coutts

spectroscopy. After a description of the methods used for the preparation of the materials, we present the results of the different characterization techniques.

PREPARATION OF CERAMIC POWDERS

Co-Precipitation-Synthesis of Yb:Y₂O₃-Powders

The synthesis of ytterbium-doped yttria powders was carried out according to procedures outlined in detail in the literature [1]. $YbCl_3 \cdot 6\,H_2O$, purity 99.999 %, and $YCl_3 \cdot 6\,H_2O$, purity 99.9 %, both supplied by Aldrich, were used as starting materials. The rare earth (RE) content of the starting materials was assayed by complexometric titration. Mixed solutions of the chlorides were prepared using deionized water. An appropriate amount of urea (Aldrich, A.C.S. grade) was dissolved in the solution. The solutions were heated to 85°C in a five-neck flask equipped with a stirrer, a thermocouple, a dropping funnel, and connections to a peristaltic pump. After an aging period of 2 h at the final temperature, the suspension was pumped into a graduated cylinder, where it was left standing overnight. Then, it was filtered through a 0.2 μm cellulose membrane using a single-use sterilized filter unit made of polystyrene. The filter cake was dried at room temperature. A calcination step is necessary to decompose the precipitated ytterbium doped yttrium hydroxide carbonate to the oxide.

In order to determine an optimized calcination temperature, a small piece of a sample containing 10 mol% ytterbium was placed in a heatable microscope stage (Linkam, TS1500). Using Raman spectroscopy under in situ conditions [2], the calcination process was monitored up to a temperature of 1700 K. Raman spectra were measured with a Raman microscope (LabRam, DILOR) using the 413.1 nm line of a Kr^+-ion laser (INNOVA 302, Coherent Inc.) for excitation. Spectra recorded are displayed in Fig. 1. Calcination of the material is noted first for a temperature of 1073 K, when the intensity of the carbonate stretching vibration at 1100 cm^{-1} decreased considerably while the sample was held at this temperature. The result of this experiment was verified by a thermogravimetric analysis performed for the same sample. To ensure complete conversion to the oxides, the filter cakes were calcined at 900 °C in an air flow for 5 h.

Synthesis of polycrystalline Yb:YAG-Powders

Polycrystalline Yb:YAG-powders were prepared using a modification of the Chick-Pederson method [3] for the preparation of oxides. For safety reasons, a total amount of 40 mmol metal nitrate per batch should not be exceeded.

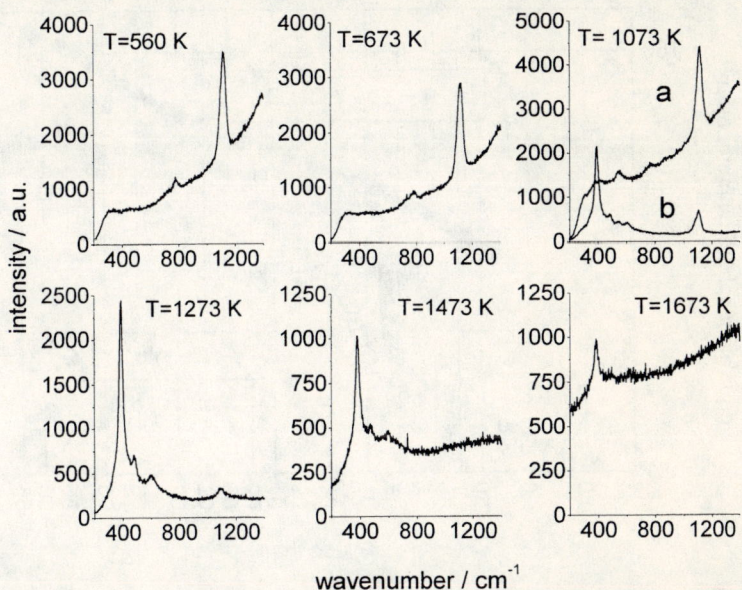

FIGURE 1. Raman spectra of sample $Yb_{0.2}Y_{1.8}O_3$, recorded in an air flow at the temperatures indicated. Traces measured at T=560 K and T=673 K show signals typical for carbonate groups. The calcination reaction sets in at 1073 K, as inferred from comparison of the spectra recorded while holding the sample at this temperature: trace a) was recorded at the beginning of the hold time, trace b) was recorded at the end of the hold time of 60 min. Trace a) shows Raman bands of carbonate groups, while trace b) is dominated by the Raman bands typical for the ytterbium-yttrium mixed oxide. Beyond T= 1273 K, only signals due to the mixed oxide are observed. Measurement conditions: 413.1 nm excitation, 5 mW, 120 s exposure time.

PARTICLE SIZE DISTRIBUTION ANALYSIS

Particle size distribution analysis was performed with a laser diffraction particle size distribution analyzer (LA500, Horiba). Raw data were processed to calculate the 0th and 3rd moments of the distribution. Typical curves for samples prepared are displayed in Fig. 2. Data shown in Fig. 2 are weighted with the volume of the particles. Additionally, the BET surface was determined by nitrogen adsorption for samples containing 20 mol% ytterbium using a Micromeritics ASAP2000 instrument. Results of both methods are displayed in Table 1.

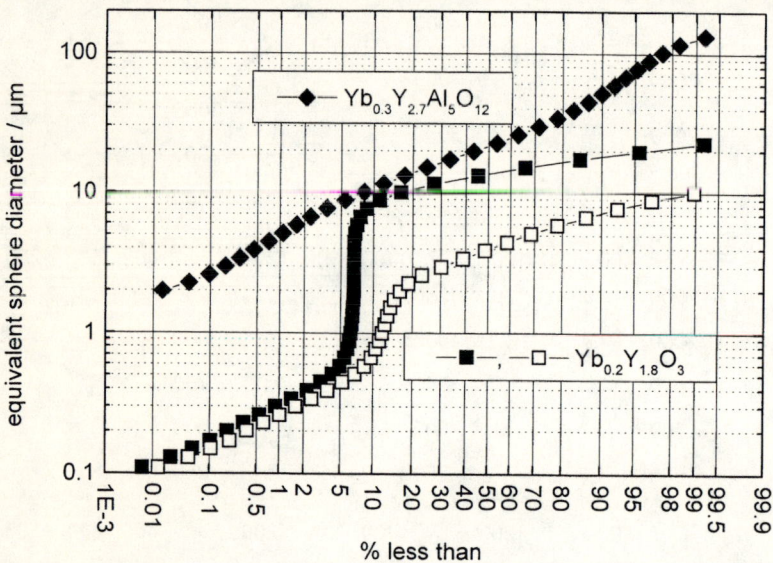

FIGURE 2. Particle size distribution of selected samples. □: $Yb_{0.2}Y_{1.8}O_3$, filter cake; ■: $Yb_{0.2}Y_{1.8}O_3$, calcined at 1173 K; ◆: $Yb_{0.3}Y_{2.7}Al_5O_{12}$. The data for the ytterbium-yttrium mixed oxide particles can be interpreted in terms of a bimodal distribution. In contrast, the ytterbium-yttrium aluminum garnet particles are distributed according to a log-normal distribution, approximately.

The preparation method used strongly influences the morphology of the particles. Particles prepared by the precipitation method are approximately of globular shape [1], whereas the particles prepared by the Chick-Pederson method are of irregular shape that may be compared with the constitution of fly ash particles. This is reflected in the particle size distribution data displayed in Fig. 2. In Table 1, average particle size is given for the samples prepared. Generally, ytterbium-yttrium mixed oxides obey a bimodal size distribution. The smaller value given in Table 1 reflects the size of the primary particles, whereas the larger diameter is characteristic for the size of agglomerates composed of these particles. Contrary, the ytterbium-yttrium aluminum garnet samples are distributed according to a log-normal distribution, with mean particle sizes ranging from 20 to 60 μm.

Interestingly, the surface of the particles accessible to nitrogen adsorption, measured for samples where 20 mol% ytterbium was substituted for yttrium, is larger for the garnet samples, though this samples are characterized by a considerable larger particle diameter, Again, this reflects the different morphology of the garnet samples.

X-RAY DIFFRACTION

X-Ray Diffraction (XRD) was performed on a Philips X'pert powder diffractometer using Cu K_α radiation. The data were compared to the respective reference patterns in order to check the homogeneity of samples prepared.

According to the results of the XRD characterisation, the samples prepared by the precipitation method are solid solutions of ytterbia and yttria, whereas the samples of ytterbium doped yttrium aluminum garnet are solid solutions of ytterbium aluminum garnet and yttrium aluminum garnet. This was concluded from the continous shift in peak position observed by increasing the ytterbium content of the samples.

TABLE 1. Particle Size Distribution of Samples Prepared, Weighted by Particle Volume

Sample	Type of Distribution	Average Diameter[a]	BET Surface[b]
$Yb_xY_{2-x}O_3$, Mixed Oxides, Calcined @1173 K			
x= 0.04	bimodal	2 ± 1 14 ± 3	
0.10	bimodal	1.5 ± 0.7 14 ± 3	
0.20	bimodal	1.5 ± 0.7 15 ± 3	
0.40	bimodal	1.8 ± 0.9 24 ± 7	3.4 ± 0.2
1.00	bimodal	2.8 ± 1.2 27 ± 7	
2.00	bimodal	1.6 ± 0.8 13 ± 4	
$Yb_xY_{3-x}Al_5O_{12}$ Garnets			
x= 0.06	log-normal	20 ± 8	
0.15	log-normal	20 ± 9	
0.30	log-normal	24 ± 12	
0.60	log-normal	22 ± 11	11.5 ± 0.3
1.50	log-normal	60 ± 29	
3.00	log-normal	60 ± 25	

[a] in µm
[b] in square meter / gram, selected samples only

FLUORESCENCE SPECTROSCOPY

Fluorescence spectra were obtained with the Raman microscope mentioned above. An external laser beam delivered by a Kr^+-ion laser (INNOVA 302, Coherent Inc.) was coupled into the microscope. The excitation wavelength was 413.1 nm. The results of the characterization by optically excited fluorescence are shown in Figure 3 and Figure 4. Figure 3 shows a selection of fluorescence spectra recorded in the spectral region attributed to the $^2F_{5/2} \rightarrow {}^2F_{7/2}$ transition of trivalent ytterbium [xxx]. Fewer bands are observed for the case of the ytterbium-yttrium mixed oxide samples when compared to the garnet samples. In addition, we found that the observed fluorescence intensity of the sample calcined in air at 1673 K is about 30 times the intensity observed of the sample calcined at 1173 K (compare left and middle trace of Figure 3). This observation may be explained by assuming that the particle size increases by sintering of the primary particles. After the thermal treatment at 1673 K, the fluorescence intensity recorded is of the same order as the intensity emitted by the corresponding garnet sample (compare middle and right trace of Figure 3).

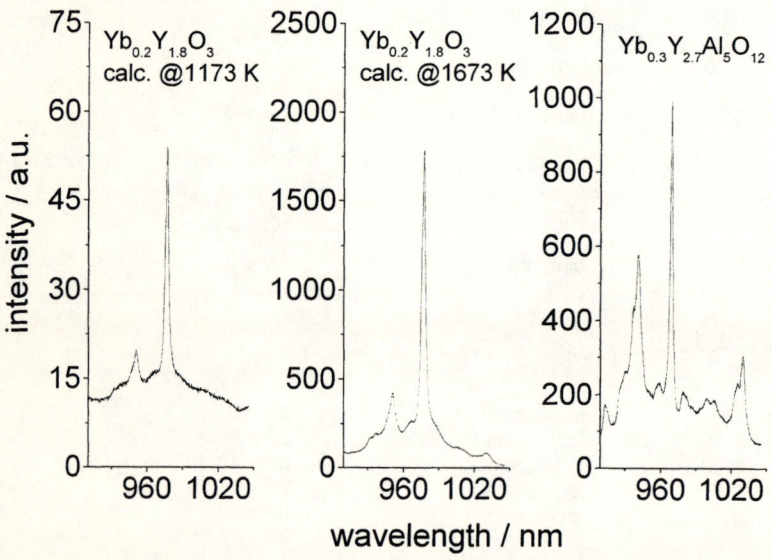

FIGURE 3. Fluorescence spectra (excitation wavelength: 413.1 nm) recorded for three samples doped with 10 mol% ytterbium. Intensity data were normalized with respect to exposure time. Note the effect of the calcination temperature on the fluorescence emission of the ytterbium-yttrium mixed oxide, as shown in the left and middle trace.

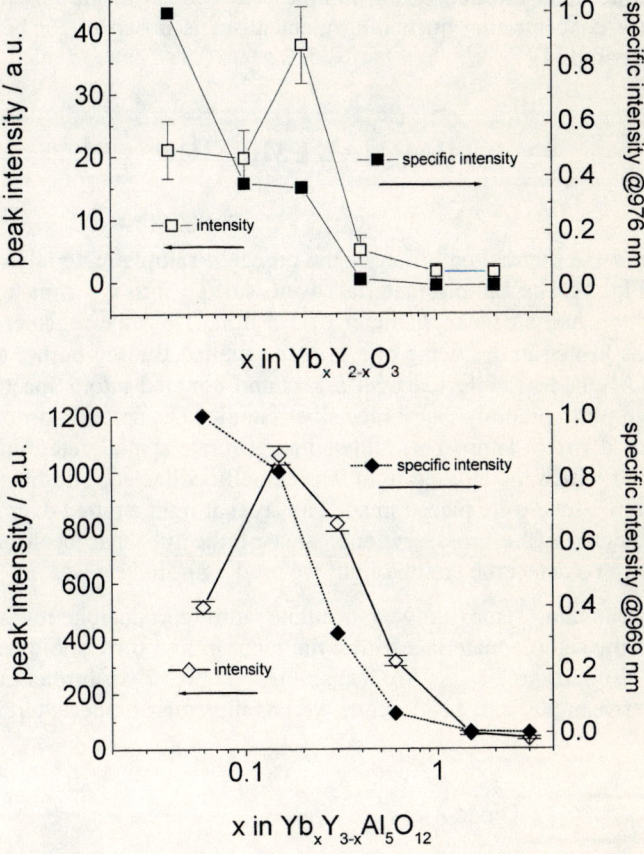

FIGURE 4. Fluorescence intensity of the central $^2F_{5/2} \rightarrow {}^2F_{7/2}$ transition of the ytterbium centres in ytterbium-yttrium mixed oxides (upper diagram) and ytterbium doped yttrium aluminum garnet. Solid symbols refer to intensity per unit time, and open symbols refer to specific intensity per ytterbium center.

Turning to Figure 4, where the results of an analysis of the fluorescence experiment are given, we note that a maximum of fluorescence intensity per unit time is observed for both series of samples. This maximum occurs at 10 mol% ytterbium substitution in case of the ytterbium-yttrium mixed oxides, and at 5 mol% dopant level in case of the ytterbium doped yttrium aluminum garnet samples. By plotting the specific intensity (defined by the ratio peak intensity / mole fraction of ytterbium) at the peak wavelength against the ytterbium concentration in the samples, it is shown that this parameter never reaches a constant value upon dilution of the luminescent centers in the host lattice. This means that a concentration quenching mechanism is observed for both series of samples prepared [4].

THERMAL EMISSION

To measure the thermal emission of the prepared sample materials we used the setup shown in Fig. 5. The sample materials were stuffed into a 5 mm long tube made of alumina which had an inner diameter of 1.5 mm. The tube together with the sample material was heated in the flame of a propane fuelled Bunsen burner to about 1050 °C. The emitted light was collected with a lens and coupled into a spectrometer with one arm of a two port randomly distributed fiber bundle. The second port of the fiber bundle was connected with a lamp. This allowed an accurate spatial determination of that part of the tube of which the emitted light was actually collected. The fiber bundle, the lens and the ceramic tube were placed in such a way that light emitted over a spot with about 1 mm diameter on the cross sectional area of the tube has been analysed with the spectrometer. To detect the emitted light we used a Si-diode array.

This experimental setup allowed accurate and reproducible measurements of the emission of the sample materials within the tube. In Fig. 6 we show recorded spectra of the $Yb_xY_{2-x}O_3$ and the $Yb_xY_{3-x}Al_5O_{12}$ samples. In Fig. 7 we further compiled the total emission between 600 and 1180 nm as well as the emission per ytterbium center in the

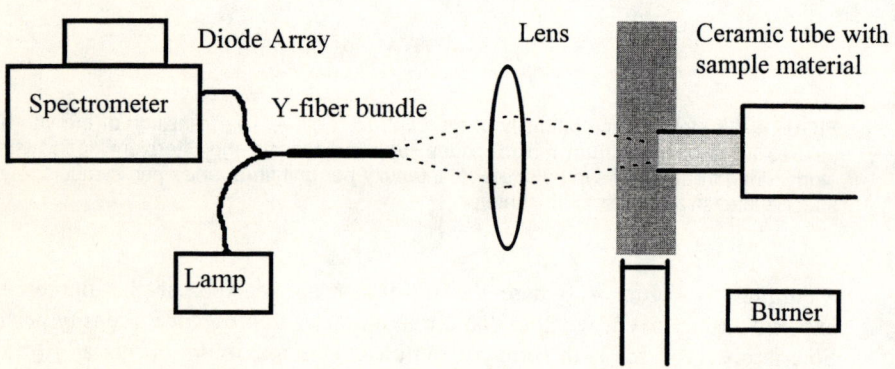

FIGURE 5. Setup used to record the thermal emission of the sample materials.

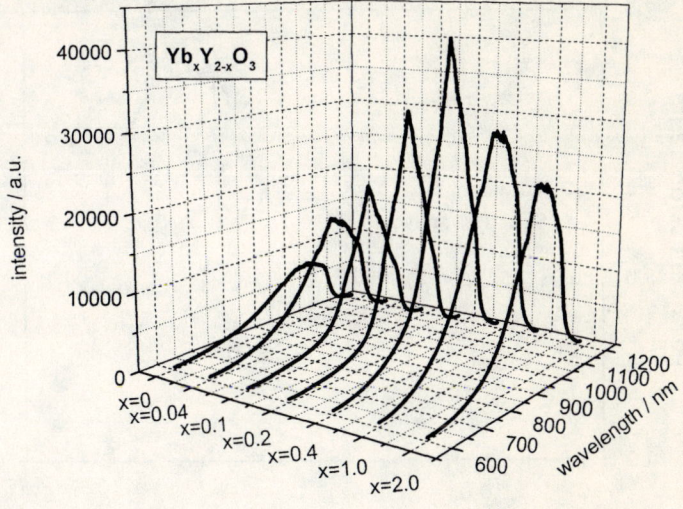

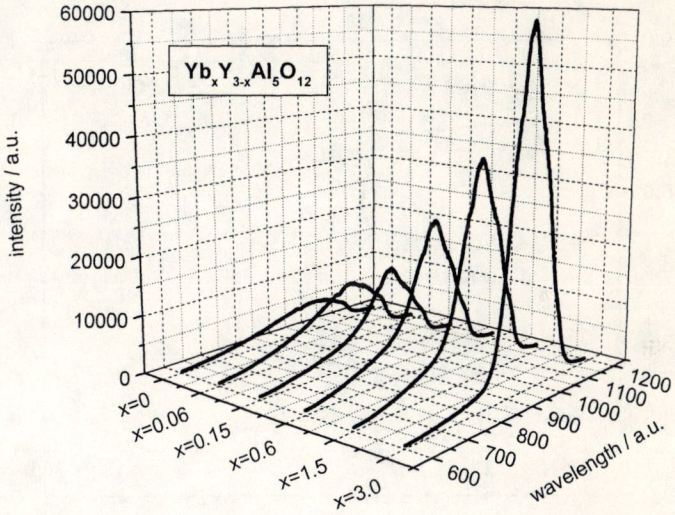

FIGURE 6. Thermal emission of samples prepared when heated in a propane gas flame. Upper diagram: ytterbium-yttrium mixed oxides, lower diagram: ytterbium doped yttrium aluminum garnets.

273

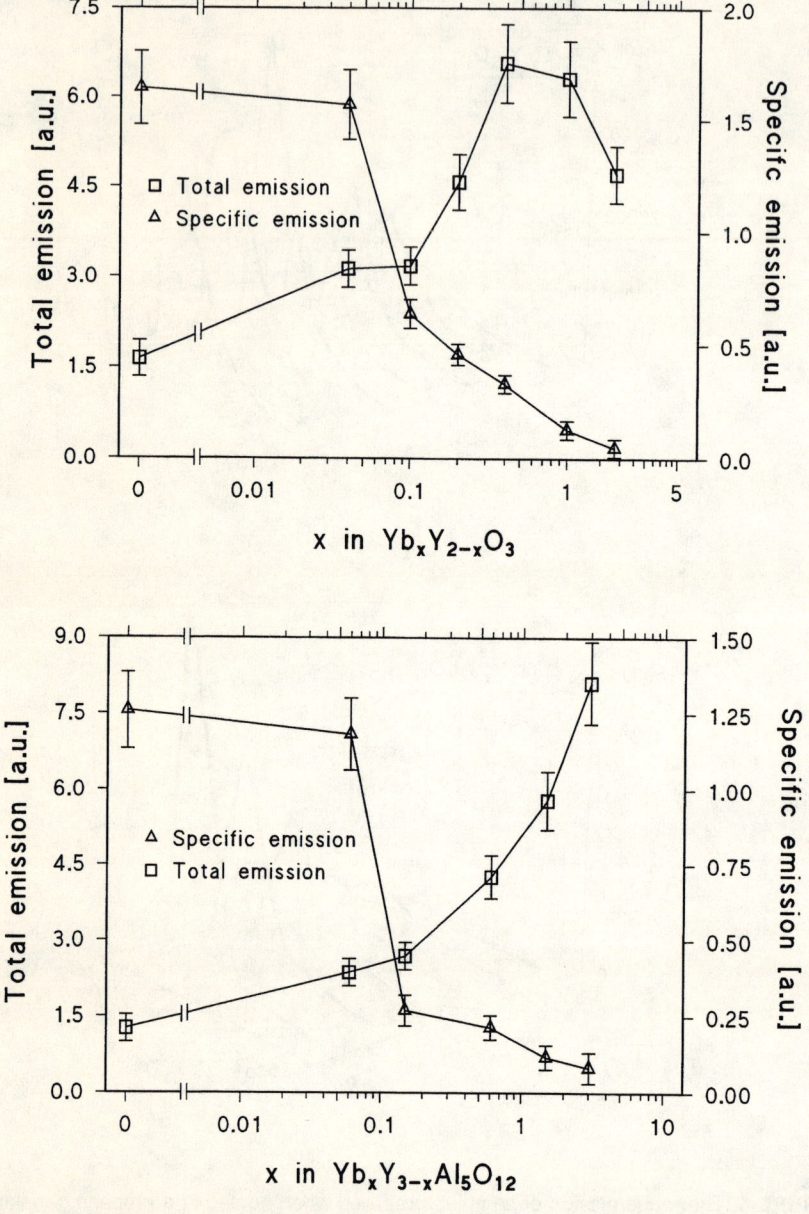

FIGURE 7. Thermal emission intensity, plotted against the ytterbium content of the samples.

prepared materials. It can be seen, that in case of $Yb_xYb_{2-x}O_3$ a maximum emission occurs at about x=0.4 wheras in the doped garnets maximum emission coincides with ytterbium concentration. In both sample material classes the specific emission decreases with increasing ytterbium concentration which is a consequence of enhanced self-quenching at higher concentrations.

The shape of the thermal emission band of ytterbia (cf. Figure 6) is in good agreement with the data published by Guazzoni [5] and Nelson [6]. So far we do not have a comprehensive explanation for the different band profiles of the spectra observed for different concentrations of ytterbium in the yttrium oxide host. But it should be noted that the shoulder at 1180 nm is more distinct at low ytterbium content. Then, a tentative interpretation is that these different band shapes are due to the background emission of the host material.

ELECTRICAL CHARACTERISATION

The electrical characterisation of TPV generators can be performed at PSI's test facility for solar photovoltaic components [7]. About 40 current/ voltage data points are recorded within 120 seconds. From the current/voltage pairs, the power/voltage curve is calculated. By coating a commercial mantle with different emitter materials, it is possible to investigate the effects of emitter composition on the electrical performance of the silicon photocells. In Figure 8, we show a typical curve obtained under operating conditions in a TPV test generator.

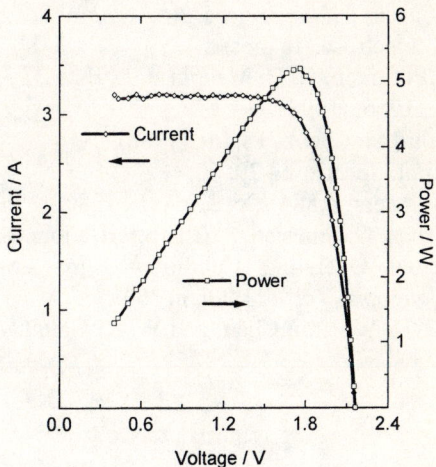

FIGURE 8. Current and power output against voltage of a TPV test generator. A mantle coated with ytterbia heated by a butane flame ($P_{thermal}$ ~1.2 kW) was used as emitter. Silicon photocells with an area of 400 cm^2 are mounted in a cylindrical geometry around the emitter. It is estimated that about 50% of the emitted radiation reaches the photocells.

CONCLUSIONS

The present work has shown that ytterbium doped polycrystalline yttrium aluminum garnets show a strong emission in a spectral band suitable for TPV systems based on silicon photocells. As compared to pure ytterbia, the ytterbium doped garnets deliver a comparable selective output

Further investigations will be aimed at the detailed study of the influence of particle size on thermal emission intensity. Also, we will investigate the effects of temperature on the concentration quenching observed, which is of interest in relation to a letter published recently by Golevlev et al. [8], who proposed models for two- and three-level selective emitters.

ACKNOWLEDGEMENTS

This work has been supported by the Research Foundation of the Swiss Gas Industry (FOGA). Thanks are due to Luiz Carlos De Sousa for technical assistance with the particle size distribution analysis and to Alwin Frei for providing the thermal analysis data.

REFERENCES

1. B. Aiken, W. P. Hsu and E. Matijevic, J. Am. Ceram. Soc., **71**, 845, (1988).
2. J.-C. Panitz, Appl. Spectrosc., in press.
3. L. A. Chick, L. R. Pederson, G. D. Maupin, J. L. Bates, L. E. Thomas, and G. J. Exarhos, Mater. Lett., **10**, 6, (1990).
4. G. Blasse, Prog. Solid St. Chem., **18**, 79, (1988).
5. G. E. Guazzoni, Appl. Spectrosc., **29**, 60, (1972).
6. R. E. Nelson, U. S. Patent, 4,584,426, 22. April 1986.
7. W. Durisch, J. Urban, G. Smestad, "Characterisation of Solar Cells and Modules under Actual Operating Conditions", in: *Proceedings of World Renewable Energy Congress*, Vol. 1, 1996, p. 359.
8. V. V. Golovlev, C. H. Winston Chen, and W. R. Garett, Appl. Phys. Lett., **69**, 280, (1996).

TPV Power Generation Prototype Using Composite Selective Emitters

Peter L. Adair, Zheng Chen, and M. Frank Rose

Space Power Institute
231 Leach Center, Auburn University, AL 36849

Abstract. Research in the field of thermophotovoltaics at Auburn University's Space Power Institute has been conducted since 1992. One of the main focus points of the research has been centered on emitter structures. The composite emitter structure, which was developed at the Space Power Institute, incorporates fibrous rare earth oxides held in a structural material fiber matrix and has been implemented into a breadboard prototype TPV system. Various aspects of the prototype system will be evaluated. The system consists of a diffusion type burner which uses propane and forced air as the inlet fuels. This will heat a cylindrical geometry composite emitter which illuminates 0.75 eV InGaAs lattice-matched to InP photovoltaic cells. The combustion products flow through the highly porous (over 90%) composite emitter structure. This results in more efficient coupling to the combustion flame. Quartz tubes prevent the combustion products from interacting with the cell array. Thermal losses will be due to end and quartz heating. The fuel input to radiant output efficiency will be determined as well as the radiant to electrical output efficiency. Hence, a total system efficiency can be determined. Emitter radiant uniformity will be discussed for several different emitter geometries.

INTRODUCTION

High system conversion efficiency is of primary importance in TPV. The three main components of a TPV system are the heat source, radiant emitter, and photoconversion device. The overall system efficiency depends on how these three key components function as a system. The total system efficiency can be determined from the products of these individual component efficiencies.

The first component listed was the heat source. For the purpose of this research, a combustion flame was used as the heat source. The heat source or thermal efficiency can be found by determining how much of the heat of combustion of the fuel is successfully transferred to the radiant emitter. In an ideal system, all of the energy in the combustion flame would be used to heat the emitter to incandescence. Unfortunately, in real systems there are significant energy losses due to heating of non-emitter surfaces, incomplete combustion, poor coupling of flame to emitter surfaces, etc. It has also been shown that the thermal efficiency decreases at a rate of 0.05% per Kelvin increase in radiation temperature (1). A typical number for thermal efficiency with no recuperation is 30% for a radiator temperature of 1475 K (1). This number can be significantly increased by recuperation of the exhaust heat. It has been shown that by using

CP401, *Thermophotovoltaic Generation of Electricity: Third NREL Conference*,
edited by Benner/Coutts

recuperation, the system thermal efficiency can be improved from 10% to 70% for a radiator operating at 2075 K (1).

The second component efficiency is the emitter efficiency. The emitter efficiency refers to the ability of the spectral control element to emit only photoconvertible radiation to the photovoltaic device. This can be determined by measuring the amount of photoconvertible radiation to that of the total radiation generated by the emitter. For a filter system, this refers to how much of the out-of-band radiation is successfully reflected back to the source and how much of the in-band radiation is transmitted through the filter. For a selective emitter system, it refers to the ratio of the in-band radiation to total radiation. The emitter efficiency is directly dependent upon the type of photovoltaic device used. A device which can photoconvert more of the emitted radiation will increase the value of the emitter efficiency. It has been shown theoretically, at maximum temperature, as much as 70% of the total radiant energy is centered in a narrow band for some rare earth oxides (2). It has been shown experimentally at lower temperatures that as much as 42% of the radiant energy is centered between 1.0 and 2.0 μm in wavelength for specific rare earth oxide combinations (3).

The last component efficiency is the photoconversion efficiency which refers to the ability of the photovoltaic cell to photoconvert radiation to electrical power. There are many factors which influence the conversion efficiency some of which are cell bandgap energy, cell temperature, radiant intensity, series resistance, and front surface metal grid and optical reflection. Because TPV requires the use of relatively low temperatures (less than 2000 K), low bandgap materials are required for efficient photoconversion. Two of the more popular material systems for TPV cells are InGaAs and GaSb. A total radiant power to electrical power conversion efficiency for an InGaAs device illuminated by a rare earth oxide emitter has been reported to be 16.2% (4).

The Auburn University TPV research focuses on designing and building a small laboratory breadboard unit for power generation applications for the U. S. Army. The basic system includes the use of a combustion heat source, a selective emitter, and InGaAs photovoltaic cells. Previous research centered on determining the appropriate selective emitter system and matching the emitter to a corresponding photovoltaic cell. Current research has focused on designing a viable TPV system and implementing the design into a working breadboard prototype unit as well determining possible system improvements and their predicted results.

EXPERIMENTAL

Combustion Burner

The burner for the TPV system constructed for this research is shown in Figure 1. A Fisher Blast Burner was modified for the TPV system. The modified burner system consisted of a base section and emitter housing section. Two separate inlets into the burner base, for propane and air, were equipped with flow meters for accurate metering of the combustion gases.

The original mixing chamber of the Blast Burner was removed and replaced with a brass water-cooled mixing chamber. A circular metal wire screen was

278

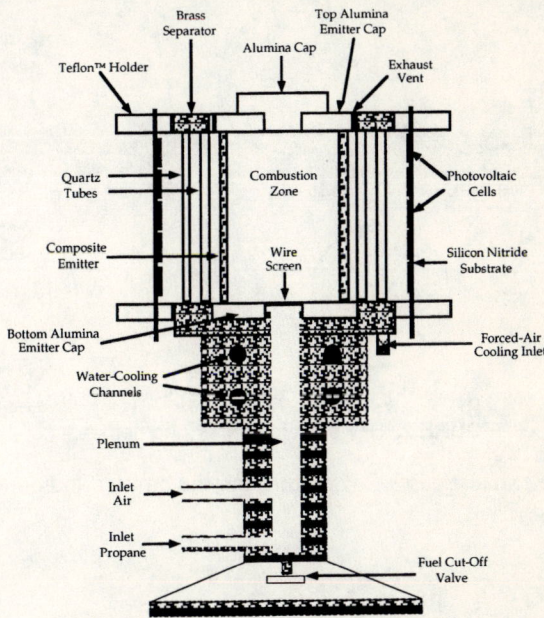

FIGURE 1. Diagram of TPV system burner.

attached to the brass mixing chamber. The screen was used to anchor the flame to the base. The water cooling and screen were used to prevent flashback of the flame into the mixing chamber. The emitter housing section consisted of two optical grade quartz tubes, a brass air manifold, and a brass separator. The emitter was placed inside the inner quartz tube. The tubes were placed concentrically into the brass air manifold used to cool the quartz tubes. There was a single pressurized air inlet and 40 small diameter air outlets uniformly spaced around the manifold in a circular pattern between the two quartz tubes. The cooling gases flowed out of the manifold between the quartz tubes through the brass separator.

TPV System Emitters

The emitters for the TPV burner system were fabricated using a patented process (5). The cylindrical emitters were 7.62 cm in height and 5.5 cm in diameter. Three different emitter thicknesses were studied to determine the optimum thickness for maximum radiant output. Several different emitter configurations were also studied to determine the best geometry for the system based upon radiant output. The various emitter configurations studied are shown in Figure 2.

The cylindrical emitters were cemented between two alumina insulation disks. Eighteen small diameter holes were drilled into the top disk around the outside to allow the combustion gases which flowed through the emitter to exhaust. The emitter was cemented to the disks with alumina adhesive. The emitter/disk assembly fit snugly into the inner quartz tube of the emitter housing. In this

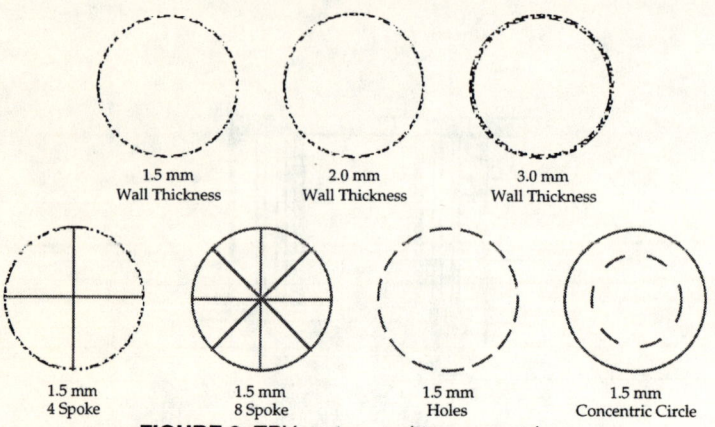

FIGURE 2. TPV system emitter geometries.

manner, all of the exhaust gases flowed through the emitter and out of the holes in the top disk.

Photovoltaic Cell Array

To obtain a significant amount of power, photovoltaic cells must form an array around the TPV system emitter. To project possible system performance, 0.75 eV InGaAs photovoltaic cells were purchased from the Research Triangle Institute. The 0.75 eV bandgap was chosen based upon previous test results using single InGaAs cells obtained from the NASA Lewis Research Center (4). Three cells with differing bandgaps of 0.60, 0.66, and 0.75 eV (6) were illuminated by various rare earth oxide composite emitters. Each cell had a total area of 1 cm². Five cells were placed in series on an alumina substrate. A diagram of the photovoltaic cell array is shown in Figure 3.

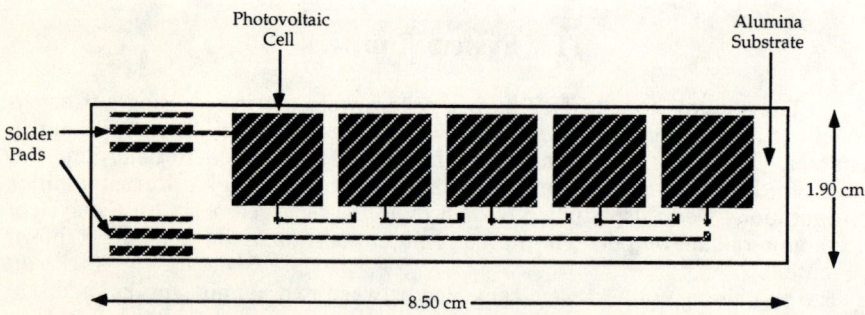

FIGURE 3. Diagram of photovoltaic cell array.

TPV System Setup

After all of the TPV system components had been designed and fabricated, the system as a whole was tested. Initial tests were performed on the burner system to determine the best emitter geometry to be used in the final system based upon radiant output. Once the final emitter geometry was chosen, the heat of combustion from fuel to radiant power conversion efficiency was determined as well as the radiant power to electrical power conversion efficiency for the system. Finally, the total system efficiency was calculated from heat of combustion to electrical power output.

The emitter geometry test involved heating various emitters by placing them inside the emitter housing section in the presence of combustion flame. Three different fuel flow rates were used to heat the emitters corresponding to three different emitter temperatures. A thin metal baffle with 5 mm diameter hole was placed between the emitter housing and thermopile sensor head of a broadband power meter. The sensor head was positioned between the top and bottom of the emitter housing section.

From initial results, the three best emitter candidates, based on highest radiant output power, were then studied for uniformity of emission. The thermopile sensor head was positioned 1.27 cm from the base of the emitter housing at normal incidence. The burner base was mounted on a rotary table and a preliminary scan around the emitter was performed to determine uniformity of radiant output as a function of azimuthal angle. It was determined that the radiant output had a 90° symmetry. Therefore, the burner was rotated only through 100° at 5° increments. The radiant power was recorded at each interval. The sensor head was then raised 1.27 cm and the procedure repeated for the entire height of the emitter. In this manner, the radiant uniformity of each emitter with given fuel flow rates was established.

The fuel flow into the burner was measured using flow meters allowing the equivalent fuel "power" to be determined. The radiant power emitted from the system emitter was determined using a power meter. The quartz tubes were removed from the emitter housing to prevent filtering of the radiant emissions. In this manner, a conversion efficiency for fuel to radiant power could be determined by dividing the fuel "power" into the radiant power of the emitter.

The total radiant power of the emitters was determined again with the power meter, however, the quartz tubes, which act as filters and as insulators of the combustion gases, were placed back into the emitter housing. The photovoltaic array was placed at normal incidence to the emitter housing. It is necessary to assume that little to no radiation is reflected back from the array to the emitter which would cause a localized increase in emitter temperature. The electrical power was determined by measuring the current and voltage of the photovoltaic array for different load resistances. The maximum power point was determined from a plot of these currents and voltages. The total radiant power could then be divided into the total electrical output power and the radiant to electrical power conversion efficiency established. Projected total electrical output was calculated based upon the electrical output of the photovoltaic array at a fixed distance from the emitter. In this manner, assuming an entire array was positioned completely around the emitter, the total electrical power output could be calculated for the array.

The temperature of the emitter was estimated by placing a thermocouple through one of the exhaust ports onto the surface of the emitter. The overall system conversion efficiency was determined by simply measuring the inlet fuel flow rate (fuel "power") and dividing it into the projected electrical output power of the photovoltaic array. This was done for several different fuel flow rates.

RESULTS AND DISCUSSION

Initial Emitter Geometry Test Results

The first goal of this experiment was to determine the three best emitter geometry candidates based on radiant output power for the burner system. It was determined that the best emitter geometries, based on radiant output, are the emitters with a wall thickness of 1.5 mm. There is a decrease in the radiant output of the emitter with increasing wall thickness. Unfortunately, a wall thickness less than 1.5 mm results in an emitter that is fragile and prone to breakage. Therefore, a compromise must be made between emitter structural integrity and radiant output. The 1.5 mm wall thickness, 1.5 mm wall thickness/4 spoke, and 1.5 mm wall thickness/8 spoke emitters were determined to be the three best emitter geometries.

To determine the radiant output as a function of azimuthal angle and height for the three emitters, the burner was mounted on a rotary table as described in the experimental section. The resultant scan for the 1.5 mm wall thickness emitter is shown in Figure 4. Analysis of the results show that the 1.5 mm wall thickness emitter demonstrated the most uniform output radiation. The 4 Spoke and 8 Spoke emitters both had higher relative intensities but also had non-uniform output radiation. Both of these geometries had considerably more radiation in the bottom center of the emitter and also varied in radiant output azimuthally. These problems are due to the inside spoke structures which interfere with the

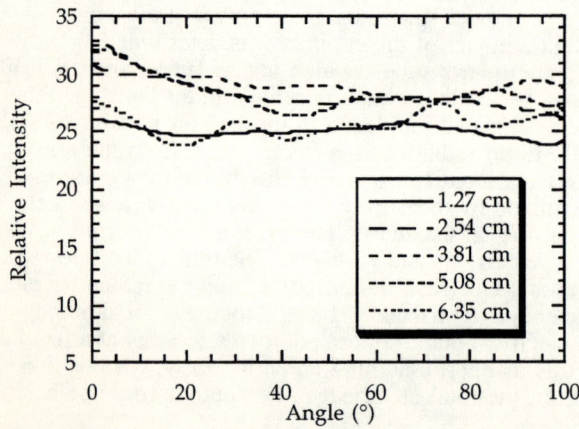

FIGURE 4. Radiant output as a function of azimuthal angle and height for 1.5 mm wall thickness emitter.

TABLE 1. Emitter radiant power density, total power, and fuel to radiant conversion efficiency for three input fuel powers for the TPV system

Input Fuel Power (W)	Radiant Power Density (W/cm^2)	Total Emitter Radiant Power (W)	Conversion Efficiency (%)
800	1.06	128	16.0
1070	1.48	179	16.7
1330	1.83	221	16.6

combustion gases flowing evenly through the emitter surface. Non-uniform radiant output will present system problems for the photovoltaic array due to the series connection of the cells. Hence, the 1.5 mm wall thickness emitter was chosen to be used in the final thermophotovoltaic system configuration.

Fuel to Radiant Power Conversion Efficiency Results

Fuel flow into the burner was measured with flow meters so that the equivalent input power could be determined. Three different powers of 800, 1070, and 1330 W were input into the burner. The radiant power was determined by measuring the radiant power of the system emitter without any quartz tubes present in the emitter housing section. The thermopile detector was positioned level with the base of the emitter at a fixed distance normal to the emitter surface to measure the radiant power. The results are listed in Table 1. Table 1 also lists the total power exiting the emitter as well as the fuel to radiant power conversion efficiency for each flow rate. Some assumptions that were made in this analysis were that the emitters radiate uniformly both with respect to height and azimuthal angle and that the emitters radiate diffusely.

The results show that the inlet fuel to radiant output conversion efficiencies are approximately 16% for the TPV system. One of the losses that can be attributed to lowering the conversion efficiency is end loss. This refers to the two insulating disks which are located at the top and bottom of the emitter structure. The two disks heat up during the combustion process and in turn lose energy by convection and radiation. For example, at an input fuel power of 1070 W, the top insulating disk reached a temperature of 755 K and the bottom insulating disk reached a temperature of 555 K. The bottom insulating disk was partially cooled by the water-cooled brass mixing chamber. At these two temperatures, assuming a disk emittance of 1.0 (worst case scenario), the top disk radiates 106 W and the bottom disk 26.6 W. An additional loss mechanism is convective heat loss which is estimated to be almost half as much as the radiant losses. These losses do not account for the stored energy in the two disks which is yet another end loss mechanism.

Radiant to Electrical Power Conversion Efficiency Results

In the fuel to radiant power conversion efficiency analysis, the quartz tubes in the emitter housing section were not present. For the radiant to electrical power conversion efficiency analysis, the quartz tubes were replaced in the emitter

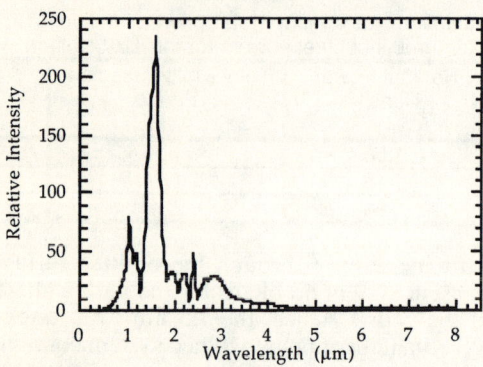

FIGURE 5. Erbia TPV system emitter spectrum.

housing section. To determine the effect of the quartz tubes on the radiant emissions, the emitter housing section of the TPV system was placed in front of a calibrated spectrometer system. The resultant radiant spectrum is shown in Figure 5.

As shown in Figure 5, the quartz tubes act as absorption filters which absorb all of the radiant emissions beyond 4.5 μm in wavelength. The peak at 1.5 μm in wavelength is due to erbia emission and the small bump at 2.9 μm is from propane and air combustion products. Unfortunately, the quartz tubes present two problems. The first is that the tubes must be cooled to prevent quartz radiation. The second problem is that each quartz tube only transmits approximately 95% of the in-band radiation. This results in attenuation of almost 10% of the in-band radiation when both quartz tubes are present in the TPV system. However, the benefits of using the quartz tubes outweigh the disadvantages. Without the quartz tubes, the cells would have to be placed further away from the system emitter to prevent interaction with hot combustion gases. This would result in a lower power density. There is only a slight decrease of in-band radiation whereas the out-of-band radiation decreases by almost 45%.

To determine the radiant to electrical power conversion efficiency, the radiant power density incident on the photocell surface must be measured. The projected photovoltaic array would be cylindrical in shape with a diameter of 10.0 cm and a height of 6.7 cm resulting in an irradiated surface area of 210 cm^2 for the TPV system. The circumference of this irradiated cylinder is approximately 31 cm. Therefore, space for approximately 30 photovoltaic arrays is available at this distance from the emitter surface. The radiant power density incident on the projected array as well as the total radiant power is listed in Table 2 for three different input fuel powers.

TABLE 2. Radiant power incident on projected array for TPV system

Input Fuel Power (W)	Radiant Power Density at Array Surface (W/cm^2)	Total Radiant Power Incident on Array (W)
800	0.121	18.2
1070	0.200	30.0
1330	0.256	38.4

TABLE 3. Projected electrical output power for TPV system

Input Power (W)	V_{oc} (V)	I_{sc} (A)	P_{max} (W)	Total Power (W)
800	1.91	0.122	0.16	4.80
1070	1.93	0.203	0.249	7.47
1330	2.12	0.221	0.324	9.72
1570	2.12	0.238	0.343	10.3
1800	2.13	0.290	0.400	12.0

The photovoltaic array was normal to the emitter and the burner heated to different input powers. The current-voltage curves for the array were generated by varying a load resistance and measuring the resultant currents and voltages. There was an improvement of electrical output with increasing illumination. This is due to an increase in the array's output current whereas the voltage remains fairly constant with increased illumination. Table 3 lists the results of the projected output power based on 30 arrays for 5 different input powers as well as the open-circuit voltage, short-circuit current, and maximum power for the array for the TPV system.

Table 3 shows that there is an increase in output electrical power with increasing input power. Unfortunately, at higher input powers the emitter structure showed signs of cracking due to the excessive amount of combustion gases inside the emitter. At the high flow rates required for high input powers, there is a large pressure drop between the inside and outside of the emitter even though the emitter is highly porous. There is also a tremendous strain on the cement which bonds the emitter to the insulating disks located at the top and bottom of the emitter.

The final step in determining the radiant to electrical conversion efficiency was to divide the radiant power incident on the projected array into the total projected electrical output. Table 4 lists the radiant to electrical power conversion efficiencies for the first three input powers for the TPV system. A current-voltage characteristic curve is shown in Figure 6 for the 0.75 eV InGaAs array with an input power of 1070 W. As can be seen in the figure, there is a significant series resistance problem with the array which leads to a reduction in the fill-factor and overall conversion efficiency.

Analysis of Table 4 shows that at least 25% of the incident radiation on the cells is converted to electrical energy, which is a very high value. As stated previously, the highest reported conversion efficiency to date using composite emitters illuminating InGaAs cells was 16.2%. It should be noted that none of these cells had an anti-reflection coating which would increase the conversion efficiency. Similarly, a reduction of the series resistance in the array interconnects would increase the conversion efficiency.

TABLE 4. Radiant to electrical conversion efficiencies

Input Fuel Power (W)	Radiant Power on Cells (W)	Projected Electrical Output Power (W)	Projected Conversion Efficiency (%)
800	18.2	4.80	26.5
1070	30.0	7.47	24.9
1330	38.4	9.72	25.3

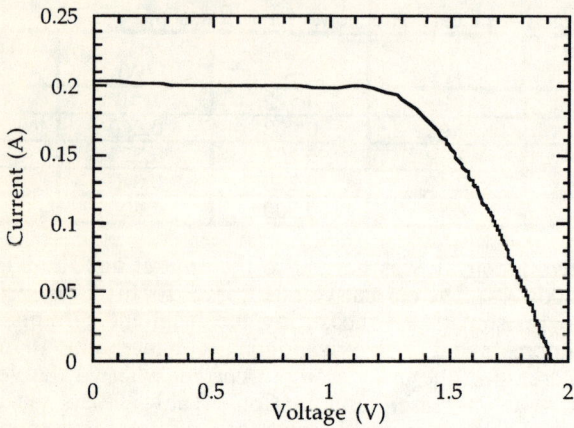

FIGURE 6. Current-voltage curve for photovoltaic array under erbia system emitter illumination.

TPV System Results

Now that the individual component efficiencies have been determined, the total TPV system efficiency can be easily calculated. The TPV system efficiency is simply the ratio of the input fuel power to output electrical power. Table 5 lists the results for the projected total system conversion efficiency calculations based upon the photovoltaic array results.

Although these conversion efficiencies are low, using very reasonable assumptions, the total system conversion efficiency can be expected to improve greatly as will be discussed in the next section. As can be seen for this TPV system, there is an improvement in system conversion efficiency up to a point with increase in input power. However, after a certain input power, the system efficiency does start to decrease as would be expected based on the decrease in thermal efficiency.

SYSTEM IMPROVEMENTS

The various system conversion efficiencies were discussed in the previous section. Due to the nature of the system configuration however, high system

TABLE 5. Predicted TPV system conversion efficiencies

Input Fuel Power (W)	Predicted Output Electrical Power (W)	Conversion Efficiency (%)
800	4.80	0.60
1070	7.47	0.70
1330	9.72	0.73
1570	10.3	0.66
1800	12.0	0.67

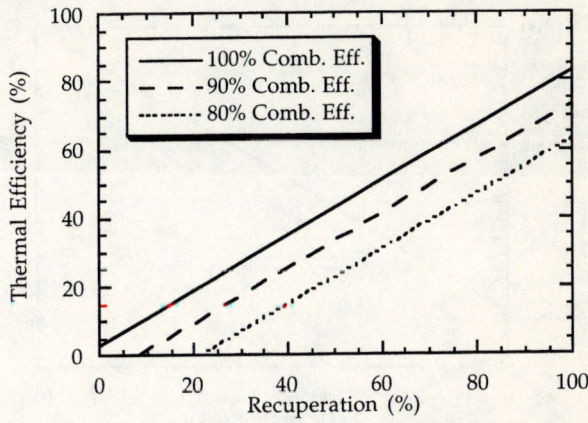

FIGURE 7. Effect of heat recuperation on system thermal efficiency (1).

conversion efficiencies were not expected. The TPV system designed and fabricated for this research allowed for ease of handling and versatility of design. Once the system was studied in operation, improvements could be suggested and their results predicted. The following section will discuss improvements which could be made to the system.

The fuel to radiant conversion efficiency for this system without heat recuperation was approximately 16% for the three input powers used. To increase this efficiency, heat recuperation must be used. This would in turn lower the required amount of inlet fuel and air which would heat the emitter to the same temperature without recuperation. Schroeder, et. al. (1), shows the dramatic results of heat recuperation for a system operating with JP-8 fuel at three different combustion efficiencies, using 110% theoretical air, and an emitter temperature of 2075 K. The emitter used in this analysis was ytterbia illuminating silicon photovoltaic cells. Figure 7 shows the thermal efficiency as a function of heat recuperation for three different combustion efficiencies (1).

Using a TPV system analysis program developed by Schroeder (7), the effect of recuperation on the fuel to radiant conversion efficiency for the system used in this research was calculated. The inlet fuel flow rate was set at 0.75 SLPM propane and 20.24 SLPM air for this calculation. Also, the emitter temperature was set at 1625 K and thermal losses were neglected. Figure 8 shows the calculated effect of heat recuperation on fuel to radiant conversion efficiency for an input power of 1070 W.

Figure 8 shows the dramatic effect of heat recuperation on the theoretical maximum fuel to radiant conversion efficiency. The maximum theoretical conversion efficiency is only 32.83% with no recuperation but is already 70.23% with only 50% recuperation. Of course this is a theoretical number which assumes a high emitter temperature and does not account for end losses which, as shown earlier, are considerable.

Reducing thermal losses is a necessity for improved system performance. There is a significant amount of energy which is not utilized by the radiant emitter. As shown earlier, more than 20% of the 1070 W input power is lost radiantly and convectively from the two insulating disks. This is due to the fact

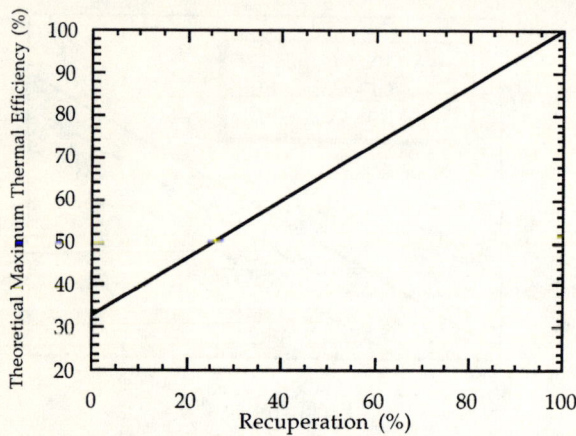

FIGURE 8. Effect of heat recuperation on fuel to radiant conversion efficiency with 1070 W input at 1350 K and no thermal losses.

that the total radiating emitter surface area is 121 cm^2 and the radiating area of the insulators is 109 cm^2. To reduce the ratio of the insulator surface area to emitter surface area, the emitter dimensions should be modified. The new emitter should be greater in height and smaller in diameter which would keep the emitter surface area constant and greatly reduce the area of the insulators. One problem that can be foreseen with this new geometry is maintaining the uniformity of emitter radiant output.

Another significant problem is that 683 W of thermal energy remains in the exhaust products for a 1070 W input, assuming an exhaust temperature equal to that of the outside emitter surface (1155 K) (7). Better coupling between the combustion products and emitter surface will ensure higher thermal to radiant efficiencies and reduce recuperation requirements.

The thermal losses can be quantified by reviewing the values of thermal input, radiant output, and exhaust temperatures. For this system, with 1070 W fuel equivalent input power, the measured radiant output is 179 W. The power remaining in the exhaust products was calculated using reference (7) and was determined to be 683 W. The remaining power, 208 W, was lost through end loss mechanisms such as radiation and convection. All of these losses must be significantly minimized to improve the fuel to radiant power conversion efficiency.

Another significant factor concerns the view factor of the photovoltaic cells. This refers to the percentage of radiation from the emitter which reaches the photovoltaic cell surface. In the present configuration, the photovoltaic cells are approximately 2.54 cm from the emitter surface. Because of this distance, the power radiated from the emitter surface is not the same as is incident on the photovoltaic array surface. This is in part due to attenuation of the quartz tubes. However, as stated previously, less than 10% of the photoconvertible radiation is actually absorbed in the quartz. Also, the quartz tubes are needed to prevent hot combustion gases from coming into contact with the photovoltaic array. A possible design improvement would be to increase the incident power on the photocells by decreasing the distance between emitter and cell surface or by concentration of the radiant energy. The key will be to maintain adequate cell cooling.

The cells used in the TPV system did not have an anti-reflection coating on the surface. Therefore, photoconvertible radiation reflected off of the cell surface. It is estimated that there would be an increase of at least 5% for the radiant to electrical conversion efficiency. This is based on analyzing the electrical output of the cells from the NASA Lewis Research Center which have an anti-reflection coating and the cells from the Research Triangle Institute which do not under the same composite emitter illumination. The 5% increase would raise the predicted radiant to electrical conversion efficiencies from approximately 25% to 30% based on the photovoltaic array for the three input powers.

It is also possible to improve array performance by reducing the series resistance. The effect of series resistance losses can be seen in the fill factor of the array. A lower fill factor indicates a higher series resistance in the cells. Series resistance occurs both inside the cell and through the interconnects. At present, the cells are connected with electrically conductive epoxy, wire bonds, and screen printed patterns on an alumina substrate. To reduce series resistance losses, monolithically integrated structures are currently being investigated at the NASA Lewis Research Center (8).

Incorporating these improvements into the system would result in a system with much greater output power and improved system efficiency. For example, assuming a conservative fuel to radiant conversion efficiency improvement from 16% to 70%, a 17% increase in radiant power incident on the array by improving the view factor, and a 5% increase in the radiant to electrical conversion efficiency would result in a modest system efficiency of 6.3%. The main problem that will be encountered will be successful integration of the recuperator into the TPV system.

CONCLUSIONS

The basic components of a combustion TPV system have been studied and assembled into a working system to determine how these components work together as a unit. Composite selective emitters used in TPV power generation were shown to be a feasible way to generate electrical power from a gaseous fuel source. By using selective emitters, several advantages are gained. One advantage is that less fuel is required to heat a selective emitter to a given temperature compared to a blackbody emitter, which in turn reduces the heat source requirements (9). Similarly, the photoconversion process of the photovoltaic device is improved by using selective emitters because almost all of the radiation is centered in a narrow band which is matched to the bandgap of the photocell. Thus, there is minimal lattice heating which in turn reduces the photoconversion efficiency.

The various TPV system components were studied to determine their respective conversion efficiencies as well as the total system efficiency. Because of the nature of the TPV system designed for this research, the component efficiencies and total system efficiency were low. However, high radiant to electrical conversion efficiencies (approaching 30%) were demonstrated using selective emitters illuminating 0.75 eV InGaAs cells. Insights into future system improvements were established including the use of heat recuperation, decreasing end losses, view factor improvements, and the use of anti-reflection coatings on the photovoltaic cells.

TPV offers a viable alternative to traditional power generation sources. It is clean, quiet, and offers many potential applications for both commercial and military needs. As shown previously, using very reasonable assumptions, modest system conversion efficiencies of over 6% can be expected with system modification. The input fuel requirements for this system should be reduced by more than one-half simply by using recuperation (1).

The main thrust of this research was to design and build a small breadboard system which would illustrate TPV power technology incorporating composite emitters. Upon completion of the TPV system, improvements were illustrated and discussed for future designs.

ACKNOWLEDGEMENTS

The authors would like to thank the Army Research Office for their funding and support under research grant DAALO39260205-1, as well as the Department of Energy under research grant DE-FG05 95ER12156.

REFERENCES

1. K. L. Schroeder, M. F. Rose, and J. E. Burkhalter, "A Parametric Study of TPV Systems and the Importance of Thermal Management in System Design and Optimization," 30th Intersociety Energy Conversion Engineering Conference, Vol. 1, 1995.

2. Chubb, D. L., "Reappraisal of Solid Selective Emitters," *NASA Tech. Memo 103290,* 1990.

3. Z. Chen, P. L. Adair, and M. F. Rose, "Selective Emitters for Thermophotovoltaic Energy Converters-A Study Based on Material Prospects," 31st Intersociety Energy Conversion Engineering Conference, Vol. 2, 1996, pp. 1013-17.

4. P. L. Adair, Z. Chen, and M. F. Rose, "Photoelectric Conversion Efficiencies for InGaAs Photovoltaic Cells Illuminated by Composite Emitters," 31st Intersociety Energy Conversion Engineering Conference, Vol. 2, 1996, pp. 1018-22.

5. M. F. Rose and P. Adair, "Selective Infrared Line Emitters," U. S. Patent No. 5,447,786. Filed May 25, 1994. Issued September 5, 1995.

6. D. M. Wilt, N. S. Fatemi, R. W. Hoffman, Jr., P. P. Jenkins, D. J. Brinker, D. Scheiman, R. Lowe, M. Fauer, and R. K. Jain, "High Efficiency Indium Gallium Arsenide Photovoltaic Devices for Thermophotovoltaic Power Systems," Applied Physics Letters, Vol. 64, No. 18, May 2, 1994.

7. K. L. Schroeder, "A Model for Predicting Performance and Optimizing Thermophotovoltaic Systems," Dissertation to be Submitted for Partial Requirements for the Degree of Doctor of Philosophy, 1997.

8. M. F. Rose, editor, Prospector VIII: Thermophotovoltaics--An Update on DoD, Academic, and Commercial Research, July 14-17, 1996.

9. P. L. Adair and M. F. Rose, "Increased Thermophotovoltaic System Efficiency Using Selective Emitters," 30th Intersociety Energy Conversion Engineering Conference, 1995.

Effect of Temperature Gradient on Thick Film Selective Emitter Emittance

Donald L. Chubb,* Brian S. Good,* Eric B. Clark*
and Zheng Chen[+]

NASA Lewis Research Center, Cleveland, Ohio 44135
[+]Auburn Space Power Institute, Auburn, Alabama 36849-5320

Abstract. A temperature gradient across a thick ($\geq$ .1 mm) film selective emitter will produce a significant reduction in the spectral emittance from the no temperature gradient case. Thick film selective emitters of rare earth doped host materials such as yttrium-aluminum-garnet (YAG) are examples where temperature gradient effects are important. In this paper a model is developed for the spectral emittance assuming a linear temperature gradient across the film. Results of the model indicate that temperature gradients will result in reductions the order of 20% or more in the spectral emittance.

INTRODUCTION

Emission from thick films is not a surface phenomenon as is usually assumed when discussing emissive materials. It depends on the geometry of the material, which for the film emitters means the film thickness. Thus radiation leaving the film originates at various depths within the film.

To model these film emitters we use a macroscopic approach. That is we solve the radiative transfer equation that applies for Boltzmann equilibrium of excited state densities and includes stimulated emission and absorption, as well as, spontaneous emission and scattering of radiation. These atomic processes manifest themselves on the macroscopic scale through the extinction coefficient, α_λ.

The product of the extinction coefficient, α_λ, and the film thickness, d, $\alpha_\lambda d = K_d$, which is usually called the optical depth, will determine the spectral emittance if the temperature is a constant through the film. However, for thick films ($\geq$ .1 mm) the temperature gradients are not negligible (> 100°K) so the emittance model must

CP401, *Thermophotovoltaic Generation of Electricity: Third NREL Conference*,
edited by Benner/Coutts
© 1997 The American Institute of Physics 1-56396-734-0/97/$10.00

include a variable temperature through the film. In the analysis to follow we assume a linear temperature variation across the film. This is the result that will occur if thermal conduction dominates radiative energy transfer. In the case where $d \leq 1$ mm this is a good assumption for the rare-earth selective emitters we are considering (3).

In the following section the emittance model will be developed. Following that, two approximate expressions for the spectral emittance, ε_λ, that apply when scattering is neglected and the temperature gradient is small will be presented. The first approximation is applicable for large optical depth, K_d, and the second approximation applies for small optical depth. Both of these approximations are compared to the exact result for ε_λ, neglecting scattering but for any temperature gradient, obtained by a numerical solution of the governing equations. Following that a discussion of the optimum film thickness to obtain maximum emittance will be presented. Finally, spectral emittance results will be compared to experimental results obtained for an erbium oxide (Er_2O_3) selective emitter that has an emission band centered at a photon wavelength, $\lambda = 1.5 \mu m$.

THICK FILM EMITTANCE MODEL

The emittance model for the thick film emitter has been previously developed for the case of no temperature gradient (1, 4). This model can be extended to include a temperature gradient across the film. The model is based on the radiative transfer equation (5), which is macroscopic in nature. Thus the emissive, absorptive and scattering properties of the material, which depend on the atomic structure, are expressed through the extinction coefficient, α_λ. The key parameter in determining the spectral emittance, ε_λ, is the optical depth, $K = \alpha_\lambda d$.

Consider Figure 1 which is a schematic drawing of a thick film emitter. Thermal energy enters through the film substrate. Part or all of the thermal input leaves the film at $x = d$ as radiation flux, $Q_\lambda(K_d)$. To determine ε_λ, $Q_\lambda(K_d)$ must be calculated since ε_λ is defined as follows.

$$\varepsilon_\lambda \equiv \frac{Q_\lambda(K_d)}{e_{bs}(\lambda, T_s)} \tag{1}$$

Where $e_{bs}(\lambda, T_s)$ is the blackbody emissive power and T_s is the substrate temperature.

$$e_{bs} = \pi i_{bs} = \frac{2\pi h c_o^2}{\lambda^5 \left[\exp(hc_o / \lambda k T_s) - 1\right]} \qquad (2)$$

Here h is Plank's constant, k is Boltzmann's constant, c_o is the vacuum speed of light, and i_{bs}, is the blackbody intensity. Notice that ε_λ has been defined in terms of the substrate temperatures, T_s. The spectral emittance could be defined in terms of the film surface temperature, T_f, or some combination of T_f and T_s. However, defining ε_λ in terms of T_s means $\varepsilon_\lambda \leq 1$ in all cases since $e_{bs}(\lambda, T_s) \geq Q_\lambda(K_d)$. This definition agrees with the usual concept of emittance.

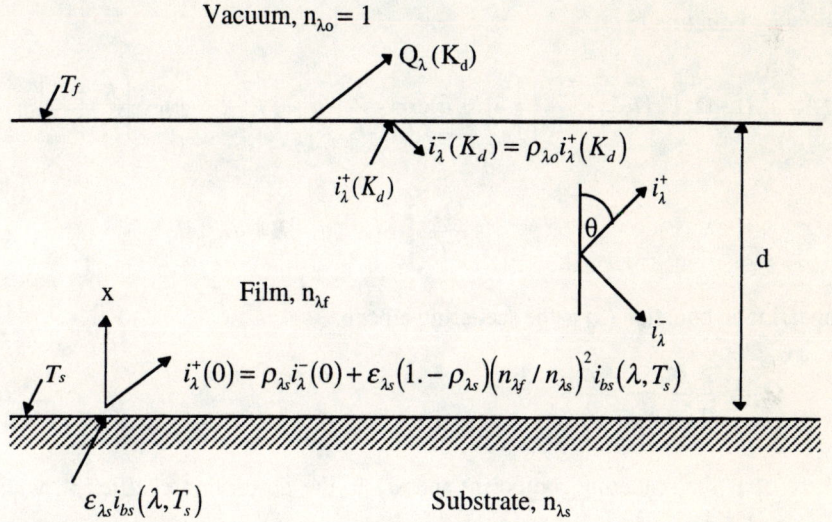

n_λ = index of refraction
$\rho_{\lambda o}$ = reflectance at film-vacuum interface
$\rho_{\lambda s}$ = reflectance at film-substrate interface
$\varepsilon_{\lambda s}$ = emittance of substrate
$i_{bs}(\lambda, T_s)$ = blackbody intensity for $T = T_s$

FIGURE 1. Schematic Diagram of Thick Film Emittance Model

To calculate Q_λ we require the radiative transfer equations for radiation intensity moving in the + x direction, $i_\lambda^+(K,\cos\theta)$, and intensity in the - x direction, $i_\lambda^+(K,\cos\theta)$, (5).

$$i_\lambda^+(K,\mu) = i_\lambda^+(0,\mu)\exp\left[-\frac{K}{\mu}\right] + \int_0^{K_d} S(K^*,\mu)\exp\left[\frac{K-K^*}{\mu}\right]\frac{dK^*}{\mu} \tag{3}$$

$$0 \le \mu = \cos\theta \le 1$$

$$i_\lambda^-(K,\mu) = i_\lambda^-(K_d,\mu)\exp\left[-\frac{K_d-K}{\mu}\right] - \int_0^{K_d} S(K^*,\mu)\exp\left[\frac{K^*-K}{\mu}\right]\frac{dK^*}{\mu} \tag{4}$$

$$-1 \le \mu = \cos\theta \le 0$$

In using these equations we are assuming that y and z variation of intensity can be neglected. Appearing in equations (3) and (4) is the so-called source function, $S(K,\mu)$, which in the case of isotopic scattering ($S(K,\mu) = S(K)$) satisfies the following equation (5).

$$S(K) = n_{\lambda f}^2(1-\Omega_\lambda)i_{\lambda b}(T,\lambda) + \frac{\Omega_\lambda}{2}\left\{\int_0^1 i_\lambda^+(0,\mu)\exp\left[-\frac{K}{\mu}\right]d\mu + \int_0^1 i_\lambda^-(K_d,-\mu)\exp\left[\frac{K_d-K}{-\mu}\right]d\mu\right\}$$

$$+\frac{\Omega_\lambda}{2}\int_0^{K_d} S(K^*)E_1\left(|K^*-K|\right)dK^* \tag{5}$$

Appearing in equation (5) is the scattering albedo.

$$\Omega_\lambda = \frac{\sigma_\lambda}{\sigma_\lambda + a_\lambda} = \frac{\sigma_\lambda}{\alpha_\lambda} \tag{6}$$

Where σ_λ is the scattering coefficient and a_λ is the absorption coefficient, which have the dimensions, cm^{-1}. The sum of σ_λ and a_λ is the extinction coefficient, α_λ. Also appearing in equation (5) is the film index of refraction, $n_{\lambda f}$, and the exponential integral, $E_1(x)$.

The general exponential integral, $E_n(x)$, is defined as follows.

$$E_n(x) \equiv \int_0^1 v^{n-2}\exp\left[-\frac{x}{v}\right]dv \tag{7}$$

Note that we are assuming isotropic scattering. As a result, S is independent of $\mu = \cos\theta$. Therefore, assuming diffuse boundary intensities, $i_\lambda^+(0,\mu) = i_\lambda^+(0)$ and $i_\lambda^+(K_d,\mu) = i_\lambda^+(K_d)$ we see from equations (3) and (4) that i_λ^+ and i_λ^- are also independent of μ.

The diffuse (independent of μ) boundary conditions at $K = K_d$ and $K = 0$ are the following.

$$i_\lambda^-(K_d) = \rho_{\lambda o} i_\lambda^+(K_d) \qquad\qquad \text{at } K = K_d \qquad (8a)$$

$$i_\lambda^+(0) = \rho_{\lambda s} i_\lambda^-(0) + (1 - \rho_{\lambda s}) \varepsilon_{\lambda s} (n_{\lambda f}/n_{\lambda s})^2 i_{bs}(\lambda, T_s) \qquad \text{at } K = 0 \qquad (8b)$$

Equation (8a) states that the radiation leaving the film-vacuum interface in the -x direction is equal to the reflected radiation at that interface. For the film-substrate interface equation (8b) states that $i_\lambda^+(0)$ is the sum of the reflected radiation and the radiation emitted by the substrate that is transmitted $(1-\rho_{\lambda s})$ through that interface. The $(n_{\lambda f}/n_{\lambda s})^2$ term accounts for refraction at the interface (5, pg 738). The reflectance at the film-vacuum interface is $\rho_{\lambda o}$ and the reflectance at the film-substrate interface is $\rho_{\lambda s}$. In the previous studies (1,3,4) the transmittance, $(1-\rho_{\lambda s})$, at the film-substrate interface was assumed to be 1 and the refraction term $(n_{\lambda f}/n_{\lambda s})^2$ was neglected. We approximate $\rho_{\lambda o}$ and $\rho_{\lambda s}$ by the reflectance for normal incidence, (5)

$$\rho_{\lambda o} = \left(\frac{n_{\lambda f} - 1}{n_{\lambda f} + 1}\right)^2 \qquad\qquad (9)$$

$$\rho_{\lambda s} = \left(\frac{n_{\lambda s} - n_{\lambda f}}{n_{\lambda s} + n_{\lambda f}}\right)^2 \qquad\qquad (10)$$

Where, $n_{\lambda s}$ is the substrate index of refraction.

At the film-substrate and film-vacuum interfaces there is the possibility of total reflection occuring. At an interface between a material with an index of refraction, n_ℓ, and a material with index of refraction n_m, where $n_\ell > n_m$, radiation moving from ℓ into m with an angle of incidence, $\theta > \theta_{\ell m}$, where $\theta_{\ell m}$ is given by Snell's law will be totally reflected. This will be taken into account when calculating $Q_\lambda(K_d)$. At the film-substrate interface refraction has been taken into account by including the

$(n_{\lambda f}/n_{\lambda s})^2$ term in equation (8b). However, the possibility of total reflection is not included. Therefore, by using equation (8b) as the boundary condition we are assuming that $n_{\lambda f} > n_{\lambda s}$ so that total reflection does not occur for radiation entering the film from the substrate.

Now consider $Q_\lambda(K_d)$, which is the radiation flux leaving the film. Since $n_{\lambda f} > 1$ the radiation leaving the film will be refracted and some of the radiation that reaches the film-vacuum interface will be totally reflected at the interface. Therefore,

$$Q_\lambda\left(K_d\right) = 2\pi \int_{\theta=0}^{\theta_M} \left[i_\lambda^+\left(K_d, \cos\theta\right) - i_\lambda^-\left(K_d, \cos\theta\right)\right] \cos\theta \, \sin\theta \, d\theta \qquad (11a)$$

and using equation (8a) and letting $\mu = \cos\theta$ this becomes the following.

$$Q_\lambda\left(K_d\right) = 2\pi \left(1 - \rho_{\lambda o}\right) \int_{\mu_M}^1 i_\lambda^+\left(K_d, \mu\right)\mu \, d\mu \qquad (11b)$$

Where μ_M is given by Snell's Law.

$$\mu_M^2 = \cos^2\mu_M = 1 - n_{\lambda f}^{-2} \qquad (12)$$

Substituting (3) in (11b) yields the following.

$$Q_\lambda\left(K_d\right) = \left(1 - \rho_{\lambda o}\right)\left[2\pi i_\lambda^+(0)h_- + \Phi_+ - \Phi_M\right] \qquad (13)$$

Where,

$$h_- = E_3\left(K_d\right) - \mu_M^2 E_3\left(\frac{K_d}{\mu_M}\right) \qquad (14)$$

$$\Phi_+ = 2\pi \int_0^{K_d} S(K) \, E_2 \, (K_d - K) \, dK \qquad (15)$$

$$\Phi_M = 2\pi\mu_M \int_0^{K_d} S(K) \, E_2 \left(\frac{K_d - K}{\mu_M}\right) dK \qquad (16)$$

298

Equation (13) gives $Q_\lambda(K_d)$ in terms of the source function S(K) and $i_\lambda^+(0)$. The $i_\lambda^+(0)$ intensity is obtained by using equations (3) and (4) to get two simultaneous equations for $i_\lambda^+(K_d)$ and $i_\lambda^-(0)$. These can then be solved for $i_\lambda^-(0)$ and the result used in equation (8b) to obtain $i_\lambda^+(0)$ (4).

$$\pi i_\lambda^+(0) = q^+(0) = \frac{1}{D}\left[\left(\frac{n_{\lambda f}}{n_{\lambda s}}\right)^2 (1-\rho_{\lambda s})\varepsilon_{\lambda s}e_{bs}(\lambda,T_s)\right. \tag{17}$$

$$\left. +2\rho_{\lambda 0}\rho_{\lambda s}E_3(K_d)\Phi_+ + \rho_{\lambda s}\Phi_-\right]$$

Where,

$$D = 1 - 4\rho_{\lambda 0}\rho_{\lambda s}E_3^2(K_d) \tag{18}$$

$$\Phi_- = 2\pi\int_0^{K_d} S(K)\,E_3(K)\,dK \tag{19}$$

Now substitute equation (17) in (13).

$$Q_\lambda(K_d) = \frac{1-\rho_{\lambda 0}}{D}\left\{2\left[\varepsilon_{\lambda s}\left(\frac{n_{\lambda f}}{n_{\lambda s}}\right)^2 (1-\rho_{\lambda s})e_{bs}(\lambda,T_s)+\rho_{\lambda s}\Phi_-\right]h_- + \Phi_+ h_+ - \Phi_M D\right\} \tag{20}$$

Where,

$$h_+ = 1 - 4\rho_{\lambda 0}\rho_{\lambda s}\mu_M^2 E_3(K_d)E_3\left(\frac{K_d}{\mu_M}\right) \tag{21}$$

Equation (20) can be substituted in equation (1) to obtain the spectral emittance, ε_λ, in terms of the source function, S(K). In the general case where scattering exists the source function must be obtained by solving equation (5). In the case of no scattering, $\Omega_\lambda = 0$, and equation (5) reduces to the following.

$$S(K) = n_{\lambda f}^2 i_b(\lambda,T) \tag{22}$$

If we also assume T is a constant through the film, $T = T_s$, then the integrations in Φ_+, Φ_-, and Φ_M, can be carried out to yield the following.

$$\varepsilon_{\lambda o} = \frac{n_{\lambda f}^2(1-\rho_{\lambda 0})}{D}\left\{2h_-\left[\frac{\varepsilon_{\lambda s}(1-\rho_{\lambda s})}{n_{\lambda s}^2}+\rho_{\lambda s}(1-2E_3(K_d))\right]+h_+\left[1-2E_3(K_d)\right]\right.$$
$$\left.-\mu_M^2 D\left[1-2E_3\left(\frac{K_d}{\mu_M}\right)\right]\right\}$$

constant temperature, no scattering   (23)

Thus ε_λ is determined by the optical depth, K_d, the indices of refraction, $n_{\lambda f}$ and $n_{\lambda s}$ and the substrate emittance, $\varepsilon_{\lambda s.}$ In the case when scattering is important ε_λ will also be a function of the scattering albedo, Ω_λ.

Now consider the case where a temperature gradient exists. To demonstrate the temperature gradient effects in the simplest manner we consider the no scattering case since in that case the source function has the simple solution given by equation (22). We also assume a linear temperature gradient across the film. As discussed in the introduction this is a good approximation for the rare earth selective emitters. As a result, the temperature across the film is given by the following expression.

$$\frac{T}{T_s} = 1-\Delta T\left(\frac{x}{d}\right) = 1-\Delta T\left(\frac{K}{K_d}\right)$$ (24)

Where , the temperature gradient is defined as follows.

$$\Delta T \equiv \frac{T_s - T_f}{T_s}$$ (25)

Using equations (24), (22) and (2) in the expressions for Φ_+, Φ_-, and Φ_M, yields the following.

$$\Phi'_+ = \frac{\Phi_+}{2n_{\lambda f}^2 e_{bs}(\lambda, T_s)} = (e^u-1)K_d\int_{\upsilon=0}^1 \frac{E_2[K_d(1-\upsilon)]}{\exp\left[\dfrac{u}{1-\upsilon\Delta T}\right]-1}\,d\upsilon$$ (26)

$$\Phi'_- = \frac{\Phi_-}{2n_{\lambda f}^2 e_{bs}(\lambda, T_s)} = (e^u-1)K_d\int_{\upsilon=0}^1 \frac{E_2(K_d\upsilon)}{\exp\left[\dfrac{u}{1-\upsilon\Delta T}\right]-1}\,d\upsilon$$ (27)

$$\Phi'_M = \frac{\Phi_M}{2n_{\lambda f}^2 e_{bs}(\lambda, T_s)} = \mu_M (e^u - 1) K_d \int_{\upsilon=0}^{1} \frac{E_2 \left[\dfrac{K_d}{\mu_M}(1-\upsilon) \right]}{\exp \left[\dfrac{u}{1-\upsilon\Delta T} \right] - 1} \, d\upsilon \qquad (28)$$

Where,

$$u = \frac{hc_0}{\lambda k T_s} \qquad (29)$$

$$\upsilon = \frac{K}{K_d} \qquad (30)$$

Equations (26) - (28) can be used in equations (20) and (1) to obtain ε_λ.

$$\varepsilon_\lambda = \frac{2n_{\lambda f}^2 (1-\rho_{\lambda o})}{D} \left\{ \left[\frac{\varepsilon_{\lambda s}(1-\rho_{\lambda s})}{n_{\lambda s}^2} + 2\rho_{\lambda s}\Phi'_- \right] h_- + \Phi'_+ h_+ - \Phi'_M D \right\} \qquad (31)$$

no scattering, with temperature gradient

As equations (26) - (28) indicate Φ'_+, Φ'_-, and Φ'_M, are functions of ΔT. The integrations in equations (26) - (28) must be carried out numerically. However, for small ΔT approximations to the integrals can be made. In most cases of interest for selective emitters, $(\lambda \le 7\mu m, T_s \le 2000K)$, the dimensionless photon energy, u, is greater than 1. Therefore, the following approximations can be made.

$$\left[\exp\left(\frac{u}{1-\upsilon\Delta T} \right) - 1 \right]^{-1} \approx \exp\left[\frac{-u}{1-\upsilon\Delta T} \right] \qquad e^u \gg 1 \qquad (32a)$$

$$e^u - 1 \approx e^u \qquad e^u \gg 1 \qquad (32b)$$

In addition for $\Delta T \ll 1$ and $0 \le \upsilon \le 1$;

$$\exp\left[\frac{-u}{1-\upsilon\Delta T} \right] \approx e^{-u} e^{-u\Delta T\upsilon} \qquad e^u \gg 1, \Delta T \ll 1 \qquad (33)$$

With the approximations given by equations (32) and (33) equations (26) - (28) become the following after a change in the integration variables.

$$\Phi'_+ \approx e^{-u\Delta T} \int_0^{K_d} \exp\left[\frac{Ku\Delta T}{K_d}\right] E_2(K)dk \qquad (34)$$

$$\Phi'_- \approx \int_0^{K_d} \exp\left[\frac{-Ku\Delta T}{K_d}\right] E_2(K)dk \qquad (35)$$

$$\Phi'_M \approx \mu_M^2 e^{-u\Delta T} \int_0^{\frac{K_d}{\mu_M}} \exp\left[\frac{K\mu_M u\Delta T}{K_d}\right] E_2(K)dk \qquad (36)$$

For a selective emitter the optical depth, K_d, will be large ($K_d>1$) in the emission band and small ($K_d<<1$) outside the emission band. Therefore, consider the two limiting cases; $\dfrac{u\Delta T}{K_d}<<1$ and $\dfrac{K_d}{u\Delta T}<<1$. For the case where $\dfrac{u\Delta T}{K_d}<<1$, integration by parts using

$$E_{n-1}(x) = -\frac{dE_n(x)}{dx} \qquad (37)$$

results in the following to first order in $\dfrac{u\Delta T}{K_d}$.

$$\Phi'_+ \approx \frac{1}{2}e^{-u\Delta T} - E_3(K_d) - \left[E_4(K_d) - \frac{1}{3}e^{-u\Delta T}\right]\frac{u\Delta T}{K_d} \qquad (38)$$

$$\Phi'_- \approx \frac{1}{2} - e^{-u\Delta T}E_3(K_d) + \left[e^{-u\Delta T}E_4(K_d) - \frac{1}{3}\right]\frac{u\Delta T}{K_d} \qquad \frac{u\Delta T}{K_d}<<1 \qquad (39)$$

$$\Phi'_M \approx \mu_M^2\left\{\frac{1}{2}e^{-u\Delta T} - E_3\left(\frac{K_d}{\mu_M}\right) - \left[E_4\left(\frac{K_d}{\mu_M}\right) - \frac{1}{3}e^{-u\Delta T}\right]\frac{u\Delta T}{K_d}\right\} \qquad (40)$$

Since we are interested in showing the effect of temperature gradient on ε_λ we define the following quantity.

$$\Delta\varepsilon_\lambda \equiv \varepsilon_{\lambda 0} - \varepsilon_\lambda \qquad (41)$$

Where $\varepsilon_{\lambda 0}$ is the emittance for no temperature gradient and is given by equation (23). By using $\Delta\varepsilon_\lambda$ to demonstrate the temperature gradient effect the dependence

302

on substrate emittance, $\varepsilon_{\lambda s}$, is removed. Therefore, using equations (38) - (40) in (31) and equation (23) for $\varepsilon_{\lambda 0}$ results in the following.

$$\Delta\varepsilon_\lambda \approx \frac{2n_{\lambda f}^2(1-\rho_{\lambda 0})}{D}\left\{\frac{1}{2}\left[h_+ - 4\rho_{\lambda s}h_- E_3(K_d) - \mu_M^2 D\right]\left[1-e^{-u\Delta T}\right]\right.$$

$$+\left[2\rho_{\lambda s}\left(\frac{1}{3}-e^{-u\Delta T}E_4(K_d)\right)h_- - \left(\frac{e^{-u\Delta T}}{3}-E_4(K_d)\right)h_+ \right.$$

$$\left.\left.-\mu_M^3\left(\frac{e^{-u\Delta T}}{3}-E_4\left(\frac{K_d}{\mu_M}\right)D\right)\right]\frac{u\Delta T}{K_d}\right\}$$

$$\text{no scattering, } \Delta T \ll 1, \ e^u \gg 1, \ \frac{u\Delta T}{K_d} \ll 1 \qquad (42)$$

Notice that if $u\Delta T \ll 1$ then $\Delta\varepsilon$ as given by equation (42) will be a linear function of ΔT. However, if $u\Delta T$ is not small then $\Delta\varepsilon_\lambda \sim (1-e^{-u\Delta T})$ provided $\frac{u\Delta T}{K_d} \ll 1$.

Also, note by looking at equation (31) that $\Delta\varepsilon_\lambda$ is independent of the substrate emittance.

Now consider $\Delta\varepsilon_\lambda$ for the case where $K_d \ll 1$. In that case $E_2(K)$ can be expanded in a power series and the integrations in equations (34) - (36) performed. To first order in $\frac{K_d}{u\Delta T}$ the results are the following.

$$\Phi'_+ = \Phi'_- \approx (1-e^{-u\Delta T})\frac{K_d}{u\Delta T} \qquad (43)$$

$$\frac{K_d}{u\Delta T} \ll 1$$

$$\Phi'_M \approx \mu_M^2(1-e^{-u\Delta T})\frac{K_d}{\mu_M u\Delta T} \qquad (44)$$

If equations (43) and (44) are used in (31) and equation (23) for $\varepsilon_{\lambda 0}$ then $\Delta\varepsilon_\lambda$ becomes the following when the approximation $E_3(K_d) \approx \frac{1}{2}-K_d$ is used.

$$\Delta\varepsilon_\lambda = \frac{2n_{\lambda f}^2(1-\rho_{\lambda 0})}{1-\rho_{\lambda 0}\rho_{\lambda s}}\left\{1+(1-\mu_M^2)\rho_{\lambda f}-\mu_M[1-\rho_{\lambda 0}\rho_{\lambda s}(1-\mu_M)]\right\}K_d\left[1-\frac{1-e^{-u\Delta T}}{u\Delta T}\right]$$

$$\text{no scattering, } \Delta T \ll 1, e^u \gg 1, \frac{K_d}{\mu\Delta T} \ll 1. \qquad (45)$$

303

Again, if $u\Delta T \ll 1$ then $\Delta\varepsilon_\lambda$ will be approximately a linear function of ΔT just as in the case of $\dfrac{u\Delta T}{K_d} \ll 1$. Also note that $\Delta\varepsilon_\lambda$ is a linear function of K_d and that $\varepsilon_{\lambda s}$ has no effect on $\Delta\varepsilon_\lambda$.

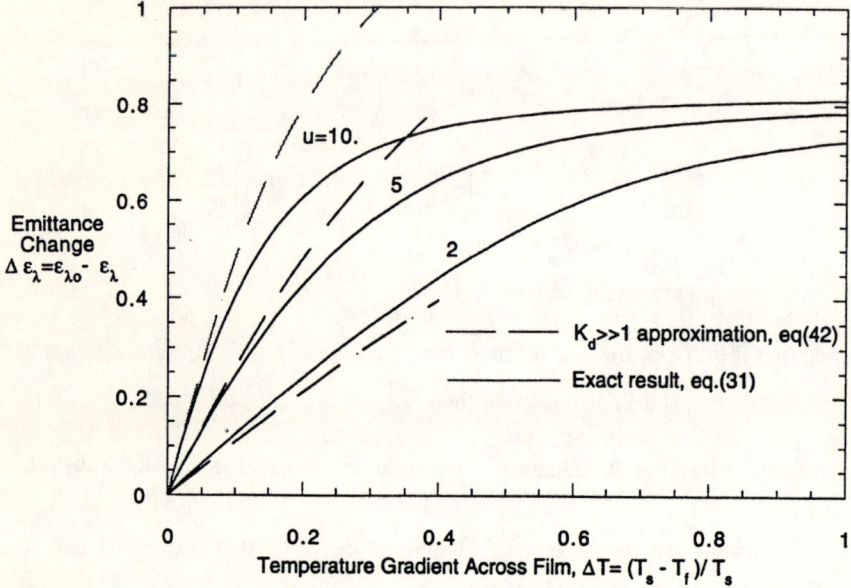

FIGURE 2. Emittance change as a function of temperature gradient at large optical depth, $K_d = 2$, for several dimensionless photon energies, $u = hc_0/\lambda kT_s$ with $n_{\lambda s} = 10$ and $n_{\lambda f} = 1.9$.

TEMPERATURE GRADIENT EFFECT ON SPECTRAL EMITTANCE FOR NO SCATTERING

Comparison of Exact and Approximate Solutions for Spectral Emittance

With the results developed in the previous section we can now illustrate the effect of ΔT on ε_λ. In Figure 2 $\Delta\varepsilon_\lambda$ is shown as a function of ΔT for large optical depth $\left(K_d = 2\right)$ at several values of u. The exact result for $\Delta\varepsilon_\lambda$ is obtained using equation (31) for ε_λ and numerical integration to obtain Φ'_+, Φ'_- and Φ'_M. Also, the $\dfrac{u\Delta T}{K_d} \ll 1$ result for $\Delta\varepsilon_\lambda$ (equaiton (42)) is shown in Figure 2.

As Figure 2 indicates $\Delta\varepsilon_\lambda$ changes rapidly at small ΔT with the slope increasing for increasing u. Thus even for $\Delta T \leq .1$ there will be a significant reduction in the spectral emittance for $u \geq 5$. In most cases, for the emission bands of rare earth selective emitters where $K_d > 1$ the dimensionless photon energy, $u > 5$. Therefore, even a small temperature gradient will result in a significant reduction in the spectral emittance in the emittance band of the rare earth selective emitters. Obviously, making the emitter as thin as possible will reduce ΔT. However, the optical depth will also be reduced, if the thickness, d, is reduced, resulting in decreased ε_λ. As a result, there will be an optimum thickness, d, to obtain maximum ε_λ. This will be discussed in the next section. Note also that the approximate solution (equation (42)) is in the close agreement with the exact results when $\Delta T < .1$.

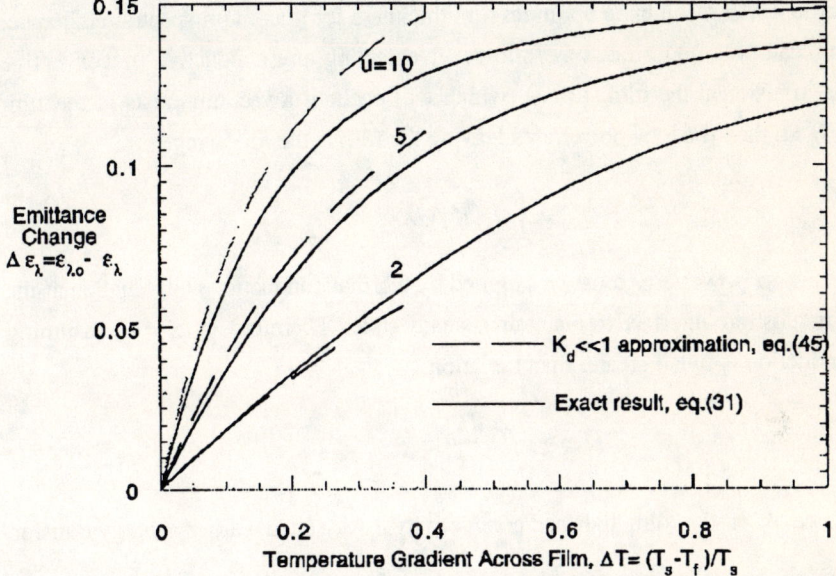

FIGURE 3. Emittance change as a function of temperature gradient at an optical depth, $K_d = .1$, for several dimensionless photon energies, $u = hc_o/\lambda kT_s$ with $n_{\lambda f} = 1.9$ and $n_{\lambda s} = 10$.

Results in Figure 2 are for large optical depth ($K_d = 2$). However, similar results occur for small optial depth and are illustrated in Figure 3 where $K_d = .1$. Again there is good agreement between the approximate solution (equation (45)) and the exact solution when $\Delta T < .1$. The range of values for $\Delta \varepsilon_\lambda$ is much smaller for the case where $K_d \ll 1$ than for $K_d > 1$. Thus the temperature gradient has only a small effect on ε_λ when $K_d \ll 1$. Therefore, for a selective emitter the emittance outside the emission band will not be greatly effected by ΔT.

Optimum Thickness for Maximum Spectral Emittance

As already stated, the counteracting effects of increasing spectral emittance with optical depth and decreasing spectral emittance with increasing temperature gradient will result in an optimum film thickness for maximum spectral emittance. This can be demonstrated as follows. Neglecting any conductive or convective heat transfer at the film surface (which will occur if a vacuum exists at the film surface) then the total power/area leaving the film is the following.

$$Q_{out} = \int_0^\infty Q_\lambda (K_d) d\lambda \tag{46}$$

This same power/area must be supplied by thermal conduction and radiation at the film-substrate interface to maintain a steady state. Therefore, at $x = 0$, assuming conduction is much greater than radiation,

$$Q_{out} = -\beta_f \left. \frac{dT}{dx} \right|_{x=0} \tag{47}$$

Where β_f is the film thermal conductivity. As stated earlier, energy transfer through the film is dominated by thermal conduction so that, equation (24) applies and $-\left. \frac{dT}{dx} \right|_{x=0} = \left(\frac{T_s - T_f}{d} \right)$. Therefore, from equations (46) and (47) the following is obtained.

$$\Delta T = \frac{T_s - T_f}{T_s} = \frac{Q_{out}}{\beta_f T_s} d \tag{48}$$

To calculate Q_{out}, equation (31) for ε_λ, which is a function of ΔT must be used to determine $Q_\lambda(K_d)$ (equation (1)). However, since ε_λ is a function of ΔT, equations (46) and (48) must solved simultaneously in order to obtain ΔT as a function of Q_{out}. This has been done in ref. 3. But to illustrate how an optimum thickness occurs we can write Q_{out} as follows.

$$Q_{out} = \varepsilon_T \sigma_{sb} T_s^4 \tag{49}$$

Where ε_T is the total emittance of the film and will be a function of T_s and σ_{sb} is the Stefan-Boltzmann constant (5.67×10^{-12} w/ cm^2 K^4). By using equation (49) in equation (48) the following results.

$$\Delta T = \tau_f d \tag{50}$$

Where,

$$\tau_f = \frac{\varepsilon_T \sigma_{sb} T_s^3}{\beta_f} \quad cm^{-1} \tag{51}$$

The quantity $\tau_f d$ is the ratio of radiation to thermal conduction (3). Thus equation (50) shows that ΔT will be small as long as thermal conduction dominates.

For selective emitters of interest, $\varepsilon_T < .2$, $\beta_f > .02$ w/cmK and $T_s < 2000$K, so that $0 < \tau_f < 5mm^{-1}$. If equation (50) is used for ΔT in equation (31) and since $K_d = \alpha_\lambda d$ the results for ε_λ when $\alpha_\lambda = 100$cm^{-1} shown in figure 4 are obtained. An extinction coefficient $\alpha_\lambda = 100$ cm^{-1} is representative of the emission band of a selective emitter. The first thing to note from figure 4 is that for $\Delta T > 0 (\tau_f > 0)$ there is an optimum thickness for maximum ε_λ. For the case of no temperature gradient $(\tau_f = 0)$ there is no optimum d. The larger the temperature gradient the more pronounced the optimum d becomes. For small τ_f large values of ε_λ occur over a broad range of thicknesses. Note that the curve for $\tau_f = 2$ mm^{-1} and $\tau_f = 5$ mm^{-1} have been truncated at d = .5 mm and d = .2 mm since $\Delta T = \tau_f d \leq 1$. Also notice that the optimum d becomes smaller as τ_f increases (larger ΔT). Based on the results of figure 4 it appears that the optimum selective emitter thickness to obtain maximum emittance in the emission band for $\alpha_\lambda = 100$cm^{-1} is in the range

$.15 \leq d \leq .4$ mm. For $\alpha_\lambda = 100 \text{cm}^{-1}$ this corresponds to an optical depth range, $1.5 \leq K_d \leq 4$.

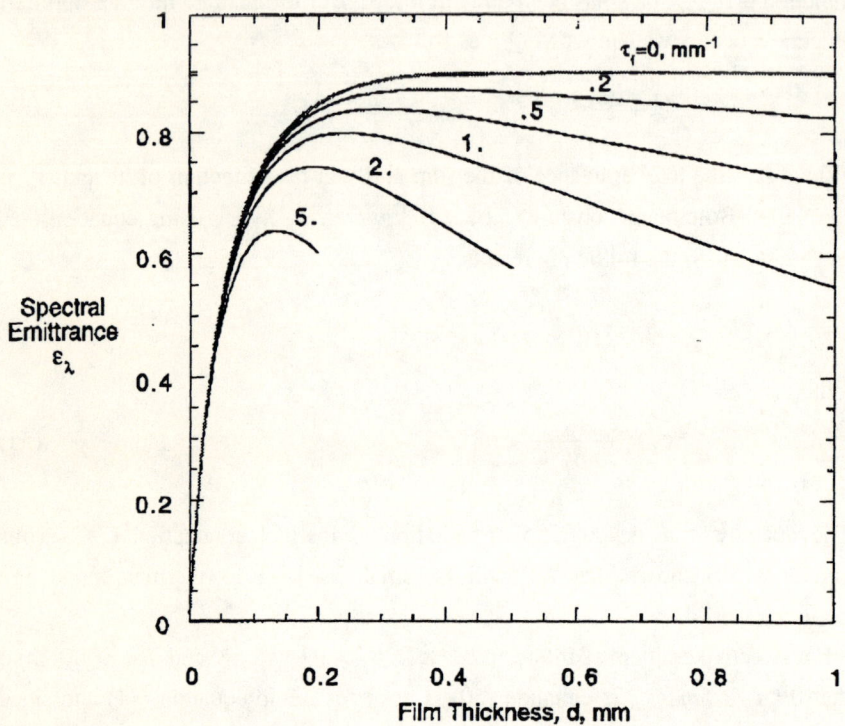

FIGURE 4. Effect of temperature gradient on spectral emittance for large extinction coefficient, $\alpha_\lambda = 100 \text{ cm}^{-1}$, at several values of the temperature gradient parameter, τ_f. Also, u=5., $\varepsilon_{\lambda s}$=.1, $n_{\lambda f}$=1.9, $n_{\lambda s}$=10.

Now consider the case of small extinction coefficient, which is representative of the wavelength region outside the emission band of a selective emitter. Spectral emittance results for $\alpha_\lambda = 1 \text{ cm}^{-1}$ are shown in figure 5. In this case, ε_λ does not attain a maximum value even for thicknesses over 1mm. Because α_λ is small much larger thicknesses (1cm to obtain K_d=1) as required before ε_λ will approach its maximum value.   For d<.4mm, the region where maximum ε_λ occurs for large α_λ, the spectral emittance is nearly independent of τ_f.

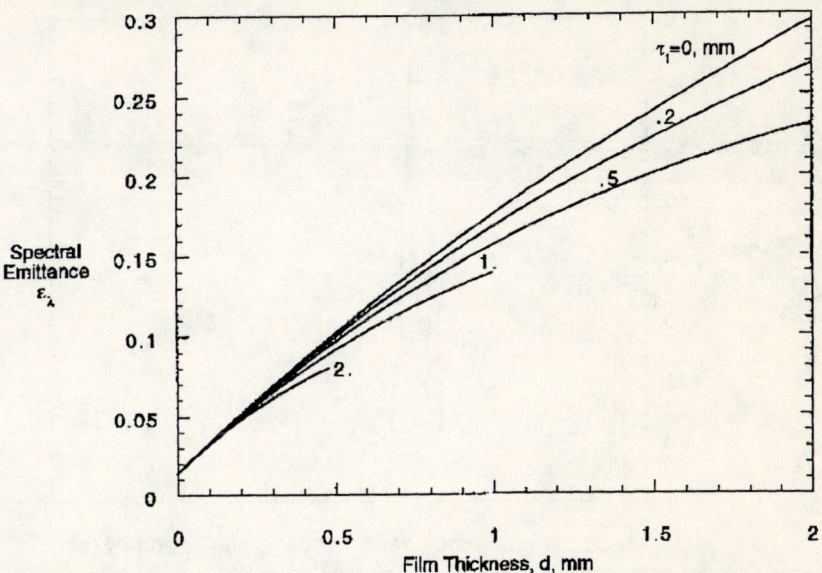

FIGURE 5. Effect of temperature gradient on spectral emittance for small extinction coefficient, $\alpha_\lambda = 1$ cm^{-1}, at several values of the temperature gradient parameter, τ_f. Also, u=5., $\varepsilon_{\lambda s}$=.1, $n_{\lambda f}$=1.9, $n_{\lambda s}$=10.

Based on the results displayed in figures 4 and 5 several conclusions can be made about the efficiency of a thick film selective emitter. The emitter efficiency (1,3,4) depends on the ratio of the emittance within the emission band ε_b to the emittance outside the emission band, ε_e. Obviously it is desirable for $\varepsilon_b/\varepsilon_e$ to be as large as possible. For the emission band, where α_λ is large, there will be an optimum thickness, d_{opt}, (corresponding to $1.5 \leq K_d \leq 4$.) to maximize ε_b. Outside the emission band, where α_λ is small, the spectral emittance increases at a much slower rate with d than for the emission band for d < d_{opt}. For d<.4mm figure 5 shows that ε_λ increases nearly at the same linear rate regardless of the temperature gradient. Thus it appears that maximum emitter efficiency will occur for the thickness, d_{opt}, corresponding to maximum emittance within the emission band. As stated earlier this thickness corresponds to $1.5 \leq K_d \leq 4.0$ when $\alpha_\lambda = 100$cm^{-1}.

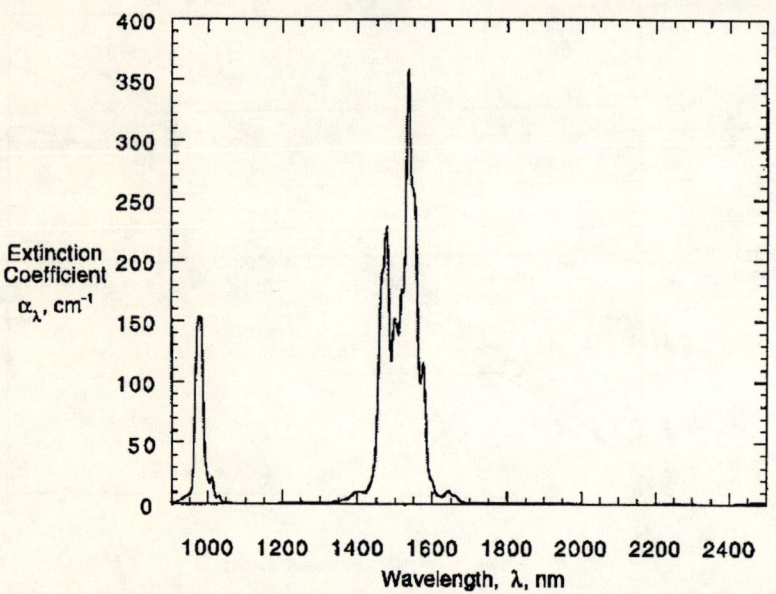

6a. Extinction Coefficient

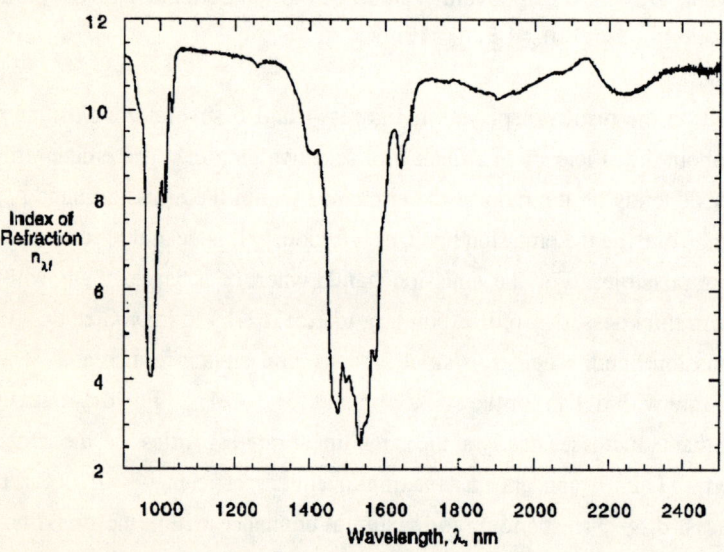

6b. Index of refraction, $n_{\lambda f}$

FIGURE 6. Extinction coefficient and index of refraction for Er_2O_3-Al_2O_3 selective emitter

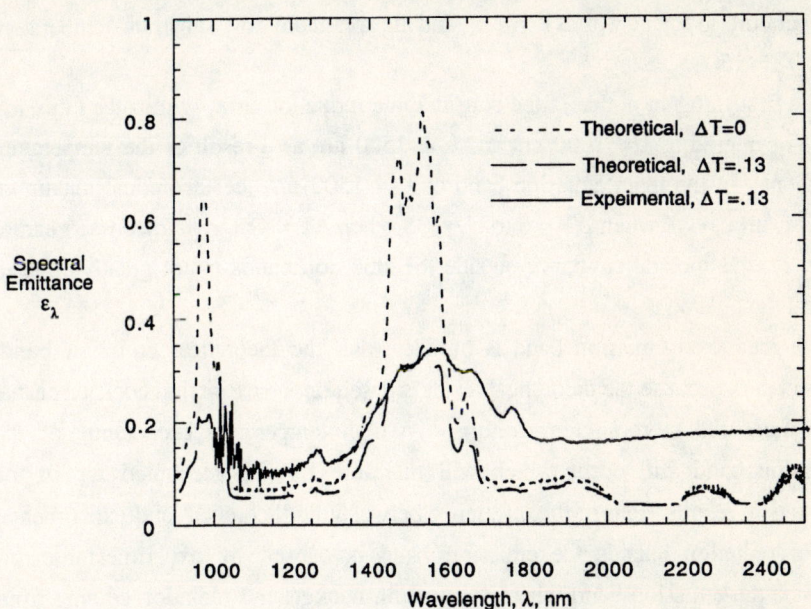

FIGURE 7. Comparison of theoretical and experimental spectral emittance for Er_2O_3-Al_2O_3 thick film selective emitter. Film thickness, d = .36 mm, $n_{\lambda s}$= 1., $\varepsilon_{\lambda s}$ = .2, T_s=1500K.

Comparison of Experimental and Theoretical Spectral Emittance

To complete this study we compare the measured spectral emittance of a selective emitter made of erbia (Er_2O_3) reinforced with alumina (Al_2O_3) with the spectral emittance calculated using equation (31). This emitter was fabricated at the Auburn Space Power Institute.(6) The calculated ε_λ is based on the extinction coefficient, α_λ, and index of refraction, $n_{\lambda f}$, shown in figure 6. These quantities were obtained using measured transmittance and reflectance data (6).

Figure 7 shows the experimental and theoretical ε_λ for an emitter of thickness, d = .36mm. This emitter had a platinum foil substrate. A constant substrate emittance $\varepsilon_{\lambda s}$ = .2 was used for the platinum foil. However, since there is an air gap between the foil and the film the appropriate index of refraction for the film-substrate interface is $n_{\lambda s}$ = 1.0, which was used in the calculation. The measured temperature gradient was $\Delta T = .13$ and the platinum foil substrate temperature was $T_s = 1500K$.

The first thing to notice is the considerable reduction in ε_λ within the emission bands centered at $\lambda \approx 1000$ nm and $\lambda \approx 1500$ nm as a result of the temperature gradient. In the main emission band at $\lambda \approx 1500$ nm the theoretical maximum goes from $\varepsilon_\lambda \approx .8$ when $\Delta T = 0$ to $\varepsilon_\lambda \approx .35$ when $\Delta T = .13$. As discussed earlier (fig. 3), the spectral emittance outside the emission bands is not greatly affected by ΔT.

The measured emission band is broader than the theoretical emission band. This occurs because the theoretical result is based on the extinction coefficient that was measured at room temperature. At high temperature broadening of the emission band will occur which will therefore not be accounted for in the theoretical results. Part of the difference between the theoretical and experimental ε_λ for radiation outside the emission band is caused by experimental error. Outside the emission band where ε_λ is small, background radiation coming from sources other than the emitting film result in the measured ε_λ being larger than the actual value, (2).

CONCLUSION

The no scattering theoretical spectral emittance model shows the importance of even small $(\Delta T \approx .1)$ temperature gradients on ε_λ. For both small $(K_d \ll 1)$ and large $(K_d \gg 1)$ optical depths, approximations for ε_λ were developed that give good agreement with the exact results as long as $\Delta T \leq .1$.

Because of the opposite dependence of ε_λ on temperature gradient and optical depth there will an optimum film thickness for maximum, ε_λ. The model predicts that the optimum optical depth has the range, $1.5 \leq K_d \leq 4.0$, depending on the temperature gradient.

Finally, there is good agreement between the theoretical spectral emittance and experimental spectral emittance for a Er_2O_3-Al_2O_3 selective emitter fabricated at the Auburn Space Power Institute.

REFERENCES

1. Chubb, D.L. and Lowe, R.A., *Journal of Applied Physics* 74, 5687 (1993).
2. Lowe, R.A., Chubb, D.L., Farmer, S.C. and Good, B.S., *Applied Physics Letter.*, 64, 3551 (1994)
3. Good, B.S., Chubb, D.L. and Lowe, R.A., "Temperature-Dependent Efficiency Calculations for a Thin-Film Selective Emitter," presented at the First NREL Conference on Thermophovoltaic Generation of Electricity, *AIP Conference Proceedings* 321, 1995.
4. Chubb, D.L., Lowe, R.A. and Good, B.S., "Emittance Theory for Thin Film Selective Emitter," presented at the First NREL Conference on Thermophovoltaic Generation of Electricity, *AIP Conference Proceedings* 321, 1995.
5. Siegel, R. and Howell, J.R., "Thermal Radiation Heat Transfer," *2nd edition Washington DC, Hemisphere*, 1981, Ch. 14.
6. Chen, Zhen, Rose, M.F., Clark, E.B. and Chubb, D.L., "Reinforced Solid Erbium Oxide Emitters for TPV Applications," presented at the 1996 International Mechanical Engineering Congress and Exposition, Atlanta, GA, November 17-22.

Superemissive Light Pipe for TPV Applications

Mark K. Goldstein, Larry G. DeShazer, Alexandr S. Kushch and
Steven M. Skinner

Quantum Group, Inc.
11211 Sorrento Valley Rd., San Diego, California 92121

Abstract. The construction and operation of a selective thermal emitter is described when attached to a light pipe or optical waveguide. The emitters studied are single crystals of rare-earth aluminum garnets involving ytterbium, erbium and holmium, and the light pipes are undoped single crystals of yttrium aluminum garnet (YAG). Typical total optical power output (0.25-12 μm) from these emitters is near 12 W/cm^2 exhibiting low graybody background emission. Emission spectra measured by a spectroradiometer are shown from 0.4 to 14.5 μm, with weak mid-IR (8-14.5μm) emission being only the blackbody emission due to crystal lattice vibrations at the near ambient temperatures of the guide's output end. Temperatures of the distal (output) and proximal ends of the light pipe are estimated by comparing the measured emission spectra to Planck's blackbody formula in the range 9-12 μm. Self-reversal (bifurcation) of the emission peaks of erbium and holmium garnet emitters is observed in the thermal emission and is due to self-absorption within the emitters. Preliminary tests of the thermophotovoltaic capability of Yb and Er garnet emitters using Si and GaSb cells, respectively, yield 1.04 and 1.56 W/cm^2 of electrical power density, respectively.

INTRODUCTION

Thermally powered light sources based on dielectric solids have been developed to power thermophotovoltaic (TPV) electrical generators. By material engineering of refractory rare-earth oxides, the thermally excited dielectric emitters are selectively matched to the spectral properties of the semiconductor cell radiation collectors. This paper reports on thermal light sources involving single crystal rare-earth garnets with their operation principle based on Kirchhoff's Law of thermodynamics.

A hot solid emits thermal radiation following Kirchhoff's Law. As usually expressed, Kirchhoff's Law states that the radiant emittance (Watts/cm^2) of a body at any temperature is equal to the radiant emittance of a blackbody at that temperature multiplied by the absorptivity of the body. Expressed in a more meaningful way for our application, an absolutely transparent body will not radiate, no matter how high its temperature. Therefore, if spectral bands of absorption (intrinsic or doped) are present in an otherwise transparent hot solid, thermal emission will only occur at the wavelengths of the absorption bands.

CP401, *Thermophotovoltaic Generation of Electricity: Third NREL Conference,*
edited by Benner/Coutts

Consequently, a hot, nearly transparent solid emits a restricted spectrum, and not the entire blackbody spectrum.

R. W. Wood (1) in 1931 first demonstrated this "selective thermal radiation" for neodymium sesquioxide (Nd_2O_3), a high melting rare-earth oxide. As is well known, rare earth oxides have sharp absorption bands and are ideal for selective thermal emitters. Nevertheless, no substance can emit a more intense thermal radiation in any region of the spectrum than a blackbody. Sometimes selective thermal emitters seemingly violate this blackbody limit, and consequentially are called "superemissive" sources. Fig. 1 presents the explanation of this situation, comparing an ytterbia (Yb_2O_3) selective emitter to blackbody emitters. When the thermal energy content of a blackbody heated to 1400°C is compacted into the narrow emission band of ytterbia centered at 980 nm, the peak emission of ytterbia becomes higher than the blackbody limit for 1400°C. This apparent violation of

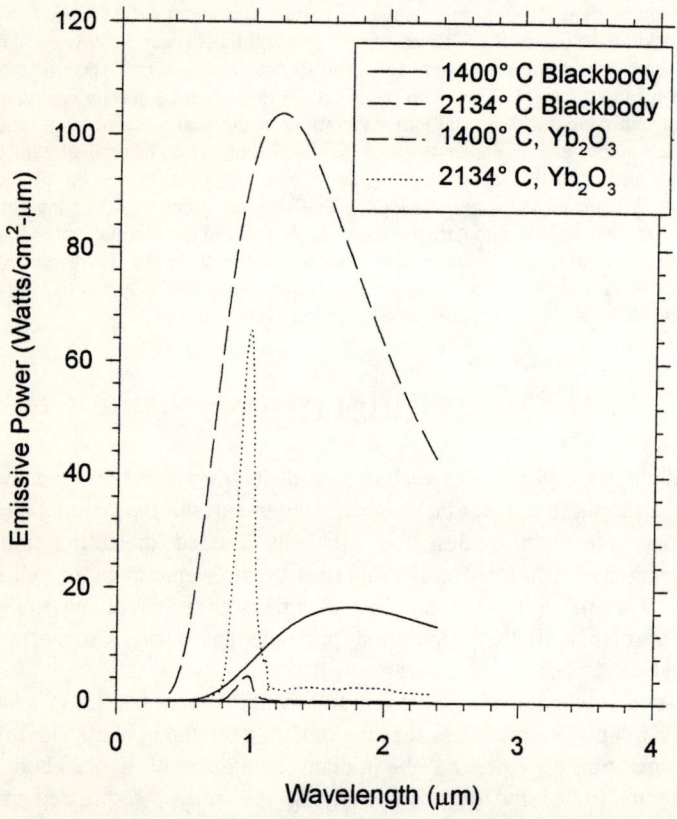

FIGURE 1. Ytterbia selective thermal emission compared to thermal radiation from a 1400°C blackbody having equal energy content. Estimated temperature of ytterbia emitter in this situation is 2134°C arising from its reduced radiative loss. For this calculation, the emissivity of ytterbia is estimated to be 0.65 at 980 nm.

the blackbody limit labels ytterbia as a "superemitter." However, there is no violation as the temperature of the ytterbia emitter becomes higher than 1400°C due to the reduced radiative loss from the ytterbia emitter in other spectral regions because of the lack of absorption in these regions. The temperature of the ytterbia is estimated to be 2134°C, and therefore the ytterbia emission should be rightfully compared to blackbody emission at 2134°C. Fig. 1 shows that the ytterbia emission peak at 980 nm, with an estimated emissivity of 0.65, is in fact below the blackbody emission for 2134°C, so the blackbody limit is not broken.

EXPERIMENT

The "SuperEmissive Light Pipe" (SELP) design attaches the superemissive source to a non-absorbing optical waveguide in order to achieve efficient collection of the thermal emission. Also, thermal insulation is provided by the light pipe where it insulates the photovoltaic cell, or other targets, from the high temperature of the thermal light source by its relatively low thermal conductivity. The configuration of the SELP device is very conformable to its particular application. Because it is a dielectric solid not depending on electrical conduction for its performance, the superemitter can be formed in many different shapes, using rods or ribbons, round or rectangular plates, as well as tubes. The SELPs reported here have the rod configuration made from single crystals, although polycrystalline and ceramic solids could perform just as well. The SELP scheme was developed and patented in 1996 by Quantum Group Inc. in response to the need of an efficient light source for TPV (2).

The SELP rods for this experiment are constructed from high purity, single crystal garnets grown by Ralph Hutcheson of Scientific Materials Inc., Bozeman, MT. Single crystals are chosen for this experiment because they can be made with the high purity (<10 ppm) necessary for achieving low intensity for the graybody background emission. The melting point of YAG is 1940°C, which is lower than the near 2400°C melting points of rare-earth sesquioxides R_2O_3, but is still more than adequate for thermal emitter operation.[1] The emitters were fabricated from rare-earth aluminum garnets ($Yb_3Al_5O_{12}$, $Er_3Al_5O_{12}$, $Ho_3Al_5O_{12}$) with cylindrical dimensions 5 mm in diameter and 25.4 mm long and polished both on the ends and sides.

The light pipe, 5 mm in diameter and 20 cm long, is made from transparent undoped single-crystal YAG, with a polished circumference as well as ends. The infrared absorption edge of YAG is at 6 μm (where the absorption coefficient α is 5 cm⁻¹), and the ultraviolet edge is at 190 nm (again, where α is 5 cm⁻¹) (3). YAG is ideal for a light pipe in SELP by having Knoop hardness of 1380 kg/mm² and room-temperature thermal conductivity of 10.3 W/m-K. The rare-earth aluminum

---

[1] In one test, however, we did unintentionally manage to melt an erbium aluminum garnet emitter.

garnet emitter crystals are attached to 20-cm long YAG light pipes using diffusion bonding by Scientific Materials Inc.

Four SELP rods having 2.54-cm long YbAG ($Yb_3Al_5O_{12}$), ErAG ($Er_3Al_5O_{12}$) and HoAG ($Ho_3Al_5O_{12}$) emitters, and a 2-mm long ErAG emitter, were tested for spectral and total power output. In addition, a 20-cm long YAG light pipe, without any emitters attached, was heated at one end to study its thermal and broadband emission effects. All data reported are for single crystal components. The heater used for most of these tests is two methane-oxygen torches at 2643°C located 9 cm from the emitter's side, with MAPP gas torches and an electric furnace used in other selected tests.

The instrumentation for the tests includes a thermopile power meter and a spectroradiometer. The power meter (Molectron model PM 30) has a spectral range of 0.25-12 μm with a flat response over the range of $\pm$ 2%. The spectroradiometer (CI Systems model SR 5000) uses a circular variable filter with a range of 0.4-14.5 μm having resolution of 3% (*e.g.*, 60 nm at 2 μm) with an internal blackbody reference. The detectors used with the spectroradiometer are liquid-nitrogen cooled HgCdTe/InSb for the long wavelengths (>1.5 μm), and Si/PbS sandwich, thermoelectrically cooled to -30°C, for short wavelengths (< 2 μm). The spectroradiometer was calibrated with a 1100°C blackbody supplied by CI Systems (model SR-2-33-SA-1").

OPTICAL POWER

The total power outputs from three SELP rods measured with the PM30 power meter are listed in Table 1. These measurements compare within 15% to the integrated spectral distributions obtained with the spectroradiometer. The power measured for the YbAG/YAG SELP is low because the YbAG crystal was fractured[1] during its testing before full power was achieved. The IR absorption of the YAG guide prevents the mid-IR (6-14.5 μm) emission from the superemitter reaching the output end of the light pipe. In a previous measurement, the ErAG/YAG SELP achieved 17.3 W/cm^2 output when heated by four MAPP

Table 1. Total Optical Power Output from SELPs

SELP	Heat Source	Power (W)	Power Density (W/cm^2)
YbAG / YAG	MAPP gas torches	0.72	3.7
ErAG / YAG	CH_4/O_2 torches	2.35	12.0
HoAG / YAG	CH_4/O_2 torches	2.32	11.8

---

[1] The fracturing was initiated by touching the side of the YbAG crystal with a Pt/Rh-Pt thermocouple having 25-μm sized wire. The temperature indicated by the thermocouple at the time was 1540 ^{o}C.

torches, but no spectral distribution was obtained for that particular measurement.

SPECTRAL DISTRIBUTION

Spectroradiometry

The spectral power distributions for the YbAG, ErAG and HoAG SELPs are shown in Figs. 2, 3 and 4 respectively. The flame spectra usually within the 2..5-4 μm band are carefully avoided in taking these thermal emission spectra. All emitter crystals are 25.4 mm long for these measurements, and the emitter temperatures are estimated to be near 1800°C. The spectra inset in these three Figures are the thermal emissions due to the strong absorption originating from the garnet's lattice vibrations at wavelengths longer than 6 μm. These broad band mid-IR spectra are about 400X lower than the selective thermal emission and are due to ambient temperature graybody ($\varepsilon \sim 0.9$) spectra of the distal (output) end of the YAG light pipe. The equivalent high-temperature mid-IR spectra from the rare-earth garnet emitters are absorbed by the YAG light pipe, never reaching the distal end. The low level thermal emission within the 2-5 μm region is due to crystal impurity and color-center absorption in the rare-earth garnets.

The peaks of the thermal emission of the three rare-earth SELPs match their room-temperature absorption peaks fairly well, with some wavelength shifting.

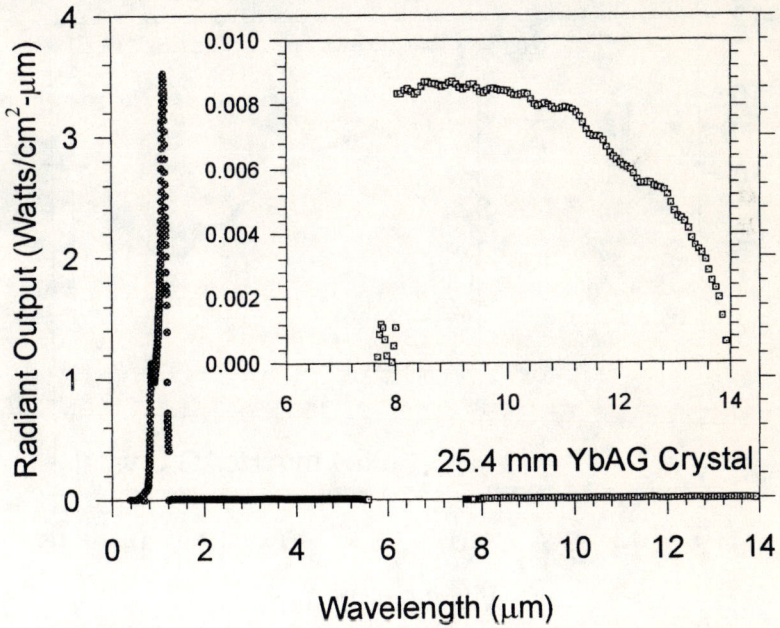

FIGURE 2. Thermal emission spectra of ytterbium aluminum garnet SELP

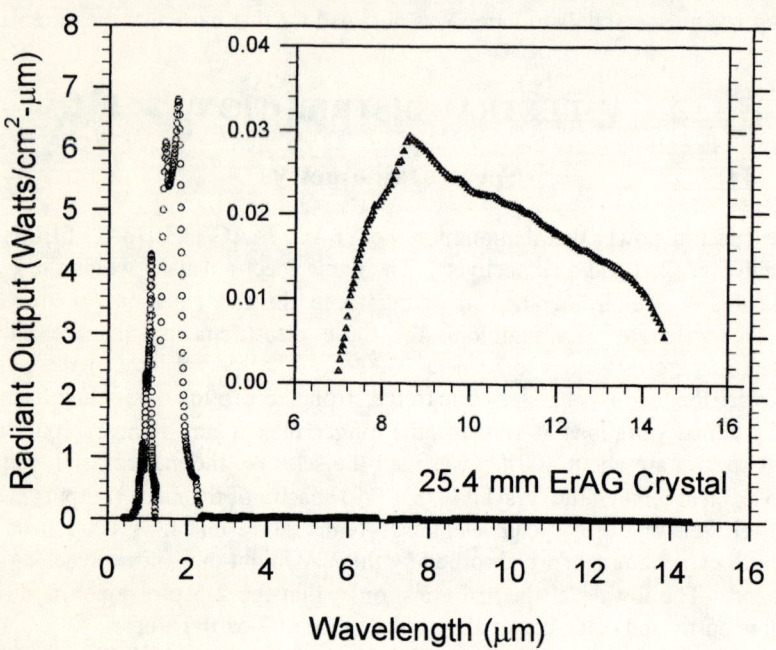

FIGURE 3. Thermal emission spectra of erbium aluminum garnet SELP showing selective emission at 0.79, 0.97 and 1.5 μm.

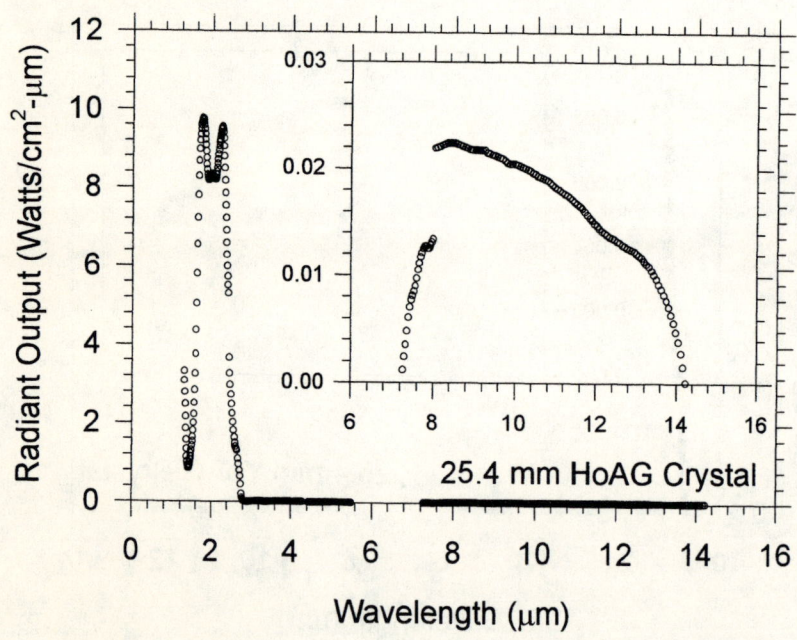

FIGURE 4. Thermal emission spectra of holmium aluminum garnet SELP.

For example, the peak of the room-temperature absorption for Yb:YAG is at 940 nm whereas the peak at high temperature is near 980 nm. The peak absorption coefficient of YbAG is determined to be 120 cm^{-1} from the Yb^{3+} peak absorption cross-section of 8×10^{-20} cm^2 measured at room temperature (4). The magnitude of this absorption coefficient indicates that 95% (= $1-e^{-3}$) of the thermal emission originates within a 250-μm depth (3/120 cm) of the Yb crystal. The fine line structure of the Yb absorption group at room temperature is smoothed out at high temperatures due to the thermal line broadening. No rare-earth fluorescence is observed at high temperatures >1500°C as it is quenched by crystal phonon interactions producing fluorescence lifetimes less than 1 psc. The peak absorption coefficient of ErAG at 1.5 μm is estimated to be 40 cm^{-1} from absorption spectra taken with a 40% Er:YAG crystal. This spectroscopy of the rare-earth garnet thermal emitters is a work in progress.

Self-Absorption

Splitting or bifurcation of the strong thermal emission peaks is observed for the erbium and holmium emitters. This bifurcation is particularly evident for the holmium peak at 2 μm (Fig. 4). The spectral shape of the emission is modified by absorption within the light source itself by the same species (*i.e.*, rare-earth ions) causing the emission. This phenomenon is called self-absorption and has been readily observed and analyzed in spectroscopy (5). The absorption at the peak of the emission is stronger than at the wings of the emission, and so can produce a relative suppression of the emission at the peak to the wings, splitting the emission into two lobes. This bifurcation is called self-reversal and occurs when the peak absorption coefficient α_o follows the condition $\alpha_o L > 1$.

Three stages of self absorption for erbium thermal emitters are shown in Fig. 5. The left-hand and center spectra in the figure show the 1.5-μm Er thermal emission from 2-mm and 25.4-mm long ErAG crystals, respectively, both in the SELP configuration. It is observed that the 25.4-mm thick crystal exhibits a near self reversal situation, while the 2-mm crystal does not appear to be self reversed. Because the Er 1.5-μm emission is a group of many (fifty-six) lines of varying peak absorptions, and not a single line, it is difficult to give a simple interpretation of the self-absorption modification here. The right-hand spectrum illustrates the emission from an erbia (Er_2O_3) mantle which has effectively a very thin thickness, demonstrating a sharply peaked emission indicative of weak self absorption.

Spectral Background

The weak, broadband, background emission up to 4 μm is due to graybody emission from impurity and color-center absorption in the light source. The reduction of this graybody emission is of primary concern in making an efficient selective thermal emitter. Single crystal garnets can be grown with high purity and low density of crystalline structural defects, primarily as a result of the well-developed laser technology which has grown high optical quality garnet crystals in

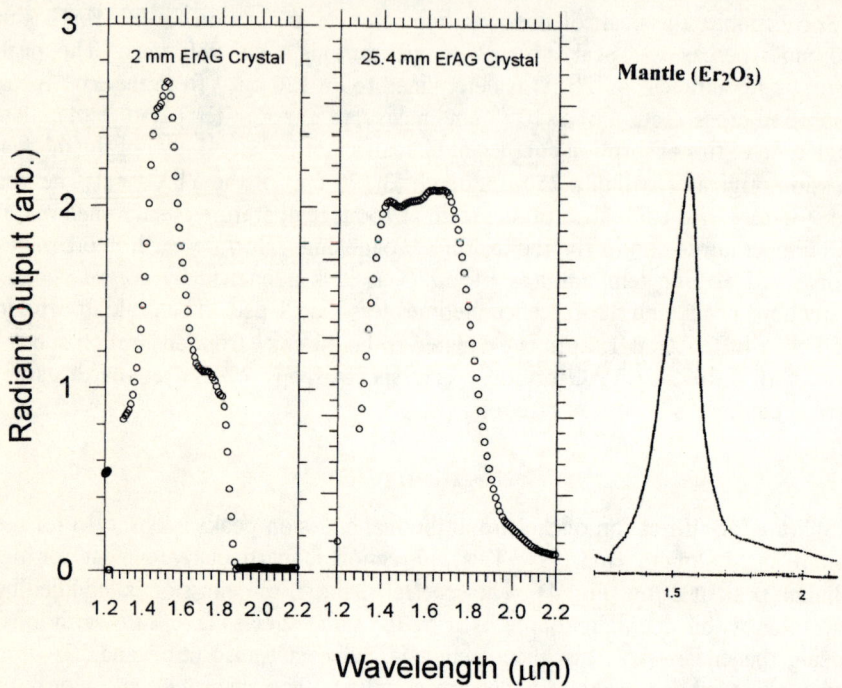

FIGURE 5. Effect of emitter thickness on modification of Er emission by self absorption.

large sizes for over thirty years. Therefore, garnets are chosen as selective emitters in this investigation to reduce the graybody background. The dominate impurities in garnets giving potential graybody emission are trivalent chromium (Cr^{3+}), divalent rare earths (such as Yb^{2+}), and other unwanted trivalent rare earths (such as Tm^{3+}). For example, for the YbAG/YAG SELP, we observed a bluish color due to Yb^{2+} content in what should have been a water-white YbAG crystal. By temperature cycling, the blue color was annealed out of the YbAG, and no divalent absorption was subsequently identified.

Color-center absorption due to as-grown structural defects give rise to a broadband absorption. The crystal defects trap electrons which produce broadband absorption both in the visible and infrared regions. Ion mobility occurs for O^{-2} in garnets at temperatures above 1300°C (6), which also lead to broadband absorption. Finally, color centers can be formed by phase changes, both solid-solid and solid-liquid, near the garnet's melting point. However, it is believed that the garnet emitter temperatures are too low in the current experiment to observe such phase-change absorption. Fig. 6 compares the graybody background emission for heated YAG (undoped) and YbAG. Both garnet crystals have nearly identical graybody backgrounds, both in magnitude and shape. The YAG spectrum on the left-hand side shows emission possibly arising from chromium impurities in the visible region and broadband emission from as-grown color-center defects which

322

peak near 1.6 μm. No thulium impurity is detected which has a strong emission signature at 1.75 μm.

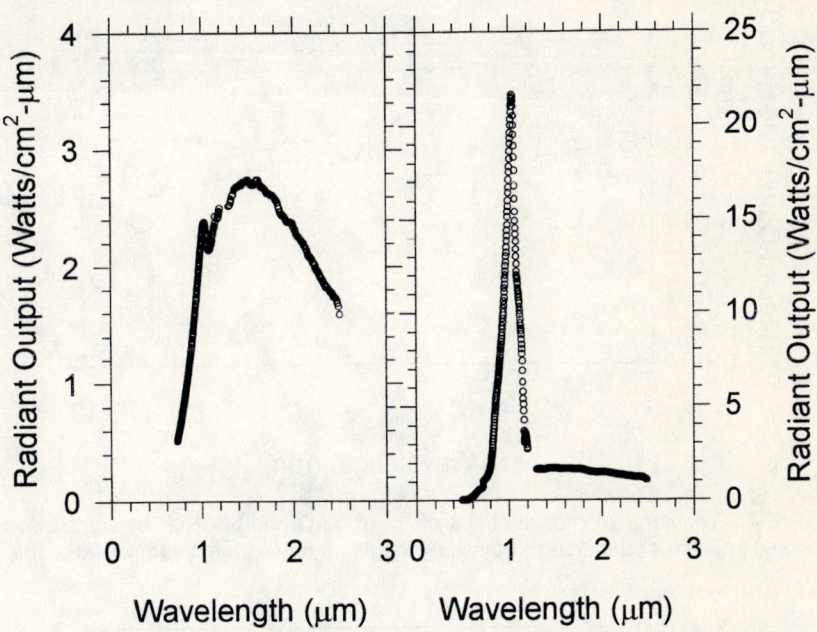

FIGURE 6. Graybody background emission from undoped YAG and YbAG emitters.

TEMPERATURE MEASUREMENT OF SELP

Temperature of thermal emitters has been difficult to measure. Use of a platinum-rhodium thermocouple has not proven successful in this investigation on rare-earth garnets. The temperature of the thermal emitters are estimated by comparing the mid-IR distribution in the 9-12 μm spectral region to the Planck blackbody distribution. Fig. 7 compares the mid-IR emission from the "cold" (distal) end of the YAG light pipe to Planck distributions at 160 and 200°C, where ε ~ 1. This determination is approximate but it certainly gives the correct order of magnitude of the temperature. It was checked by attaching a thermocouple to the side of the distal end of a YbAG/YAG SELP rod where the YbAG emitter is heated inside an electric furnace. The thermocouple measured 40°C at 1100°C furnace temperature, and 76°C at 1500°C furnace temperature.

The temperature of the garnet emitters is estimated by measuring the spectral distribution in the 9-12 μm region of the "hot" (proximal) end with the spectroradiometer and comparing it to Planck distributions from 1500 to 3000°C. For undoped YAG, the temperature is near 2000°C (see Fig. 8) and for ErAG, 1800°C. Again, these temperatures are approximate. Accuracy of this technique in determining temperatures of selective emitters will be further investigated.

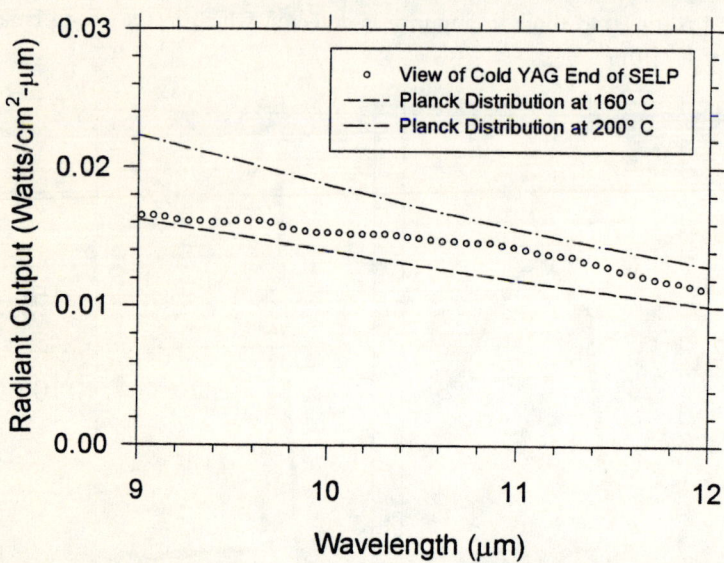

FIGURE 7. Temperature measurement of "cold" YAG end of SELP by comparison of measured spectral distribution to Planck distributions having temperatures from 160 to 200°C.

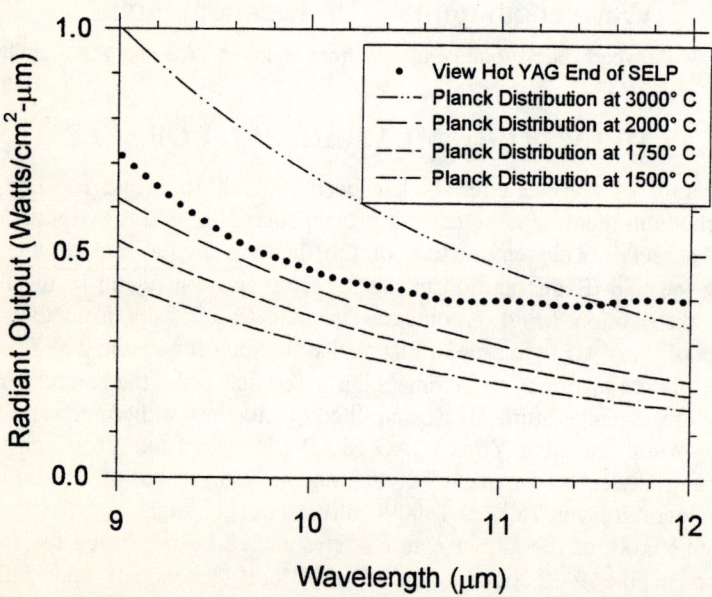

FIGURE 8. Temperature measurement of "hot" YAG end of SELP by comparison of measured distribution to Planck distributions having temperatures 1500 - 3000°C.

THERMOPHOTOVOLTAIC TESTS

Two TPV tests were conducted with the YbAG/YAG and ErAG/YAG SELPs. Fig. 9 shows the photovoltaic electrical output with the YbAG emitter heated by an electric furnace from 1000 to 1700°C. The photovoltaic cell used for this test was Si cell rated at 250 suns (Sunpower) with active cell area of 1.56 cm². No cell cooling was used in this test and the distal end of the SELP was 0.7 mm from the PV cell. The electrical power density generated is 1.04 W/cm². When the YbAG emitter was heated by MAPP gas torches, the electric power produced compared within a factor of 2 of the electric furnace test; the rod fractured before further tests could be made.

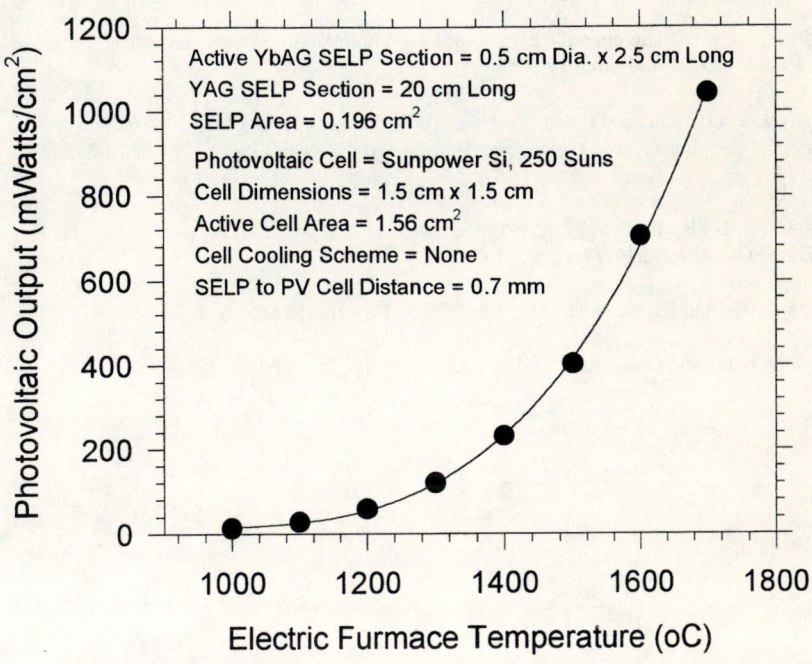

FIGURE 9. TPV test of YbAG SELP with Si photovoltaic cell.

The ErAG/YAG SELP produced electrical power density of 1.56 W/cm² using a GaSb photovoltaic cell, 5 mm in diameter. The ErAG was heated by three MAPP torches producing a total optical output of 1.05 W (5.35 W/cm²). The electrical output was 0.306 W with an electrical-optical conversion efficiency of 29%.

CONCLUSION

The use of single crystal thermal emitters coupled to optical waveguides or light pipes leads to efficient collection of thermal emission, low graybody background emission, and removal of mid-infrared emission by the waveguide absorption. Availability of high purity single crystals as emitters for TPV is made possible by the advanced state-of-the-art of crystal growth accomplished for laser technology.

REFERENCES

1. Wood, R. W., Phys. Rev. **38**, 487-490 (1931).

2. Goldstein, M. K., "Superemissive Light Pipes and Photovoltaic Systems Including Same," U.S. Patent 5,500,054, March 19, 1996.

3. DeShazer, L. G., Rand, S. C., and Wechsler, B. A., "Laser Crystals," in *CRC Handbook Series of Laser Science and Technology*, Boca Raton, FL: CRC Press, 1986, ch. 1.5, pp. 281-338.

4. DeLoach, L. D., Payne, S. A., Chase, L. L., Smith, L. K., Kway, W. L., and Krupke, W. F., IEEE J. Quant. Elect. **29**, 1179-1190 (1993).

5. Cowan, R. D., and Dieke, G. H., Rev. Mod. Phys. **20**, 418-455 (1948).

6. Schwartz, K B., and Duba, A. G., J. Phys. Chem. Solids **46**, 957-962 (1985).

SESSION 6:
TPV SYSTEMS

Development of a Portable Thermophotovoltaic Power Generator

Frederick E. Becker, Edward F. Doyle, and Kailash Shukla

Thermo Power Corporation, Tecogen Division
Waltham, Massachusetts 02254-9046

Abstract. A 150 Watt thermophotovoltaic (TPV) power generator is being developed. The technical approach taken in the design focused on optimizing the integrated performance of the primary subsystems in order to yield high energy conversion efficiency and cost effectiveness. An important aspect of the approach is the use of a selective emitter radiating to a bandgap matched photovoltaic array to minimize thermal and optical recuperation requirements, as well as the non-recoverable heat losses. For the initial prototype system, fibrous ytterbia emitters radiating in a band centered at 980 nm are matched with high efficiency silicon photoconverters. The integrated system includes a dielectric stack filter for optical energy recovery and a ceramic recuperator for thermal energy recovery. The system has been operated with air preheat temperatures up to 1350K. The design of the system and development status are presented.

INTRODUCTION

In both military and commercial markets there exists a need for portable electric power units for a variety of applications. Examples of these applications include 20 to 100 watt person-portable power systems for communication equipment and soldier systems, 100 to 500 watt power systems for grid independent appliances, and greater than 500 watt systems for remote site instrumentation power, camping and construction site power generation, and emergency power source during net power outages. Many of these applications are currently powered by engine generators or batteries. Existing engine generators, however, are noisy, polluting, and of limited life. Batteries, on the other hand, have problems related to low power density, limited shelf life, minimal repair capability, high cost and disposal. Advanced power systems are, therefore, needed that do not have the limitations of motor generators or batteries and are quiet, have high power density, and are fueled systems (replenished from a fuel source) with long shelf life and low maintenance.

CP401, *Thermophotovoltaic Generation of Electricity: Third NREL Conference,*
edited by Benner/Coutts
© 1997 The American Institute of Physics 1-56396-734-0/97/$10.00

Thermophotovoltaics (TPV) is an attractive approach for meeting these requirements. The continuous combustion used for TPV permits a greater degree of control of exhaust products and noise relative to internal combustion engines. The absence of moving parts in the main power stream is expected to provide vibration-free operation and lower maintenance requirements than internal combustion engines. TPV is also potentially the least costly of the direct conversion processes because of its simplicity and the availability of cost effective photoconverters. If all these advantages of TPV can be exploited, the range of potential applications for TPV systems can be quite diverse.

Individual components needed to produce TPV power systems such as PV cells, emitters, and optical filters are actively being developed. However, the key issue in realizing commercially viable TPV systems is the integrated performance of the major components and its impact on overall system efficiency and cost. In this regard, Thermo Power Corporation, with funding support from DARPA/NASA-Lewis, is developing completely integrated TPV power sources (1) in the range of 150 to 500 watts to demonstrate the technological pathways for achieving system efficiencies and cost effectiveness consistent with the needs of military and commercial markets. Initially, the TPV generator will use gaseous fuels with future capability for operating on liquid hydrocarbon fuels.

TECHNICAL APPROACH

The technical approach taken in this project is the use of wavelength selective fibrous emitter radiating to a bandgap matched photovoltaic array to minimize thermal and optical recuperation requirements, as well as non-recoverable heat losses. While a well matched emitter-photoconverter has significant design advantages, high PV cell efficiency and high optical and thermal recuperation remains a critical element for high system performance. It is also important that the design minimize other potential loss mechanisms such as photoconverter fill factor and radiation view factor, ancillary power requirements and non recoverable thermal losses from optical windows and insulating surfaces.

For the initial prototype systems, fibrous ytterbia emitters radiating in a wavelength selective band centered at 980 nm are matched with high efficiency silicon photovoltaic converters. The ytterbia selective emitter system is based on Thermo Power's patented large, supported continuous fiber radiant structure (SCFRS) (4,5) that can operate up to temperatures of 2100K with good thermal shock resistance and rapid response time. The emitter consists of continuous fibers woven into a porous ceramic base. As such, the emitter preserves the advantages of gas light mantles that have traditionally operated at high temperatures for long times. Unlike gas mantles, the Thermo Power emitter can be made in planar,

15 cm x 15 cm or larger tiles without becoming fragile. Although the initial prototype TPV generators are being developed for operation with fibrous ytterbia emitters and silicon photoconverters, they can be easily adapted for operation with other combinations of fibrous ceramic emitter materials and bandgap matched PV cells, for example erbia and gallium antimonide, by just changing the emitter fiber material and the PV cells for the arrays.

SYSTEM DESIGN

Two configurations of TPV power modules are being developed - a single emitter module and a dual emitter module. In the single emitter TPV system module, shown in Figure 1, a single emitter/photovoltaic array is close-coupled to a ceramic recuperator. This arrangement provides a compact design and can be scaled-up for higher power output by adding more modules. In the dual emitter TPV system module, shown in Figure 2, two emitters are coupled to a single very high effectiveness recuperator. This arrangement provides easier access and more space in the rear of the emitter substrate for fuel delivery system and is the preferred configuration, especially for liquid fuels. In operation, the combustion products pass through the ceramic recuperator to preheat the combustion air to about 1350K. The combustion air flows through the back of the porous ceramic substrate which supports the ytterbia emitter in a filament form. Because the preheated air is far in excess of the auto-ignition temperature of the fuel, and in order to minimize the temperature of the substrate, the fuel is micromixed with the air just above the substrate surface. This enables the ytterbia filaments to see the maximum gas temperature and associated radiative output potential.

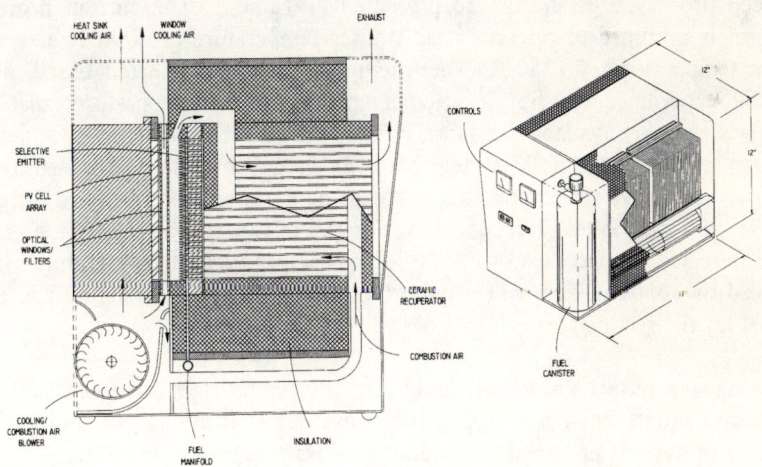

FIGURE 1. Single-Emitter TPV System Prototype

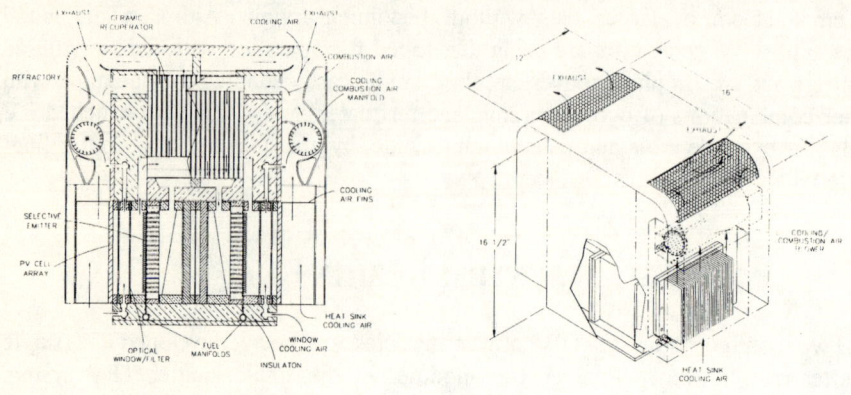

FIGURE 2. 150 W Dual-Emitter TPV System Prototype

To protect the PV array from direct contact of hot combustion gases and to reflect any non convertible radiation back to the emitter, two quartz windows with optical filters are used. Cooling air passes between the two windows to maintain them at temperatures consistent with the capability of the optical coatings. The PV array is shown mounted to a forced convection cooled heat sink, although a water cooled heat sink is used for initial prototypes. A common blower (one per emitter/PV array) is used to provide the cooling air to the heat sink and quartz windows, and for combustion air.

The design specifications for the current TPV prototype generator and for an advanced prototype generator are presented in Table 1. The current prototype generator is designed to operate at an emitter temperature of 2000 K and an air preheat temperature of 1350 K. These temperatures were selected based on our operating experience with fibrous ytterbia emitters and the temperature capabilities of materials readily available for fabrication of the ceramic recuperator. A detailed computer model of the TPV system has also shown that the power density and efficiency of the system continues to increase as these temperatures are increased when an efficient optical filter is used. For the advanced prototype, the emitter temperature will be increased to 2100 K and the air preheat temperature will be increased to 1650 K. This will require the use of a higher temperature ceramic material for the recuperator.

The planar emitters and PV arrays in the prototype system are essentially equal in area and square in shape. With a total emitter/PV array area of 290 cm^2, the current prototype is designed to produce a gross power of 245 Watts at a gross

332

efficiency of 6.5%. The net power and efficiency are 170 Watts and 4.5% respectively, after allowing power for the ancillary equipment.

With the same total emitter/PV array area, the advanced prototype with higher temperature capability will be able of producing a gross power of 372 Watts at a gross efficiency of 8.8%. The net power and efficiency will be 315 Watts and 7.5% respectively, after allowing power for the ancillary equipment.

TABLE 1. TPV Prototype Design Specifications

	Current Prototype	Advanced Prototype
Emitter Temperature – K	2000	2100
Air Preheat Temperature – K	1350	1650
Average Emitter Exitance – W/cm^2	3.50	5.00
Average Exitance View Factor	0.80	0.80
Average Exitance @ Array – W/cm^2	2.80	4.00
Emitter Area (2 Emitters) – cm^2	291	291
PV Array Area – cm^2	288	288
PV Cell Efficiency	38%	38%
PV Array Uniformity Factor	80%	85%
Gross Power (2 Emitters) – W	245	372
Ancillary Power – W	75	57
Net Power – W	170	315
Gross System Efficiency	6.5%	8.8%
Net System Efficiency	4.5%	7.5%

DEVELOPMENT STATUS

To facilitate development of the overall TPV system, a "pathfinder" emitter system, shown in Figure 3, was fabricated and tested to characterize the emitter, optical filter, and PV cell. Spectrally resolved power density measurements of the emitter were made by Essential Research Inc. with a spectral radiometer, that made use of a dual grating monochrometer and three detectors: silicon and germanium operating in the photovoltaic mode and a lead sulfide detector operating in the photoresistive mode. Figure 4 shows the spectral exitance of the ytterbia emitter as viewed through quartz windows. The emitter has a strong emission peak at 980 nm, silicon-convertible power of 3.24 W/cm^2, and total radiative power of 10.67 W/cm^2. Also shown in Figure 4 are the efficiency vs. wavelength characteristics for two prototype PV cells. The open circuit voltage, short circuit current, fill factor, peak power, responsivity and overall efficiency for each of these cells are listed in Table 2. The "initial PV cells", which were selected at the start of the program for use in the prototypes, had a peak monochromatic cell efficiency of 35% at 920 nm. At the ytterbia emission peak of 980 nm, the monochromatic cell efficiency dropped to 33% and continued to drop at higher wavelengths where the ytterbia emitter was still emitting strongly. As a result, the overall efficiency of this cell in converting radiation below 1180 nm, was reduced to 27% when matched with the ytterbia emitter. The "new PV cells", which will be used in the next generation prototype, are much better matched to the ytterbia emitter characteristics as shown in Figure 4. For this PV cell, the peak monochromatic cell efficiency of 46% occurs at the ytterbia peak of 980 nm. With the better matching characteristics and the higher peak efficiency, the new PV cells have an overall efficiency of 38% when matched with the ytterbia emitter.

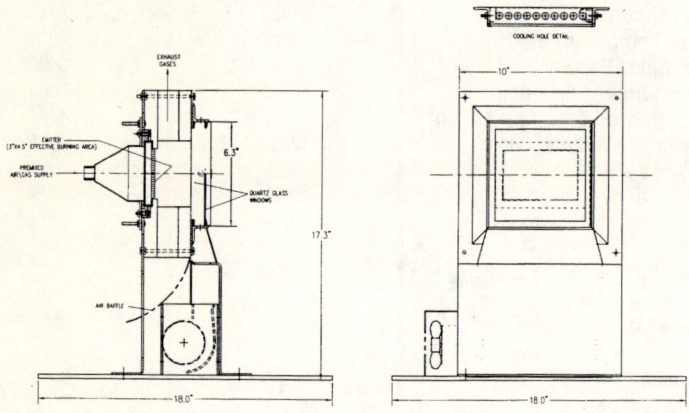

FIGURE 3. "Pathfinder" Emitter System

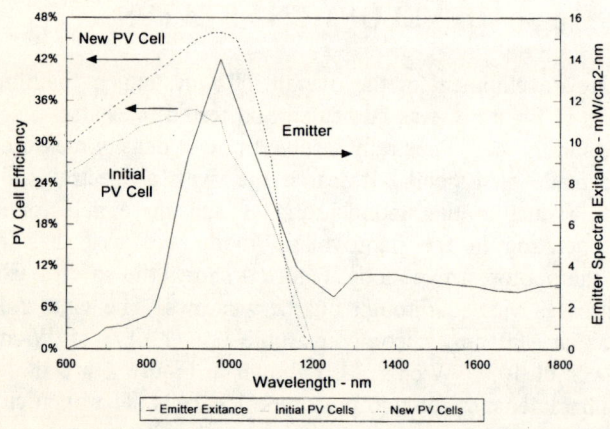

FIGURE 4. PV Cell Efficiency and Emitter Spectral Exitance vs. Wavelength

334

TABLE 2. PV Cell Characteristics

	Initial PV Cells	New PV Cells
Voc – Volts	0.79	0.77
Isc – A/cm^2	1.59	2.04
FF	0.72	0.80
Pm – A/cm^2	0.90	1.26
Responsivity – A/W	0.47	0.61
Efficiency	27%	38%

Two types of filters are being investigated for optical performance and mechanical stability by Essential Research Inc. One is a multilayer dielectric stack filter and the other is a thin "transparent" conductive coating also known as a solar control film. The dielectric stack filter performance is shown in Figure 5 by comparing the spectral content of the power radiated for two operating conditions: two quartz windows, and the combination of one quartz window and a dielectric stack filter. The first two bars show the effect on total power: the dielectric stack reduced output power to 60% of the unfiltered output. The next two bars show that the silicon convertible radiation dropped by 2% with the use of a filter. In the near IR, the wavelength range between 1180 and 3500 nm, output power was reduced by a factor of four: just 24% of the unfiltered power level. With the use of a filter, the ratio of in-band to out-of-band radiation (convertible fraction) thus increased from 30% to 50%.

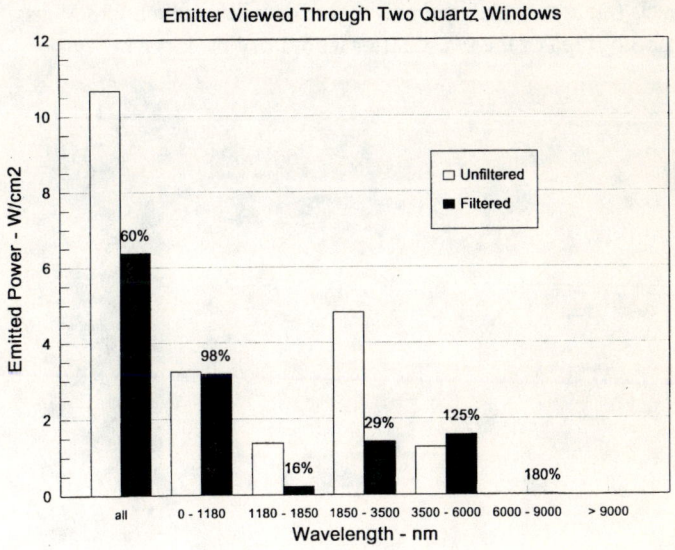

FIGURE 5. Dielectric Stack Filter Performance

A compact, monolithic, high-efficiency, counter-flow, ceramic recuperator, shown in Figure 6 was fabricated and tested to provide air preheat temperatures of up to 1350K. The desired air preheat temperature is the key determinant of recuperator material. However, temperature related effects such as material differential expansion, thermal shock, thermal conductivity must also be considered and accounted for in recuperator design. Table 3 compares the relevant thermal properties of potential recuperator materials. The baseline material chosen for the initial prototypes is cordierite (2MgO-2Al2O3-5SiO2). Future recuperators will be fabricated using higher temperature materials like silicon carbide and whisker reinforced alumina.

The two prototype TPV generators have been tested extensively in the laboratory and development of the technology with the prototypes continues. We have operated at emitter temperatures of 2000 K and air preheat temperature of 1350 K for many hours and have demonstrated the feasibility of the micromixed surface combustion concept needed to operate at these conditions. A convertible exitance of 3.7 W/cm^2 has been measured at the emitter and, with the optical filter, a 55% convertible radiation fraction has been achieved.

Figure 7 shows the measured power output as a function of the convertible exitance at the emitter for the initial PV arrays, with the poorly matched and low efficiency PV cells, and the projected power for arrays with the new high efficiency PV cells. A peak power of 150 watts was measured with the initial array at an emitter convertible exitance of 3.7 W/cm^2. At the same emitter convertible exitance level, the arrays with the new PV cells will produce 228 Watts. Figure 8 shows the photograph of the two-emitter laboratory prototype.

FIGURE 6. Ceramic Recuperator

TABLE 3. Ceramic Recuperator Material Options

Recup. Material	Max. Use Temp.,	Air Preheat Temp., °C		Coeff. of Thermal Expans.	Thermal Conductivity		Thermal Shock Resistance	
	°C	$\varepsilon = 0.8$[1]	$\varepsilon = 0.9$[1]	10^{-6}/°C	W/m-K @20°C	Rating	ΔT_c,[2] °C	Rating
Cordierite	1250	1000	1125	1.5	5	Fair	500	Good
Silicon Nitride	1250	1000	1125	3	20	Good	700	Very Good
Aluminum Titanate	1350	1075	1200	0.5	8	Fair	1000	Excellent
Silicon Carbide	1700	1350	1500	4.0	100	Excellent	375	Good
Whisker Reinforced Alumina	1750	1400	1575	7	35	Very Good	700	Very Good
Alumina	1850	1475	1650	7.5	35	Very Good	250	Fair

[1] ε : Recuperator effectiveness
[2] ΔT_c : Critical temperature differential at which thermal stress exceeds material fracture strength (Reference: Coors Ceramic Company)

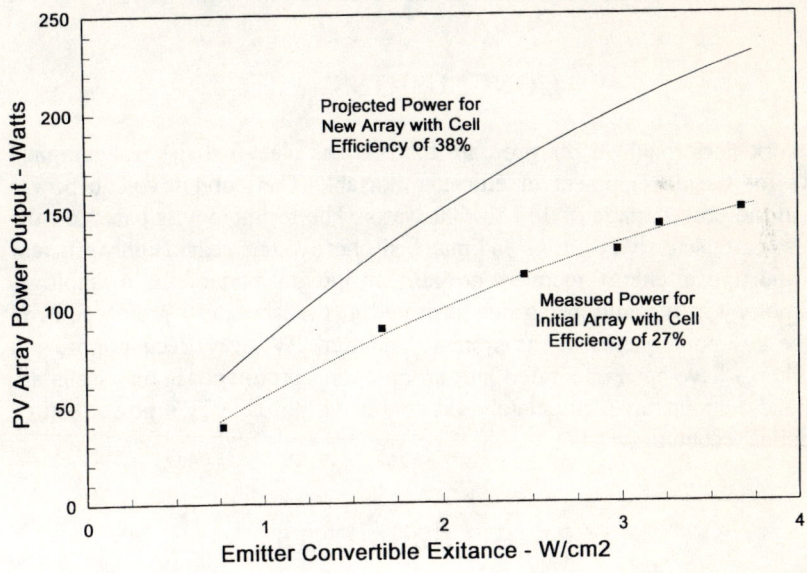

FIGURE 7. PV Array Power vs. Emitter Convertible Exitance for Current Prototype

FIGURE 8. 150 W Dual Emitter Laboratory Prototype

CONCLUSIONS

The work performed on the program to date has identified the technological pathways for the development of efficient, portable Thermophotovoltaic power sources in the power range of 150 to 500 watts. The technology is based on the use of rare earth selective emitters and matched photovoltaic cells. Highly efficient thermal and optical energy recovery are also an integral part of the technology. The technology is currently being demonstrated at the laboratory prototype level where the key components of the system – emitter, PV array, recuperator, and optical filter – have been integrated into an operating prototype. Future plans are to build and demonstrate a completely self contained, portable TPV power source based on this technology.

ACKNOWLEDGMENTS

This work has been supported by the U.S. Department of Defense through DARPA and NASA-Lewis. The authors acknowledge the guidance provided by Dr. Bob Rosenfeld of DARPA and Mr. Dave Wilt of NASA-Lewis. We also acknowledge Mr. Phil Jenkins of Essential Research for contributing technical insights and designing some figures used in the paper.

REFERENCES

1. Becker, F.E., Doyle, E.F., Mastronardi, R., Shukla, K., Linder, E.B., and Garverick, L.M., "Development of a 500 Watt Portable Thermophotovoltaic Power Generator," Twenty-Fifth IEEE Photovoltaic Specialists Conference, Washington, D.C., 1996, pp. 1413-1416.

2. Becker, F.E., "Recuperators for Thermophotovoltaic Energy Conversion Systems," Prospector VIII: Thermophotovoltaics – An Update on DOD, Academic, and Commercial Research, sponsored by Space Power Institute. Auburn University, Alabama and Army Research Office, Research Triangle Park, North Carolina, July 14-17, 1996, Edited by M. Frank Rose, pp. 173--194.

3. Doyle, E.F., "System Aspects of TPV Energy Conversion," Prospector VIII: Thermophotovoltaics – An Update on DOD, Academic, and Commercial Research, sponsored by Space Power Institute. Auburn University, Alabama and Army Research Office, Research Triangle park, North Carolina, July 14-17, 1996, Edited by M. Frank Rose, pp. 173-194.

4. Nelson, R. E., "Grid Independent Residential Power Systems," The Second NREL Conference on Thermophotovoltaic Generation of Electricity, Colorado Springs, 1995, AIP Conference Proceedings #358, The American Institute of Physics, New York © 1996, pp. 221-237.

5. Nelson, R. E., "Thermophotovoltaic Emitter Development," The First NREL Conference on Thermophotovoltaic Generation of Electricity, Copper Mountain, 1994, AIP Conference Proceedings #321, The American Institute of Physics, New York © 1995, pp. 80-95.

Multifuel (Liquid Hydrocarbons) TPV Generator

Guido Guazzoni* and Malachy McAlonan**

*CECOM - Advanced Systems Directorate, Fort Monmouth, New Jersey 07703
**TELEDYNE BROWN ENGINEERING - Energy Systems, Hunt Valley, Maryland 21031-1325

Abstract . A system configuration study of a TPV power source based on a liquid hydrocarbon burner, a SiC emitter and low bandgap PV cells is presented together with projected performance characteristics. The TPV power source design is scaleable to output power levels ranging from few hundreds watts (Man-portable TPV battery charger) to few Kilowatts (TPV Power Generator).

INTRODUCTION

The TPV Power Source system analysis presented in this paper is based on the size and characteristics of existing TPV components used in an experimental investigation conducted by CECOM RDEC, Power Sources Division as part of the TPV Generators/ Battery Charges development effort. The results of the experimental phase, which is planned to be completed in early 1998, should provide both, verification of the system analysis projections in terms of system characteristics and performance, and guidance on the selection of critical components, structural materials and final design configuration.

The TPV Power Source under study is scaleable from power output levels of few hundred watts (power range of interest for the development of manportable, TPV battery chargers) to 2-3 kW output for possible development of portable, lightweight, multifuel power generators.

In both cases the TPV power source must be capable of operating with liquid hydrocarbon fuels available in the field with major focus on operation with diesel because of the military decision to proceed with the development of ground vehicles and portable power generators which utilize the diesel family of fuels. Many of the properties of diesel fuel such as flash point, vapor pressure, and flame speed make it an attractive fuel from the stand point of safety. However, these

CP401, *Thermophotovoltaic Generation of Electricity: Third NREL Conference,*
edited by Benner/Coutts
© 1997 The American Institute of Physics 1-56396-734-0/97/$10.00

same properties also make diesel difficult to ignite and burn at low firing rates and low temperatures.

DISCUSSION

During the development of military Thermoelectric Generators in the 1970-1985 period, the Army has extensively evaluated the performance of ultrasonic atomization for the conditioning of liquid fuels to meet multifuel operation requirement . Even if ultrasonic atomizer burners have been successfully developed and have demonstrated capability of meeting low heat requirement dictated by the small size of some military applications (120 W TEG) , problems such as undesirable crystal wetting , overheating, brittleness, high fuel viscosity/surface tension, and smoke during startup at low temperature has cast doubts on the long time reliability of this technology.

The multifuel capability of the burner to be used in this study is based on a different principle of fuel atomization; the Babington atomizer. In this burner (Fig. 1), which is a modified version of a highly reliable commercial unit, the fuel is supplied to the exterior surface of a hollow convex atomizer that contains a small aperture. The liquid fuel spreads out and covers the entire atomizing surface, aperture included, with a continuous thin film of fuel. Low pressure compressed air is introduced to the interior of the hollow atomizer. As the compressed air escapes through the small orifice, it ruptures a portion of the thin flowing film of fuel creating a continual dispersion of liquid particles that are smaller in diameter than those produced by conventional nozzles. Some of these burners were extensively tested and met many military requirements including startup and operation at -53 °F. The firing rate capability of the burners to be used in this investigation is listed in Table I.

TABLE I.   Burner Firing Rates.

Burner No 1:
Maximum Firing Rate = 5.8 liter/hour = 191,000 BTU/hour = 55,848 watts(t)

Burner No 2:
Maximum Firing Rate = 2.8 liter/hour =   79,200 BTU/hour = 23,158 watts(t)

The atomized fuel is carried away from the atomizer orifice along the axial section of the burner tube by a stream of air maintained at room temperature. The burner tube is made of two coaxial cylindrical sections, spaced approximately 1 cm

FIGURE 1. Burner and cylindrical mantle.

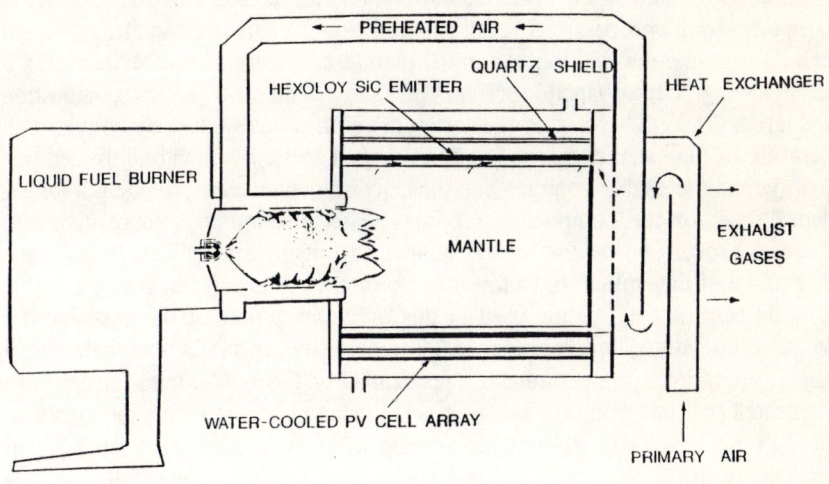

FIGURE 2. Cross sectional view of TPV test unit design.

apart, which form an annular space used to channel the primary air for combustion down the burner tube. The internal cylindrical section is provided with openings, properly spaced and distributed along circumferential rings, that allow the primary air to mix with the atomized fuel at specified locations.

Fig. 2 is a cross sectional view of the design of the TPV test unit prototype. The emitter is an open end cylindrical structure made of Hexoloy SA SiC, 21.5 cm long , with an inside diameter of 15.5 cm. and a wall thickness of 0.6 cm. This cylindrical structure is supported by high temperature insulation walls that separate it from the burner tube on one side and from the primary air pre-heating system (heat exchanger) on the other side. Fig. 3 is a picture of the burner-emitter assembly. Inside the emitter, and coaxial with it, there is a closed end cylindrical mantle, made of Hastelloy, which is spaced 1 cm apart from the inside surface of the emitter. The burner tube feeds directly into the cylindrical mantle, which is 22 cm long and has a inside diameter of 13.3 cm. The mantle is provided with perforations that cover its entire cylindrical surface and that are grouped in sections of different holes size. Combustion fully develops inside the mantle and is distributed over the entire inside surface of the SiC emitter by the mantle perforations that have the function to equally distribute the thermal energy provided by the combustion over the inside emitter surface. The combustion products move along the annular space formed by the mantle and the emitter and leave the combustion chamber through a ring of openings located at the end of the mantle structure. Here the exhaust gases interface with a heat exchanger, made of Inconel 610, that pre-heat the primary air for combustion.

The Hastelloy mantle used in the actual combustion chamber is the result of a long optimization process on both combustion "characteristic length" and uniformity of flame distribution. An initial mantle configuration , provided with a flame diverting cone-shaped section, resulted in a too short combustion "characteristic length" for the gas stream with consequent excessive high temperature of the exhaust gases. The cone-shaped section, which represented one third of the mantle usable volume, was subsequently eliminated providing a longer residence time for the combustion products. This resulted in a more complete combustion process inside the mantle volume, a longer flame dwell time on the inside surface of the emitter and a lower temperature of the exhaust gases.

The utilization of the mantle used in this preliminary part of the experimental phase is temperature limited because Hastelloy, the mantle's material, has a maximum operational temperature of approximately 1350 °C. Initial tests have demonstrated that operating the burner at a set firing rate to maintain the mantle at 1250-1300 °C, results in the outside surface of the SiC emitter stabilizing at approximately 1000 °C. At this time this seems to be the maximum emitter operational temperature achievable with this mantle material.

The first part of this analysis is therefore based on the SiC emitter radiating at 1000 ° C. For higher operational emitter temperatures (maximum operating temperature of Hexoloy SA is 1650 ° C) the present mantle and burner tube will be replaced with similar components made of Hexoloy SA. However, before starting fabrication of a Hexoloy SiC mantle which will require considerable effort, it is planned to utilize several Hastelloy mantles to experimentally optimize the mantle design (with respect to perforations size and distribution), to achieve the highest possible temperature uniformity over the entire emitter radiating surface.

Fig. 2 shows that the TPV cells will be mounted on the inside surface of a cylindrical, water cooled, annular barrel to face the emitter over its entire length. The cell array is spaced 2.5 cm apart from the emitter surface. Without cells, the annular barrel can be used as a calorimeter that collects more than 95% of the energy radiated by the emitter. The primary air for combustion is pre-heated by a simple three-pass, cross-flow heat exchanger, made of Inconel 610, at the exit of the mantle-emitter assembly. Every pass comprises three layers of tubular structure for the passage of the primary air which is driven by a small dc blower. The exhaust gases cross flow through the tubular structure, preheating the incoming air. Preliminary tests indicate that this simple design heat exchanger can preheat to 450-500 ° C primary air for the combustion of a fuel rate needed for emitter operation at 1000 ° C. The pre-heated primary air is recycled via an insulated duct to the burner tube, enters the tube annular space and finally is distributed inside the burner tube at specific locations to mix with the atomized fuel-air stream.

Two TPV cells (GaSb and GaInAs) are considered in this study. Both cell types can convert the emitted energy below 1.7 μm with slightly different conversion efficiency values.

The normal, spectral emittance of Hexoloy SA (α - SiC) is well characterized over a wide spectral range (up to 8 μ m) and for temperatures up to 1500 °C (1) Using Hexoloy's available spectral emittance data (Fig.4), which do not vary substantially for time exposure to an oxygen rich environment in the 1000 - 1300 °C temperature range, the amount of energy radiated (H) by an Hexoloy emitter operating at 1000 °C is equal to 11.97 W/cm^2 with the following distribution of the radiated energy in three specific spectral regions : 0 - 1.7μm, 1.7 μm - 4.0 μm and 4.0 μm -∞ .

H(0-1.7 μm)	=	8.2 % H	=	0.99 W/cm^2
H(1.7-4 μm)	=	52.3 % H	=	6.26 W/cm^2
H(4 μm - ∞)	=	39.5 % H	=	4.72 W/cm^2

FIGURE 3. Burner-emitter assembly.

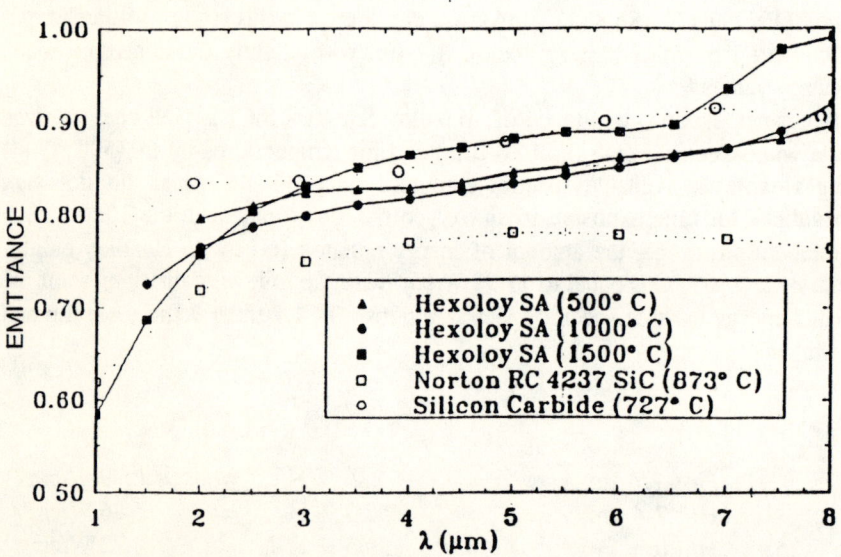

FIGURE 4. Normal, spectral emittance of Hexoloy and other SiC-based materials.

A blackbody radiator at the same temperature emits 15 W/cm^2 over the entire spectrum with the following energy spectral distribution .

Hbb(0-1.7 µm)	=	9.4 % H	=	1.41 W/cm^2
Hbb(1.7-4 µm)	=	55.6 % H	=	8.34 W/cm^2
Hbb(4.µm-∞)	=	35.0 % H	=	5.25 W/cm^2

The approach envisioned to eliminate the radiation above 1.7 µm, which is unusable by the cell, is to use a combination of treatments available from the optical coating industry to tailor the radiation emitted by the SiC Hexoloy emitter. For maximum effect, the treatment must be optimized for the bandgap of the PV cell. Ideally, in our case, we want the coating to be totally transparent below 1.7 µm to leave unaffected the emittance of the Hexoloy emitter at wavelengths just above the cutoff, and beyond the cutoff reduce the emittance to zero. Combinations of optical coatings have been successfully applied on SiC emitters, and tested to 1000 °C without apparent problems. It is felt that these coatings would perform up to 1200 ° C. Figure 5 shows the actual reflection data obtained on the coated SiC. For higher temperature application , the silicon would be replaced by thorium and coating of tantalum pentoxide and silicon dioxide would be dominant.

Figure 5 indicates that the emitted energy is allowed to pass through the coating with a transparency of approximately 0.85 in the desired wavelength band (1 to 2 µm) and that the emittance is essentially shutdown beyond 2 µm. Furthermore, it seems to be possible to treat the SiC emitter to obtain different wavelength cutoffs without degrading the emittance in the desired range. The emittance window can be moved along the wavelength scale by about ±15 % from 2µm, to 2.3 µm or to 1.7 µm. However, at this point the long wavelength (above 4-5 µm) spectral emittance of a coating-treated SiC is unknown. It seems that beyond 4-5µm the coating-treated SiC emits energy in the direction of the cell array with an average emittance value of approximately 0.5. With this assumption and operating at 1000 °C the coating-treated Hexoloy cylindrical emitter radiates 3849 W distributed in the three wavelength bands as follow:

H(0-1.7 µm) = 0.9 x 0.99 W/cm^2→ Total Emitter Surface radiates 962 W
H(1.7-4 µm) = 0.05 x 6.26 W/cm^2→ Total Emitter Surface radiate 335 W
H(4 µm-∞) = 0.5 x 4.72 W/cm^2→ Total Emitter Surface radiates 2,552 W

347

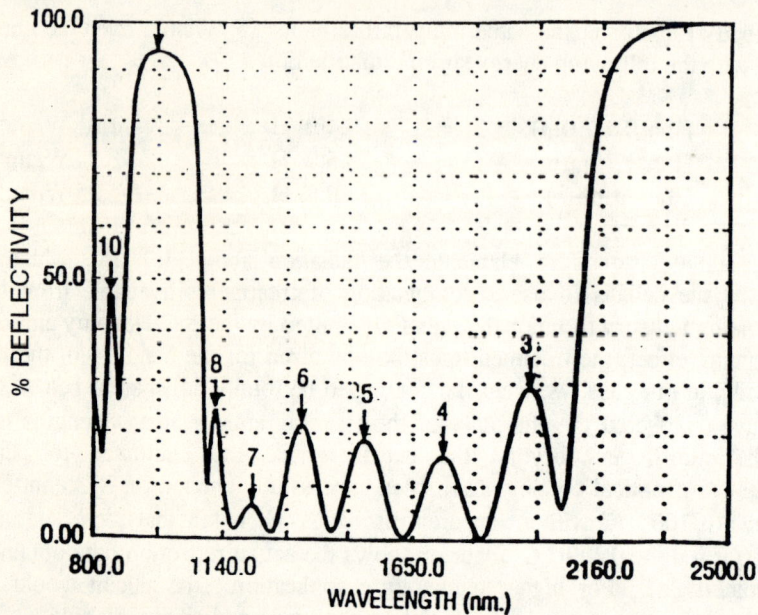

FIGURE 5. Reflectivity of coating-treated SiC.

The water cooled cylindrical structure that supports the cell array has an inside diameter of 21.8 cm and a surface of 1,403 cm². With a cell packing factor of 0.8 (1 mm spacing between adjacent cells) it can accommodate 930 GaSb cells , each with a surface of 1.2 cm² (1 cm x 1.2 cm), for a total cell area of 1,116 cm².

Assuming a 4 % end losses (energy lost on the two end belts between emitter and cell array) for all the energy radiated by the emitter, the cell usable radiation impinging on the total cell surface is computed as:

Cell Usable Radiation = 962 W x 0.96 x 0.8 = 738.8 W

The GaSb cell that we plan to use in the experimental investigation converts the energy in the cell usable band (for a gray body emitter operating at 1000 °C) into electric power with a conversion efficiency of approximately 39 % (and of 36 % for the same emitter at 1500 °C)(3). Fifteen percent of the usable energy impinging on the cell is reflected by the cell grids and the balance, between 46 % and 49 %, results in thermal heating of the cell and must be removed by the cooling system.

With the SiC Hexoloy emitter operating at 1000 °C and a cell efficiency of 39%, the 930 cell array will provide an electric output of 288 W.

Heat is transferred from the combustion gases to the emitter inner surface mainly by convection. At equilibrium the emitter radiates all the heat transferred from the gases. The combustion products exit the emitter area much hotter than the emitter and in even a well-designed heat transfer arrangement, the required gas temperatures are estimated to be about 150 °C hotter than the emitter at 1000 °C and as much as 450 °C hotter to run the emitter at 1400 °C, and most of the combustion heat remains in the exhaust gases. A recuperator can extract heat from the exhaust and the pure counterflow exchanger approach offers the most effectiveness. In combustion processes the exhaust products have a higher heat capacity than the inlet air, because the exhaust mass flow is greater and the specific heats are species and temperature dependent. The ratio of the heat capacities (about 0.8 to 0.85) governs the ratio of temperature change in the fluid flows (by

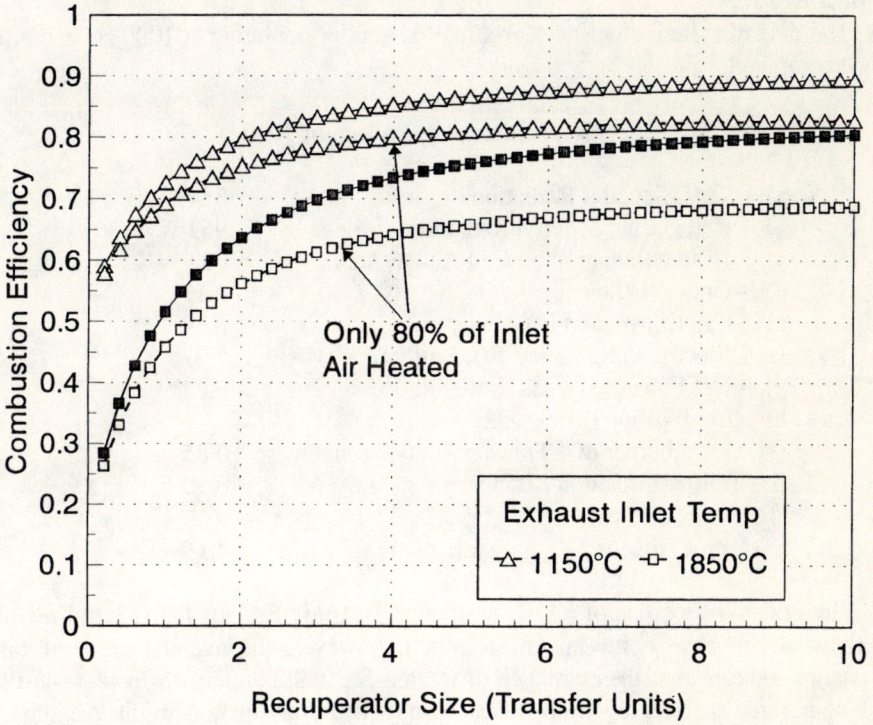

FIGURE 6. Combustion efficiency for ideal burner and for air-atomizing burner.

conservation of energy) and consequently the exhaust temperature drop through the exhanger is about 80% of the inlet temperature rise and this in turn is limited to the exhaust inlet temperature and effectiveness of the recuperator. Therefore, at higher emitter temperatures, there is a greater exhaust heat loss. An air-atomizing burner compounds this problem because the air used for atomizing is not available to the recuperator and so the exhaust leaves the recuperator at a higher temperature than it would normally. The heat transfer size of a recuperator controls its effectiveness and affects its ability to reduce exhaust loss. Fig. 6 shows the net effect on combustion efficiency for the estimated exhaust inlet temperatures corresponding to emitter temperatures of 1000 °C and 1300 °C for both an ideal burner and for an air-atomizing burner.

The power requirement for the burner (presently around 68 W because of the use of commercial components) can be reduced to 28 W. The cell array water cooled system will require approximately 52 W, leaving a TPV system net power output of 208 W.

The overall system efficiency for a treated emitter operating at 1000 °C can now be computed.

Treated Emitter operating at 1000 °C

a)	- Total Spectral Radiation	3849 W
b)	- Cell Usable Spectral Radiation	962 W
c)	- Usable Radiation on Cell Active Area	738 W
d)	- Optical Efficiency (c/a)	0.19
e)	- Cell Conversion Efficiency	0.39
f)	- Electric to Radiation Eff. Ratio (d x e)	0.074
g)	- Net Power to Gross Power Ratio	0.722
h)	- Combustion Efficiency:	
	(Ideal burner & 4 Transfer Units Recuperator	0.85
i)	-Burner Insulation Efficiency	0.90

System Efficiency (f x g x h x i)   **4.1%**

The optical properties of a coating-treated Hexoloy SiC for the spectral region above 4 μm have not been characterized, however, the manufacturers of this coating indicate that the emittance of treated SiC will continue to increase in the long infrared spectral region to values close to 1. In this analysis we have assumed that the coating on the Hexoloy emitter radiates with emittance equal 0.5 over the entire spectral region beyond 4μm. This radiation amounts to 2,552 W.

In thermal equilibrium the emitter absorptivity is equal to the emitter emittance (assuming complete opacity of the emitter). Therefore, in accordance with its absorptivity value the emitter can reabsorb a fraction of the radiation with wavelength above 4 µm if this radiation is reflected back. If a quartz window (which has a transmissivity = 0 for wavelengths above 4 µm) is placed directly on the cell surface, it will reflect back a fraction (estimated at approximately 20% = 510 W) of the impinging radiation with wavelength above 4 µm. The quartz, at approximately the same temperature of the cell, absorbs the balance (2,042 W) of the impinging radiation that the cell cooling system removes from the cell-quartz assembly. However, if a quartz shield is placed away from the cell surface, at some location between the emitter and the cell array, the quartz will reach an equilibrium temperature and then it will reemit the absorbed radiation, in approximately same amounts, in two directions, toward the cell array (1,021 W) and back to the emitter (1,021 W). In this condition the treated Hexoloy emitter receives back from the quartz shield 1,531 W (510 W reflected and 1,021 W reradiated). With an absorptivity of 0.5 the emitter reabsorbs half of this energy (765 W) and, to keep radiating at the same operating temperature, requires from the combustion only 3,849W - 765 W = 3,084 W.

The quartz has a transmission of 0.95 for the radiation in the cell usable spectral region (< 1.7 µm), therefore the quartz shield reduces the usable radiation on the cell active area to 701 W.

The overall system efficiency for the same emitter operating conditions and with the addition of a quartz shield between emitter and cell array results as follows :

Treated Emitter operating at 1000°C plus Quartz Shield

a)	- Total Spectral Radiation	3,084 W
b)	- Cell Usable Spectral Radiation	962 W
c)	- Usable Radiation on Cell Active Area	701 W
d)	- Optical Efficiency (c / a)	0.22
e)	- Cell Conversion Efficiency	0.39
f)	- Electric to Radiation Eff. Ratio (d x e)	0.088
g)	- Net Power to Gross Power Ratio	0.708
h)	- Combustion Efficiency:	
	(Ideal burner & 4 Transfer Unit Recuperator)	0.85
i)	- Burner Insulation Efficiency	0.90

System Efficiency (f x g x h x i)   **4.8 %**

By replacing the Hastelloy mantle with a same design mantle made of SiC Hexoloy SA the TPV test unit can be operated at higher fuel firing rate. With a mantle working at 1550-1600°C the Hexoloy emitter's surface temperature will be at approximately 1300°C radiating 9,646 W with the following radiation spectral distribution:

$H(0-1.7 \mu m) = 0.9 \times 4.8$ W/cm^2 Total Emitted Radiation = 4,672 W
$H(1.7-4\mu m) = 0.05 \times 16$ W/cm^2 Total Emitted Radiation =   865 W
$H(4\mu m-\infty) = 0.5 \times 7.6$ W/cm^2 Total Emitted Radiation = 4,109 W

At this higher emitter temperature the usable radiation on the cell active area amounts to 4,672 W x 0.96 x 0.8 = 3,588 W, reduced to 3,408 W with the addition of the quartz shield, and with a cell conversion efficiency of 37%, the 930 cell array will provide a gross electric output of 1,261 W (a net electric output of 1,159 W for a projected auxiliary power requirement of 102 W).The system efficiencies for both, treated emitter and treated emitter plus quartz shield , are as follows :

Emitter operating at 1300°C	Treated Emitter	Treated Emitter plus Quartz Shield
a) - Total Spectral Radiation	9,646 W	8,414 W
b) - Cell Usable Spectral Radiation	4,672 W	4,672 W
c) - Usable Radiation on Cell Active Area	3,588 W	3,408 W
d) - Optical Efficiency (c / a)	0.37	0.40
e) - Cell Conversion Efficiency	0.37	0.37
f) - Electric to Radiation Eff. Ratio (d x e)	0.137	0.15
g) - Net Power to Gross Power Ratio	0.92	0.91
h) - Combustion Efficiency: (Ideal burner & 4 Transfer Units Recuperator)	0.745	0.745
i) - Burner Insulation Efficiency	0.88	0.88
System Efficiency (f x g x h x i)	**8.2 %**	**9.0 %**

Fig. 7 plots the system efficiency obtainable for treated emitter temperatures ranging from 1000 °C and 1300 °C with and without quartz shield.   This operational temperature range can be cover by the smaller of the two burners (Burner # 2) available for the experimental phase of this investigation.

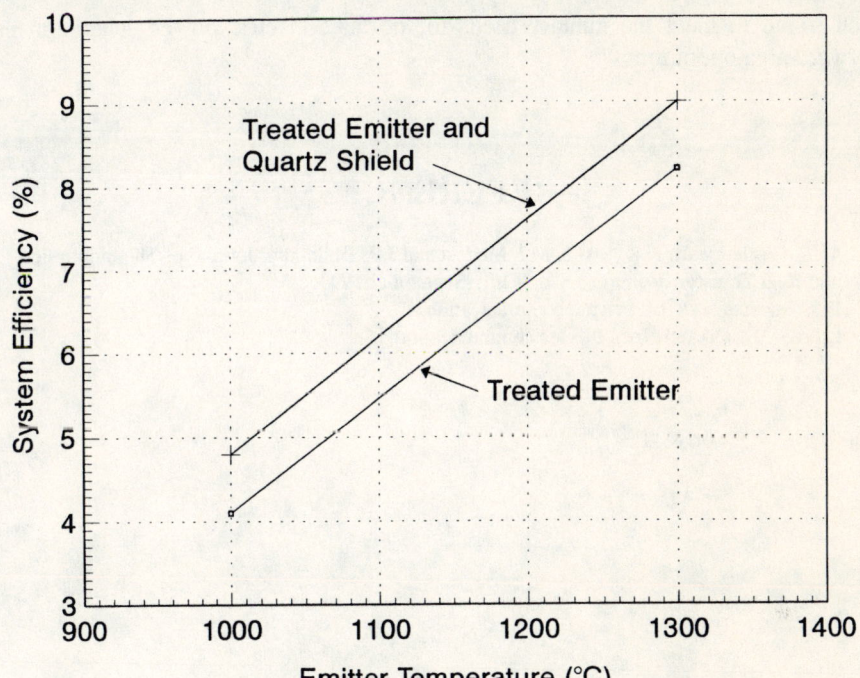

FIGURE 7. System efficiency for emitter's temperature in the 1000°C - 1300°C range, with and without quartz shield.

CONCLUSIONS

A coating-treated SiC emitter in conjunction with an array of low bandgap (GaSb) PV cells could represent the state-of-the-art most viable and practical approach for the development of military TPV power sources capable of operating from logistically available Hydrocarbon fuels.

The overall system efficiency can reach values around 9% at reasonable operational temperatures of both burner and emitter.

The experimental phase of this investigation should demonstrate the liquid fuel operation, confirm the validity of the assumptions taken, and verify the projected system efficiency. It will also provide the basis for a realistic projection of the system power density in terms of power output per unit weight (W/kg) and power output per unit volume (W/m^3). We expect that this TPV approach will yield system power densities substantially higher that those of a system based on an ytterbia selective emitter with an array of silicon cells and thus make this approach well suited to meet the military needs for portable electric power generation in forward area operations.

REFERENCES

1. M.A. Postlethwait, K.K. Sikka, M.F.Modest and J.R. Hellmann, Journal of Thermophysics and Heat Transfer, Volume 8, No.3, July-September 1994.
2. P. Beauchamp, OCLI, Private communication.
3. L.Fraas, JX Crystals Inc., Private communication.

Conceptual Design of 500 Watt Portable Thermophotovoltaic Power Supply Using JP-8 Fuel

Crispin L. DeBellis and Mark V. Scotto
McDermott/Babcock & Wilcox Research and Development Division
Alliance, Ohio

Stephen W. Scoles
Babcock & Wilcox Naval Nuclear Fuel Division
Lynchburg, Virginia

Lewis Fraas
JX Crystals
Issaquah, Washington

Abstract. Babcock & Wilcox (B&W) and JX Crystals (JXC) have developed an innovative design for a compact, 500 watt net electric (We), 24-VDC thermophotovoltaic (TPV) power supply using JP-8 fuel. As currently envisioned, the TPV generator will be approximately 20 cm (8 inches) in diameter and 50 cm (20 inches) high, not including a fuel tank and controls. The total system may weigh as little as 7.5 kg (16.5 lb) without fuel. This system will achieve high efficiency and high power density relative to its size through the use of low bandgap gallium antimonide (GaSb) PV cells and a matched emitter. A thermally integrated fuel vaporizer and recuperator will boost system efficiency by transferring the unused energy in the exhaust stream to the incoming fuel and combustion air. At rated conditions and 500 We output, the system is expected to have an overall efficiency of 8% to 10%. This paper examines the trade-offs between system efficiency, power density, and weight required in the selection and configuration of the major system components.

CP401, *Thermophotovoltaic Generation of Electricity: Third NREL Conference,*
edited by Benner/Coutts
© 1997 The American Institute of Physics 1-56396-734-0/97/$10.00

INTRODUCTION

The major components of a TPV system are shown in Figure 1. Fuel and air are mixed and burned in the burner. The heat of combustion is transferred via convection and radiation to an emitter, which is typically a ceramic material. As a result, the emitter is elevated to a high temperature (1500° to 2000°K). Because of its temperature and composition, the emitter intensely radiates energy to the PV cell assembly. Optical filters may be used to reflect unusable radiation back to the emitter. The PV cells convert a portion of this energy into DC power, which can be altered to suit a particular need with a power conditioner. The portion of the energy absorbed by the PV cells that is not converted to electricity must be removed as waste heat. The PV cells cannot be allowed to heat up, because their conversion efficiency decreases with increased temperature. The combustion products leaving the emitter cavity still contain a significant amount of energy. To achieve a high system efficiency, a recuperator is required to transfer the energy in the combustion products to the incoming combustion air.

The basic design of a TPV system is governed by a few key technology choices:
- What is the operating temperature of the emitter?
- What kind of PV cells are used?
- What are the radiation characteristics of the emitter?

Emitter temperature directly controls power density and indirectly impacts production costs and system efficiency. Higher emitter temperatures produce higher power densities. However, the emitter temperature is limited by material properties and heat transfer. Material properties can limit operating temperature, because the few high-temperature materials suitable for use in oxidizing environments are prone to thermal shock failure. This can be alleviated by the use of ceramic composites. However, ceramic composite technology is not mature, and data on long-term stability are sparse. On the other hand, thermodynamics suggest using lower emitter temperatures to maximize the heat transfer from the combustion gas to the emitter. These competing factors must be considered in the system design process.

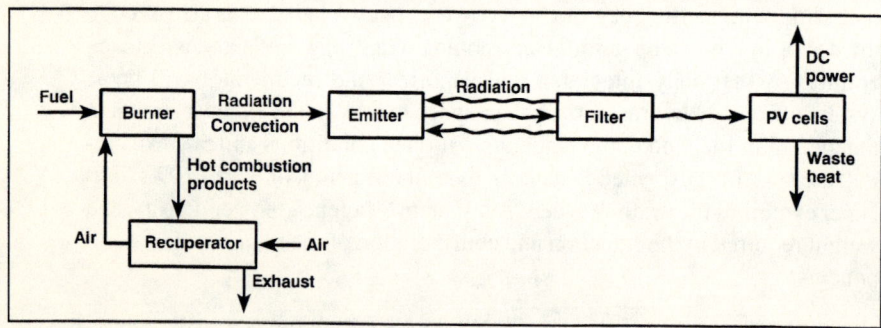

FIGURE 1. TPV system components.

PV cell selection must be based on power density, system efficiency, cell availability, and projected manufacturing cost. Silicon cells are currently available at reasonable costs but have a high bandgap of 1.12 ev (1.11 micron wavelength). Emitter temperatures required for efficient silicon cell use (> 2000°K) generally result in rapid degradation and failure of the emitter and other high-temperature components. GaSb cells are currently more expensive but have a bandgap of 0.73 ev (1.70 micron wavelength). As shown in Figure 2, the cell bandgap is important, because photons striking the cell with energies lower than the bandgap (longer wavelengths) do not produce photo electrons. Also shown in Figure 2 is the energy emitted by a near black-body radiator, such as silicon carbide (SiC), as a function of temperature and the amount of energy falling below the bandgap of the cells.

GaInAs and other III-V PV cells have been fabricated with bandgaps below 0.68 ev (1.84 micron wavelength), but these cells are significantly more expensive than GaSb cells and do not significantly increase the amount of usable energy. Even with GaSb cells, a large fraction of the energy radiated by a black-body emitter is below the bandgap of the cell (longer wavelengths) and therefore unusable. Much of this energy can be reflected back to the emitter by a dielectric or other bandpass filter.

As an alternative to black-body emitters, materials that selectively radiate most of their energy at useful wavelengths can be employed. Rare-earth oxides such as erbium and ytterbium oxide have spectral lines at wavelengths below the bandgaps of common PV cells. However, these spectral lines are narrow (0.2 to 0.3 microns wide) and significantly reduce the power density of the emitter. Other selective emitters emit over a broader band. One such emitter under development at JXC is referred to as a "matched" emitter. As shown in Figure 3, the measured emission of this material at 1700°K closely matches the GaSb cell quantum efficiency curve. It

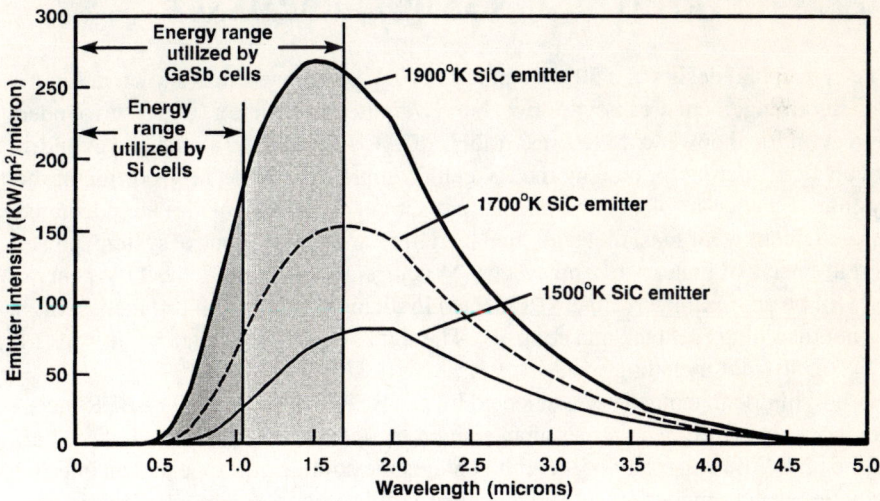

FIGURE 2. Black-body emission and PV cell bandgaps.

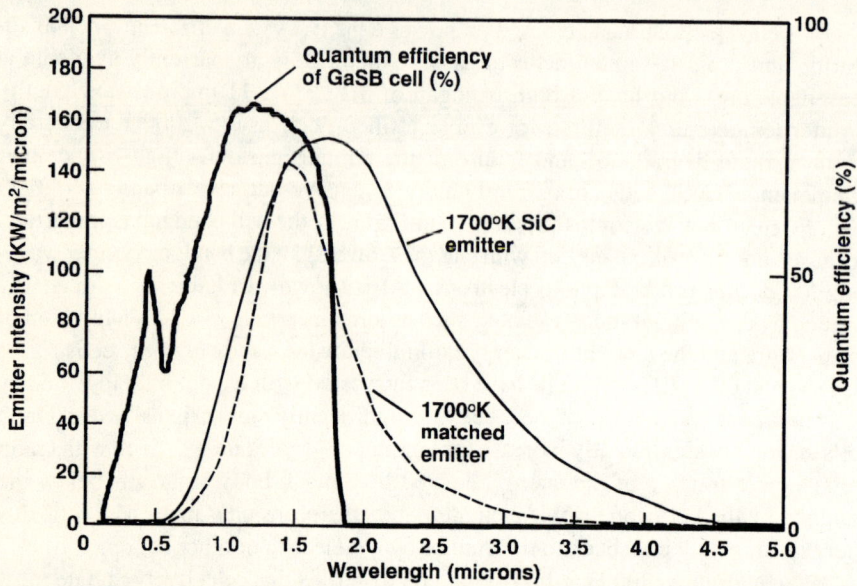

FIGURE 3. Characteristics of GaSb cells, 1700°K SiC emitter and "matched" emitter.

also radiates almost as much energy within useful wavelengths as a SiC black-body emitter but radiates significantly less energy at longer wavelengths. This allows high power density and high-efficiency operation with simplified band-pass filters.

SYSTEM DESIGN AND PERFORMANCE

The conceptual design of a 500-We portable TPV power system is shown in Figure 4. This arrangement uses a cylindrical burner/emitter/recuperator (BER) surrounded by a cylindrical power converter assembly (PCA). A horizontal air fan and cylindrical air flow ducting for cooling the PV cells complete the basic TPV portion of the system. Balance-of-plant components (not shown on the schematic) include housing and ducting, air fans, fuel tank, fuel feed lines and valves, control system, power conditioner, and battery start-up system. As currently envisioned, the TPV generator will be approximately 20 cm (8 inches) in diameter and 50 cm (20 inches) high — not including fuel tank and controls. The total system may weigh as little as 7.5 kg (16.5 lb), not including fuel.

A cylindrical geometry was selected for the BER and the PCA. The BER incorporates a vaporizer in the center of an annular recuperator located in the bottom half of the unit. The upper half of the unit contains the combustion zone and an internal radiating wall to make the emitter (the outer wall) isothermal. The emitter is impervious to gas flow through its walls. This potentially eliminates the need for a win

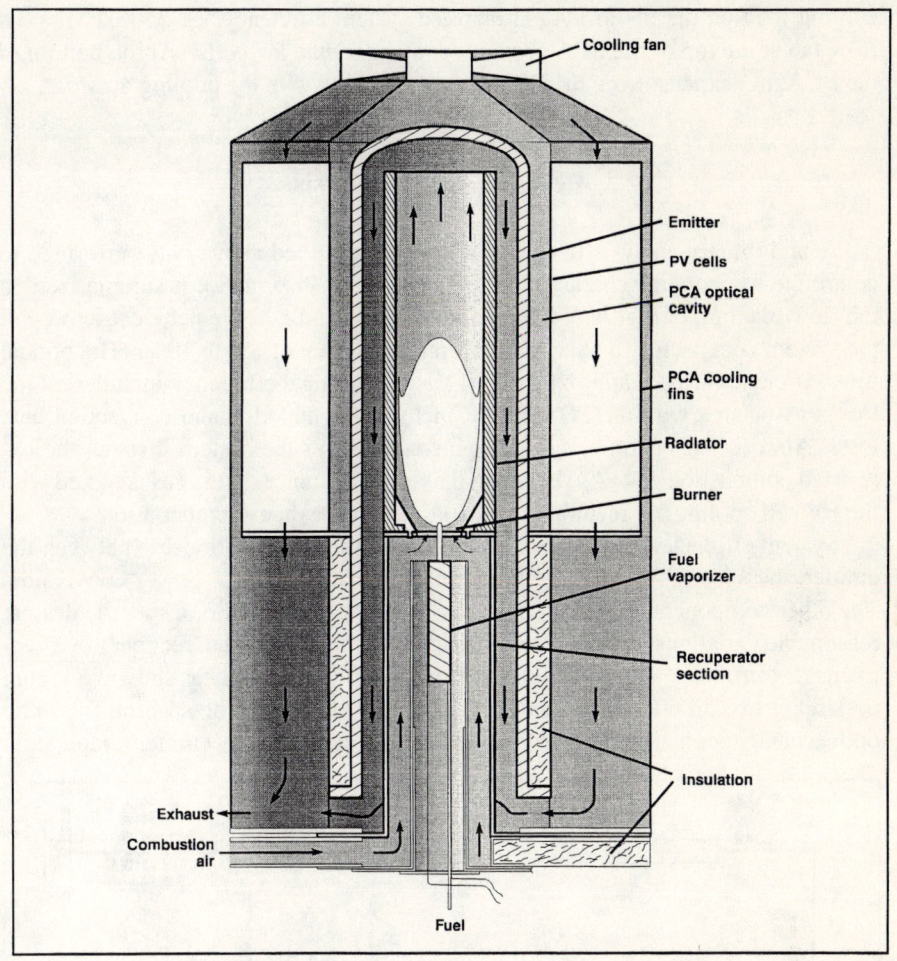

FIGURE 4. Schematic of portable TPV generator.

dow (quartz) to protect the cells from the hot combustion products. However, if vapor phase transport of emitter material degrades PV cell performance, a window may still be required. The emitter is isolated from the PV cells by an air gap that is designed to eliminate natural convection, thus minimizing convective losses in the cavity.

The PCA was also designed with a cylindrical geometry. It is composed of 16 interlocking plates with heat transfer fins on the outside and the PV cells on the inside. Each circuit contains a string of GaSb PV cells producing approximately 34 We. The PV cells are covered with a spectral filter to reflect out-of-band radiation. The interconnects between the PV cells are covered with a highly reflective material to reflect radiation back into the system. Both the filter and the reflectors reduce the cooling load on the PV array and increase system efficiency. A 250-CFM, axial flow fan at the top of the PCA provides cooling for the PV cells. At the bottom of the PCA, hot exhaust from the BER is vented to mix with the cooling air flow.

Performance Analysis

A preliminary analysis of the TPV system described above was performed to determine its operating characteristics. The system performance is summarized on the flow diagram in Figure 5. At rated conditions and 500 We net electric output, the system is expected to have a net overall efficiency of 8% to 10%. The optical train efficiency will be approximately 14%. The optical efficiency includes all the losses associated with the PCA cavity, including conduction and convection heat loss. Almost half of the waste heat generated leaves the system through the exhausted combustion gas. As shown on the flow diagram, exhaust gas is mixed with the PV cell cooling air, resulting in a 330 K average exhaust temperature.

System efficiency is primarily controlled by the inter-relationship between the emitter, the spectral filter, and the type of photocells used for power conversion. The other component affecting efficiency is the recuperator. For a specific design, reasonable variations in recuperator size (directly proportional to recuperator effectiveness) can change system efficiency from 5% to 10% overall, as shown in Figure 6. Higher overall efficiencies can be achieved by changing the spectral filter and optical cavity geometry. System power density is controlled by emitter temperature

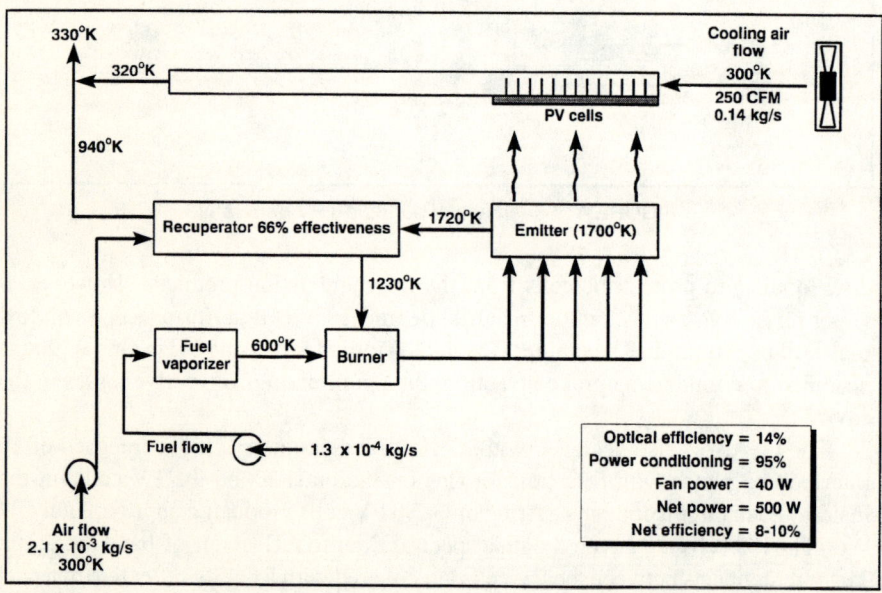

FIGURE 5. Process flow diagram at rated conditions.

360

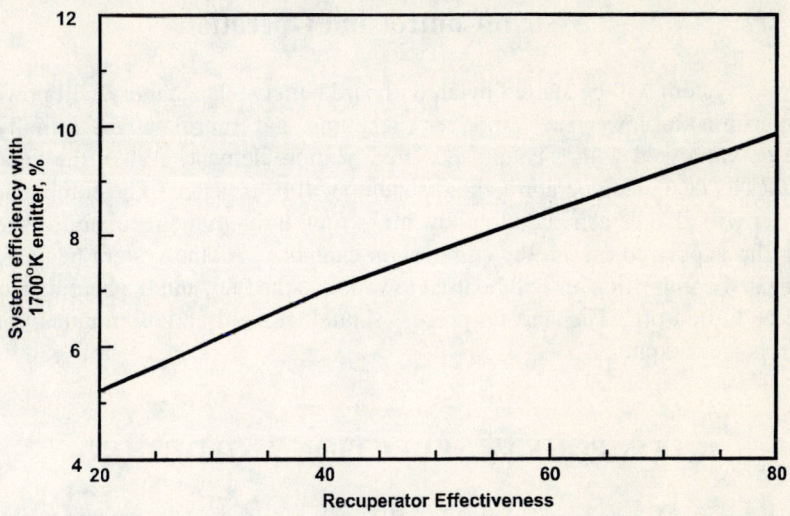

FIGURE 6. System efficiency vs. recuperator effectiveness.

and, to a lesser extent, by filter characteristics. As shown in Figure 7, when emitter temperature increases, power density increases, reducing total system size and mass. Because power density directly affects total PV cell area, system cost is also directly impacted by emitter temperature.

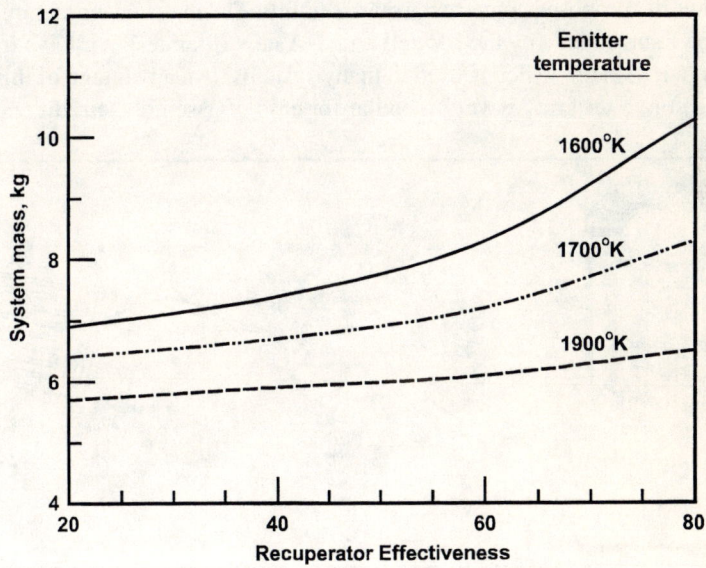

FIGURE 7. System weight vs. recuperator effectiveness and emitter temperature.

System Control and Operation

The system will be started by an on-board battery. The battery will power the combustion air blower, fuel vaporizer, fuel pump, and control valves. Initially, fuel will be vaporized with a battery-powered heating element. When the vaporizer reaches its design temperature, the fuel pump will be actuated. The combustion air blower will also be activated to allow air to flow through the recuperator and mix with the vaporized fuel in the combustion chamber. As the system heats up, the preheated combustion air will be used to vaporize the fuel, and the heating element will be turned off. The start-up process should take only a few minutes to reach rated power output.

COMPONENT SELECTION AND DESIGN

In the development of a portable TPV power system, a number of key issues must be addressed in the component selection and design. These are described below for the PCA and BER subassemblies.

Power Converter Assembly (PCA)

The power converter assembly is a critical subsystem in the TPV generator. Figure 8 shows the proposed converter array assembly. The matched emitter cylinder is at the center surrounded by the PV cell array. A new filter design allows for eliminating a quartz shield, which is used in many systems. An advantage of this design is that the converter array is very modular for ease of assembly, circuit encapsula-

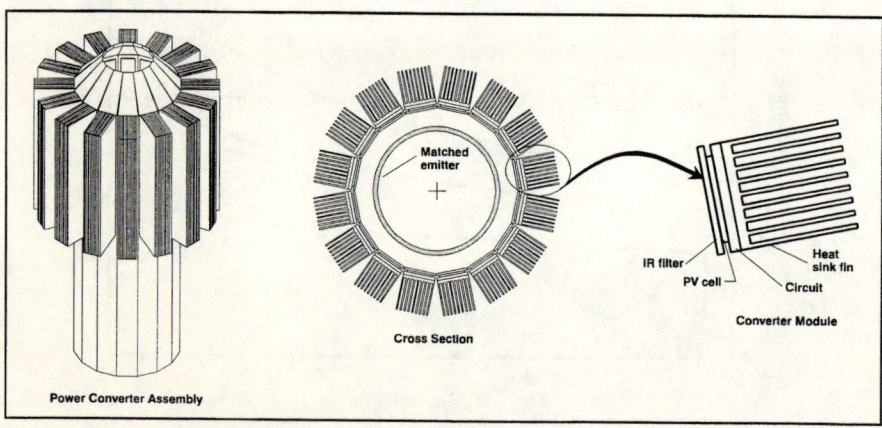

FIGURE 8. Schematic of Power Converter Assembly.

tion, and generator converter assembly repair. A major challenge for the development of this system will be to build and optimize the PCA. The PCA design determines the optical efficiency of the system. The optical efficiency can be improved by recycling as much unusable energy as possible back into the system. This can be accomplished by maximizing the view factor between the emitter and PV cells, reducing end losses, utilizing highly reflective materials on non-PV cell surfaces, and minimizing conduction/convection losses. This converter assembly consists of the GaSb PV cells and circuits, IR filter, and matched emitter.

GaSb PV Cells and Circuits

Over the past three years, JXC has established a pilot manufacturing line for producing high-quality GaSb cells.[1] It has fabricated thousands of GaSb cells for PV applications. Figure 9a shows the current/voltage response of a cell measured in a flash simulator, which is an ideal test condition. Figure 9b shows a similar curve for a GaSb cell under continuous operation in front of a 1650°K black-body source. The latter curve represents what a practical TPV system will produce. Cells are solder mounted into circuits. A typical TPV circuit may consist of multiple cells in series producing 36 We. Spade connectors on the positive and negative terminals are used to draw the current just as in a small battery. While JXC has fabricated both cells and circuits, process automation and volume production are required for cost reduction.

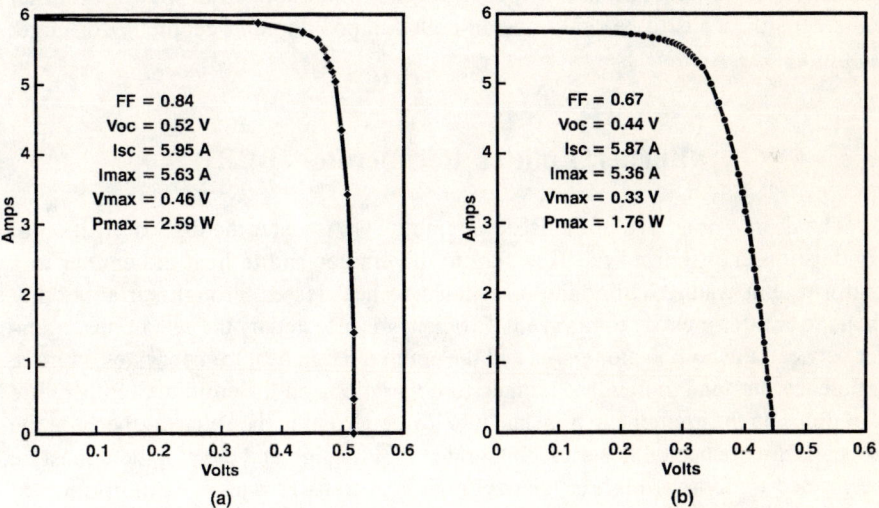

FIGURE 9. Characteristics of JXC's 1.0 CM² GaSb cell (a) Flash test and (b) 1650°K black-body.

Previous work at JX Crystals has focused on SiC black-body emitters. For these emitters, a simple dielectric filter has been used to return the large amount of energy in the 1.8 to 4-micron band back to the emitter. However, the new matched emitter has a much smaller amount of energy in this band, which now allows a redesign of the filter to address the whole 1.8 to 15-micron band. For the conceptual design, a simple metal film / two-dielectric layer filter designed by W. Biter at Sensortex and described by G. Guazzoni and M.F. Rose[2] was used. While this filter was designed for use with a selective emitter and silicon cells, it was modified for use with the matched emitter and GaSb cells.

Matched Emitter

In 1996, JXC developed an infrared emitter that is nearly perfectly matched to the GaSb cell,[3] as shown in Figure 3. This new material emits strongly for wavelengths where the GaSb cell responds and weakly for longer wavelengths where the cell does not respond. Unlike previous rare-earth oxide selective emitters, the bandwidth of this new emitter closely matches the cell bandwidth rather than being much narrower. The narrow bandwidth associated with rare-earth oxide selective emitters severely reduces power density. However, this new emitter, promises to make the TPV generator more powerful, more efficient, smaller, and lighter. While the matched emitter is very exciting, it has only recently been incorporated in small gas burner units. It still needs to be optimized in both composition and geometry for larger liquid fueled systems.

Burner, Emitter, Recuperator (BER)

The design objective of the BER assembly is to transfer the highest fraction of combustion energy from the JP-8 fuel to the emitter and to heat the emitter to a uniform temperature. In addition, conductive heat losses through surfaces other than the emitter must be minimized. Efficiency is affected by the rate of heat transfer between the combustion gases and the emitter. High heat transfer rates improve efficiency but tend to increase temperature variations on the emitter. Temperature variations on the emitter can translate into lower current in the PV array, because the PV cells are connected in series. Integrated BER design and development must be performed to achieve high efficiency, high heat transfer rates, and uniformity of emitter temperature.

Burner

The conceptual design for a JP-8 fueled BER using an impervious emitter and a diffusion flame is illustrated in Figure 10. The primary burner is a gas nozzle that rapidly entrains combustion air. The gas nozzle generates several jets that spread the fuel into the combustion chamber. The fuel jets are ignited with a spark ignitor. A combination of air swirl, jet geometry, and refractory geometry will establish a well back-mixed flame stabilized near (but not too close to) the burner face. A fuel vaporizer precedes the gas nozzle. The vaporizer will operate in the 600°K temperature range. The recuperator surrounds the fuel vaporizer. Heat from the hot combustion air and internal surfaces of the recuperator heats the vaporizer. The liquid fuel supply requires a fuel pump to deliver liquid fuel at moderate pressures to the vaporizer. During start-up electrical heating is provided to vaporize the fuel. The preheater would initially be run off battery power until the recuperator can supply the necessary heat.

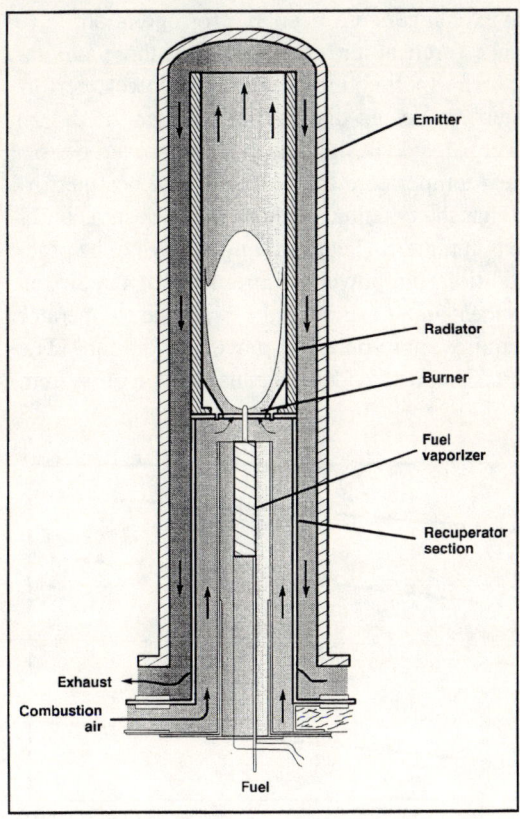

Emitter

Radiator

Burner

Fuel vaporizer

Recuperator section

Exhaust

Combustion air

Fuel

FIGURE 10. Schematic of JP-8 fueled burner / emitter / recuperator.

Emitter

In addition to the energy transfer benefits of the matched emitter, the successful development of the TPV generator depends largely on the structural capabilities of the materials of construction. The emitter, in addition to other critical generator components, including the burner, recuperator, and thermal insulation, will operate at high temperatures (1000° to 2000°K) and will be subject to thermal and mechanical shock. These components must exhibit long-term mechanical, chemical, and dimensional stability. For all these components, low mass will be important, as will manufacturability at reasonable cost. To survive operating conditions, the emitter and other key components will be fabricated from ceramic materials. Ceramic composites will be used if they can meet requirements for high-temperature operation at reasonable cost. Ceramic composites are preferred because of their excellent thermal shock resistance and resistance to catastrophic failure.

Recuperator

A high-temperature recuperator will be required to achieve high overall system energy efficiency. B&W has extensive experience in the design and development of high-temperature heat exchangers.[4] Ideally, ceramic materials that can withstand the high combustion gas temperature would be used in the recuperator. However, reliable (long life) compact ceramic heat exchangers may not be available in the near term. Therefore, a combination of simple monolithic ceramic components to recover the highest temperature heat near the burner exit, followed by a high-temperature metallic alloy heat exchanger, will be used.

Commercial, compact alloy heat exchangers are available for operation up to 1300°K. In Figure 11 the performance (preheat air temperature) of three annular heat exchanger design concepts is shown as a function of size. The lowest performance is from a plain annular counterflow design that cannot produce the design preheat temperature of 1200°K. The corrugated design has much improved performance and reaches the desired preheat temperature in 1.2 units. The best performance is from an annular plate/fin. It has the potential of meeting the design conditions in 0.5 units. However, a full-length plate/fin heat exchanger may not be practical with available materials. In addition, the power requirements of the air fan must be considered. The higher the preheat air temperature, the larger the recuperator and air-side pressure drop, which requires more parasitic power for the fan. The recuperator is a key component, and its design provides a means to trade off system efficiency, weight, and cost.

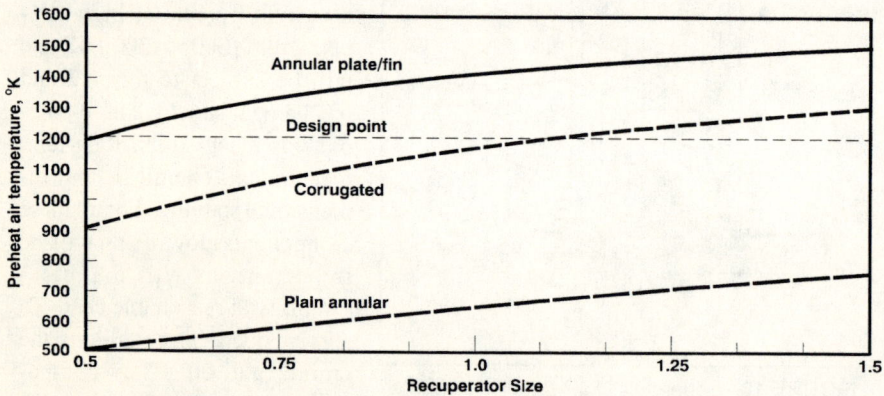

FIGURE 11. Recuperator performance

CONCLUSIONS

B&W and JXC believe that this design for a 500-We, JP-8 fueled, portable TPV power system appears to be superior to other technologies in meeting the requirements of a portable military power generation system.

This TPV system should have the highest power conversion efficiency of any static device in this size range that uses JP-8 fuel directly.

- The expected conversion efficiency of the optical power train is 14%, with a net efficiency of 8% to 10%.
- There will be few moving parts in the system, which should significantly increase the mean time between equipment failures and simplify repairs.
- Steady-state combustion will produce significantly less noise than an internal combustion engine.
- With proper design, the noise level should be inaudible at short distances.
- Military logistics should be improved with this system because it uses JP-8 fuel.

The team also believes that the proposed system is superior to competing TPV approaches. The choice of GaSb cells with a matched emitter produces higher power densities at lower emitter operating temperatures than silicon cells; this increases system efficiency and dramatically improves system reliability. Furthermore, GaSb cells, fabricated using processes similar to silicon cells, are expected to be cost competitive when manufactured at high volumes.

REFERENCES

1. L. M. Fraas, H.H. Xiang, R. Ballantyne, J. Avery, J.X. Fraas, S. Hui and Y. Shi-Xhong, "Low Cost, Low Bandgap Thermophotovoltaic Cells," AIP Conference Proceedings, Third NREL Conference on Thermophotovoltaic Generation of Electricity, Colorado Springs, Colorado, 1997, AIP Press, Woodbury, New York.
2. G.E. Guazzoni and M.F. Rose, "Extended Use of Photovoltaic Solar Panels," AIP Conference Proceedings 358, Second NREL Conference on Thermophotovoltaic Generation of Electricity, Colorado Springs, Colorado, 1995, AIP Press, Woodbury, New York, 1996
3. L. Ferguson, R. Ballantyne, W. Connelly, J. Samaras, M. Sea, and L. M. Fraas, "Matched Infrared Emitters for Use with GaSb TPV Cells," AIP Conference Proceedings, Third NREL Conference on Thermophotovoltaic Generation of Electricity, Colorado Springs, Colorado, 1997, AIP Press, Woodbury, New York.
4. D.L. Hindman and C.L. DeBellis, "Performance of an Advanced Heat Exchanger Using Ceramic Composite Tubes in a Hazardous Waste Incinerator," Advances in Enhanced Heat/Mass Transfer and Energy Efficiency, ASME HTD-Vol. 320, November 1995.

2-Amp TPV Cogenerator Using Forced-Air Cooled Gallium Antimonide Cells

Lewis Fraas, James Avery, Russ Ballantyne, Paul Custard,
Luke Ferguson, Huang Han Xiang, Jason Keyes, Bill Mulligan,
John Samaras, and Doug Williams

JX Crystals, Inc., 1105 12ᵗʰ Ave NW, Suite A2, Issaquah, WA 98027

Abstract. We will describe a wall mounted TPV cogenerator for use as a battery trickle charger and 5,000 BTU/hr room heater on boats, in remote cabins, and in recreational vehicles. Propane is used to heat a proprietary matched emitter, and the emitter is surrounded by a photovoltaic conversion array consisting of 48 GaSb cells connected in series. Warm air generated by forced-air cooling of the array cooling fins is used for room heating, while combustion exhaust gases are vented to the outside. The generator will be demonstrated at the conference. Beta site test units are presently being assembled, and production units are expected to be available this fall

Introduction

JX Crystals has developed and built a wall mounted TPV cogenerator for use as a battery trickle charger and 5,000 BTU/hr room heater on boats, in remote cabins, and in recreational vehicles. This cylindrical unit is approximately 6 inches in diameter and 24 inched tall. Our prototype unit, which will be demonstrated at the conference, generates approximately 1.6 amperes of charging current at 12.5 V. With minor engineering improvements, the production model is expected to generate over 2 amperes of charging current. Figure 1 shows a photograph of the prototype unit.

Despite its modest peak power output, when operated continuously this unit generates about 500 watt-hours of electricity

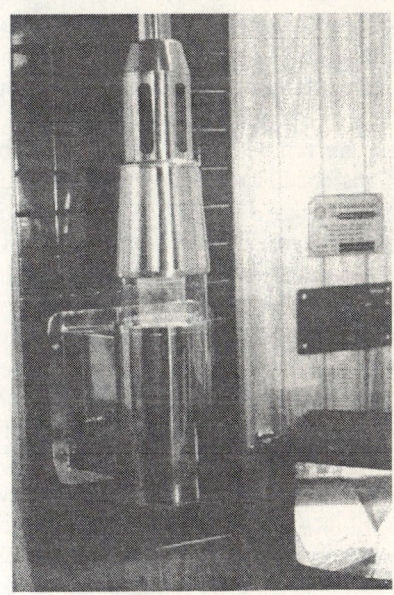

Figure 1. JX Crystals' 2-amp TPV Cogenerator prototype.

CP401, *Thermophotovoltaic Generation of Electricity: Third NREL Conference,*
edited by Benner/Coutts

per day which, for example, is approximately equivalent to the daily output of a 200-watt flat-panel photovoltaic system operated in the northern U.S. during the winter. In cold climates, room heat cogeneration is a clear advantage over traditional PV systems. In fact, we believe that the TPV cogenerator is very complementary to flat panel solar systems in northern climates. In cold, cloudy winter months the TPV cogenerator can be used to generate the bulk of the electric power plus room heat, while in warm summer months with high solar insolation the solar panels will be used predominately. Our market research indicates that there are at least 15 thousand potential customers per year purchasing solar panels in off-grid environments such as mountain cabins, pleasure boats, and recreational vehicles. Most of these solar customers are potential TPV cogenerator customers, and these customers are accessible through solar panel distributors. Considering additional features of the TPV cogenerator such as small size, ease of installation, quiet operation, and pleasing appearance, we expect this initial product to be very attractive.

Results

A schematic of the 2-amp TVP cogenerator is shown in figure 2. The TPV cogenerator assembly consists of a fuel/air mixing tube, a combustion chamber, and an infrared emitter surrounded by a TPV cell circuit with attached cooling fins. Propane is used as a fuel source. The 2-inch diameter matched emitter operates between 1200 and 1400°C. Details of the matched emitter used in the cogenerator are reported elsewhere in these same conference proceedings [1]. The TPV circuit contains 48 single-crystal GaSb cells in series. The assembly is enclosed in a cylinder with two fans at the bottom, one used for forced air-cooling of the array, and another to supply combustion air. Clean air heated by the circuit cooling fins is used for room heat, while combustion exhaust gases are vented through an exhaust tube to the outside. A heat exchanger is used to transfer additional heat from the combustion gas stream to the clean air for room heating. The TPV cogenerator will be demonstrated at the conference in a typical system configuration, with the TPV array output connected to a Trace Engineering Model C12 charge controller. A deep-cycle marine 12V battery is used for energy storage. A low cost, commercially available card and electrodes are used for electronic ignition and flame sense. Should a flame-out condition occur, a solenoid valve is used to shut off gas flow to the combustion chamber.

At operating temperature, the TPV array delivers over 15 V at its maximum power point. An I-V curve for the TPV array operating in the prototype system is shown in figure 3. In this prototype array/system configuration, high series resistance leads to a relatively low fill factor. Note that the fill factor of typical GaSb cells is about 75% [2]. With minor modifications, we expect to be able to reduce the series resistance substantially, increasing the current delivered to the battery at charging voltage.

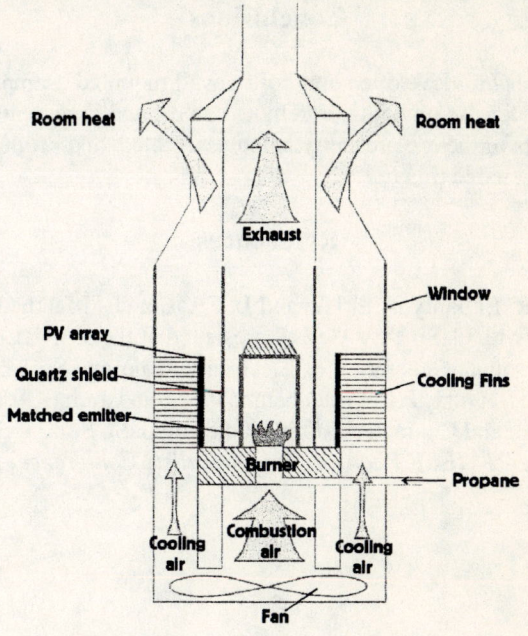

Figure 2. Cross-section of JX Crystals 2-amp TPV Cogenerator

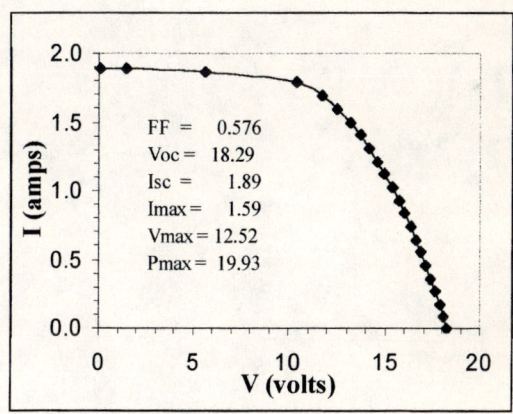

Figure 3. I-V curve of the PV array operating in the cogenerator system.

Conclusions

JX Crystals has developed and built a wall mounted 2-amp, 5,000 BTU/hr TPV cogenerator for use on boats, in remote cabins, and in recreational vehicles. Beta site test units of the are presently being assembled, and production units are expected to be available this fall.

References

1. L. Ferguson, R. Ballantyne, S. Hui and L. Fraas, and, "Matched Infrared Emitters for Use with GaSb TPV Cells," presented at the 3rd NREL Conference on Thermophotovoltaic Generation of Electricity, Colorado Springs, CO, 1997.
2. L.M. Fraas, J.E. Avery, P.E. Gruenbaum, V.S. Sundaram, K. Emery, and R. Matson, "Fundamental Characterization Studies of GaSb Solar Cells," in *Proceeding of the 21st IEEE Photovoltaic Specialists Conference*, Las Vegas, NV, 1991, p80.

Development of a Cogenerating Thermophotovoltaic Powered Combination Hot Water Heater/Hydronic Boiler

Aleksandr S. Kushch
Steven M. Skinner
Richard Brennan
Pedro A. Sarmiento

Quantum Group, Inc.
11211 Sorrento Valley Road; San Diego, CA 92121

ABSTRACT

A cogenerating thermophotovoltaic (TPV) device for hot water, hydronic space heating, and electric power generation was developed, designed, fabricated, and tested under a Department of Energy contracted program. The device utilizes a cylindrical ytterbia superemissive ceramic fiber burner (SCFB) and is designed for a nominal capacity of 80 kBtu/hr. The burner is fired with premixed natural gas and air. Narrow band emission from the SCFB is converted to electricity by single crystal silicon (Si) photovoltaic (PV) arrays arranged concentrically around the burner. A three-way mixing valve is used to direct heated water to either the portable water storage tank, radiant baseboard heaters, or both. As part of this program, QGI developed a microprocessor-based control system to address the safety issues, as well as photovoltaic power management. Flame sensing is accomplished via the photovoltaics, a technology borrowed from QGI's Quantum Control™ safety shut-off system.

Device testing demonstrated a nominal photovoltaic power output of 200 W. Power consumed during steady state operation was 33 W, with power drawn from the combustion air blower, hydronic system pump, three-way switching valve, and the control system, resulting in a net power surplus of 142 W. Power drawn during the ignition sequence was 55 W, and a battery recharge time of 1 minute 30 seconds was recorded. System efficiency was measured and found to be more than 83%. Pollutant emissions at determined operating conditions were below the South Coast Air Quality Management District's (California) limit of 40 ng/J for NOx, and carbon monoxide emissions were measured at less than 50 dppm.

CP401, *Thermophotovoltaic Generation of Electricity: Third NREL Conference,*
edited by Benner/Coutts
© 1997 The American Institute of Physics 1-56396-734-0/97/$10.00

INTRODUCTION

The primary objectives of this project were to, 1) develop TPV subsystems capable of producing enough electric power to run a combustion blower, water pump, controls and other electric devices, 2) integrate the subsystems into proof-of-concept prototype combination (combo) water/space heating system, 3) evaluate the performance of combo appliance, 4) develop cost-effective burner fabrication process, and 5) investigate the potential for performance improvement.

Combination appliances, or combo appliances, were first introduced into the market in 1980. The first systems relied on a storage water heater to provide for portable hot water and space heating requirements, as well as external power for auxiliaries such as the hydronic pump. One of the many possible TPV powered combo configurations is schematically presented in Figure 1.

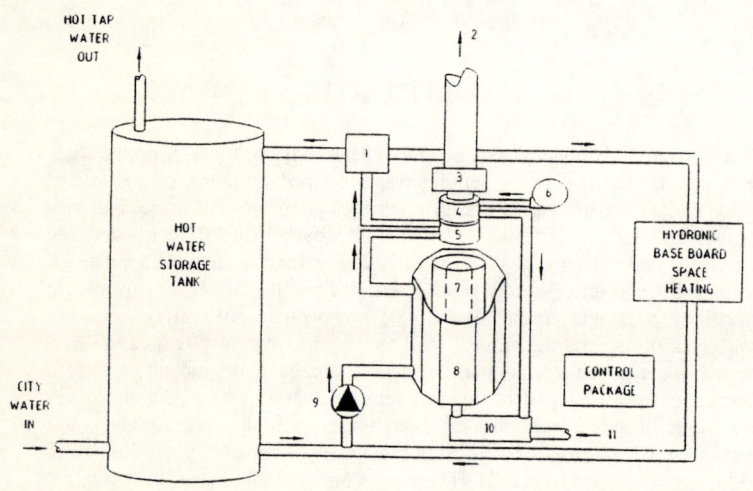

[1] Water Valve, [2] Exhaust, [3] Flue Damper, [4] Air Preheater, [5] Convective Heat Exchanger, [6] Blower, [7] Superemissive Burner, [8] PV Cooling System [9] Water Pump, [10] Mixing Chamber

FIGURE 1.   Schematic diagram of TPV powered combo water heater/space heater

374

Gaseous fuel is introduced into the mixing chamber (10) where it mixes with the combustion air. A recuperator (4) can be utilized to preheat the combustion air supplied by the blower (6). Recuperation was not incorporated in this proof-of-concept prototype. The combustible mixture is then delivered to a radiant burner that converts the fuel energy into thermal radiation and sensible heat of the exhaust products. Following ignition, the burner heats up to become incandescent. Photovoltaics are arranged concentrically around the burner. Radiant energy leaving the burner is incident on the photovoltaic arrays that generate adequate electricity to power the entire system. Hydronic loop return water driven by pump (9) is directed to the radiant heat exchanger that serves as a PV cooling system. This arrangement ensures that the PV cell temperature is within design limits. A transparent pyrex or quartz tube separates the combustion exhaust from the photovoltaics to reduce thermal load on the cell arrays.

An additional benefit of the pyrex or quartz tube is the filtering of long wavelength (over 3 micron), radiation resulting in further reduction of PV thermal load. Water exiting the radiant heat exchanger is directed into the convective heat exchanger where it is heated to the final temperature by the combustion exhaust products. Hot water exiting the convective heat exchanger is directed to either a storage tank or hydronic heating loop, or both, by way of a three-way mixing valve. Exhaust products are emitted into the atmosphere through a flue vent (2).

According to a Gas Research Institute (GRI) Report (Ref. 1), by 1994, approximately 750,000 Combo units were installed in the residential market. It is expected that the market for the Combo appliance will grow by 5 to 30% per year, depending on marketing strategy. This will result in the installation of approximately 600,000 additional units by the year 2000, with gas consumption of about 69.8 billion cubic feet per year.

The most significant advantages of Combo systems, as indicated by GRI (Ref. 1), are as follows:

- Improved system reliability, i.e., fewer components in the system
- Reduced footprint (one unit functions as two appliances)
- Simpler installation; one unit, one vent, one gas line
- Lower installation cost compared to installing two separate units
- Lower operating cost, i.e., fewer maintenance requirements

TPV powered Combo systems will provide additional benefits, such as:

- Greater system reliability due to electric grid independence

- Net electric power surplus allowing for emergency lighting or communication during blackouts

- Greater fuel utilization (when viewed from source)

- Reduced pollutant emission via the application of powered low NOx burners

COMBO SYSTEM REQUIREMENTS

The proof-of-concept combo system prototype design started with a survey conducted by QGI to determine typical requirements for existing combo system components. One objective of this study was to define component power consumption, thereby providing adequate power generation capable of supplying combo subsystems. It was found that the typical fuel input for combo appliances range from 55-130 kBtu/hr. The major power consumer is the hydronic water pump, which requires from 48 to 168 W of electric DC power to provide water flow rates from 3.25 to 9.3 gpm. It has a system head of 20 feet, which is typical for residential installations. Another electric power consumer is the combustion air blower. A premixed burner fired at 80,000 Btu/hr requires about 15 cfm of combustion air flow to provide 15% excess air. The combustion air blower utilized in the current prototype is capable of supplying 23 cfm of air at a pressure drop of about 1 inch WC, and requires only 4.2 W of electric DC power. The other electric power consumers include the ignitor (22 W), 3-way mixing valve (6 W), and safety/power management controls (< 1 W). Considering component power consumption, the total system power requirement was calculated to be between 100 and 200 W.

COGENERATOR COMPONENTS DEVELOPMENT

Eight major cogenerator components were designed, developed and integrated into the system during this project. Figure 2 shows the undressed prototype unit. The components are as follows:

- Superemissive burner
- PV array
- Convective heat exchanger
- Ignition, safety, flame controls

- Fuel/air mixing/delivery system
- PV cooling system
- Heat storage/delivery system
- Power management system

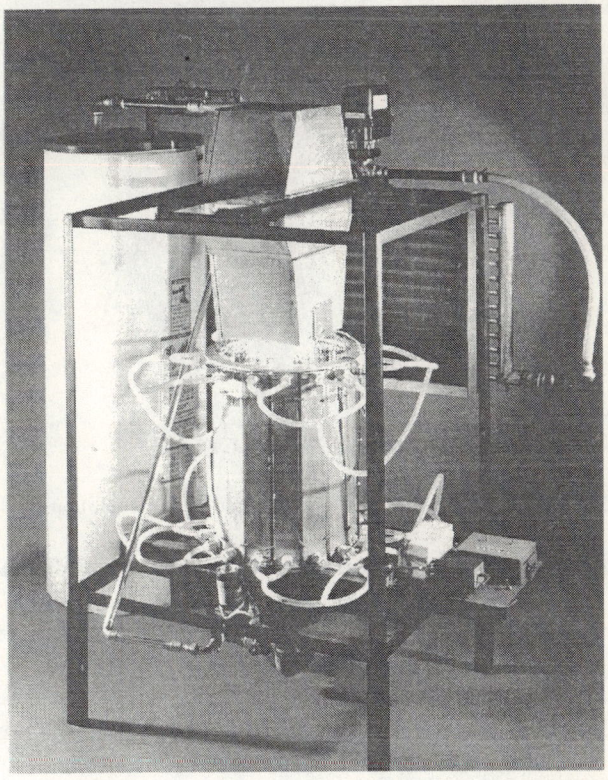

FIGURE 2 . Cogeneration System for Combo Appliances

SUPEREMISSIVE BURNER

The superemissive premixed burner with the nominal capacity of 80,000 Btu/hr is presented in Figure 3.

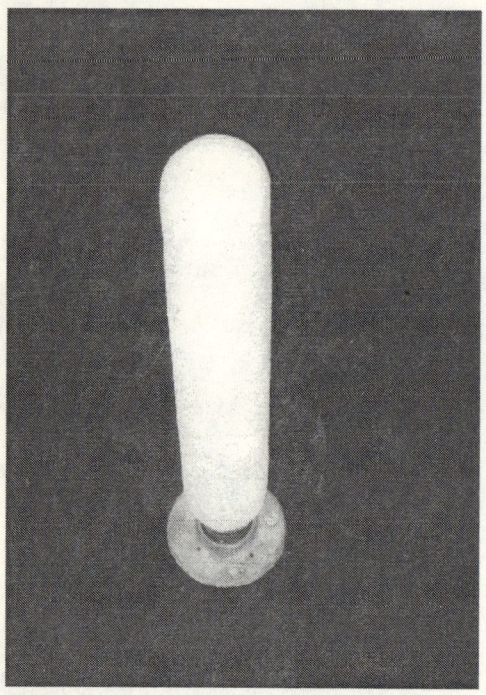

FIGURE 3. Superemissive Burner for 80,000 Btu/hr Cogeneration device

The burner is composed of a ceramic fiber matrix of alumina fabricated by Quantum's low cost vacuum forming technology. Burners are subsequently coated with the ytterbia (Yb_2O_3) superemitter. It is well known that ytterbia provides a photon emission peak centered at 980 nm that ideally fits the spectral response silicon photovoltaics. The burner was 18 inches in height and about 3 inches in diameter, and exhibited a pressure drop of about 0.5-0.7 inches of WC. Extensive testing revealed several desirable burner features, such as easy ignition, stability, and complete combustion with low NOx and CO emission when excess air was adjusted to about 10%. No evidence of flashback has been recorded during the test program. In order to prevent the contact between hot exhaust products and PV cell and to minimize PV heat rejection requirement, the burner was separated from the PV arrays with a pyrex tube. As noted earlier, this also cuts off longwave radiation that is not silicon convertible, thus reducing the PV cell heat rejection requirements.

FUEL/AIR MIXING/DELIVERY SYSTEM

The ceramic fiber matrix burner (CFMB) is powered by a DC blower (23 cfm, 4.2 W) which delivers the combustion air into fuel/air mixing chamber, then drives the combustible mixture into the burner. The mixing chamber was designed to provide effective gas/air mixing with a minimal pressure drop.

PV ARRAYS AND COOLING SYSTEM

One sun silicon single crystal PV cells, manufactured by SunPower, were selected for this project. These cells are reported to have a 20.5% light-to-electricity conversion factor based on the solar spectrum. A single PV array is shown in Figure 4 and measures 50 cm by 5 cm. Each array consists of 12 PV cells, each measuring 24 mm by 73 mm. The PV arrays were mounted on a water cooled heat sink designed to ensure cell temperature within operating limits.

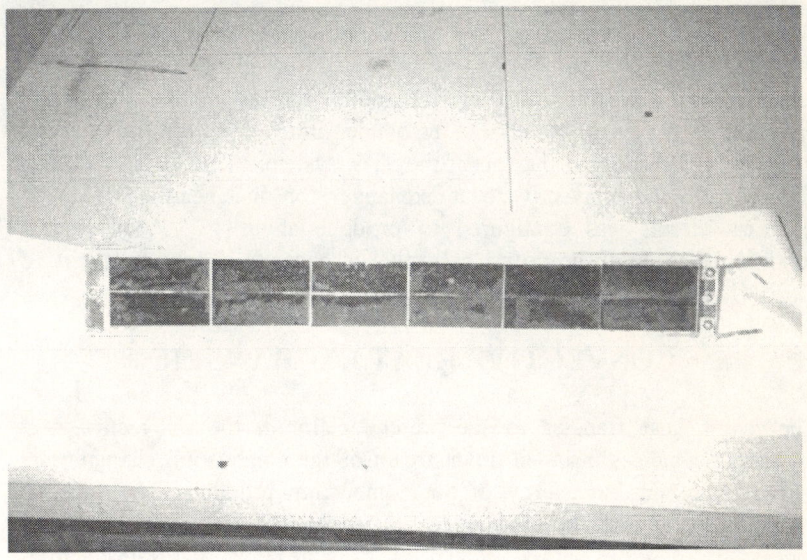

FIGURE 4. Single PV Array

Sixteen PV arrays were integrated into the TPV power generation subsystem as depicted in Figure 5.

FIGURE 5. TPV Section of Cogenerator

The PV section was about 50 cm tall and 33 cm in diameter. Incoming (hydronic loop return) water was introduced into the bottom of each PV array and rejected from the top of the TPV system. This water was then directed into the convective heat exchanger for final heating. The PV electrical circuit was configured to produce about 18.5 VDC and is conditioned through the power controller in order to supply electricity to 12 VDC electric components.

CONVECTIVE HEAT EXCHANGER

The major heat transfer to the water occurs in the convective heat exchanger which is installed downstream of the combustion chamber and extracts the sensible energy of the combustion products. In order to minimize the size of the convective heat exchanger section, QGI utilized four copper fin tubes with a 5/8 inch ID, 13 inches long, and 8 fins/inch. It was determined that such a design provided for efficient heat transfer with low aerodynamic resistance to the flue gases.

HEAT STORAGE/DELIVERY SYSTEM

The TPV cogenerator prototype system was comprised of the TPV/combustion subsystem and the heat storage/delivery subsystem which included a water pump, a water storage tank, a three-way switching valve, and a copper finned radiator that simulates radiant baseboard space heating. A typical residential water heater tank (50 gallons) was used as the hot water storage device. The copper finned radiator was used to provide heat transfer from hot water to the heating space. The system was capable of simultaneously directing hot water to the water storage tank, the space heat radiator, or both. The system was designed to provide about 2 gpm of water flow with a temperature differential of 65° F.

IGNITION, SAFETY, FLAME CONTROLS, POWER MANAGEMENT SYSTEM

The burner subsystem is composed of two major components -- the control and the TPV subsystems. The control system consists of a microprocessor, electronic components, and a circuit board. The control system initiates the battery (during startup), as well as the hot surface ignitor, gas valve, blower, water pump and switching valve. The control system is also responsible for flame sensing/safety shutoff, the recharge circuit with general power management, as well as system activation following a call for hot water or space heating.

The flame sensing and safety shutoff feature of this system utilizes the approach of the patented Quantum Control® (Ref. 2, 3, 4). Combustion devices must include a flame sensing shutoff system to achieve a high degree of safety. Typically, flames are monitored in large appliances by flame rectification, ionization, or optical sensors. These systems are externally powered and expensive. Low cost thermocouple and thermopile systems are used throughout the small appliance industry and, although self-powered, they are slow to respond to flame failure due to the inability of the thermoelectric generator and its supporting structure to dissipate heat. The Quantum Control® system can replace thermocouple and thermopile powered safety shutoff systems, as well as flame rectifier devices in gas appliances and provide system shutoff within two seconds. The Quantum Control® responds faster to flame failure than a thermoelectric powered valve and is less expensive than flame rectification systems.

COGENERATOR PERFORMANCE EVALUATION

The completely integrated combo system was tested in order to evaluate the major performance characteristics and to optimize operating conditions. The results of the tests are presented in Figures 6, 7, 8 and 9. PV power versus excess air and fuel input are presented in Figure 6.

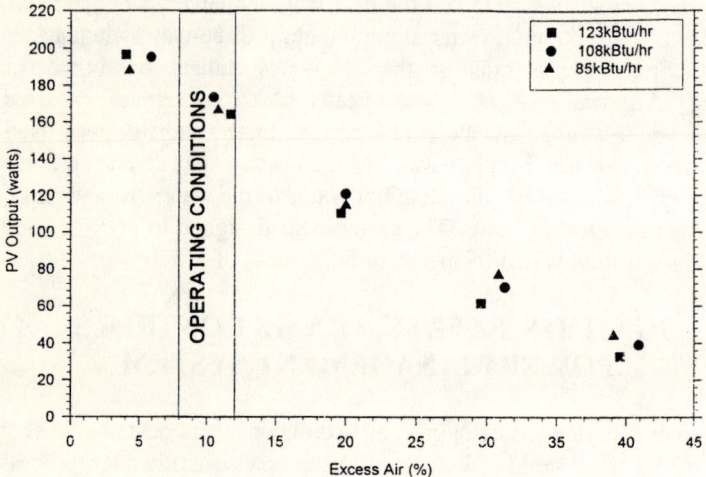

FIGURE 6. Electric Power Generation

This data indicates that the target PV power output of 100 to 200 W was achieved, and is capable of self-powering the entire system and provides a surplus of electric power. Maximum power output is achieved with low excess combustion air where the flame temperature reaches the maximum value. The fuel input rate does not affect PV power output within the range tested (85 to 123.5 kBtu/hr). The operating point of about 10% excess air was selected based on pollutant emission considerations. Figure 7 presents NOx emissions versus the excess air for the same range of fuel input rates. The South Coast Air Quality Management District's (SCAQMD) limit on NOx emissions is 40 ng/J and less (5). The lowest fuel input rate (85 kBtu/hr) results in low NOx emissions at all excess air values tested. Greater fuel input causes slightly higher NOx emissions at excess air less than 10%. Excess air of 10% or more results in the cogenerator operating with low NOx emission at all fuel inputs rates tested.

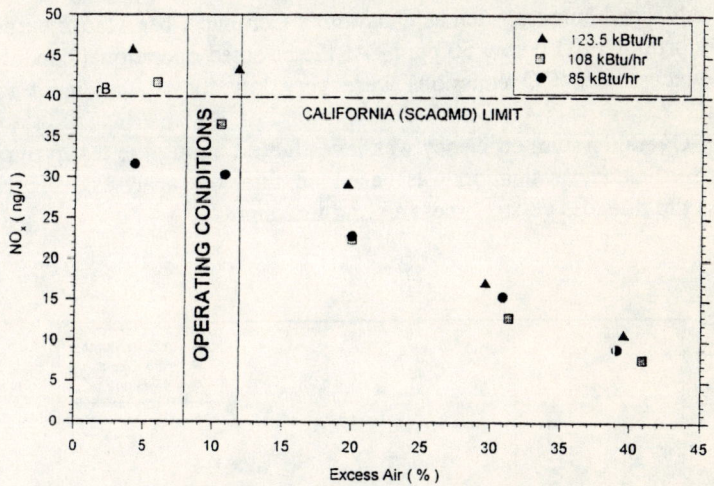

FIGURE 7. NOx Emission Characteristics

Carbon Monoxide (CO) is another pollutant that must be addressed when gas appliances are considered. CO formations versus the combustion variables are presented in Figure 8.

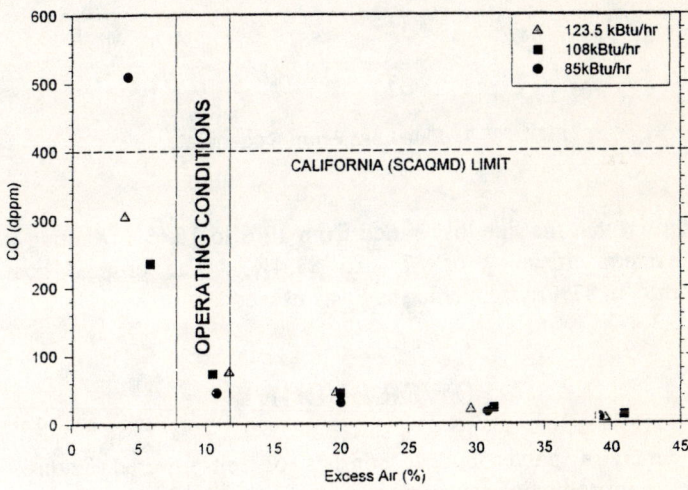

FIGURE 8. CO Emission Characterization

CO emissions exhibit rapid increase at excess air levels of less than 7%, which is typical for such combustion systems. SCAQMD requires a CO emission limit from industrial and commercial boilers of under 400 ppm (6).

Most advanced burners, such as Quantum's Ceramic Fiber Matrix burners, usually produce CO below 50 ppm. At the selected operational conditions (10% excess air), CO emissions were very low for all fuel input rates tested.

Overall system efficiency was investigated during the test program. System efficiency evaluation was based on flue loss analyses. Figure 9 depicts the flue loss versus excess air and fuel input.

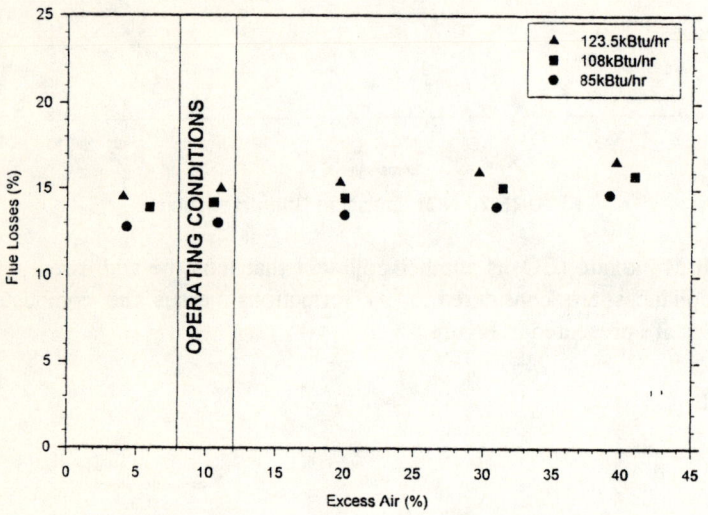

FIGURE 9. Flue Loss From Cogenerator

It was shown that the flue loss varied from 12.6 to 16.6%, resulting in an overall system efficiency of 87.4 to 83.4%. The original goal to demonstrate an 83% system efficiency was exceeded.

POWER BUDGET

The Cogenerator power budget consists of self-powered feature and surplus power production that can be utilized in home electrical devices.

Power requirements of the Combo unit are as follows:

SYSTEM COMPONENT	POWER REQUIREMENTS
Hydronic pump	20 W
Combustion blower	4.2 W
Three-way valve	6 W
Igniter	22 W
Gas Valve	less than 1 W
Controls	less than 1 W
Total start up	55 W
Total steady operation	33 W
Maximum power produced	197 W
Power surplus	142 W

CONCLUSION

A proof-of-concept TPV Cogeneration device for combination water heating and space heating was developed, fabricated, and tested. PV power output of about 200 W was demonstrated during the test program. This power is adequate to supply electric components for typical combo appliances. High system efficiency and low pollutant emissions were confirmed by tests. The presented system features about 142 W power surplus that could be useful in residential electric devices.

ACKNOWLEDGMENTS

The authors would like to take this opportunity to thank the Department of Energy for their funding and support under Research Grant DE-FG03-91ER81136.

REFERENCES

1. Combination Space Heating and Water Heating Systems Market Analysis. Prepared by DeLima Associates, Gas Research Institute, June 1995.

2. Goldstein, M.K., et al, "Photovoltaic Control system", U.S. Patent No. 4,793,799.

3. Goldstein, M.K., et al, "Photosensitive Control of Electrically Powered Emissive Ignition Devices", U.S. Patent No. 4,898,531.

4. Dolnick, E., et al, "Self-Powered Intermittent Ignition and Control System for Gas Combustion Appliances", U.S. Patent No. 4,778,378.

5. Rule 1121, "Control of Nitrogen Oxides from Residential Type, Natural Gas-Fired Water Heaters", SCAQMD, 1978.

6. Rule 1146, "Emissions of Oxides of Nitrogen from Industrial, Institutional and Commercial Boilers, Steam Generators, and Process Heaters", SCAQMD, 1988.

SESSION 7:
NOVEL CONCEPTS

Minority-Carrier Transport in InGaAsSb Thermophotovoltaic Diodes

R. U. Martinelli, D. Z. Garbuzov, H. Lee, N. Morris, T. Odubanjo,
G. C. Taylor, and J. C. Connolly

Sarnoff Corporation, Princeton, NJ 08543-5300

Abstract

Uncoated InGaAsSb/GaSb thermophotovoltaic (TPV) diodes with 0.56 eV (2.2 µm) bandgaps exhibit external quantum efficiencies of 59 % at 2 µm. The devices have electron diffusion lengths as long as 29 µm in 8-µm-wide p-InGaAsSb layers and hole diffusion lengths of 3 µm in 6-µm-wide n-InGaAsSb layers. The electron and hole diffusion lengths appear to increase with increasing p- and n-layer widths. At 632.8 nm the internal quantum efficiencies of diodes with 1- to 8-µm-wide p-layers are above 89 % and are independent of the p-layer width, indicating long electron diffusion lengths. InGaAsSb has, therefore, excellent minority carrier transport properties that are well-suited to efficient TPV diode operation. The structures were grown by molecular-beam epitaxy.

Introduction

There has been a recent interest in thermophotovoltaic (TPV) electric power generation from blackbody sources at about 1000 °C [1, 2]. Candidate III-V compound materials for the TPV diode active region are InGaAs grown on InP and InGaAsSb grown on GaSb. Since the bandgap of the optimal TPV diode is about 0.55 eV, the InGaAs diodes have a compositionally graded layer interposed between the InP substrate and the InGaAs layer to accommodate the lattice mismatch. The InGaAsSb diodes are lattice-matched to the substrate. Lattice-matched devices may offer more possibilities in terms of bandgap engineering and the variety of material choices, but these advantages are not all clear at present. Nonetheless, while InGaAs diodes currently offer superior performance in terms of quantum efficiency and saturation current, owing in part to their longer-term development [3], InGaAsSb is demonstrating excellent minority-carrier transport characteristics that make it a very attractive material for TPV diodes.

In this paper we show quantum efficiencies of 59 % at 2 µm wavelength from an InGaAsSb/GaSb diode. The inferred electron diffusion length in this device is 29 µm. Our spectral quantum efficiency (SQE) curves compare favorably with recently reported results for both InGaAsSb and InGaAs TPV diodes [3].

CP401, *Thermophotovoltaic Generation of Electricity: Third NREL Conference,*
edited by Benner/Coutts
© 1997 The American Institute of Physics 1-56396-734-0/97/$10.00

Experimental Results

The TPV diodes discussed in this paper are designed to generate electrical power from 1000 °C blackbody radiation. The active region is a nominally 0.55 eV InGaAsSb p/n homojunction. To optimize the power-conversion efficiency, radiation up to the bandgap cut-off wavelength must be maximally absorbed, and the photo-generated carriers must be transported to the p/n junction with minimal loss. This places a tradeoff between widening the active region to increase optical absorption at near-bandgap wavelengths, where the absorption constant is relatively small, and decreasing the p- and n-layer thicknesses to maximize carrier collection at the junction. The values of the electron and hole diffusion lengths, L_e and L_h, are critical to the optimization of the TPV structure. In general, long diffusion lengths allow wide active regions, which lead to high quantum efficiencies. We undertook the study of the TPV diode SQE to determine the values of L_e and L_h.

The TPV structures were grown by solid-source, molecular-beam epitaxy (MBE) and were lattice-matched to the GaSb substrates in a manner very similar to that used for the growth of antimonide-based lasers [4, 5]. To study the transport properties of the diodes, 500-μm-diameter mesas were etched into the wafers, and individual mesa diodes were evaluated. Figure 1 schematically shows the vertical structure of the TPV diode.

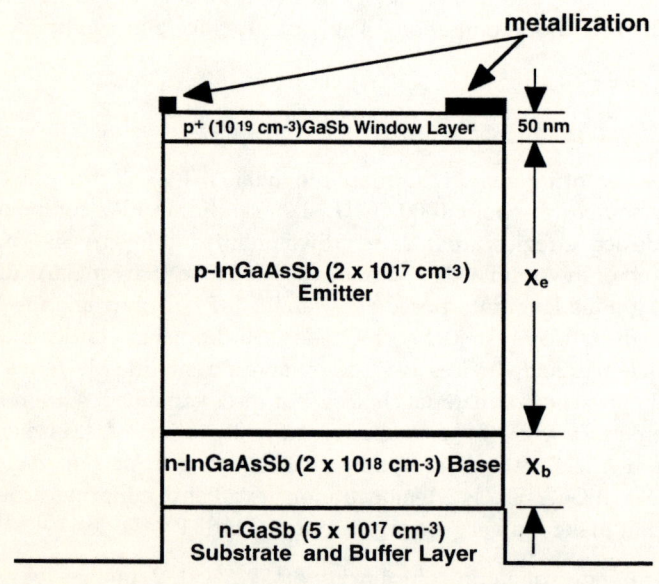

Figure 1. Vertical Layer structure of a p-on-n InGaAsSb TPV diode.

After growing a 500-nm-thick n-GaSb buffer layer on the n-GaSb substrate, the so-called base region is grown x_b μm thick. For most of the devices used in this work, x_b is 1 to 6 μm. The n-type base layers are Te-doped to about 2×10^{18} cm^{-3}. On top of the base the p-type emitter, Be-doped to 2×10^{17} cm^{-3}, is grown to thicknesses, x_e, ranging from 1 to 8 μm. A 50-nm-thick p$^+$ GaSb window layer, Be-doped to 10^{19} cm^{-3}, is grown on the emitter, completing MBE growth of this so-called p-on-n diode. The GaSb window layer insures very low electron recombination at the p$^+$-GaSb/p-InGaAsSb window-layer/emitter interface. While we have grown and characterized n-on-p TPV diodes, this work describes results from p-on-n devices.

Following growth, the wafer is thinned to 150 μm and metallized with standard Au/Ge/Ni/Au. The epitaxial surface is patterned and metallized with Cr/Au, and the mesas are etched several microns into the substrate using a bromine/methanol solution.

Relative SQE curves in the wavelength interval of 1 to 2.5 μm were obtained using a monochrometer with a known relative output-power spectrum. The diode's absolute quantum efficiency was measured using a 1.575 μm fibered diode laser with a known output-power-versus-laser-current characteristic. The relative SQE data were then adjusted to agree with the 1.575 μm absolute data. Absolute quantum efficiency measurements were also made at 632.8 nm using a HeNe laser.

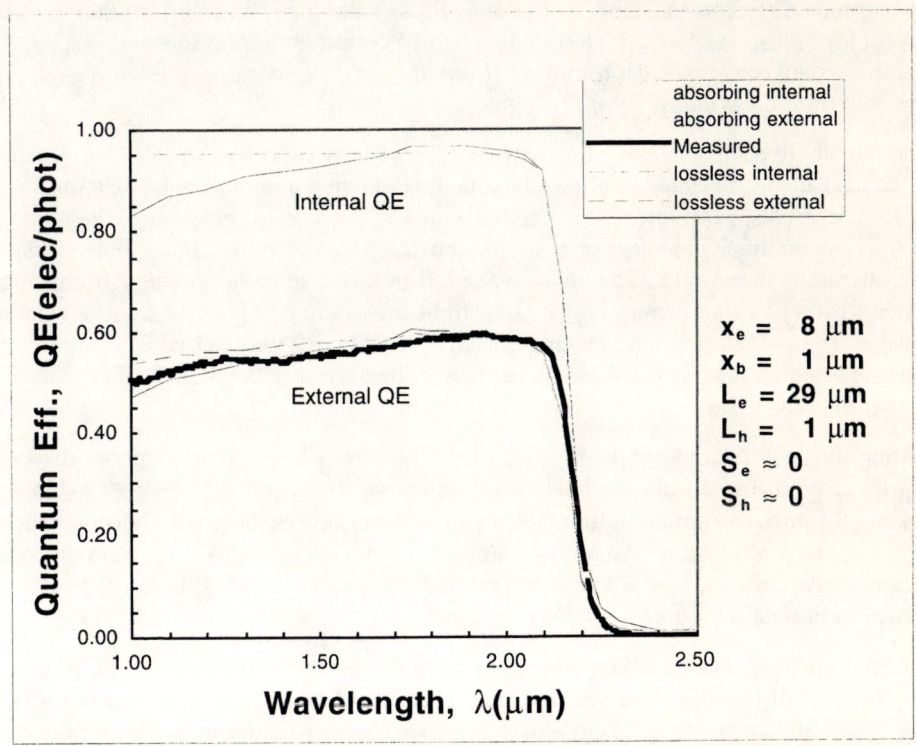

Figure 2. The spectral quantum efficiency of a p-on-n TPV diode having an 8-μm emitter and a 1-μm base. Model curves show external and internal SQE for absorbing and lossless GaSb window layers.

Figure 2 shows the SQE of a TPV diode having an 8-μm emitter and a 1-μm base, along with model curves for external and internal SQE. The measured SQE rises sharply at wavelengths shorter than the bandgap wavelength of 2.2 μm (0.56 eV), reaching a maximum QE of 59 % at 2 μm. As the wavelength decreases, the SQE slowly decreases to 50 % at 1 μm.

The model used to fit the SQE data is a simple, one-dimensional minority-carrier diffusion model commonly employed in SQE data analysis [6]. The measured reflectivity was used to calculate the photon density within the diode. In some cases we used the measured absorption constant of InGaAsSb, but mostly we used semi-empirical absorption data computed through a method similar to that of Borrego [7]. Accounting for absorption by the 50-nm-thick GaSb window layer, two sets of SQE curves were calculated: one set assuming that the window layer contributes all electrons photo-generated within it to the emitter (labeled "lossless internal (or external)" in Figure 2), and the other set assuming that the window layer contributes no photo-generated electrons to the emitter (labeled "absorbing internal (or external)" in Figure 2).

Electron and hole diffusion lengths, L_e and L_h, characterize minority-carrier transport. The base/substrate and the window-layer/emitter interfaces are characterized by recombination velocities S_b and S_e, respectively. Note that we normalize these recombination velocities to the "bulk" recombination velocity D/L, where D is the appropriate diffusion constant. In this way $S_b \gg 1$ describes an interface that acts like a sink for holes, and $S_b \ll 1$, describes a nearly perfect reflecting interface. The total photo-current comprises contributions from the emitter, base, depletion-region, and possibly from the window layer.

The model curve of external SQE with an absorbing window describes the measured SQE curve in Figure 2 reasonably well using the following parameters: $L_e = 29$ μm, $S_e \approx 0$, $L_h = 1$ μm, and $S_b \approx 0$. Note that in diodes where $x_e > x_b$, as in this device, the influence of the base on the SQE is negligibly small, and the parameters describing hole transport are, therefore, inaccurate. The values we used in this case were obtained from diodes having 1-μm bases and emitters. The best fit to the external SQE for a lossless window layer gives $L_e = 25$ μm, still much longer than x_e. In Figure 2 the external SQE curve for a lossless window layer is calculated with $L_e = 29$ μm, which gives a curve that is slightly higher than the data.

Setting the spectral reflectivity to zero gives the internal-SQE model curves shown in Figure 2. From 1 to 2.1 μm the internal QE is above 80 %, and it is over 90 % between 1.4 and 2.1 μm. Absorption in the GaSb window layer causes the gradual decrease in the SQE beginning at 1.7 μm. Assuming complete photo-electron collection from a lossless window layer, the internal SQE is constant at 95 % for wavelengths shorter than about 2 μm, as indicated in Figure 2.

Remarkably long electron diffusion lengths have also been inferred from the SQE data of diodes with different emitter widths. The following relation describes the majority of our observations: $L_e \geq x_e$. Figure 3 shows our inferred values of L_e for diodes with emitter widths of 1 to 8 μm. The dashed line in Figure 3 denotes $L_e = x_e$. Twenty-one of the twenty-four data shown in Figure 3, or 88 %, lie on or above the line, strongly suggesting that the relation $L_e \geq x_e$ is valid.

The model also fits the SQE data reasonably well by assuming non-zero values of S_e, but consequently, even larger values of L_e must be used. Our approach has been to assume that the smallest value of L_e, consistent with the data, is the most prudent estimation.

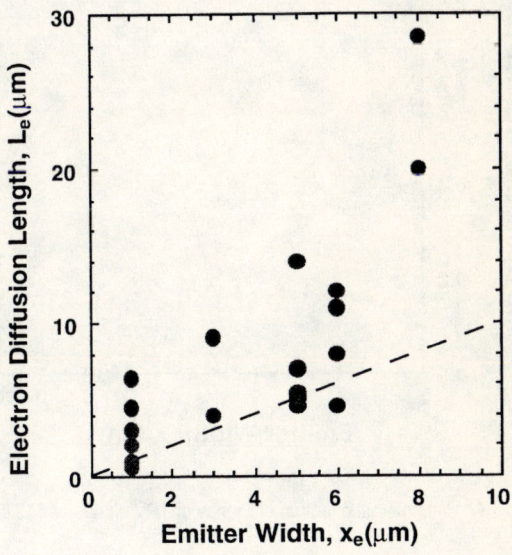

Figure 3. The inferred electron diffusion length for TPV diodes with emitter widths of 1 to 8 μm.

Measurements of the QE at 632.8 nm of diodes with different emitter widths corroborates the above results. Figure 4 shows the inferred internal QE at 632.8 nm for diodes with 1-, 3-, 6-, and 8-μm emitters. Light at 632.8 has an estimated absorption constant in 0.56-eV InGaAsSb greater than 10^5 cm^{-1}, so that photo-electrons are generated very close to the GaSb/InGaAsSb interface. It is also energetic enough to create more than one electron-hole pair in the InGaAsSb [8], which would increase the measured QE. The measured internal QE is essentially independent of emitter width. Its high values between 89 and 98 % strongly imply that $L_e >> x_e$, in basic agreement with the conclusions reached from our elementary analysis of the SQE data. They also suggest that the GaSb window is lossless.

To a lesser extent, we have observed the same phenomenon with respect to hole transport. Not as many wide-base diodes were measured, since efficient p-on-n TPV diodes will have wider emitters than bases, owing to the fact that electrons have the longer diffusion length. Figure 5 shows the dependence of L_h on base thickness. As in the case of electrons, the values of L_h inferred from the model increase with increasing base width, but the magnitudes of L_h for a given x_b are only about one tenth that of electrons. This result may reflect the lower mobility and non-radiative lifetime of holes.

The base layers of these diodes were doped an order of magnitude higher than were the emitter layers, which might decrease the holes' non-radiative lifetime.

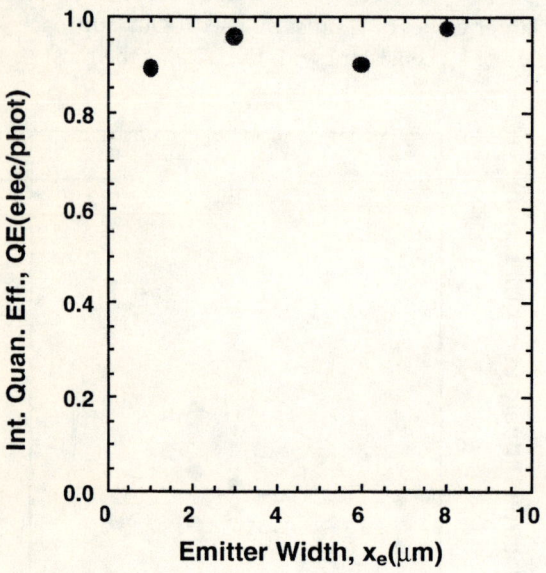

Figure 4. High internal quantum efficiency of TPV diodes at 632.8 nm.

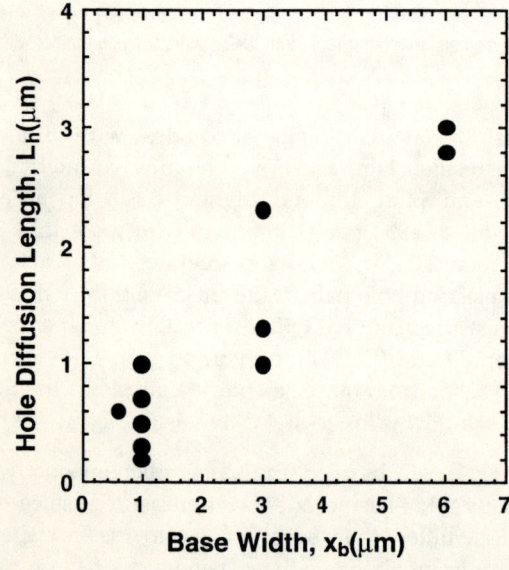

Figure 5. The inferred hole diffusion length for TPV diodes with base widths of 1 to 6 μm.

Conclusions

In conclusion, we have shown that the transport of photo-generated electrons in p-InGaAsSb is very efficient. p-on-n TPV diodes with bandgaps of 0.56 eV have external quantum efficiencies of 59 % at 2 μm and exhibit electron diffusion lengths of 29 μm in an 8-μm-wide emitter. The electron diffusion length appears to increase with increasing emitter width. Similarly, the hole diffusion length appears to increase with increasing base width, although the hole diffusion lengths are about one tenth those of the electron. These results were corroborated by the internal quantum efficiencies greater than 89 % that were measured at 632.8 nm. Given these results, an optimized InGaAsSb/GaSb p-on-n TPV diode with a 6- to 8-μm-wide emitter will have an excellent spectral quantum efficiency.

References

1. 1st NREL Conference on the Thermophotovoltaic Generation of Electricity, Copper Mtn, CO, AIP Conf. Proc. **321** (1994).

2. 2nd NREL Conference on the Thermophotovoltaic Generation of Electricity, Colorado Springs, CO, AIP Conf. Proc. **358** (1995).

3. G. W. Charache, J. L. Egley, L. R. Danielson, D. M. Depoy, P. F. Baldasaro, B. C. Campbell, S. Hui, L. M. Frass, and S. J. Wojtczuk, "Current Status of Low-Temperature Radiator Thermophotovoltaic Devices," presented at the 25th Photovoltaic Specialists Conference, 1996.

4. H. Lee, D. Z. Garbuzov, R. U. Martinelli, R. J. Menna, and J. C. Connolly, "Molecular Beam Epitaxy of AlGaAsSb/InGaAsSb/GaSb Mid-Infrared High-Power Separate-Confinement Quantum-Well Lasers," to be presented at the Electronic Materials Research Conference, Ft. Collins, CO, June 25-27, 1997.

5. H. Lee, P. K. York, R. J. Menna, R. U. Martinelli, D. Garbuzov, and S. Y. Narayan, "2.78 μm InGaAsSb/AlGaAsSb Multiple Quantum Well Lasers with Metastable Wells Grown by Molecular Beam Epitaxy," J. Crystal Growth, **150**, 1354 (1995).

6. S. M. Sze, "Physics of Semiconductor Devices," 2nd Ed., John Wiley and Sons, New York, 1981, pp. 802 - 805.

7. J. Borrego, M. Zierak, and G. W. Charache, "Parameter extraction for TPV Cell Development," 1st NREL Conference on the Thermophotovoltaic Generation of Electricity, Copper Mtn, CO, AIP Conf. Proc. **321**, pp. 371-378 (1994).

8. T. Kobayashi, "Average energy to form hole-electron pairs in GaP diodes with alpha particles," Appl. Phys. Lett., **21**, 150 (1972).

Electric Characteristics Of Germanium Vertical Multijunction (VMJ) Photovoltaic Cells Under High Intensity Illumination

Vadim A. Unishkov

Quantum Group, Inc.
11211 Sorrento Valley Road; San Diego, CA 92121

ABSTRACT

This paper presents the results of the performance evaluation of Vertical Multijunction (VMJ) germanium (Ge) photovoltaic (PV) cells. Vertical Multijunction Germanium Photovoltaic cells offer several advantages for Thermophotovoltaic (TPV) applications such as high intensity light conversion, low series resistance, more efficient coupling to lower temperature sources, high output voltage, simplified heat rejection system as well as potentially simple fabrication technology and low cost photovoltaic converter device.

INTRODUCTION

Vertical Multijunction Photovoltaic cells are very promising for various applications. There are several major advantages of VMJ PV devices over the planar type PV cell. One of the most important features is that VMJ cells are capable of accepting an extremely high radiant flux, up to 100 W/cm^2 (Ref. 1). In the last two years, Quantum Group, Inc. (QGI) has developed and demonstrated narrow band emission technologies which are capable of providing a radiant flux up to 27 W/cm^2. High concentrator type PV cells are expected to be an important component for TPV

CP401, *Thermophotovoltaic Generation of Electricity: Third NREL Conference,*
edited by Benner/Coutts

applications. Another advantage of VMJ cells is better photon collection because there is no shadowing caused by the current collecting grid common to planar PV cells. By selecting the appropriate cell thickness (which depends on the incident radiant flux), total photon absorption is expected in a VMJ cell without a sophisticated surface treatment. In addition, the VMJ cells simplify TPV system design by providing a higher voltage output which is proportional to the number of junctions in the assembly. The detailed analysis of VMJ advantages has been reported (Ref. 2). Early investigation of the potentials and development of VMJ cells was conducted in Russia and results discussed (Ref. 1) The test results of the germanium (Ge) VMJ cell are presented in this paper.

Germanium was selected to be a PV material because it's absorption spectrum is well matched to the emissive characteristic of erbia. As previously presented (Ref. 3), a Ge PV collector is suitable for incorporating into a TPV power generator with an erbia emitter. The competing PV materials such as gallium-antimonide (GaSb) and other III-V semiconductors are significantly more expensive. Although the intrinsic carrier concentration of Ge is about five times that of InGaAs (Ref. 4) Ge is commercially available and it is expected that the cost of Ge PV cells will be close to the cost of silicon (Si) cells because the fabrication technologies are similar.

Typically, with increasing light intensity the PV current and power rates of rise decrease due to the influence of series resistance. For VMJ PV cells with p-n junctions, located parallel to the direction of the incident light, the series resistance is influenced generally by volumetric resistance of the base region, in other words by resistivity of the semiconductor.

It is known that the lowering of the resistivity of a semiconductor causes an increase of the open circuit voltage (V_{OC}) because of a decrease of the reverse saturation current. At the same time the lowering of the resistivity reduces the mobility and lifetime of the minority charge carriers which causes a reduction of the PV current.

TEST CONDITIONS AND RESULTS DISCUSSION

We investigated the electrical characteristics of germanium VMJ cells with resistivities of 0.1, 1.0 and 10 Ω cm. The first sample was assembled of 15 junctions, the second consisted of 14 junctions and the third was comprised of 18 junctions. All the junctions of the VMJ cell had the same thickness of 200 microns. The results of the tests were calculated for a single PV junction. The characteristics of the samples are presented in Table 1.

Table 1.   Characteristics of the Tested Samples

Sample ##	Resistivity Ω cm	Length cm	PV Base Thickness cm	Electric Characteristics at Illumination of 0.1 W/cm^2		
				Voltage of p-n junction, V_{OC}	SC Current per 1 cm, mA	Efficiency %
1	0.1	0. 8	0.02	0.16	1.2	5 4
2	1.0	0. 8	0.02	0.12	1.75	4.8
3	10.0	0.85	0.02	0.09	2.5	6.0

A 600 W Tungsten electric bulb was used as the light source during the experiment. The investigation of germanium VMJ PV cell performance was conducted under an incident radiant flux from 0.1 to 5.0 Watts/cm^2. High intensity incident radiation was provided by a Fresnel lens allowing for the control of radiant flux intensity via relative positioning of the cell along the optical axis. The area of the focal spot was significantly larger than the area of the PV cell ensuring uniform illumination. The incident radiant flux on the PV face was measured by the standard measurement device which is a Ge VMJ cell with the accurate linear dependence of the short circuit current (I_{SC}) versus radiant flux intensity in the above mentioned range of illumination power. A pulsed illumination source was utilized to avoid overheating of the PV cell. An electronic interruption device was used to control the illumination duration and all measurements were made at the peak of the light pulse.

Figure 1 presents the open circuit voltage (V_{OC}), short circuit current (I_{SC}) and efficiency (η) versus radiant flux intensity of the tested Ge VMJ samples with resistivities of 0.1 Ω cm (1), 1.0 Ω cm (2) and 10.0 Ω cm (3).

The behavior of V_{OC} versus radiant flux is identical for all the samples; V_{OC} increases in the radiant flux range from 0.1 to 2.5 W/cm^2 and exhibits a tendency for saturation when radiant flux raises above this range. The calculations indicate that the maximum V_{OC} per single cell at radiant flux of 5.0 W/ cm^2 is expected to be 0.25, 0.2 and 0.16 V for the samples of Ge with the resistivities of 0.1, 1.0, and 10.0 Ω cm, respectively. The samples of Ge with low resistivity (0.1 Ω cm) demonstrated linear dependence of I_{SC} from radiant flux within whole range of tested light intensity.

The PV sample with the highest resistivity demonstrated nonlinear behavior of the I_{SC}-radiant flux characteristic when the radiant flux exceeded 3.0 W/cm^2. This phenomenon is caused by a voltage drop in base regions of

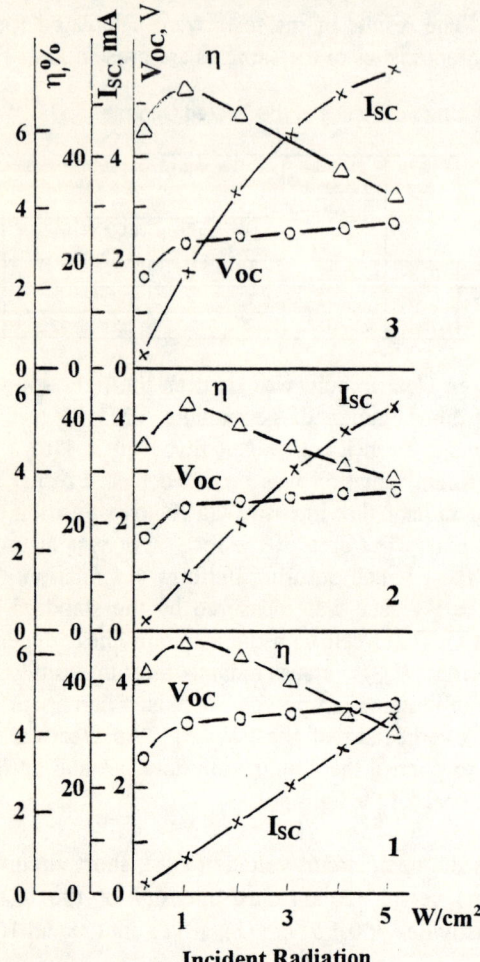

Figure 1. Electrical characteristics of Ge VMJ cells versus light intensity

the individual junctions of the Ge VMJ cells. This voltage drop (V) is calculated by equation (1).

$$V = I\, n\, \rho\, d/s \qquad (1)$$

Where: **n** is the number of base regions of the individual junction connected in the series

    ρ is the resistivity

    d is the thickness of base regions of the individual junction

    s is the area of p-n junction

    I (of course) is the PV current

The results of the calculations of the voltage drop in the base and series resistance (R_n) which were done by the Handy method (Ref. 5) and, from an ideal I-V characteristic (corrected to 1 cm^2 of p-n junction), are presented in Table 2.

Table 2. Comparison of the V and R_n Calculated by Handy Method and I-V characteristic

Sample ##	Voltage Drop, V	R_n, Ohm/cm^2 (by Handy)	R_n, Ohm/cm^2 from I-V chart
1	0.012	0.25	0.32
2	0.16	0.56	0.5
3	2.4	1.03	1.15

The values of R_n, calculated by the Handy method, are well correlated with the data obtained from an ideal I-V characteristic. The values of I_{SC} were calculated per 1 cm of junction length at the light intensity of 5.0 W/cm^2. These values are 40, 53 and 67 mA for the samples of Ge, with ρ of 0.1, 1.0, and 10.0 Ω cm, respectively.

All of the diagrams that represent the efficiency versus radiant flux indicate the maximum efficiency at illumination is approximately W/cm^2. At higher illumination, the efficiency decreases. The Ge sample with the highest resistivity of the base regions of the individual junction demonstrated the highest efficiency drop, due to non-linear dependence of I_{SC} versus radiant flux. The relatively low efficiency of the tested VMJ PV cells can be explained by the mismatch of the emission spectrum of the light source and the absorption characteristic of the Ge PV cell. Better performance is expected from a Ge VMJ cell integrated into a TPV device with a QGI high power density burner-emitter system which is tuned to the emission peak of 1.55 micron.

REFERENCES

1. Landsman, A. P., Strebkov, D. S., and Unishkov, V. A., "High Voltage Photovoltaic Power Source for ELP" in Proceedings of Conference on IV All Union Conference of ELP and Photovoltaic, Moscow, 1971, pp. 200-208.

2. Sater, B. L., Vertical Multi-Junction Cells for Thermophotovoltaic Conversion. The First NREL Conference on Thermophotovoltaic Generation of Electricity, Copper Mounting, CO, 1994, pp. 165-176.

3. Kittle, E., and Guazzoni, G. E., "Design Analysis of TPV Generator System" 25th Annual Proceedings Power Sources Conference, May 1972.

4. Gray, J. L., and El-Husseini, A., "A Simple Parametric Study of TPV System Efficiency and Output Power Density Including a Comparison of Several TPV Materials" The Second NREL Conference on Thermophotovoltaic Generation of Electricity, Colorado Springs, CO, 1995, pp. 3-15.

5. Handy R.I. Solid State Electronics, 1967, **10**, No 8.

Thin-Film Polycrystalline Ga$_{1-x}$In$_x$Sb Materials

Miguel Contreras, H. Wiesner, J. Webb

National Renewable Energy Laboratory
1617 Cole Blvd.
Golden, CO 80401

Abstract. Polycrystalline Ga$_{1-x}$In$_x$Sb thin films with $0<x<1$ have been grown on 7059 Corning glass by evaporation from the elemental sources. X-ray diffraction data reveal single-phase materials with cubic structure (space group F $\overline{4}$ 3 m). The addition of In to the GaSb matrix results in an increase in lattice parameter and consequently a reduction in optical bandgap. Also, increased In content in the films produces a change in the preferred orientation on films grown on bare 7059 glass. In this contribution, materials of interest are limited to those with optical bandgaps between 0.7 and 0.5 eV.

The as-deposited, near-stoichiometry films with Sb/(In+Ga)$\leq$1 show p-type electrical conductivity. Addition of small amounts of Sn provides an effective extrinsic doping effect. In this material, Sn behaves amphoterically providing both donor and acceptor states depending on which element is substituted. Sn incorporation in materials with Sb/(In+Ga)<1 yields p-type conductivity, whereas materials with Sb/(In+Ga)>1 show n-type conductivity.

INTRODUCTION

The motivation to study polycrystalline thin-film materials for thermophotovoltaic (TPV) applications is the same as for traditional polycrystalline thin-film photovoltaic (PV) technologies: the potential for lower cost in the overall cost of manufacturing generators. Some promising TPV compounds have already been mentioned in the literature[1,2]. Indeed, there are several alloys that can provide the flexibility to tailor a bandgap value for a specific radiant energy source. But ultimately, all potential TPV materials must answer the questions of stability, reliability and degradation. The development of single-crystal, GaSb-based TPV cells [3,4], epitaxial (Ga,In)As [5], and (Ga,In)Sb thin-film materials are important steps that already provide some of the answers in advancing TPV toward a larger scale of applications and electrical power generation.

CP401, *Thermophotovoltaic Generation of Electricity: Third NREL Conference,*
edited by Benner/Coutts
© 1997 The American Institute of Physics 1-56396-734-0/97/$10.00

In this paper, we evaluate the prospects and feasibility of polycrystalline (Ga,In)Sb thin films for TPV applications. Such a compound semiconductor system was chosen mainly because of the practical demonstration of single-crystal GaSb as TPV converters and the ease in evaporating its elemental components. All three elements (In, Ga, and Sb) are stable at room temperature. The only health concern comes from potential airborne Sb dust particles that can cause respiratory disease. This risk is minimized by careful handling and storage of Sb and the use of protective equipment while working in the evaporator reactor.

The objectives of this work are to determine structural, electrical, and optical properties of evaporated $Ga_{1-x}In_xSb$ polycrystalline thin films and how growth conditions affect some of those properties. Materials of interest are limited to those with optical bandgaps between 0.7 and 0.5 eV.

THIN-FILM GROWTH

Film growth is carried out in a multisource evaporator equipped with an electron impact emission spectroscopy (EIES) system (Sentinel III by Inficon) for rate monitoring and controlling of In and Ga. The Sb source is monitored and controlled by a standard quartz-crystal oscillator instrument (XTC by Inficon). Substrate heating is achieved by an array of infrared quartz lamps. The vacuum chamber was always evacuated to pressures of $<2 \times 10^{-6}$ torr before film growth. Because of the vapor pressure of the constituent elements and the substrate heating, the chamber pressure during evaporation rises to $\sim 10^{-5}$ torr.

Substrates used in this preliminary study are 7059 Corning glass. The bare glass was first washed in a soap solution, rinsed in DI water, and blow-dried with a high purity nitrogen jet before being loaded into the chamber.

Due to the physical configuration of the evaporator system (sources and substrate geometry), all samples grown present a gradient in their composition as a function of position in the plane of growth. Most of this gradient comes from the nonuniform distribution of Sb in the substrate. This feature is used to our advantage in that changes of physical properties with deviations from stoichiometry can be evaluated in a single run. Further details are given in the following paragraphs.

EXPERIMENTAL RESULTS

Composition of all relevant samples has been determined by inductively coupled plasma spectroscopy (ICP). The compositional gradient mentioned above has been quantified by this technique, and for this purpose, several pieces from a 5-cm x 5-cm sample were analyzed and the results mapped. The atomic composition/distribution of Ga and In are fairly uniform in all areas of the sample (<1 at%). On the other hand, the Sb distribution varies slowly and unidirectionally (from one side to the other) of the sample. The variation can be as much as 6 at% (in Sb content) from one side to the other. This situation allows us to evaluate changes in physical properties as a function of deviation from stoichiometry in a single run.

Structural Analysis

The III-V semiconductors GaSb and InSb can crystallize in more than one phase. The JCPDS-ICDD database shows that orthorhombic, hexagonal, tetragonal, and cubic phases exist for both materials. Because it is the cubic phases that are of interest for TPV, film growth efforts were concentrated in attaining such a structure (space group F$\bar{4}$3m). This was possible for runs where the substrate temperatures were above 400°C during film growth. Fig. 1 displays the XRD pattern of selected $Ga_{1-x}In_xSb$ films, where peak position and indexing follows that of the cubic structure.

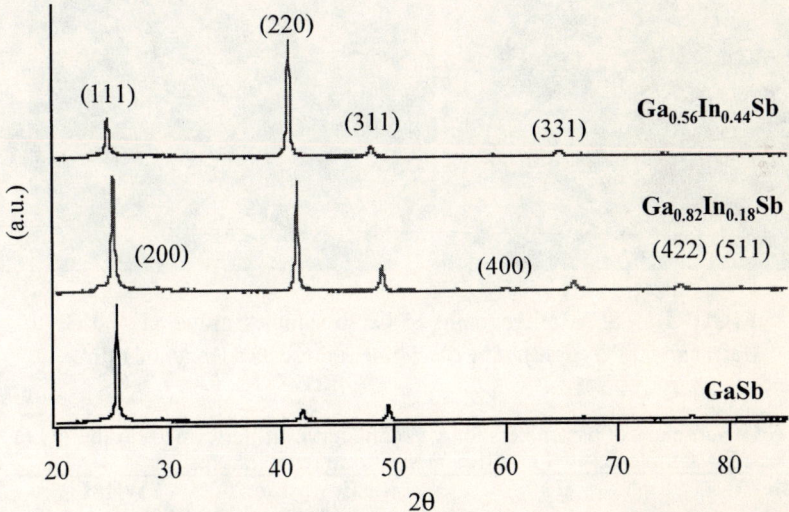

FIGURE1. XRD patterns of cubic $Ga_{1-x}In_xSb$ materials grown by coevaporation at a substrate temperature ~400°C on 7059 Corningglass.

Note from the XRD patterns in Figure 1 that the introduction of In leads to an apparent change in the preferred orientation of the films. Specifically, GaSb films present a (111) orientation, but In incorporation with x>0.2 changes the texture to a (220) orientation. This phenomenon is not surprising in that similar changes in unidirectional texture (preferred orientation relative to the plane of growth, but random orientation relative to the crystal axis) have been observed in other semiconductor systems—see, for instance, the case of the $Cu(In_{1-y}Ga_y)_3Se_5$ semiconductor system in ref. [6].

In general, substrate temperature has a significant effect on crystal quality and morphology. To look at such effects, GaSb samples were grown at three different substrate temperatures (namely 400°, 450°, and 500°C). Scanning electron micrographs taken from this set of samples reveal an increase in grain size with the increase of substrate temperature (Fig. 2). The increase in grain size with temperature is also supported by the analysis of peak broadening of XRD data. Results from a Voight curve-fitting routine performed in the (111) peak of the samples in question are summarized in Table 1.

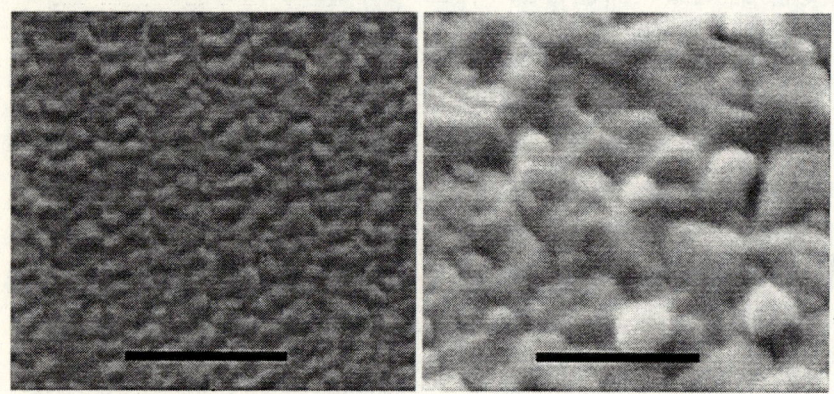

FIGURE2. SEM micrographs of GaSb samples grown at 400°C (left) and 500°C (right). The black line represents a length of 1 μm.

TABLE 1. Summary of parameters for a Voight curve fit performed on the (111) XRD peak for samples grown at different substrate temperatures.

Sample	Ts (°C)	peak position (deg)	FWHM (deg)
C649	400	25.259	0.155
C652	450	25.278	0.100
C653	500	25.292	0.076

From the curve fitting of XRD data we observe that the relative intensity (cps) of the (111) peaks increase with increased substrate temperature and that the full width at half maximum (FWHM) of the fitted peaks decreases. Such changes and observations can be associated with an enhancement of grain size and film quality. However, the grain size in all samples is still much less than 1 μm, that is, columnar and continuous films across the thickness of a sample have not yet been obtained. Columnar grains with grain sizes comparable to the thickness of a film (~2μm) are desirable because they present a continuous path for charge carriers diffusing/drifting away from the junction (probability of recombination is decreased).

Optical Measurements

From the XRD data, we see that the lattice parameter increases with In incorporation (peak positions shift to lower 2θ values); therefore, a change in the optical absorption is anticipated. Optical properties have been studied by reflection (R) and transmission (T) measurements performed in a Cary 6000 spectrophotometer. We must point out here that the noise level in R and T is increased at longer IR wavelenghts, particularly those in the range of interest to TPV. This is due to the poor performance of the IR source lamp at those wavelengths (λ>2000 μm). The absorption characteristics of $Ga_{1-x}In_xSb$ films can be seen in Fig. 3. An absorption coefficient was not obtained because of the noise in the optical data; nevertheless, the graph in Fig.3 shows clearly the change in the absorption edge with increased In content.

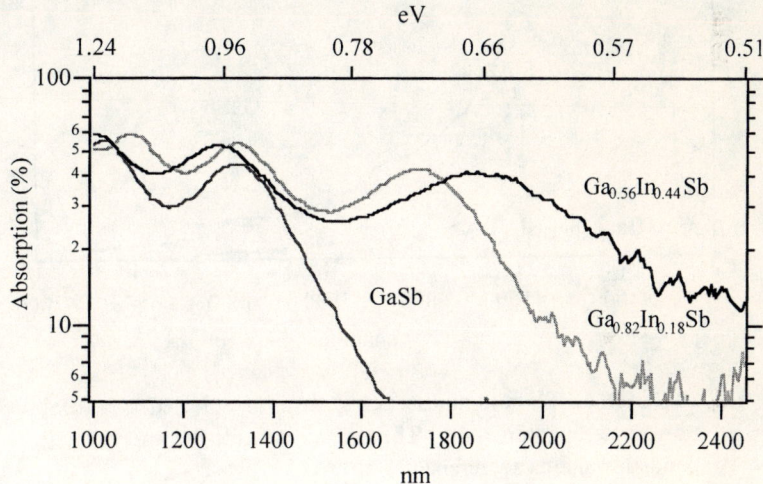

FIGURE 3. Absorption curves for ~ 5000-Å-thick $Ga_{1-x}In_xSb$ films grown on 7059 glass at a substrate temperature ~400°C.

In addition to the reflection and transmission measurements, we have performed photoluminescence (PL) measurements to gain some knowledge about film quality (gap defects/states) and its viability for device fabrication. It has been reported that there is a strong correlation between PL spectra (peak shape and intensity) and PV quality of films [7]. Room-temperature photoluminescence spectra were obtained using a Fourier transform-Raman spectrophotometer and a methodology described in ref. [8].

The PL data corroborate the R and T measurements in that a clear shift of optical emission with In content is observed. However, the PL spectra in all samples are characterized by broad and low-intensity peaks (see Fig. 4 for two $Ga_{1-x}In_xSb$ films). Such characteristics of the PL spectra indicate a poor film quality, perhaps not suitable for devices. The broad peak is a result of a variety of radiative transitions resulting from defect states present within the bandgap of the materials. Such gap states could behave as effective recombination or trap states that hinder PV action.

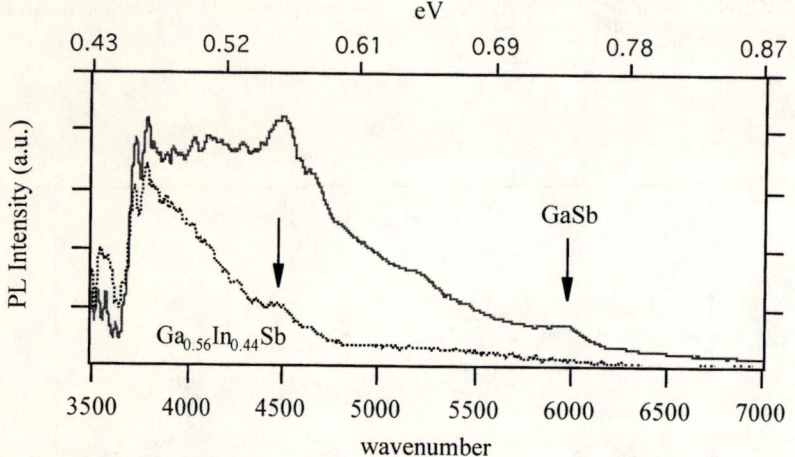

FIGURE 4. Photoluminescence spectra of selected (Ga,In)Sb polycrystalline thin-films. The arrows indicate the approximate bandgap value as a reference.

For comparison, and as a reference for the reader, it has been shown that high-quality polycrystalline PV cells show PL spectra characterized by relatively intense PL peaks (almost Gaussian in shape), with a narrow FWHM (see refs. 7 and 8). The samples we consider here do not show such desirable properties or characteristics.

Hall Measurements

All as-grown samples where Sb/(Ga+In)≤1 show p-type conductivity. It is well known that even single-crystal GaSb materials show p-type conductivity without the need of extrinsic doping. The mechanism responsible for this effect is a native Sb antisite defect [5]. We assume that the same mechanism is responsible for the polycrystalline thin-film case presented here. Because of the need of both types of conductivity, extrinsic doping has been explored using Sn.

Sn incorporation was accomplished by depositing thin layers (100-200Å) of this element onto the substrate prior to the growth of ~2 μm-thick $Ga_{1-x}In_xSb$ films. Because film growth is carried out at relatively high temperatures (>400°C), diffusion and incorporation of Sn into the $Ga_{1-x}In_xSb$ matrix is very likely to occur.

P-type conductivity is enhanced by Sn in samples with Sb/(Ga+In)<1. In this case, Sn atoms will most likely substitute for Sb; conversely, for samples with Sb/(Ga+In)>1, the resulting conductivity is n-type. Table 2 shows a summary of Hall measurements on relevant samples.

TABLE 2. Summary of room-temperature Hall measurements on selected polycrystalline thin-film $Ga_{1-x}In_xSb$ materials

$\frac{Sb}{Ga+In}$	$\frac{In}{Ga+In}$	cond. type	R (Ω-cm)	μ_H (cm^2/V-s)	carrier conc.	comments
1.00	0	p	13.7	2.45	1.9×10^{17}	
0.97	0.18	p	14.0	2.38	1.9×10^{17}	
1.00	0.44	p	8.7	0.59	1.2×10^{18}	
1.01	0	n	1.19	0.54	9.7×10^{18}	200Å Sn added
0.96	0	p	0.11	1.15	4.8×10^{18}	200Å Sn added
0.92	0	p	0.01	6.00	6.1×10^{19}	200Å Sn added
0.96	0	p	1.22	1.93	2.7×10^{18}	100Å Sn added

CONCLUSIONS AND FINAL REMARKS

Polycrystalline (In,Ga)Sb thin-film materials have been successfully grown and crystallized in the cubic phase. Incorporation of In into the GaSb matrix lowers the bandgap and leads to a change in preferential orientation on films grown on 7059 Corning glass. As-deposited films are p-type, and extrinsic doping (both p- and n-type) is possible using Sn. We note that even though it is relatively simple to grow these types of materials by coevaporation, it is rather difficult—at this stage—to grow device-quality materials. Further optimizations are needed.

The optimization needed for (In,Ga)Sb polycrystalline thin films are similar to that of other state-of-the-art polycrystalline PV materials: (i) attaining large columnar grains in the absorber layer with grain sizes in the same order as the thickness of the film or larger; (ii) finding a suitable absorber-layer contact that is not only ohmic, but also, can be stable under the high processing temperatures during absorber growth; (iii) finding suitable or improved window layer materials for heterojunction type devices.

Perhaps the most significant improvement to attain acceptable device performance can come from grain-size enhancement by growing the absorber film on single crystal substrates (Ge, Si, others). This has been the case for polycrystalline GaAs solar cells, where the conversion efficiency has reached acceptable levels only when the grain size is in the millimeter range.

ACKNOWLEDGMENTS

M. Contreras wishes to thank T.J. Coutts, M.W. Wanlass, J.S. Ward and J. Carapella for discussions, feedback and technical support during the course of this work. Also, Rick Matson for SEM work. This work was supported by the U.S. department of Energy under contract DE-AC36-83CH10093.

REFERENCES

1. N. Dhere, *Proceedings of the Second NREL Conference on Thermophotovoltaic Generation of Electricity*, AIP Conference Proceedings 358, p. 409, (1995)
2. T.J. Coutts, M.W. Wanlass, J.S. Ward, and S. Johnson, *Proceedings of the 25th IEEE Photovoltaic Specialist Conference*, May 13-17, 1996, Washington D.C., p. 25
3. M. W. Wanlass, J.S. Ward, K.A. Emery, and T.J. Coutts, *Proceedings of the 1st World Conference in Photovolatic Energy Conversion*, Hawaii, Dec 5-9 1994. (24th IEEE Photovoltaic Specialist Conference, Vol. II, p. 1685)
4. S. Wojtczuk, P. Colter, G. Charache and B. Campbell, *Proceedings of the 25th IEEE Photovoltaic Specialist Conference*, May 13-17, 1996, Washington D.C., p. 77
5. H. Ehsani, I. Bhat, C. Hitchcock, J. Borrego and R. Gutmann, *Proceedings of The Second NREL Conference on Thermophotovoltaic Generation of Electricity*, AIP Conference Proceedings 358, p. 423, (1995)
6. Miguel A. Contreras, John Webb, Andrew Tennant and Rommel Noufi, *Mat. Res. Soc. Symp. Proc. Vol. 378*, 1995 Materials Research Society, p. 803
7. M. Tanda, S. Manaka, A. Yamada, M. Konagai and K. Takahashi, *Jpn. J. Appl. Phys., Vol 32* (1993) pt. 1, No. 5A, p. 1913
8. J.D. Webb, M. Contreras, and R. Noufi, *Proceedings of the 1st World Conference in Photovolatic Energy Conversion*, Hawaii, Dec 5-9 1994. (24th IEEE Photovoltaic Specialist Conference, Vol. I, p. 275)

Advantages of Quantum Well Solar Cells for TPV

Paul Griffin[1], Ian Ballard[1], Keith Barnham[1], Jenny Nelson[1],
Alexander Zachariou[1], Chris Button[2], Mark Hopkinson[2],
Malcom Pate[2]

[1]EXSS Group, Department of Physics, Imperial College, London SW7 2BZ, UK.
[2]EPSRC III-V Facility, University of Sheffield, Sheffield, UK

Abstract. We discuss the advantages of quantum well solar cells (QWSCs) for thermophotovoltaic (TPV) applications and illustrate them with a test cell grown in GaInAsP/InGaAs lattice-matched to InP but not optimised for TPV. It is shown that a GaInAsP quaternary cell with a bandgap of 1.1μm with 60 InGaAs QWs has an open circuit voltage of (1.7 ± 0.1) times that of a homogeneous InGaAs cell under a narrow band ytterbia-like illuminating spectrum. Similar enhancements are observed under an erbia-like illumination and in a broad band illumination approximating a black body at 3000K. The quaternary cell absorbs a similar range of wavelengths close to the InGaAs cell. Also, better temperature coefficients for the QWSC than the control cell are observed in a spectrum approximating a black body at 3000K. A comparison is made between the QWSC and two other cells, one InGaAs and one Si, using published spectral response and illuminated current/voltage data. It is shown that whilst InGaAs has a higher power output in black body spectra and Si a higher power output in an ideal ytterbia spectra the QWSC has the highest output in a combination of 2000K black body and ytterbia spectra where 56% of the power output is in the ytterbia emission band. Possible improvements to this non-optimised QWSC are discussed.

INTRODUCTION

Thermophotovoltaic (TPV) power production, the conversion of long wavelength (heat) radiation to electric power by photovoltaic (PV) cells, which was introduced in the 1960's [1], is currently undergoing a renewed interest due to recent advances in low bandgap photovoltaics and selective emitters. In TPV waste heat from sources such as nuclear or fossil fuel combustion is harnessed using low bandgap photovoltaic cells. The heat energy spectrum is often re-shaped using selective emitters such as the rare earth oxides ytterbia and erbia which absorb infra-red radiation and re-emit in a narrow band [2]. The re-emitted radiation may be efficiently converted to electric power using a PV cell of appropriate band gap. Low band gaps are required since the source temperature is much lower than that of the sun but low band gaps give low output voltages.

CP401, *Thermophotovoltaic Generation of Electricity: Third NREL Conference,*
edited by Benner/Coutts
© 1997 The American Institute of Physics 1-56396-734-0/97/$10.00

Materials such as Ge, Si and GaSb, amongst others, have been proposed but their fixed band gaps make spectral matching of source and cell difficult. The band gap of the ternary compound $In_xGa_{1-x}As$ (where x is the atomic percentage) can be altered but strain relaxation caused by the lattice mis-match between the substrate (usually InP or GaAs) and the $In_xGa_{1-x}As$ active layer can degrade performance reducing the output voltage.

Quantum Well Solar Cell

One alternative way of achieving PV conversion of appropriate band gap is to insert quantum wells (QWs) into the intrinsic (i) region of a p-i-n solar cell. The effective bandgap is easily varied by altering the QW width and depth during growth. The open-circuit voltage (V_{oc}) has been shown to be between that of p-i-n devices made from the well material and that of the base cell and is higher than expected from the change in the effective bandgap alone [3]. We discuss here the advantages of quantum well solar cells (QWSCs) for TPV applications and illustrate them with a GaInAsP/InGaAs QWSC which was designed as an optical modulator and is not optimised for TPV.

The paper is in two main sections. The first presents data for an InP/GaInAsP/InGaAs QWSC and a control cell of homogeneous InGaAs lattice matched to InP under a 3000K black body spectrum and simulated ideal ytterbia and erbia emitter spectra. The second section is an attempt to compare the QWSC with other cells used for TPV drawing on published data. The comparison is performed by calculating the short circuit current density under: black body temperatures of 3000K, 2000K and 1000K; the simulated ideal ytterbia and erbia spectra; a combination of a black body at 2000K and ytterbia spectra.

EXPERIMENTAL

The experiments performed are external quantum efficiency (QE) as a function of wavelength, the current density as a function of voltage (J(V)) and the power conversion efficiency as a function of temperature. The data is discussed and possible advantages of the QWSC for TPV are shown.

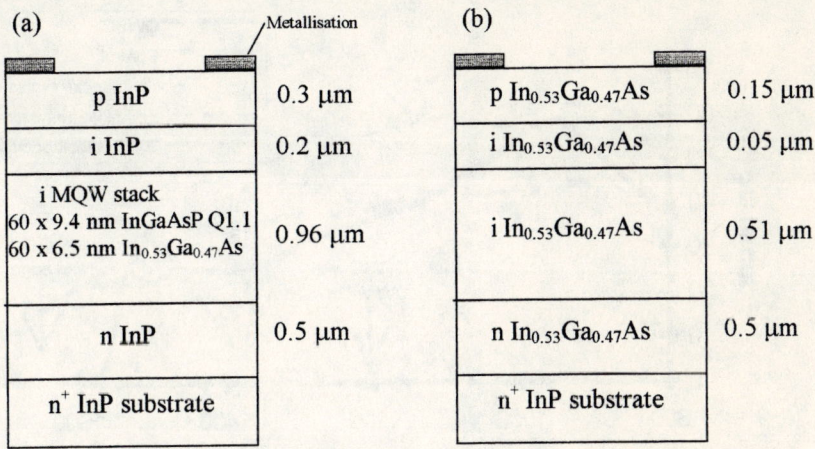

(a)			(b)	
p InP	0.3 μm		p In$_{0.53}$Ga$_{0.47}$As	0.15 μm
i InP	0.2 μm		i In$_{0.53}$Ga$_{0.47}$As	0.05 μm
i MQW stack 60 x 9.4 nm InGaAsP Q1.1 60 x 6.5 nm In$_{0.53}$Ga$_{0.47}$As	0.96 μm		i In$_{0.53}$Ga$_{0.47}$As	0.51 μm
n InP	0.5 μm		n In$_{0.53}$Ga$_{0.47}$As	0.5 μm
n$^+$ InP substrate			n$^+$ InP substrate	

FIGURE 1. Sample layer details for (a) the QWSC and (b) the control cell.

Sample Description

A wafer was grown with the quaternary $Ga_{1-x}In_xAs_yP_{1-y}$ where x and y are atomic percentages. The base cell is InP with an intrinsic region of width 0.96μm (Fig. 1) containing 60 QWs of lattice matched InGaAs 6.5nm wide separated by 9.4nm wide barriers of $Ga_{1-x}In_xAs_yP_{1-y}$ with a band gap wavelength of 1.1μm. All the layers are lattice-matched to the n-type InP substrate with a 0.2μm undoped spacer between the top of the multi-quantum well (MQW) stack and the p-layer to reduce dopant diffusion into the MQW [4]. The control cell is a comparable cell made from homogeneous InGaAs lattice matched to InP. The p-layer and i-layer are thinner than that of the QWSC but the material quality is similar. Both were processed into 1000μm diameter mesas and metallised with an optical window of 545μm diameter. No anti-reflection (AR) coating was applied to either.

Measurements

External quantum efficiency

The external quantum efficiency (QE) of all the cells was measured with a monochromator, a tungsten lamp and calibrated InGaAs and Si detectors. Both cells exhibit lower than optimum QEs due to the absence of an AR coat (Fig. 2). The QE of the QWSC above 900nm is generated from photon absorption in the

413

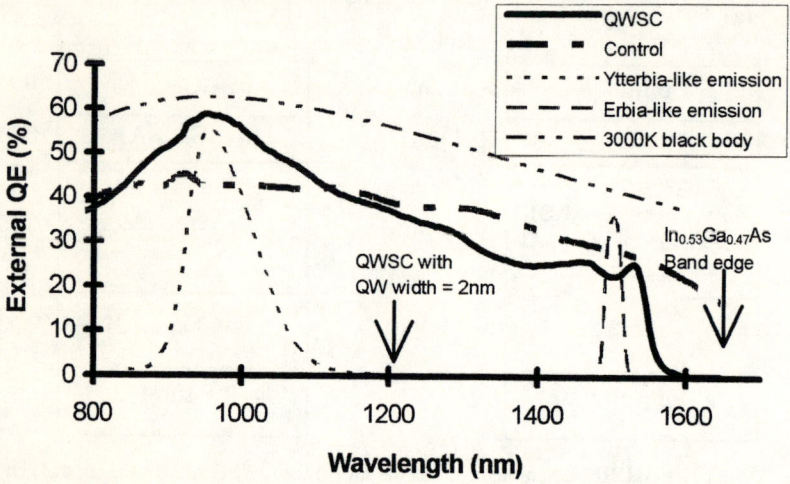

FIGURE 2. The external quantum efficiencies for the QWSC and a control cell. Also shown are the ytterbia-like and erbia-like spectra and a 3000K black body spectrum on an arbitrary scale. The arrowed lines show the position of the absorption edge of the InGaAs control on the right and a modelled InP/InGaAs QWSC.

QWs and is nearly as high as the QE of the InGaAs control. The response falls off at longer wavelengths due to fewer energy levels in the QWs and an exciton peak is seen just above 1500nm. The band edge is just lower in wavelength than that for the InGaAs cell due to quantisation of energy levels in the wells. The fall off in QE of the QWSC below 950nm is due to absorption in the thick InP p-layer.

Current/Voltage measurements

Dark. The cells were kept in the dark and the current measured as the voltage was stepped up. The temperature of the cells was kept at 25°C throughout the measurements. The current density, J, is calculated using the active area. The dark current density from the QWSC is significantly lower than that of the control (Fig. 3) suggesting that the QWSC will have superior voltage performance.

Illuminated. The cells were illuminated with a quartz halogen tungsten lamp approximating a black body at 3000K and filtered with interference bandpass filters simulating ideal ytterbia and erbia emissions [2] (Table 1 and Fig. 2). The intensity of the lamp was set to give the highest possible short circuit current density (J_{sc}) in the control cell and then kept constant for all other measurements. Using the measured spectral response of the cell and the measured short circuit

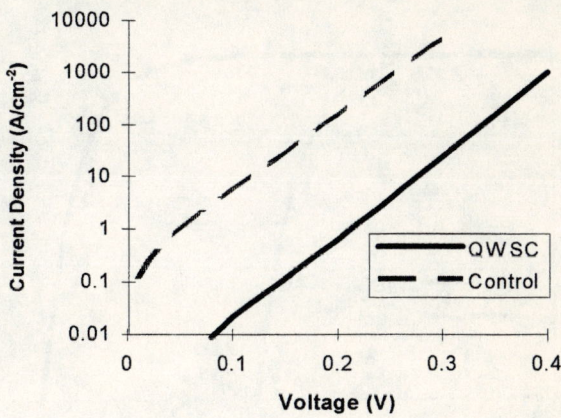

FIGURE 3. The J(V) curves of the QWSC and InGaAs control as measured in the dark and plotted on a semi-logarithmic graph.

current the incident power density on the cell was calculated to be 277mWcm^{-2} for the 3000K black body spectrum and 107mWcm^{-2} and 22mWcm^{-2} for the ytterbia-like and erbia-like spectra respectively.

The light J(V) curves for the QWSC and its control under ytterbia-like illumination are shown in Fig. 4 with J calculated using the active area. The most obvious difference is the V_{oc} of the QWSC being 1.7 ± 0.1 times that of the control (Table 2). The V_{oc} of the control is typical of a small (1mm diameter) size device of InGaAs at this level of illumination.

For the erbia simulated spectrum the V_{oc} of the QWSC is 1.8 times larger than the control (Table 2) leading to enhanced power output despite a lower current density. The fill factor (FF) is also lower for the control.

In a broad band spectrum approximating a 3000K black body the QWSC also has an enhanced V_{oc} by a factor of 1.6 (Table 2). Here the control has a higher J_{sc} but again the V_{oc} dominants.

Table 1. The peak position and full width at half maximum of the bandpass filters used above with ytterbia and erbia emitters.

	Ytterbia		Erbia	
	Filter	**Emitter**	**Filter**	**Emitter**
Peak centre (nm)	950	980	1500	1500
FWHM (nm)	90	150	25	65

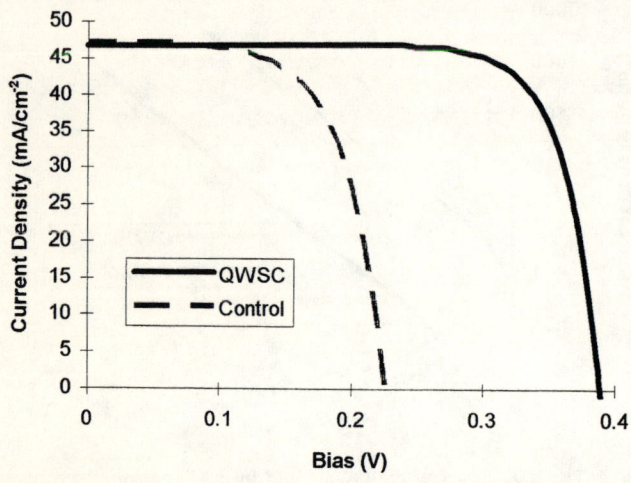

FIGURE 4. The J(V) curves of the QWSC and InGaAs control as measured in the ytterbia-like illumination.

Temperature Coefficients

The temperature coefficients of the QWSC and control cell have been measured under unfiltered tungsten light, approximating a black body at 3000K, between 20 °C and 100 °C using a Peltier temperature controlled system described in Ref. [5]. The illumination level was set to give a short-circuit current from a

Table 2. Cell characteristics under ytterbia-like, erbia-like and 3000K black body (BB) illuminating spectra.

	J_{sc} (mA/cm^2)	V_{oc} (V)	FF	P_{out} (mW/cm^2)
	Ytterbia			
QWSC	47	0.39	0.75	14
Control	45	0.23	0.63	6.5
	Erbia			
	J_{sc} (mA/cm^2)	V_{oc} (V)	FF	P_{out} (mW/cm^2)
QWSC	6.1	0.34	0.73	1.5
Control	8.9	0.18	0.59	0.95
	3000K BB			
	J_{sc} (mA/cm^2)	V_{oc} (V)	FF	P_{out} (mW/cm^2)
QWSC	50	0.39	0.75	15
Control	64	0.24	0.63	9.5

416

Table 3. Temperature coefficients under an illumination approximating a 3000K black body.

	$(dJ_{sc}/dT)/J_{sc}$ $x10^{-3}$ / K	$(dV_{oc}/dT)/V_{oc}$ $x10^{-3}$ / K	$(dFF/dT)/FF$ $x10^{-3}$ / K	$(dEff/dT)/Eff$ $x10^{-3}$ / K
QWSC	0.130	-5.1	-2.2	-6.3
Control	0.400	-9.0	-4.8	-11

calibrated homogeneous GaAs cell equal to that which the calibrated cell gave under AM1.5 1000Wm^{-2} spectrum. This light level was kept constant over all the cells as the temperature was varied. The parameters are shown in Table 3 and have been normalised by dividing by the value at 20°C. Note that the change in J$_{sc}$ is increasing whereas all the others are decreasing. Most important is that the efficiency of the QWSC is found to fall more slowly with temperature than the control. The efficiency temperature coefficient of the QWSC is half that of the control. Also the V$_{oc}$ of the QWSC becomes greater than 3 times that of the control at 100°C. This makes it less susceptible to heating which is significant in TPV power production.

Discussion

The above results show that a QWSC has an enhanced V$_{oc}$ over the control cell, made of the same material as the QWs. The result is independent of whether the spectrum is narrow or broad band. The V$_{oc}$ leads to an enhancement in the output power under ytterbia, erbia and a 3000K black body illumination which is expected to hold also for lower source temperatures.

The V$_{oc}$ is increased as a direct result of the lower dark current. Modelling of the dark currents of single QW samples has shown that the quasi-Fermi level separation in the QW and the barriers are not the same [6] and this is expected to be true for MQW systems.

The improved temperature performance of the QWSC over the control will help to maintain power output in a hot TPV environment.

COMPARISON WITH OTHER CELL TYPES

It is important to compare the performance of the QWSC to other types of cells that are candidates for TPV. Due to the open architecture of TPV systems almost any cell can be used [7]. Two of the most common cells are Si and InGaAs. Si is used with ytterbia emitters [8] and InGaAs for low temperature black body radiation with or without emitters [9]. Though the QWSC is not optimised for photovoltaic action a comparison will be made against cells optimised for TPV and solar applications. The spectra used will be a black body at 3000K, 2000K and 1000K and ideal ytterbia and erbia emissions. By calculating the expected J_{sc} from the measured SR given in the literature and measured light curves from the literature the output power of the devices is estimated.

Data

The QWSC is the same as described above and has no AR coating. The Si cell is that used in Ref. [10] as it is a high efficiency and also has no AR coating. The InGaAs cell is published in Ref. [11] and has an AR coating optimised for an AM0 spectrum. The spectral response is adjusted to give that expected from a sample with no AR coat by using the reflection data also provided in Ref. [11]. Having no AR coating on the cells creates similar reflective losses for all cells.

The black body spectra are calculated and the ytterbia and erbia spectra used are the spectra measured from the narrow band filters. Although the emitter spectra are not identical to the filter spectra (Table 1) they are similar enough to allow the comparison. A third type of spectrum is used where an ytterbia spectrum has been superimposed onto a black body spectrum and the ytterbia spectrum scaled to give 56% of the total power in the emission band. This is a more realistic spectrum from an ytterbia emitter. Theory predicts emission efficiencies higher than this [12] but in practice the efficiencies to date have been lower.

Calculations

Using the $QE(\lambda)$ from the literature the J_{sc} is calculated from the incident spectra ($IR(\lambda)$) by :

$$J_{sc} = \int QE(\lambda).IR(\lambda)d\lambda \qquad (1)$$

The measured light $J(V)$ curve from the literature is shifted to the calculated value for J_{sc} and the maximum power output calculated by recalculating the maximum power point.

Results

The open-circuit voltage of the QWSC was over 1.3 times that of the InGaAs cell in the black body spectra (Table 4). However in all cases, except the ideal ytterbia spectrum, the maximum power (P_m) from the InGaAs cells exceeded that of the QWSC due to the short-circuit current density of the cell being around twice that of the QWSC. Another interesting trend is that as the black body temperature is reduced the InGaAs cell's power output falls less rapidly than the QWSC and the QWSC's power output falls less than the Si cell. The Si cell only has the maximum power output for the ideal ytterbia spectrum.

418

Table 4. The modelled light J(V) characteristics under 192 mW/cm^2 spectrum corresponding to a black body of temperature shown in the table.

	3000K		
	J_{sc} (mA/cm^2)	V_{oc} (V)	P_m (mW/cm^2)
QWSC	35	0.51	14
InGaAs	65	0.38	18
Si	24	0.59	12
	2000K		
	J_{sc} (mA/cm^2)	V_{oc} (V)	P_m (mW/cm^2)
QWSC	19	0.50	7.6
InGaAs	42	0.37	11
Si	1.7	0.56	3.5
	1000K		
	J_{sc} (mA/cm^2)	V_{oc} (V)	P_m (mW/cm^2)
QWSC	0.11	0.44	0.35
InGaAs	0.39	0.23	0.77
Si	0.071	0.45	0.024

In the spectra of ideal emitters the V_{oc} is similarly enhanced over the InGaAs cell (Table 5). In ytterbia the incident power is 37mWcm^{-2}. The J_{sc} values are similar and the QWSC has more power output than the InGaAs cell. However the Si cell is well matched to this spectrum and has the highest power output. In erbia the incident power is 12mWcm^{-2} and the power output from the Si cell is negligible. The InGaAs cell has the higher power output due to the J_{sc}.

The mixed ytterbia and 2000K black body spectra has an incident power 53mWcm^{-2} with 56% in the ytterbia band. The power output of the QWSC

Table 5. The modelled light J(V) characteristics under ytterbia- and erbia-like spectra and a mixed ytterbia and 2000K black body spectra with 56% of the power in the ytterbia band.

	Ytterbia		
	J_{sc} (mA/cm^2)	V_{oc} (V)	P_m (mW/cm^2)
QWSC	16	0.49	6.3
InGaAs	19	0.34	4.6
Si	16	0.58	7.4
	Erbia		
	J_{sc} (mA/cm^2)	V_{oc} (V)	P_m (mW/cm^2)
QWSC	3.3	0.45	1.2
InGaAs	8.5	0.32	1.9
Si	------	------	-----
Ytterbia	+ 2000K	BB	
	J_{sc} (mA/cm^2)	V_{oc} (V)	P_m (mW/cm^2)
QWSC	14	0.49	5.4
InGaAs	20	0.34	4.7
Si	11	0.57	4.9

exceeds that of both the InGaAs and the Si cells as the increase of the V_{oc} over the InGaAs cell exceeds the loss in J_{sc} and for the Si cell the J_{sc} gain exceeds the loss in V_{oc}.

Discussion

The QWSC did not have the maximum power output in the ideal emitter or pure black body spectra but this was due to the current density. The InGaAs and Si are optimised structures whereas the QWSC is not. The QWSC can be improved by a thinner p-layer, more QWs, back mirroring and adjusting the QW dimensions. Due to the enhanced voltage output over the InGaAs cell improvement in the spectral response will enable the QWSC's power output to exceed that of the InGaAs cell.

Whilst the Si cell was best in ideal ytterbia and InGaAs best in black body the realistic situation will be a mixture of these. Modelling a mixed spectrum of ytterbia and 2000K black body which has 56% of the power in the emission band of ytterbia gives the QWSC the maximum power output.

CONCLUSION

In conclusion we have shown that QWSCs made from lattice matched GaInAsP/InGaAs have a significant voltage enhancement over cells made from InGaAs. The longer erbia-like emission can be effectively absorbed by QWs with higher voltage than single bandgap cell of comparable band edge. We have also shown that the temperature dependence of these QWSCs is better than a comparable InGaAs cell, giving advantages in the hot TPV environment. The advantages of QW converters can also be used in systems where the radiation is emitted in a broad band black body spectrum.

A comparison of the QWSC with devices from the literature also show that the voltage of the QWSC is enhanced. The spectral response from the QWs limits the current output, but this can be improved. Even with the present device characteristics under a mixed ytterbia and 2000K black body, a more likely spectrum than an ideal ytterbia spectrum, the QWSC has a higher output power than either of the comparison cells.

Further improvement to these cells can be made by optimising the AR coat for the illuminating spectrum and using a suitable front contact grid design to balance series resistance and shading losses at higher power densities. The higher operating voltages in QWSCs over homogeneous InGaAs cells reduces concerns over series resistance and shading.

420

As has been observed in this system [3] the open circuit voltage increases as the effective bandgap increases. If the wells are narrower (for example 2nm) the effective bandedge can be optimised for an ideal ytterbia emission spectrum at 1200nm (Fig. 2). The V_{oc} extrapolated from data in Ref. [3], is expected to be at least 0.7V. Such optimisation is much simpler in a QWSC than in a bulk cell. An improvement in spectral response by etching off the substrate and mirroring the back surface has recently been shown for AlGaAs/GaAs and GaAs/InGaAs cells [13] and any light not absorbed will be reflected back to the emitter increasing the *system* efficiency.

Problems in using lattice mismatched material are reduced in QWSCs as has been shown in the strained GaAs/InGaAs system [14]. Less strained material would be needed for QW cells than for equivalent conventional cells if less InGaAs is needed for the same level of quantum efficiency. Strain balancing is also potentially easy to implement in QWSCs by using an opposite strain in the barriers to the wells.

In summary the inclusion of InGaAs QWs into an InGaAsP cell leads to a voltage enhancement compared to a cell made from InGaAs and depending on the incident spectra can give more power than either a Si cell or an InGaAs cell. Moreover there is scope for improvement in the design which will lead to increased power output. The flexibility of using QWs can be used to easily match the cell to the emission spectra.

ACKNOWLEDGMENTS

We would like to thank Chee Tang, Mark Whitehead and Gareth Parry of the Semiconductor IRC and the University of Oxford for the use of their quaternary samples. We are grateful to EPSRC, the Clean Technology Unit, The Greenpeace Trust, The British Council and Tata Ltd (India) for financial support.

REFERENCES

1. Wedlock, B.D., "Thermo-Photo-Voltaic conversion", *Proceedings of the IEEE* **51**, 694 (1963).
2. Coutts, T.J., Wanlass, M.W., Ward, J.S., Johnson, S., "A Review of Recent Advances in Thermophotovoltaics", in *Proceedings of the 25th IEEE Photo-voltaics Specialist Conference,* 1996, pp. 25-30.
3. Barnham, K.W.J., Connolly, J., Griffin, P., Haarpaintner, G., Nelson, J., Tsui, E., Zachariou, A., Osbourne, J., Button, C., Hill, G., Hopkinson, M., Pate, M., Roberts, J.S., Foxon, T., *J. Appl. Phys.* **80**(2), 1201 (1996).
4. Zachariou, A., Barnham, K.W.J., Griffin, P. , Nelson, J., Button, C., Hopkinson, M., Pate, M., Epler, J., "A New Approach to p-doping and the Observation of Efficiency Enhancement in InP/InGaAs Quantum Well Solar Cells", in *Proceedings of the 25th IEEE Photo-voltaics Specialist Conference,* 1996, pp. 113-116.

5. Ballard, I., *MSc thesis*, Imperial College, London, 1995.
6. Nelson,J., Barnes, J., Ekins-Daukes, N., Kluftinger, B., Barnham, K.W.J., Button, C., Roberts, J.S., "Observation of Suppressed Radiative Recombination in Single Quantum Well p-i-n diodes", to be submitted to J. Appl. Phys. 1997
7. Gray J. L., El-Husseini, A., "A Simple Parametric Study of TPV System Efficiency and Output Power Density Including a Comparison of Several TPV Materials", in *Proceedings of the 2nd NREL Conference on Thermophotovoltaic Generation of Electricity*, 1995, pp. 3-15.
8. Holmquist, G. A., Wong, E. M., Waldman, C. H., "Laboratory Development TPV Generator", in *Proceedings of the 2nd NREL Conference on Thermophotovoltaic Generation of Electricity*, 1995, pp. 138-161.
9. Jain R. K., Wilt, D. M., Jain, R., Landis G. A., Flood, D. J., "Lattice-Matched and Strained InGaAs Solar Cells for Thermophotovoltaic Use", in *Proceedings of the 2nd NREL Conference on Thermophotovoltaic Generation of Electricity*, 1995, pp. 375-386.
10. Umeno, M., Kato, T., Yang, M., Azuma, Y., Soga, T., Jimbo, T., "High Efficiency AlGaAs/Si Tandem Cell Over 20%", in *Proceedings of the 1st World Conference on Photo-voltaic Energy Conversion*, 1994, pp. 1679-1684.
11. Wilt D. M., Brinker D. J., Fatemi N. S., Hoffman, Jr., R. W., Jenkins, P. P., Lowe, R., "Lattice Mismatched InGaAs Photovoltaic Devices for Thermophotovoltaic Power Systems", in *Proceedings of the 1st World Conference on Photo-voltaic Energy Conversion*, 1994, pp. 1738-1741.
12. Chubb, D. L., Lowe, R. A., *J. Appl. Phys.* **74**(9), 5687 (1993).
13. J. Barnes PhD thesis, University of London, 1995; J. Nelson "Physics of Thin films", Vol 21, Ed. M.H. Francome and J.L. Vossen, Academic Press, p. 311, 1995
14. P. Griffin, J. Barnes, K.W.J. Barnham, M. Mazzer, C. Zanotti-Fregonara, C. Olson, C. Rohr, G. Haarpaintner, J.P.R. David, J.S. Roberts, R. Grey, M.A. Pate *J. Appl. Phys.* **80**(10), 5815 (1996).

Appropriate Materials and Preparation Techniques for Polycrystalline-Thin-Film Thermophotovoltaic Cells

Neelkanth G. Dhere

Florida Solar Energy Center
1679 Clearlake Road, Cocoa, FL 32922-5703

Abstract Polycrystalline-thin-film thermophotovoltaic (TPV) cells have excellent potential for reducing the cost of TPV generators so as to address the hitherto inaccessible and highly competitive markets such as self-powered gas-fired residential warm air furnaces and energy-efficient electric cars, etc. Recent progress in polycrystalline-thin-film solar cells have made it possible to satisfy the diffusion length and intrinsic junction rectification criteria for TPV cells operating at high fluences. Continuous ranges of direct bandgaps of the ternary and pseudoternary compounds such as $Hg_{1-x}Cd_xTe$, $Pb_{1-x}Cd_xTe$, $Hg_{1-x}Zn_xTe$, and $Pb_{1-x}Zn_xS$ cover the region of interest of 0.50-0.75 eV for efficient TPV conversion. Other ternary and pseudoternary compounds which show direct bandgaps in most of or all of the 0.50-0.75 eV range are $Pb_{1-x}Zn_xTe$, $Sn_{1-x}Cd_{2x}Te_2$, $Pb_{1-x}Cd_xSe$, $Pb_{1-x}Zn_xSe$, and $Pb_{1-x}Cd_xS$. $Hg_{1-x}Cd_xTe$ (with x $\approx$ 0.21) has been studied extensively for infrared detectors. PbTe and $Pb_{1-x}Sn_xTe$ have also been studied for infrared detectors. Not much work has been carried out on $Hg_{1-x}Zn_xTe$ thin films. $Hg_{1-x}Cd_xTe$ and $Pb_{1-x}Cd_xTe$ alloys cover a wide range of cut-off wavelengths from the far infrared to the near visible. Acceptors and donors are introduced in these materials by excess non-metal (Te) and excess metal (Hg and Pb) respectively. Extrinsic acceptor imputities are Cu, Au, and As while and In and Al are donor impurities. $Hg_{1-x}Cd_xTe$ thin films have been deposited by isothermal vapor-phase epitaxy (VPE), liquid phase epitaxy (LPE), hot-wall metalorganic chemical vapor deposition (MOCVD), electrodeposition, sputtering, molecular beam epitaxy (MBE), laser-assisted evaporation, and vacuum evaporation with or without hot-wall enclosure. The challenge in the preparation of $Hg_{1-x}Cd_xTe$ is to provide excess mercury incidence rate, to optimize the deposition parameters for enhanced mercury incorporation, and to achieve the requisite stoichiometry, grain size, and doping. MBE and MOCVD techniques have paved the way for obtaining epitaxial $Hg_{1-x}Cd_xTe$ thin films at substrate temperatures of ~180° C with the desired crystalline perfection, stoichiometry, and doping without the necessity of further annealing for improving either the crystalline quality or dopant activity. Retaining larger mercury proportions during annealing would require heated enclosures as in isothermal VPE, hot-wall technique, vacuum evaporation, hot-wall MOCVD, or close-space sublimation. $Pb_{1-x}Cd_xTe$ thin films can be prepared by magnetron sputtering from cooled $Pb_{1-x}Cd_xTe$ targets on heated substrates. Hot-wall technique is suitable for the deposition of $Pb_{1-x}Cd_xTe$ thin films. $Hg_{1-x}Cd_xTe$ and $Pb_{1-x}Cd_xTe$ TPV cells will benefit from the substantial work on CdTe thin film solar cells. The paper reviews work on thin films of ternary and pseudoternary compounds of interest for TPV conversion and methods of their preparation with a view to choosing the appropriate materials and fabrication techniques for polycrystalline-thin-film TPV cells.

CP401, *Thermophotovoltaic Generation of Electricity: Third NREL Conference*,
edited by Benner/Coutts

INTRODUCTION

Thermophotovoltaic (TPV) cells convert thermal energy to electricity. Modularity, portability, silent operation, absence of moving parts, reduced air pollution, rapid start-up, high power densities, potentially high conversion efficiencies, choice of a wide range of heat sources employing fossil fuels, biomass, and even solar radiation are key advantages of TPV cells in comparison with fuel cells, thermionic and thermoelectric convertors, and heat engines [1-10]. Potential applications of TPV systems include: remote electricity supplies, transportation, co-generation, electric-grid independent appliances, and space, aerospace, and military power applications. TPV generators originally proposed over twenty years ago were limited by relatively immature materials technologies, especially in the two critical components viz. the thermal emitter and photovoltaic cells. Recent advances in material processing for both components have renewed the interest in TPV energy conversion.

The range of bandgaps for achieving high conversion efficiencies using low temperature (1000-2000 K) black-body or selective radiators is in the 0.5-0.75 eV range. Present high efficiency convertors are based on single crystalline materials such as $In_{1-x}Ga_xAs$, $GaSb$, $Ga_{1-x}In_xSb$, and $In_{1-x}Ga_xAs_{1-y}Sb_y$ [1-9]. Several polycrystalline thin films e.g. $Pb_{1-x}Cd_xTe$, $Pb_{1-x}Cd_xTe$, and $Hg_{1-x}Zn_xTe$, etc have great potential for economic large scale applications in cheaper, civilian, and military, terrestrial and aerospace applications [1]. Several deposition techniques e.g. vacuum evaporation, sputtering, close-space sublimation, electrodeposition, screen printing, metal-organic chemical vapor deposition (MOCVD), chemical bath deposition (CBD), etc. have been developed for preparation of polycrystalline-thin-film photovoltaic (PV) solar cells [1,11-15]. PV conversion efficiencies comparable to those of single crystalline Si solar cells have been achieved e.g. 15.8% for CdTe and 17.7% for $CuIn_{1-x}Ga_xSe_2$ [1,13,14]. TPV cell development can benefit from the more mature PV solar cell and optoelectronic (infrared detectors, lasers, and optical communications) technologies.

Effect of High Fluences

A blackbody emitter operating at a temperature of 2000 K which is approximately one-third that of the sun can provide light flux density of 90 W cm^{-2} to the cell. This flux density is ~900 times that of the one-sun air-mass one intensity of ~0.1 W cm^{-2}. Many concepts developed for high concentration PV cells are directly applicable to the TPV cells operating at high fluences [1,16]. As the intensity of radiation (with $hv > E_g$) incident on the cell is increased, there is a proportionate increase in the short-circuit-current density, an increase in the open-circuit voltage, and consequently an increase in the PV conversion efficiency. Interestingly the open-circuit voltage and the device efficiency increase more rapidly in the case of cells with an initially poor diode quality factor. Thus PV conversion efficiencies will be higher at higher fluences available in TPV applications.

Midgap recombination states affect the minority-charge-carrier diffusion length and intrinsic junction rectification. An allowable recombination-state density of 6.5×10^{15} cm^3 has been calculated for a direct-bandgap semiconductor with an intragrain mobility μ of 1000 cm^2 V^{-1} s^{-1} and an atomic-size recombination state cross-section of 10^{-14} cm^2, assuming optical absorption length of 1 μm and the diffusion length of 2 μm [1]. Using values of the dielectric constant ϵ of $12 \times 9 \times 10^{-14}$ F cm^{-1} and the built-in-voltage ϕ_B of 0.45 V, for the above semiconductor, and a short circuit current density of 3 A cm^{-2}, a recombination-state density below $\sim 3 \times 10^{16}$ cm^3 will be required to maintain an intrinsic junction rectification [1]. Thus for a TPV cell the diffusion length and intrinsic junction rectification criteria set limits to the recombination-state density of 6.5×10^{15} cm^{-3} and 3×10^{16} cm^{-3} respectively. Until recently these values could be met only by single-crystal materials. Recent progress in polycrystalline-thin-film solar cells makes it possible to consider these cells for TPV applications [1,17-19]. The best values of the product recombination-state density times the depletion region width N_t w are few times 10^{-11} A cm^{-2}, while the depletion region width w is a few thousand Å [1]. Hence the best recombination-state density N_t is in the acceptable range of 10^{-15}-10^{-16} cm^{-3}. More importantly, the low recombination-state density is in the region of interest for effective collection of photogenerated carriers.

MATERIALS AND DEPOSITION TECHNIQUES

A compilation of bandgap and lattice constants of II-VI and IV-VI compound semiconductors of interest for TPV cell fabrication shows that there are continuous ranges of direct bandgaps for the ternary and pseudoternary compounds $Hg_{1-x}Cd_xTe$, $Pb_{1-x}Cd_xTe$, $Hg_{1-x}Zn_xTe$, and $Pb_{1-x}Zn_xS$ covering the region 0.50-0.75 eV range [1]. Other ternary, pseudoternary, and pseudoquaternary compounds which show direct bandgaps in most of or all of the 0.50-0.75 eV range are $Pb_{1-x}Zn_xTe$, $Sn_{1-x}Cd_{2x}Te_2$, $Pb_{1-x}Cd_xSe$, $Pb_{1-x}Zn_xSe$, and $Pb_{1-x}Cd_xS$ [1]. Thus several material combinations present a range of bandgaps which would allow bandgap engineering for achieving the best performance. As discussed in the following, $Hg_{1-x}Cd_xTe$ (with x $\approx$ 0.21) has been studied extensively for infrared detectors. PbTe and $Pb_{1-x}Sn_xTe$ have also been studied for infrared detectors. $Hg_{1-x}Cd_xTe$ and $Pb_{1-x}Cd_xTe$ TPV cells will benefit from the substantial work on CdTe thin film solar cells. Not much work has been carried out on $Hg_{1-x}Zn_xTe$ thin films. In photovoltaic devices a p-type absorber is usually chosen because of the availability of several n-type window layers such as ZnO:Al, SnO$_2$:F, and cadmium stannate Cd$_2$SnO$_4$, etc [20]. CdS heterojunction partner layers can match several p-type absorbers. CdS can also serve as passivation layer for $Hg_{1-x}Cd_xTe$ [21]. p-type conductivity can be achieved more easily in tellurides than in sulphides. Thus it would be appropriate to choose p-type $Hg_{1-x}Cd_xTe$, and $Pb_{1-x}Cd_xTe$ absorber polycrystalline thin films with direct bandgaps in the range 0.50-0.75 eV for the development of TPV cells [1].

In most TPV applications, the main requirement is the output-power density [22]. For the TPV convertors, the optimum bandgap for achieving the maximum output-power density is approximately 0.2 eV lower than that for achieving the maximum conversion efficiency [22]. Thus a compromise may be achieved between achieving either the maximum power output and the maximum conversion efficiency using the great variety of materials. A simple parametric study of TPV system efficiency and output power density has shown that output power density would be lower at lower emitter temperatures. However, fairly high efficiencies can be expected at lower bandgaps with high-pass or bandpass filters [23].

PV modules based on polycrystalline-thin-film CdTe and $CuIn_{1-x}Ga_xSe_2$ are strong candidates for low-cost, large-scale manufacture of PV solar cells [11-15,24,25]. Efficiencies over 10% have been achieved by employing several processes such as magnetron sputtering, vacuum evaporation, etc for the deposition of CdTe thin films, the best efficiency being 15.8% by close-space sublimation (CSS) [14]. Vacuum coevaporation, selenization of sputtered metallic precursors, and rapid isothermal processing have resulted in $CuIn_{1-x}Ga_xSe_2$ solar cell efficiencies over 10%, the present best efficiency being 17.7% [25]. The key strength of CdTe and $CuIn_{1-x}Ga_xSe_2$ solar cells is their stability verified at NREL [11,24,25].

Comparison of the basic parameters of the best polycrystalline-thin-film CdTe, $CuIn_{1-x}Ga_xSe_2$, and $CuInSe_2$ as well as single-crystal Si and GaAs PV solar cells with the ideal cell which converts all the photons with energy above the absorber bandgap to electricity shows that i) the current densities are lower for the polycrystalline-thin-film cells because of their less efficient red response resulting from smaller diffusion lengths, lower blue collection limited by absorption in the CdS heterojunction partner layer, and less efficient antireflection coatings; ii) the open circuit voltages of polycrystalline-thin-film cells are ~15% lower than those of single-crystal cells, primarily because the forward current in polycrystalline cells is at least an order of magnitude larger; iii) the fill factors for polycrystalline cells are lower due to larger diode quality factors; and iv) the conversion efficiencies of polycrystalline cells are approximately two-thirds of those of single crystal cells [1,17-19]. Even though the values of the various parameters are lower for the polycrystalline thin-film cells, the overall values are respectable in view of the fabrication economics. The difference between the efficiencies of polycrystalline and single-crystal cells has been narrowing steadily over the last few years and the trend is expected to continue [1,17-19].

Wide bandgap semiconductors such as CdTe have a tendency to self-compensate shallow levels of deliberately incorporated extrinsic donor or acceptor impurities through spontaneous generation of native point defects, such as vacancies, antisite defects, host interstitials, or their complexes [26]. Self-compensation tries to minimize the free energy of the sample by bringing the Fermi level E_F to the intrinsic level E_i. Lower bandgap values (0.50-0.75 eV) for TPV applications make it easier to dope the semiconductors to appropriate concentrations. Greatly enhanced carrier mobilities can be achieved because the high dielectric permittivities of group II-VI and IV-VI binary and pseudo-ternary compound semiconductors heavily screen the field of ions and thus reduce the perturbation of potential considerably.

Losses in the heterojunction partner layer could be eliminated in a TPV cell where the absorption in the heterojunction partner and the interface layers would be negligible for the IR radiation and at the same time the junction quality could be improved considerably [15]. It would be easier to lower the series resistance by heavily doping the lower bandgap p-type $Pb_{1-x}Cd_xTe$ and $Hg_{1-x}Cd_xTe$ absorbers near the back contact. $Hg_{1-x}Cd_xTe$ has disadvantage of more stringent health and safety procedures. Hence it may be preferable to employ $Pb_{1-x}Cd_xTe$.

Low bandgaps and larger fluences employed in TPV cells result in very high current densities which would require specially developed front contact grids. The requirements of the deposition of special contact grids and conducting away heat preferably through a heat-conducting substrate would favor substrate structures as employed in $CuIn_{1-x}Ga_xSe_2$ thin-film solar cells as against superstrate structures employed in CdTe thin-film solar cells. Techniques for laser- and mechanical scribing, integral interconnection, and multi-junction tandem structures which have been fairly well developed for thin-film PV solar cells could be further refined for enhancing the voltages from TPV modules. Thin-film TPV cells may be deposited on metals or back-surface reflectors. Spectral control elements may be deposited directly on the TPV convertor.

The bandgaps E_g of absorber layers of the TPV cells would be 0.59 eV and 0.75 eV equivalent to cutoff wavelengths of 2.1 μm and 1.65 μm respectively so as to spectrally match the emission characteristics of holmia- and erbia-based selective emitters with emissions at 0.62 eV and 0.80 eV respectively [1].

$Hg_{1-x}Cd_xTe$ alloy system ranks as the most commonly used semiconductor for infrared detection in the two atmospheric windows 3-5 μm and 8-12 μm [27-29]. The binary compounds HgTe and CdTe have a zincblende type face-centered cubic structure, with a lattice parameter very close to each other, and are miscible in all proportions. HgTe is a semimetal and CdTe is a direct bandgap semiconductor with E_g of 1.5 eV at 300 K. PbTe has a NaCl type face-centered cubic structure. $Pb_{1-x}Cd_xTe$ alloy is a direct bandgap semiconductor. However, because of the different crystal structures, it may not be possible to obtain intermediate bandgaps by a simple rule of mixtures [30,31]. $Hg_{1-x}Cd_xTe$ and $Pb_{1-x}Cd_xTe$ alloys cover a wide range of cut-off wavelengths from the far infrared to the near visible. Acceptors and donors are introduced in these materials by excess non-metal (Te) and excess metal (Hg and Pb) respectively. Appropriate extrinsic dopants can also be introduced to obtain p- or n-type conductivity. Cu, Au, and As are known acceptor impurities while and In and Al are donor impurities.

$Hg_{1-x}Cd_xTe$ thin films have been deposited by isothermal vapor-phase epitaxy (VPE), liquid phase epitaxy (LPE), hot-wall metalorganic chemical vapor deposition (MOCVD), electrodeposition, sputtering, molecular beam epitaxy (MBE), laser-assisted evaporation, and vacuum evaporation with or without hot-wall enclosure [32-84]. Usually $Cd_{0.96}Zn_{0.04}Te$, InSb, GaAs, sapphire/CdTe substrates are employed to minimize the substrate-film mismatch.

The earliest developed process, isothermal vapor-phase epitaxy consists of evaporation of HgTe on to CdTe single crystals to form $Hg_{1-x}Cd_xTe$ layers [32-41]. The $Hg_{1-x}Cd_xTe$ layer thickness is proportional to the square of growth time and Hg content varies with depth from the surface showing that interdiffusion of Hg and Cd play an important role. The process requires proper control of excess mercury pressure [36].

Liquid phase epitaxy has become the established technique for the fabrication of $Hg_{1-x}Cd_xTe$ photodiode arrays [42-47]. Excellent composition reproducibility at Cd content x of 0.21 has been achieved with a LPE graphite sliding boat design and integral HgTe source for Hg overpressure [43]. n- and p-type carrier concentrations have been controlled using In and Al donor- and Cu and Rb acceptor- dopants rather than with native defects and residual impurities doping [43]. The LPE process usually results in 10-20 μm thick layers while the necessary absorber thickness is only a few microns for the direct bandgap semiconductors.

Cold- or hot-wall metalorganic chemical vapor deposition has been well developed for the deposition of $Hg_{1-x}Cd_xTe$ layers [48-56]. Growth temperatures have been reduced to 350° C, 200° C, and even 180° C by taking advantage of lower thermal stability of organotellurium compounds such as diisopropyltelluride, ditertiarybutyltelluride, and di(2-propen-1-yl)telluride with large hydrocarbon molecules [48,49]. p- and n-type conductivities have been obtained respectively with As- and I-doping [55]. Control of wall temperature was found to be important in order to transport Hg effectively which determined the quality of epitaxial films [49]. Epitaxial $Hg_{1-x}Cd_xTe$ thin films have been grown on sapphire coated with CdTe by MOCVD [57]. Lattice mismatch was minimized by converting the surface of CdTe layer to $Hg_{1-x}Cd_xTe$ by isothermal VPE. Later liquid phase epitaxy was utilized to grow $Hg_{1-x}Cd_xTe$ epitaxial layers.

$Hg_{1-x}Cd_xTe$ thin films with the Cd content x greater than 0.6 have been prepared by nonaqueous electrodeposition on indium-tin oxide transparent conducting windows [58]. Solar cells have been prepared employing $Hg_{1-x}Cd_xTe$ thin-films (with x ≈ 0.9) deposited by electrodeposition [59-61].

Polycrystalline $Hg_{1-x}Cd_xTe$ thin films have been deposited by triode sputtering in a mercury plasma from a cooled $Hg_{1-x}Cd_xTe$ target on CdTe or $Hg_{1-x}Cd_xTe$ substrates heated to 50-250° C in a vacuum system pumped with mercury diffusion pump [62-64]. $Hg_{1-x}Cd_xTe$ thin films were doped either p- or n-type by cosputtering from either a gold or an aluminum target. Both p-on n and n-on-p junction photodiodes were formed by sputtering doped $Hg_{1-x}Cd_xTe$ films onto substrates of the opposite conductivity type. The figure of merit R_oA_j comprising of the product of the photodiode dynamic resistance at zero-bias voltage R_o multiplied by the junction area A_j of the sputtered p-on-n photodiode with cutoff wavelength of 12.0 μm was found to be 1.8 Ω cm^2 which compared favorably with those obtained in implanted junctions [62]. Epitaxial $Hg_{1-x}Cd_xTe$ thin films have been deposited by dual ion beam sputtering on to heated CdTe single crystals [65-66]. Supplementary Hg was provided by floating $Hg_{1-x}Cd_xTe$ pieces in a mercury pool which was subsequently frozen. The combined target from liquid-nitrogen cooled solid $Hg_{1-x}Cd_xTe$ target was used for sputter deposition. In a few cases second Hg target was also sputtered to obtain supplementary Hg. Substrate temperatures over 100° C resulted mostly in polycrystalline films while epitaxial growth was observed for substrate temperatures below 100° C. If the target is not cooled to 77 K, excessive outdiffusion of mercury to the surface results in mostly mercury being sputtered from the target until the entire target is depleted in mercury. Because of this, sputter-deposition of $Hg_{1-x}Cd_xTe$ thin films has not been very successful. Laser-assisted evaporation has also been used to deposit $Hg_{1-x}Cd_xTe$ thin films [67].

Significant advances have been made in molecular beam epitaxy for producing $Hg_{1-x}Cd_xTe$ infrared arrays [68-82]. MBE growth at temperatures of 160-220° C (usually 180° C), improves the control of stoichiometry in grown layers. Cd content x has been varied over a wide (0.15-0.65) range. Composition and thickness control across 2 inch wafers, as well as composition uniformity across the thickness have been achieved. Substrate surface temperature changes during epilayer MBE growth have been controlled to improve the crystallinity [82]. Usually tellurium evaporates in the form of Te_2 molecules. Atomic tellurium from cracker effusion cell can enhance the sticking coefficient of Hg thus minimizing Hg vacancies. This helps in directing the extrinsic As dopant to group VI substitutional sites. Highly doped p-type $Hg_{1-x}Cd_xTe$ thin films have also been obtained by photoassisted MBE [83].

$Hg_{1-x}Cd_xTe$ thin films have been deposited in modified hot wall apparatus by coevaporation of CdTe and HgTe on to a heated substrate [84]. Mercury loss was practically eliminated by closing the quartz hot wall chamber with a conical quartz plug lubricated with graphite to improve its tightness. Small grain (0.1 μm) $Hg_{1-x}Cd_xTe$ thin films have been deposited by vacuum evaporation on to substrates heated to 100° C [85].

The challenge in the preparation of $Hg_{1-x}Cd_xTe$ is to provide excess mercury incidence rate, to optimize the deposition parameters for enhanced mercury incorporation, and at the same time to achieve the requisite stoichiometry, grain size, and doping. The sticking coefficient of mercury may be increased with atomic tellurium, photoassisted deposition, or plasma-assisted deposition. MBE and MOCVD techniques have paved the way for obtaining epitaxial $Hg_{1-x}Cd_xTe$ thin films at substrate temperatures of ~180° C with the desired crystalline perfection, stoichiometry, and doping without the necessity of further annealing either for improving the crystalline quality or dopant activity. Retaining larger mercury proportions during annealing would require heated enclosures as in isothermal VPE, hot-wall technique, vacuum evaporation, hot-wall MOCVD, or close-space sublimation. Cd content x in $Hg_{1-x}Cd_xTe$ would have to be adjusted at 0.45 and 0.65 to obtain the necessary bandgaps of 0.59 eV and 0.75 eV respectively. The actual Cd content would be higher because of a small nonlinearity in the variation of bandgap with the Cd content. Mercury metal has been employed as one of the precursors in MOCVD and MBE. Mercury telluride has less stringent health and safety requirements than mercury metal. Hence if possible, it would be better to substitute mercury metal by mercury telluride. Even though isothermal VPE, LPE, and MOCVD may not be suitable for deposition of TPV cells, some of the features of these techniques would be useful in the development of an appropriate technique.

The solar cell heterostructure of polycrystalline thin film solar cells is well-suited for TPV cells. For this structure, it is mainly essential to choose an appropriate technique for the deposition of preferably a p-type absorber on a conducting or metallized substrate. The TPV cell fabrication can then be completed by low-temperature depositions of an n-type heterojunction partner layer such as CdS by chemical bath deposition, an n-type transparent-conducting coating such as Cd_2SnO_4 or ZnO:Al by magnetron sputtering, and front-contact grids by evaporation through mechanical masks without degrading the absorber layer. CdS layers have been used even for passivation of $Hg_{1-x}Cd_xTe$ thin films [21]. Over 10% efficient CdTe thin film solar cells have been prepared with vacuum evaporated CdTe thin films [86]. Vacuum evaporation combined with hot-wall technique can satisfy the criteria of excess mercury incidence rate, enhanced mercury incorporation using atomic tellurium from cracker cell, stoichiometry, grain size, and doping. It is also amenable to scale-up for large scale manufacture. It would, therefore, appropriate to carry out deposition of $Hg_{1-x}Cd_xTe$ on to substrates heated to low temperatures (100-180° C) in heated enclosures by vacuum coevaporation from spaced and concentric sources. Dopants such as Cu can either be predeposited on the substrate for subsequent outdiffusion into the film during the deposition or may be coevaporated during deposition. $Hg_{1-x}Cd_xTe$ thin films deposited by MBE and MOCVD on heated substrates do not require post-annealing treatment. Hence it is expected that proper control of substrate temperatures and incident evaporant species will be sufficient in obtaining the desired stoichiometry and structural and semiconducting properties. However, if any annealing is found essential to improve crystallinity or dopant activity, it would be possible to carry it out in an evacuated and back-filled, sealed, graphite enclosure, similar to the close-space sublimation set-up using $Hg_{1-x}Cd_xTe$ charge to minimize the loss of Hg.

Comparatively little work has been done on $Pb_{1-x}Cd_xTe$ polycrystalline thin films because of the lack of technological impetus similar to $Hg_{1-x}Cd_xTe$ infrared detectors. Diffusion of Cd in PbTe has been studied by interdiffusion of CdTe/PbTe layers [87]. $Pb_{1-x}Cd_xTe$ thin films can be prepared by magnetron sputtering from cooled $Pb_{1-x}Cd_xTe$ targets on heated substrates (~300° C). PbTe thin films have been deposited by vacuum evaporation with or without hot-wall enclosures, the requirements being less stringent [88-91]. Hot-wall technique also seems to be suitable for the deposition of $Pb_{1-x}Cd_xTe$ thin films. Hence it would be appropriate to prepare $Pb_{1-x}Cd_xTe$ thin films by magnetron sputtering and hot-wall technique by co-evaporation of PbTe and CdTe from spaced and concentric sources. The process would have fewer problems of loss of Cd or Pb and hence the desired stoichiometries and doping should be obtained. Post-deposition annealing for grain-growth and dopant activation can be carried out using processes similar to those already developed for CdTe solar cells.

Single Crystal Wafers to Polycrystalline Thin Films

A natural progress has occurred from single crystal wafers to polycrystalline thin films in the preparation of solar cells. Solar cells are being fabricated employing semicrystalline silicon wafers cut from silicon blocks prepared by condensation of melt and also thin-film silicon deposited on insulating substrates solar cells. Their photovoltaic conversion efficiencies are comparable to those fabricated using single-crystal materials. Early CdTe and CuInSe$_2$ solar cells were based on single crystal materials. At present the highest efficiency solar cells based on these materials are being prepared with polycrystalline thin films. The superior performance of polycrystalline thin films results from improved understanding of the grain boundary passivation and alloying. Moreover, monolythic interconnection schemes are easier to implement in thin films deposited on insulated substrates.

TPV HEAT SOURCE AND EMITTER

A wide range of heat sources employing fossil fuels, biomass, and even solar radiation can be utilized for TPV applications. At present, the energy source for majority of energy conversion applications is a fossil-fueled flame because of the domestic availability of clean-burning natural gas and high density liquid hydrocarbon fuels. Very little luminous output is generated when most of these fuels are burned in efficient aerated flames [91]. Hence it is essential to position a stable emissive structure in the flame to enhance the radiant intensity or generate selective emission. Two of the most promising system configurations are filtered broadband emitter and a selective emitter. In the first, energy close to the bandgap of TPV cells radiated from a silicon carbide or metallic emitter heated to 900-1500° C is used [91-94]. The remaining radiant energy is reflected back to the emitter and readmitted as useful energy. The broadband emitters are capable of producing relatively high output power densities. However, losses in filters, cells, and reabsorption/emission of the radiant energy limit the overall system efficiencies. In the selective emitters, thermal energy is converted to radiation in a useful wavelength range with a rare-earth oxide emitter heated to 1050-1500° C. The available power densities from selective emitters are known to be low. However, the advantage for selective emitters lies in the fact that less fuel would be needed to achieve a given temperature thereby increasing the overall efficiency [95]. In a simple experiment, with the same geometry a selective line and composite silicon carbide emitters reached temperatures of 1420 K and 1070 K respectively [95].

Rare earth elements are usually trivalent while having different atomic numbers. The vacancies in the inner electronic shells permit electronic transitions in the visible and near infrared spectrum. Because of the shielding properties of valence electrons, the emissions can be very selective with the full width at half maximum of the order of 300 nm. Hence selective emitters such as ytterbia, erbia, holmia, and neodymia are available with emissions at 0.98 μm, 1.55 μm, 2.0 μm, and 2.4 μm, making them suitable for solar cells based on Si, Ge, GaSb, $In_{1-x}Ga_xAs$, $Cd_{1-x}Hg_xTe$, and $Cd_{1-x}Pb_xTe$, etc. Mostly fibrous ceramic emitters are being used at present. The spectral emittance of typical emitter consists of i) lattice vibration region at long wavelengths (6-10 μm and beyond), ii) absorption edge at short wavelengths, and iii) intermediate free-charge-carrier region [92]. Fossil-fuel flame temperatures are too low to excite electronic transitions from the valence to the conduction band in oxide ceramics with bandgap of 3-5 eV. Hence the intensity of short-wavelength radiation is low. The undesirable long-wavelength radiation due to lattice vibration is usually unavoidable. Luminous efficiency of 2 lumens watt[-1] was achieved by the Welsbach mantle consisting of a refractory ceramic filamentary mesh of a mixture of thorium and cerium oxides in the last century. Addition of ceria impurity moves the absorption edge of thoria from UV to visible to yield maximum luminous intensity. The emittance of Welsbach mantle in the intermediate free carrier absorption region is reduced considerably below the emittance of 0.4 of a monolithic thoria block because of the small (~10 μm) diameter of ceramic fibers which makes it very transparent in the infrared region [92].

Gas mantles are excellent thermal stress-tolerant structures even though they are not mechanically durable. The narrow section of a 10 μm diameter fiber is too small to develop thermal stress while the stress developed along the axis of the fiber is relieved by flexing. High convective heat transfer and lack of insulating boundary layers maintain the fibers in thermal equilibrium with the flame. Hence near highest realizable temperatures and consequently high radiant power can be achieved very rapidly. Mantles are prepared with the aid of a textile precursor e.g. rayon, impregnated with aqueous metal salt which are first converted to their hydroxides and is then ignited inside the camping lantern to convert the hydroxides to their oxide and to pyrolize the rayon. Escaping gases cause rifts (microcracks) and ruptures which cannot be sintered and which act as stress raisers that readily propagate when the fiber is mechanically stressed and cause fracture far below the intrinsic strength of the ceramic. Hence commercially available gas lighting mantles are extremely fragile. ThermoLyte Corporation has developed a process for ceramic fiber fabrication with reduced density and severity of flaws by carefully regulating the pyrolysis of metal salts to form oxides, and by allowing decomposition gases to escape without significant damage [92,93]. In this process, the textile precursor imbibed with meal salt (preferably nitrate) solution is dried and subjected to temperature ramps and soak cycles in carefully controlled gas ambients. ThermoLyte mantles can, therefore, survive two orders of magnitude greater mechanical impacts as compared to commercial mantles. It would be possible to develop a selective thermal emitter, using a ThermoLyte cylindrical selective erbia prototype emitter in conjunction with a simple, readily available, commercial Coleman camping lantern having a total thermal output of ~1,000 watts. The lantern is expected to provide ~250 W of thermal energy to the emitter, the rest of the heat being convective loss, part of which could be recuperated. It is expected that the emitter temperature will be raised to 1500-1525° C and that the line emission centered at 1.55 μm would be radiated with a total energy of ~100 W. Two concentric quartz cylinder may be used. The first would serve mainly to isolate the cells from combustion products while both would reflect long wavelength IR radiation back to the emitter and thus would minimize temperature rise of TPV cells. Cooled TPV cells can be mounted around on a third outer concentric cylinder to capture and convert IR radiation into electric power.

Another interesting approach consists of a durable powder metal selective emitter surface being developed by Thermacore Inc [96]. It would be possible to develop TPV source using a durable Thermacore holmia-based powder metal selective emitter with line emission centered at 2.0 μm configured to maximize both the efficiency and output power density of a TPV system operating at 1100° C. A durable and rugged holmia-based powder metal selective emitter can withstand thermal and mechanical shock and isolate combustion products. Simple joining techniques would permit evacuation of the volume between the emitter and the cells. It would consist of a metallic superalloy base, a layer of sintered 200-mesh nickel powder, and a layer of rare-earth oxide, holmia. The superalloy base of inconel 800 or Hayes alloy A214 can provide mechanical strength, ductility, and resistance to oxidation and flames while the rare-earth-oxide layer would provide the selective emittance. The intermediate nickel layer would absorb the thermal-expansion mismatch, promote adhesion, and minimize broadband emission. Radiant energy from a central cylindrical holmia-based powder metal selective emitter heated with a flame burner could be concentrated by parabolic reflectors on to narrow arrays of TPV cells mounted on a surrounding porous metal heat exchanger.

COST OF TPV CELLS

The most costly items in a TPV system are the emitter and the cells. Majority of the present TPV systems are utilized in small-scale aerospace and military applications. The total power output of the systems deployed at present is small. Hence the cost of TPV systems and the cells is not a major issue. It would be necessary to reduce the cost to the range of US$ 2-5 per watt so as to be competitive in small to medium size commercial applications [10]. The technologies being developed at present may be able to reach this goal. However, the cost will have to be reduced to the range of US ¢ 35-$ 1 per watt to reach the large-scale applications in the residential and consumer markets as well as the much larger application in hybrid-electric car market [10,94]. These markets are very competitive and the consumers are cost conscious. For example, it has been estimated that the cost of a 450 W TPV generator for a self-powered gas-fired residential warm air furnace would have to be limited to US$ 75 equivalent to the cell cost of ~US ¢ 17 per watt [94]. Energy efficient electric cars with an efficient fossil-fuel electric generator operating at a constant output supplemented by a TPV generator to provide the necessary excess energy for overtaking and climbing on steep slopes can also become economic only with cheaper TPV cells. It will not be possible to achieve these goals with the TPV cells based on single crystal materials. Polycrystalline thin-film $Hg_{1-x}Cd_xTe$ and $Pb_{1-x}Cd_xTe$ TPV cells have great potential to achieve such a cost reduction making them suitable for economic, large scale applications in cheaper, civilian and military, terrestrial and aerospace TPV applications.

PHOTOVOLTAIC MATERIALS LABORATORY

A well-equipped thin-film PV Materials Laboratory is available at the Florida Solar Energy Center for a TPV project. The laboratory has a semi-clean facility with double-door entry, systems for preparation of thin films by sputtering, vacuum evaporation, selenization, chemical bath deposition, and auxiliary support facilities. At present $CuIn_{1-x}Ga_xSe_2$ and CdTe polycrystalline thin-film solar cells are being developed using processes amenable to large-scale manufacture, under a project entitled "$CuIn_{1-x}Ga_xSe_2$ and CdTe polycrystalline thin-film solar cells" funded by NREL [97-99]. Durability of PV Modules is being studied with funding from the Sandia National Laboratories. Earlier, effect of lunar and orbital temperature variation on the efficiencies of single- and two-junction solar cells has been computed [100,101]. The knowledge and experience gained during the execution of these projects would be useful in the development of TPV cells and systems.

CONCLUSIONS

PV conversion efficiencies of the present $CuIn_{1-x}Ga_xSe_2$ and CdTe polycrystalline solar cells are in the range 15.8-17.7%. These values are approximately two-thirds of the single-crystal cell efficiencies and are expected to improve further in the future. The best recombination-state density N_t is in the acceptable range of 10^{-15}-10^{-16} cm^{-3} for TPV application for satisfying both the diffusion length and intrinsic junction rectification criteria.

$Hg_{1-x}Cd_xTe$ and $Pb_{1-x}Cd_xTe$, etc polycrystalline-thin films have great potential for economic applications in cheaper, civilian, terrestrial TPV applications. It would be possible to reduce the cost of TPV technologies based on single-crystal materials being developed at present to the range of US$ 2-5 per watt so as to be competitive in small to medium size commercial applications. However, these technologies would not be able to provide a further cost reduction to the range of US ¢ 35-$ 1 per watt to reach the more competitive large-scale residential, consumer, and hybrid-electric car markets. That would be possible only with the polycrystalline-thin film TPV cells.

ACKNOWLEDGMENTS

This work was supported by the National Renewable Energy Laboratory Contract # XG-2-11036-5 and the Sandia National Laboratories Contract # AP-7660. The author is thankful to Late Dr. Cynthia Carter, D.O.E. Division of Advanced Energy Projects, for encouragement, and to Dr. Tim Coutts of NREL, Dr. Robert E. Nelson of Quantum Group Inc and Dr. David Sarraf, of Thermacore Inc for useful discussions.

REFERENCES

1. N. G. Dhere, Polycrystalline-Thin-Film Thermophotovoltaic Cells, AIP Proc. 2nd NREL Conference on Thermophotovoltaic Generation of Electricity, Colorado Springs, CO, 409-422, (1995).
2. S. Wojtczuk, E. Gagnon, L. Geoffroy and T. Parodos, $In_xGa_{1-x}As$ Thermophotovoltaic Cell Performance vs. Bandgap, AIP Proc. 1st NREL Conference on Thermophotovoltaic Generation of Electricity, Copper Mountain, CO, 177-187, (1994).
3. H. Ehsani, I. Bhat, D. Marcy, G. Nichols, J. Borrego, J. Parrington,and R. Gutmann, OMVPE Growth and Characterization of InGaAs for TPV Cells, AIP Proc. 1st NREL Conference on Thermophotovoltaic Generation of Electricity, Copper Mountain, CO, 188-193, (1994).
4. L. Fraas, R. Ballantyne, J. Samaras,and M. Seal, Electric Power Production Using New GaSb Photovoltaic Cells with Extended Infrared Response, AIP Proc. 1st NREL Conference on Thermophotovoltaic Generation of Electricity, Copper Mountain, CO, 44-53, (1994).
5. R. K. Jain, D. M. Wilt, G. A. Landis, R. Jain, I. Weinberg,and D. J. Flood, AIP Proc. 1st NREL Conference on Thermophotovoltaic Generation of Electricity, Copper Mountain, CO, 202-209, (1994).
6. D. M. Wilt, N. S. Fatemi, R. W. Hoffmann, Jr., P. P. Jenkins, D. Scheiman, R. Lowe, and G. A. Landis, InGaAs PV Device Development for TPV Power Systems, AIP Proc. 1st NREL Conference on Thermophotovoltaic Generation of Electricity, Copper Mountain, CO, 210-220, (1994).
7. J. B. McNeely, M. G. Mauk, and L. C. DiNetta, An InGaAsSb/GaSb Photovoltaic

Cell by LiquidPhase Epitaxy for Thermophotovoltaic (TVP) Application, AIP Proc. 1st NREL Conference on Thermophotovoltaic Generation of Electricity, Copper Mountain, CO, 221-225, (1994).

8. R. K. Jain, D. M. Wilt, R. Jain, G. A. Landis, and D. J. Flood, Lattice-Matched and strained InGaAs Solar Cells for Thermophotovoltaic Use, 2nd NREL Conference on Thermophotovoltaic Generation of Electricity, Colorado Springs, CO, 375-386, (1995).

9. H. Ehsani, I. Bhat, C. Hitchcock, J. Borrego, and R. Gutmann, Characteristics of GaSb and GaInSb layers Grown by metalorganic Vapor phase Epitaxy, 2nd NREL Conference on Thermophotovoltaic Generation of Electricity, Colorado Springs, CO, 423-433, (1995).

10. L. J. Ostrowski, U. C. Pernisz, L. M. and Fraas, "Thermophotovoltaic Energy Conversion: Technology and Market Potential", AIP Proc. 2nd NREL Conference on Thermophotovoltaic Generation of Electricity, Colorado Springs, CO, July 17-19, 251-260, (1995).

11. H. Ullal, K. Zweibel, and B. G. Von Roedern, "Thin-Film CdTe and CuInSe2 Photovoltaic Technologies" Proc. ISES Solar World Congress, Budapest, Hungary, Aug. 23-27, 187-193, (1993).

12. N. G. Dhere, S. Kuttath, K. W. Lynn, R. W. Birkmire, and W. N. Shafarman, "Polycrystalline $CuIn_{1-x}Ga_xSe_2$ Thin Film PV Solar Cells Prepared by Two-stage Selenization Process Using Se Vapor" Proc. IEEE First World Conf. Photovoltaic Energy Conversion, Waikoloa, Hawaii, 190-193, (1994).

13. J. R. Tuttle, J. S. Ward, A. Duda, T. A. Berens, M. A. Contreras, K. R. Ramanathan, A. L. Tennant, J. Keane, E. D. Cole, K. Emery, and R. Noufi, "The Performance of $Cu(In,Ga)Se_2$-Based Solar Cells in Conventional and Concentrator Applications", Mat. Res. Soc. Symp. Proc. Vol. 426, 143-151, (1996).

14. C. Ferekides, J. Britt, and L. Killian, "High Efficiency CdTe Solar Cells by Close Spaced Sublimation", Proc. 23rd IEEE Photovoltaic Specialists' Conference, Louisville, KY, 389-393, (1993).

15. L. Skarp, E. Antilla, A. Rautiainen, and T. Suntola, "ALE-CdS/CdTe-PV-Cells", Int. J. Solar Energy, 12, 137-142, (1992).

16. L. M. Fraas, Current Topics in Photovoltaics, T. J. Coutts and J. D. Meakin (eds.), Academic Press, 168-221, (1985).

17. J. R. Sites and X. Liu, "Six-year Efficiency Gains for CdTe and $CuIn_{1-x}Ga_xSe_2$ Solar Cells: What Has Changed?", Proc. IEEE First World Conf. Photovoltaic Energy Conversion, Waikoloa, Hawaii, 119-122, (1994).

18. I. L. Eisgruber and J. R. Sites, "Status of Polycrystalline Thin Film Solar Cells", AIP Conf. Proc. 12th NREL Photovoltaic Program Review, Denver, CO, 407-413, (1993).

19. P. H. Mauk, H. Tavakolian, and J. R. Sites, "Interpretation of Thin-Film Polycrystalline Solar Cell Capacitance", IEEE Trans. Electron Devices, 37, 422-427, (1990).

20. X. Wu, W. P. Mulligan, J. D. Webb, and T. J. Coutts, TPV Plasma Filters Based

on Cadmium Stannate, AIP Proc. 2nd NREL Conference on Thermophotovoltaic Generation of Electricity, Colorado Springs, CO, July 17-19, 329-338, (1995).

21. J. P. Ziegler, J. M. Lindquist, J. Hemminger, Passivation of HgCdTe with CdS thin films: correlation of device characteristics with surface spectroscopy, J. Appl. Phys., 65, 2523-2529, (1989).

22. P. A. Iles, C. Chu, and E. Linder, "The influence of bandgap on TPV convertor efficiency", AIP Proc. 2nd NREL Conference on Thermophotovoltaic Generation of Electricity, Colorado Springs, CO, 446-457, (1995).

23. J. L. Gray and A. El-Husseini, A simple parametric study of TPV system efficiency and output power density including comparison of several TPV materials, AIP Proc. 2nd NREL Conference on Thermophotovoltaic Generation of Electricity, Colorado Springs, CO, 3-15, (1995).

24. N. G. Dhere, Recent Developments in Thin-Film Solar Cells, Thin Solid Films 194, 757, (1990).

25. N. G. Dhere, "CuIn$_{1-x}$Ga$_x$Se$_2$ and CdTe PV Solar Cells" AIP Conference Proceeding: 13th NREL Photovoltaic Program Review Meeting, Lakewood, CO, 428, (1995).

26. J. A. Van Vechten, Handbook of Semiconductors, (S. P. Keller, ed.), North Holland, Amsterdam, Vol. 3, Chap. 1, (1980).

27. T. Nguyen Duy and D. Lorans, Highlights of recent results on HgCdTe thin film photoconductors, Semiconductors Sci. & Technol., 6, C93-C95, (1991).

28. R. K. Willardson and A. C. Beer (eds.), Semiconductors and Semimetals, Academic Press, Vol. 12, (1977).

29. R. K. Willardson and A. C. Beer (eds.), Semiconductors and Semimetals, Academic Press, Vol. 18, (1981).

30. S. H. Wei and A. Zunger, Electronic Structure and Stability of II-VI Semiconductors and Their Alloys, J. Vac. Sci. Technol., A6, 2597-2611, (1988).

31. S. H. Wei and A. Zunger, Electronic and Structural Anomalies in Lead Chalcogenides, (Unpublished).

32. G. Cohen-Solal and Y. Marfaing, C.R. Acad. Sc., 260, 4190-4193, (1965).

33. G. Cohen-Solal, Y. Marfaing and F. Bailly, Croissance Epitaxique De Composes Semiconducteurs Par Evaporation-Diffusion En Regime Isotherme, Revue De Physique Appliquee, 1, 11-17, (mars 1966).

34. Y. Marfaing, G. Cohen-Solal and F. Bailly, A New Process of Crystal Growth: Evaporation Diffusion Under Isothermal Conditions, J. Phys. & Chem. Solids, 1, 549-553, (1967).

35. C. Sella and G. Cohen-Solal, Comptes Rendus des Seances de l'Academie des sciences, C.R. Acad. Sc. Paris, 264, 179-182, (1967).

36. O. N. Tufte and E. L. Stelzer, Growth Properties of HgCdTe Epitaxial Layers, J. Appl. Phys., 40, 4559-4568, (1969).

37. G. Cohen-Solal and Y. Riant, Epitaxial (CdHg)Te Infrared Photovoltaic Detectors, Appl. Phys. Lett., 19, 436-438, (1971).

38. T. Nguyen-Duy, J. C. Morland, and G. Cohen-Solal, HgCdTe Photodiode

Mosaics on Vapor Phase Epitaxial Layers, Proc. IEEE Electron Devices Meeting, Washington, 491-495, (1980).

39. M. Isshiki, Y. Maruyama, and S. Tobe, Isothermal Vapor Phase Epitaxy of $Hg_{1-x}(Cd_{1-y}Zn_y)_xTe$ Thin Films Using Semiclosed Open-Tube System, Jpn. J. Appl. Phys. 31, 1842-1844, (1992).

40. A. B. Trigubo and N.E. Walsoe De Reca, Orientation dependence of morphology of (Hg,Cd)Te films grown by isothermal vapor phase epitaxy, Materials Sci. & Eng., B27 , 87-91, (1994).

41. Z. Djuric, Isothermal vapour-phase epitaxy of mercury-cadmium telluride (Hg, Cd)Te, J. Materials Sci., 6, 187-218, (August 1995).

42. B. Pelliciari, State of the Art LPE HgCdTe AT LIR, J. Crystal Growth, 86, 146-160, (1988).

43. M. G. Astles, N. Shaw, and G. Blackmore, Techniques for improving the control of properties of liquid phase epitaxial (CdHg)Te, Semiconductor Sci. & Technol., 8, s211-s215, (1993).

44. S. P. Tobin, G.N . Pultz, E. E. Krueger, M. Kestigian, K. K. Wong, and P. W. Norton, Hall Effect characterization of LPE HgCdTe P/n heterojunctions, J. Electronic Materials, 22, 907-914, (1993).

45. H. R. Vydyanath, Status of Te-rich and Hg-rich liquid phase epitaxial technologies for the growth of (Hg, Cd)Te alloys, J. Electronic Materials, 24, 1275-1285, (1995).

46. F. T. Smith, P. W. Norton, P. Lo Vecchio, N. Hartle, M. Weiler, N. H. Karam, S. Sivananthan, and Y. P. Chen, Te-rich liquid phase epitaxial growth of HgCdTe on Si-based substrates, J. Electronic Materials, 24, 1287-1292, (1995).

47. K. P. Möllmann and H. Kissel, Optical Absorption of thin $Hg_{1-x}Cd_xTe$ epitaxial layers, Semiconductors Sci. Technol., 6, 1176-1169, (1991).

48. R. Korenstein, W. E. Hoke, P. J. Lemonias, K. T. Higa, and D. C. Harris, Metalorganic growth of HgTe and CdTe low temperatures using diallyltelluride, J. Appl. Phys. 62, 4929-4931, (1987).

49. S. Oda, Y. Tanaka, O. Sugiura, and M. Matsumura, Heteroepitaxial growth of HgTe on InSb at 200 C by Metalorganic chemical vapor deposition using ditertiarybutyltelluride, J. Appl. Phys., 65, 1808-1810, (1989).

50. R. Koenstien, P. Hallock, B. MacLeod, W. Hoke, and S. Oguz, The influence of crystallographic orientation on gallium incorporation in HgCdTe grown by metalorganic chemical vapor deposition on GaAs, J. Sci. & Technol. A 8, 1039-1044, (1990).

51. D. D. Edwall, J. Bajaj, E. R. Gertner, Material characteristics of metalorganic chemical vapor deposition $Hg_{1-x}Cd_xTe$/GaAs/Si, J. Sci. & Technol., A 8, 1045-1048, (1990).

52. I. B. Bhat, H. Ehsani, and S. K. Ghandhi, The growth and characterization of HgTe and HgCdTe using methylalylltelluride, J. Vac. Sci. & Technol., A 8, 1054-1057, (1990).

53. S. R. Hahn, J. D. Parson, Direct Growth of $Hg_{1-x}Cd_xTe$ alloys in a cold-wall,

annular reactant inlet inverted-vertical organometallic vapor phase epitaxy reactor at 300o C, J. Crystal Growth, 134, 90-96, (1993).

54. K. Shigenaka, L. Sugiura, F. Nakata, and K. Hirahara, Effects of growth rate and mercury partial pressure on twin formation in HgCdTe (111) layers grown by metalorganic chemical vapor deposition, J. Electronic Materials, 22, 865-871, (1993).

55. C. T. Elliott, N. T. Gordon, R. S. Hall, T. J. Phillips, C. L. Jones, B. E. Matthews, C. D. Maxey, and N. E. Metcalfe, Metal organic vapor phase epitaxy (MOVPE) grown heterojunction diodes in $Hg_{1-x}Cd_xTe$, Proceedings of SPIE, 2269, 648-657, (1994).

56. P. Mitra, T. R. Schimert, F. C. Case, R. Starr, M. H. Weiler, M. Kestigian, and M. B. Reine, Metalorganic chemical vapor deposition of HgCdTe for photodiode applications, J. Electronic Materials, 24, 661-668, (1995).

57. K. Ozaki, K. Fujiwara, H. Ebe, and Y. Miyamoto, Epitaxial growth of HgCdTe on sapphire for photovoltaic detectors, Proceedings of SPIE - The International Soc. for Optical Eng., 2269, 384-396, (1994).

58. C. L. Colyer and M. Cocivera, Thin-Film Cadmium Mercury Telluride prepared by nonaqueous electrodeposition, J. Electrochem. Soc., 139, 406-409, (1992).

59. B. M. Basol, Electrodeposited CdTe and HgCdTe Solar Cells, Solar Cells, 23, 69-88, (1988).

60. B. Basol and E. Tseng, Mercury cadmium telluride solar cell with 10.6% efficiency, Appl. Physics Lett., 48, 946-948, (1986).

61. B. Basol, E. Tseng, and O. Stafsudd, Mercury cadmium telluride thin film solar cells, IEEE 1745- 1746, (1985).

62. G. Cohen-Solal, A. Zozime, C. Motte, and Y. Riant, Sputtered Mercury Cadmium Telluride Photodiode, Infrared Physics, 16, 555-559, (1976).

63. G. Cohen-Solal, C. Sella, D. Imhoff, and A. Zozime, Jpn J. Appl. Phys., Suppl. 2, Part 1, 517, (1974).

64. A. Zozime, H. Drappier, C. Sella, M. Chaveau, and G. Cohen-Solal, Dopage in situ de couche minces de $Cd_xHg_{1-x}Te$ obtenues par pulvérisation catodique en atmosphère de mercure application à la réalisation de détecteurs d'infra-rouges, Le Vide, 175, 19-22, (1975).

65. S. V. Krishnaswamy, J. H. Rieger, N. J. Doyle, and M. H. Francombe, Ion-beam sputter deposition and epitaxy of CdTe and HgCdTe films, J. Vac. Sci. & Technol., A 5, 2106-2110, (1987).

66. S. V. Krishnaswamy, J. H. Rieger, N. J. Doyle, and M. H. Francombe, Dual ion beam sputter deposition of CdTe, HgTe, and HgCdTe films, MRS Soc. Symp. Proc. 90, 463, (1987).

67. S. V. Plyatsko, Y. S. Gromovoj, G. E. Kostyunin, F. F. Sizov, and V. P. Klad'ko, Laser-assisted evaporation of high-quality narrow-gap thin films, Thin Solid Films, 221, 127-131, (1992).

68. O. K. Wu and G. S. Kamath, An overview of HgCdTe MBE technology, Semiconductor Science and Technology, 6, 6-9, (1991).

69. J. M. Arias, J. G. Pasko, M. Zandian, J. Bajaj, L. J. Kozlowski, R. E. DeWames,

and W. E. Tennant, Molecular beam epitaxy (MBE) HgCdTe flexible growth technology for the manufacturing of infrared photovoltaic detectors, Proceedings of SPIE - The international Society for Optical Engineering, 2228, (1994).

70. Jean-Pierre Faurie, Molecular beam epitaxy of $Hg_{1-x}Cd_xTe$: growth and characterization, Prog. Crystal Growth and Characterization of Materials, 29, 85-159, (1994).

71. T. Sasaki and N. Oda, Highly producible HgCdTe molecular beam epitaxy growth technique using radiational substrate heating, Proc. SPIE, 2228, 86-95, (1994).

72. O. K. Wu, D. M. Jamba, G. S. Kamath, G. R. Chapman, S. M. Johnson, J. M. Peterson, K. Kosai, and C. A. Cockrum, HgCdTe molecular beam epitaxy technology. A focus on material properties, J. Electronic Materials, 24, 423-429, (1995).

73. H. R. Vydyanath, L. S. Lichtmann, S. Sivanathan, P. S. Wijewarnasuriya, and J. P. Faurie, Annealing experiments in heavily arsenic-doped (Hg,Cd)Te, J. Electronic Materials, 24, 625-634, (1995).

74. L. O. Bubulac, D. D. Edwall, S. J. C. Irvine, E. R. Gertner, and S. H. Shin, P-type doping of double layer mercury cadmium telluride for junction formation, J. Electronic Materials, 24, 617-624, (1995).

75. J. Bajaj, J. M. Arias, M. Zandian, J. G. Pasko, L. J. Kozlowski, R. E. De Wames, and W. E. Tennant, Molecular beam epitaxial HgCdTe material characteristics and device performance: Reproducibility status, J. Electronic Materials, 24, 1067-1076 (1995).

76. T. Sasaki, M. Tomono, and N. Oda, Crystallinity improvement of HgCdTe on GaAs by molecular beam epitaxy, J. Crystal Growth, 150, 785-789, (1995).

77. J. M. Arias, M. Zandian, J. Bajaj, J. G. Pasko, L. O. Bubulac, S. H. Shin, and R. E. De Wames, Molecular beam epitaxy HgCdTe growth-induced void defects and their effect on infrared photodiodes, J. Electronic Materials, 24, 521-524, (1995).

78. J. W. Han, S. Hwang, Y. Lansari, R. L. Harper, and Z. Yang, Modulation-doped HgCdTe, J. Vac. Sci. & Technol., A 7, 305-310, (1989).

79. M. Boukerche, S. Sivanathan, P. S. Wijewarnasuriya, and I. K. Sou, Electrical properties of intrinsic p-type shallow levels in HgCdTe grown by molecular-beam epitaxy in the (111) B orientation, J. Vac. Sci. & Technol., A 7, 311-313, (1989).

80. B. Peliciari, G. L. Destefanis, and L. DiCioccio, Evidence of anomalous behavior in low-n-type mercury cadmium telluride induced by extended defects, J. Vac. Sci. & Technol., A 7, 314-320, (1989).

81. R. Beam, K. Zianio, and J. Ziegler, CdTe/GaAs/Si substrates for HgCdTe photovoltaic detectors, J. Vac. Sci. & Technol. A 7, 343-347, (1989).

82. D. Rajavel, F. Mueller, J. D. Benson, B. K. Wagner, R. G. Benz II, and C. J. Summers, In situ growth surface temperature measurements for molecular beam epitaxial growth of CdTe, ZnTe, and $Cd_{1-x}Zn_xTe$ alloys. J. Sci. & Technol., A 8, 1002-1005, (1990).

83. K. A. Harris, T. H. Myers, R. W. Yanka, and L. M. Mohnkern, Microstructural

defect reduction in HgCdTe grown by photoassisted molecular-beam epitaxy. J. Sci. & Technol., A 8, 1013-1019, (1990).

84. A. Rogalski, J. Piotrowski, and J. Gronkowski, A modified hot wall epitaxy for the growth of CdTe and $Hg_{1-x}Cd_xTe$ epitaxial layers, Thin Solid Films, 191, 239-245, (1990).

85. F. F. Wu, W. Song, R. Jiang, J. Yan, and Z. Liu, Properties of the HgCdTe films, Proceedings of the SPIE, 2274, 76-80, (1994).

86. B. E. McCandless, Y. Qu, and R. W. Birkmire, A treatment to allow contacting CdTe with different conductors, Proc. 1994 IEEE First World Conf. PV Energy Conversion, Waikoloa, Hawaii, 107-110, (1994).

87. V. B. Bobruiko, T. A. Kouznetzova, M. P. Belyansky, and A. M. Gaskov, Diffusion of cadmium in lead telluride, Materials Sci. Eng. B: Solid-State Materials for Adv. Technol., B32 7-10, (1995).

88. A. B. Mandale, Transport Properties of Lead Telluride Films, Thin Solid Films, 195 15-21, (1991).

89. W. Feng and Y. Yen, Dependence of optical properties of thermal evaporated lead telluride films upon substrate temperature, Vacuum, 42, 1045-1046, (1991).

90. P. K. Parris, D. Mukherjee, and C. A. Hogarth, Electrical and Optical-properties of vacuum-deposited Lead-telluride layers, Physica Status Solidi A-Applied Research, 152, 461-466, (1995).

91. R. E. Nelson, "Grid-independent residential power systems", AIP Proc. 1st NREL Conference on Thermophotovoltaic Generation of Electricity, Copper Mountain, CO, 221-237, (1994).

92. R. E. Nelson, "Grid-independent residential power systems", AIP Proc. 2nd NREL Conference on Thermophotovoltaic Generation of Electricity, Colorado Springs, CO, 221-237, (1995).

93. U. C. Pernisz and C. K. Saha, Silicon Carbide emitter and burner for a TPV convertor, AIP Proc. 1st NREL Conference on Thermophotovoltaic Generation of Electricity, Copper Mountain, CO, 99-105, (1994).

94. L. M. Fraas, L. Ferguson, L. G. McCoy, and U. C. Pernisz, SiC emitter design for thermophotovoltaic generators, AIP Proc. 2nd NREL Conference on Thermophotovoltaic Generation of Electricity, Colorado Springs, CO, 488-494, (1995).

95. Adair, Rose 1st NREL Conference on Thermophotovoltaic Generation of Electricity, Copper Mountain, CO, 245-, (1994).

96. D. B. Sarraf and T. S. Mayer, Design of a TPV generator with a durable selective emitter and a spectrally matched PV cell, 2nd NREL Conference on Thermophotovoltaic Generation of Electricity, Colorado Springs, CO, 98-108, (1995).

97. N. G. Dhere, S. Kuttath, and H. R. Moutinho, Morphology of precursors of $CuIn_{1-x}Ga_xSe2$ thin films prepared by two-stage selenization process, J. Sci. & Technol. A, 13, 1078-1082 (1995).

98. N. G. Dhere, D. L. Waterhouse, K. B. Sundaram, O. Melendez, N. R. Parikh, and

B. K. Patnaik, "Studies of Chemical Bath Deposited Cadmium Sulfide Films by Buffer Solution Technique", J. Mat. Sci.: Mat. in Electronics, 6, 52-59 (1995).

99. N. G. Dhere, D. L. Waterhouse, K. B. Sundaram, O. Melendez, N. R. Parikh, and B. K. Patnaik, "Solution-Grown CdS Layers for Polycrystalline-Thin-Film Solar Cells", Proc. 23rd Photovoltaic Specialists' Conference, Louisville, KY, May 10-14, 566, (1993).

100. N. G. Dhere and J. V. Santiago, "Computation of Photovoltaic Parameters under Lunar Temperature Variation", Proc. 12th Space Photovoltaic Research and Technology Conference (SPRAT XII), 298-308, (1992),

101. N. G. Dhere and J. V. Santiago, "Effect of Lunar and Orbital Temperature Variation on the Efficiency of Single- and Two-Junction Solar Cells", Solar Engineering - 1993, ASME Solar Energy Conference, Washington, DC, April 25-28, 1993, American Society for Mechanical Engineers, New York, NY, 251-256, (1993).

A TPV System Using a Gold Filter with CuInSe₂ Solar Cells

W. J. Biter*, K. A. Georg* and J. E. Phillips[†]

*Sensortex, Inc., 402 D North Mill Road, Kennett Square, PA 19348
[†]Institute of Energy Conversion, University of Delaware, DE 19716-3820

Abstract Using a combination of techniques, it is possible to produce a high efficiency, low cost thermophotovoltaic system. This approach uses a Au/dielectric bandpass filter and a conventional CuInSe₂ solar cell. The Au/dielectric filter produces a response ideally suited to the 1eV optical bandgap of the CuInSe₂ solar cell. With this filter, the transmission in the bandpass region is controlled by interference while at the longer wavelengths, the reflectivity is controlled by the high value of the complex part of the index of refraction.

The Au/dielectric filter reflects lower energy photons back to the emitter, thereby changing the effective emissivity. The reflected energy is not lost, but reabsorbed by the source. In this case, the conversion efficiency can be as high as those obtained using solar cells whose optical bandgap is matched to the source temperature.

Since the bulk of the radiated energy is at the longer wavelengths, it is essential that the reflectivity approach 100% over the reflected spectrum. The optical performance of the thin metal films of Au is very good with reflectivities approaching that of bulk Au (> 99%) in the region of interest. Such filters were fabricated and measured using both silicon and hydrogenated amorphous silicon as the matching dielectrics. The high optical index (~3.6-3.7) of these materials predicts a transmission over 80%. The results of optical transmission and reflection measurements made on these filters, from the visible through the IR, will be presented.

The system efficiency depends on the transmission in the bandpass region, the reflection in the IR, collection efficiency of the PV cell and other thermal losses. Finite element modeling of total losses, which include transmission in the bandpass region, reflection in the IR, the collection efficiency of the CuInSe₂ solar cell and other thermal losses, indicate that system efficiencies well over 10% are possible, even with emitter temperatures as low as 1200° C.

INTRODUCTION

Background

The problem in fabricating a thermophotovoltaic (TPV) generator is that most energy is radiated at longer wavelengths that cannot be utilized by solar cells. Material considerations restrict surface temperatures of thermal heat limiting the

CP401, *Thermophotovoltaic Generation of Electricity: Third NREL Conference,*
edited by Benner/Coutts
© 1997 The American Institute of Physics 1-56396-734-0/97/$10.00

peak emission of the blackbody radiation to around 0.6 eV (2.1 μm), a lower energy than the bandgap of conventional solar cells.

The most direct solution is to fabricate a photovoltaic (PV) cell to match this spectrum that could be mounted in front of the emitting to convert the radiant energy into electricity.

An alternate technique is to mount a high-pass or band-pass filter in front of the source. This reflects the longer wavelengths back to the source while transmitting shorter wavelengths, changing the effective emissivity of the surface. As long as the filter is lossless, the reflected energy is absorbed by the source and hence, energy is not lost. A somewhat similar approach is the use of a spectrally selective surface to emit predominately at wavelengths better utilized by the cell. Both techniques shift the radiated energy to a higher photon energy and allow the use of solar cells with larger bandgaps. Previous work on filters attempted to use plasma filters. These are generally oxides that have been doped, e.g. indium tin oxide (ITO), but their performance in the long wavelength infrared (LWIR) is poor, rarely having IR reflectivities above 92-93%. However, thin metal films of Au are also used as transparent conductors. Used with Au thickness of 5.0-7.5 nm, their performance is superior to the oxides and the thin films of Au have been electrodeposited with IR reflectivity approaching that of bulk Au (>99.5%).

Approach

A design for a high efficiency TPV, based on the small overlap between the radiated energy from the black body source and the collection efficiency of the CuInSe$_2$ cell (Figure 1), was developed. The key to the system is the use of a filter that reflects most of the energy back to the emitting surface.

Filter Optimization

The constraints on the filter are severe and different from most filter design. It is not sufficient that the filters pass the region of interest but other bands must be almost completely reflected, ~99%. Since the energy at shorter wavelengths is decreasing rapidly, the filter may be a low pass filter with no loss in performance. The use of filters to reflect the unused energy back to the source is not a new approach (1). Metal-like filters, termed plasma filters, operate by having a sharp change in the optical constants at the plasma frequency and a high reflectivity at longer wavelengths. The most common materials used for these filters are the oxides, indium tin oxide (ITO) and SnO$_2$ (2 and 3). A somewhat similar filter is a thin layer of metal (4). The metals have high reflectivity over most wavelengths and are fairly lossless until longer wavelengths. Other approaches are possible, as listed in the following sections.

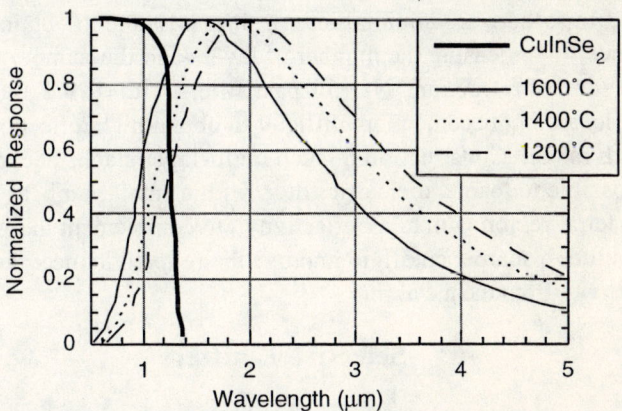

FIGURE 1. Quantum efficiency of CuInSe$_2$ cell and normalized black body emission curves for 1200°C, 1400°C, and 1600°C

Mesh Filters

Metal mesh filters are widely used at microwave frequencies. The designs are complex since they involve scattering at distances comparable to a wavelength. Generally, the dimensions of the elements are a fraction of a wavelength, requiring a filter design of 0.25 µm or less. The state of the art for conventional photolithography is 0.5 µm feature size. Beyond this requires either x-ray or direct write electron beam lithography. If the feature size could be kept above 0.5 µm, then the cost of fabricating such a filter is nominal and likely cheaper than any dielectric stacks. With the continuing advance in the integrated circuit industries, it is likely that this 0.5 µm will be pushed to about 0.25 µm within a few years. A possible problem is the behavior of this filter at the higher IR frequencies. At microwave frequencies, the patches behave as perfect conductors. This is not true in the IR, and becomes less realistic as the frequency is raised into the near IR (~1 µm). Filters are currently being fabricated in the LWIR region for the low bandgap PV cells(5). The mesh filters, since they can use the thicker electrodeposited Au as the metal, will have good IR reflectivity and can be used as the base film. Another problem is the high cost. The TPV system uses a relatively large area while the cost of filters fabricated with nanofabrication techniques are likely to be high.

Dielectric filters are based on interference effects, using multiple layers of dielectric films with different indices of refraction and different thickness. With dielectric films, there are minimal losses. It is possible to obtain virtually any performance by increasing the number of layers and the number of films with different index of refraction. Narrow pass filters with >99% transmission are easily fabricated but become more difficult to obtain high reflection over a large wavelength range. Construction of such multi-layer stacks becomes costly. It may be possible to team a dielectric filter, with a small number of layers and a narrow spectral region with high reflectivity, to complement the Au filter. This could introduce a sharper cutoff to improve the reflectivity near the on-set of the high reflectivity from the metal film.

Selective Emitters

By using a selective emitter, the surface of the heat source is modified so that its spectral emissivity is a closer match to the solar cell. These modify the emissivity by the use of rare earth oxides (e.g. Nd_2O_3, Yb_2O_3). Changes in the feature size of the surface will also modify the spectral emissivity. One problem with the use of a selective emitter is that they typically have a finite emissivity outside the range of interest. This results in a significant amount of energy being radiated in these spectral regions. A second problem is that they tend to be fragile since, to have low emissivity outside the peak, they need to be thin.

Photovoltaic Cells

$CuInSe_2$ has a bandgap close to ideal for the proposed approach. It is also a thin film cell that is ready for large volume production. This will be reflected in the cost of the system, essential for consumer applications. It has produced high efficiency in sunlight and the proposed approach concentrates the radiation near its optimum collection efficiency.

One reason it has received so much interest is the extremely high absorption coefficient. As single a crystal, its bandgap is 0.95eV. Bandgaps for thin films measured by optical absorption techniques are generally slightly higher than the single crystal values. Measurements of the change of bandgap with temperature indicate a value between 1.2 and 1.6×10^{-4} eV/K.

The design and fabrication using $CuInSe_2$ PV cells is attractive but silicon cells are also usable, with only a small change in performance. The attractive feature of the thin film cells, besides the potentially low cost, is the possibility of having the cells on a flexible substrate where they can conform to the surface of the cylindrical filter. Silicon cells would have to be cut in sections but the added optical losses would be small.

FILTER DESIGN

Filter Optimization

The Au filter is a critical component since it controls most of the reflected energy. The transmission and reflection of the Au films are dominated at shorter wavelengths by the interference produced by the various layers while at the longer wavelengths, its response is controlled by the high value of the complex part of the index of refraction. Figure 2 shows the values of n and k selected for the modeling. These are basically bulk optical constants. However, films of electrodeposited Au have been deposited whose performance (based on IR reflectivity) approach these values (6). These films are relatively thick (>0.1 μm.) but do indicate that the desired level of performance is achievable.

The filter is a simple three layer structure using a thin layer of Au sandwiched between two dielectric layers. Original calculations were based on common thin film materials and used an optical index of 3.2.

One of the thin film solar cells under development at the Institute of Energy Conversion (IEC) uses hydrogenated amorphous silicon (a-Si). This film has a high optical index, ~ 3.6-3.7, and is transparent beyond 1 μm. With the metal film filters, performance improves with a higher optical index of the dielectric layer.

The effect of a higher index can be seen in Figures 3 and 4. Figure 3 plots the transmission versus wavelength for different values of the index of the dielectric layers while plots the peak transmission versus the index of refraction of the dielectric layer. These results are based on a simple three layer structure with the center layer of Au and the outer layers kept at the same optical thickness. The higher index of the a-Si layers results in a higher transmission filter and makes the performance of this simple 3-layer filter very attractive.

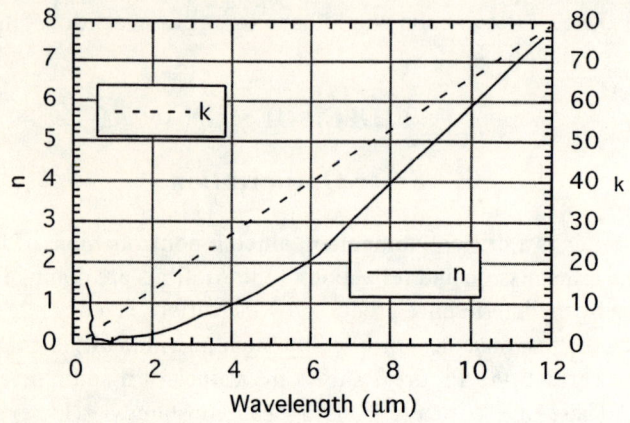

FIGURE 2. Optical constants (n and k) used for filter calculations

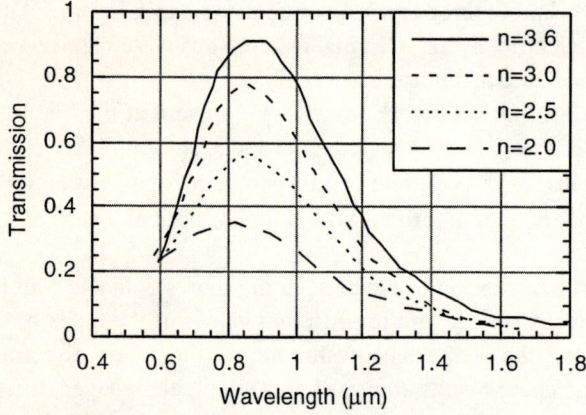

FIGURE 3. Transmission versus wavelength for the three layer induced transmission filter (300 nm Au) for various refractive indices of the dielectric layers.

The calculated reflection and transmission of a typical filter (see Table 1) are shown in Figure 5 and Figure 6 respectively. The reflectivity is relatively flat at longer wavelengths, increasing slightly as the thickness of the Au increases. Since a large amount of energy is in this region, this increase is important.

TABLE 1. Filter Design

		Thickness (nm)	Index of Refraction
Substrate	Glass	-	-
Dielectric 1	a- Si	38.9	3.6
Metal	Au	31.4	-
Dielectric 2	a- Si	38.9	3.6

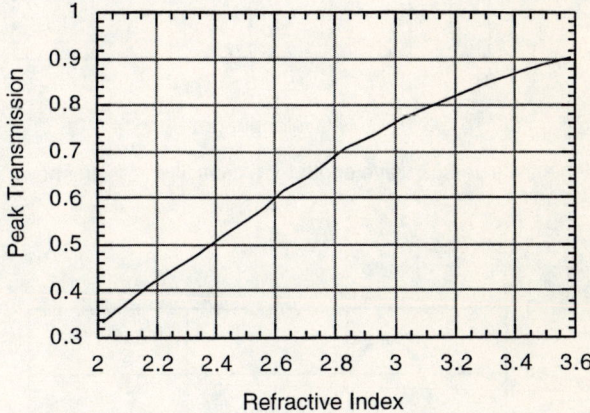

FIGURE 4. Peak transmission versus index of refraction of the dielectric layer for the induced transmission filter (30 nm Au)

With the design of the Au filter, there are two competing effects. The location of the bandpass is primarily controlled by the optical thickness of the layers but the magnitude of the peak is controlled by the thickness of the Au films. This is a result of the sharp increase in the complex part of the index k ($n = $ n - ik) with wavelength. At slightly thicker metal films, the loss becomes severe, limiting the region where the filter can be used.

As the Au thickness increases, the maximum current from the CuInSe$_2$ cell is decreasing and the peak is shifting to shorter wavelengths. However, since the reflectivity of the Au film increases slightly in the LWIR region, the efficiency can increase even though the transmission of the filter is decreasing. This effect can be seen in Figure 7, which shows the transmission for the same thickness of Au but using a thicker dielectric film that shifts the peak to longer transmission. The decrease of the peak amplitude reflects the higher loss in the Au at slightly longer wavelengths.

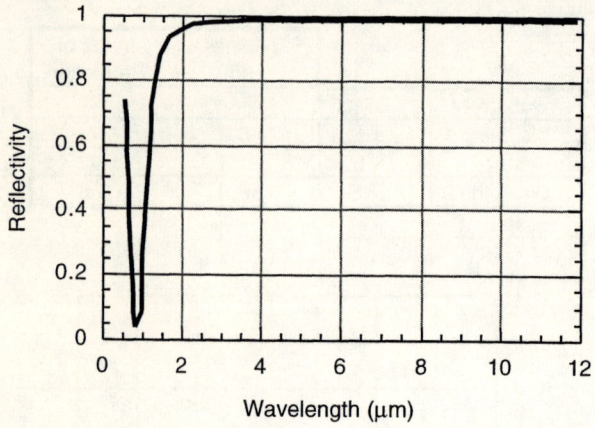

FIGURE 5. Reflectivity versus wavelength for typical filter design shown in Table 1 at a temperature of 1200°C.

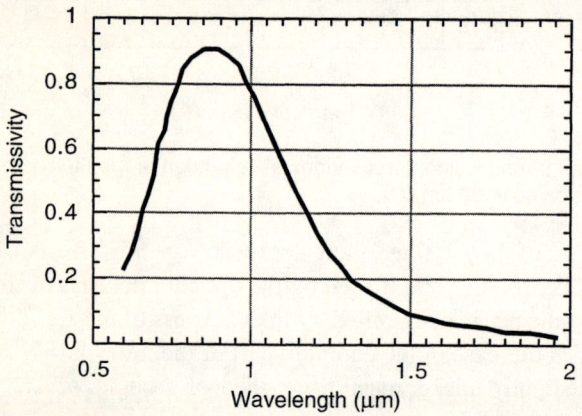

FIGURE 6. Transmissivity versus wavelength for typical filter design shown in Table 1 at a temperature of 1200°C.

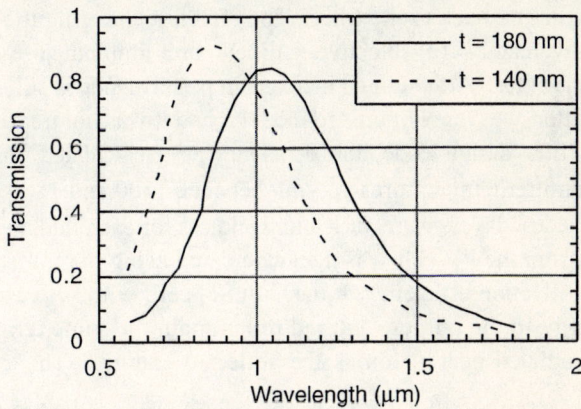

FIGURE 7. Transmission versus wavelength for the induced transmission filter (300 nm Au) at two thicknesses of the dielectric layer (n=3.6). Note decrease in transmission at longer wavelengths produced by an increase in absorption of the Au film.

This is a basic limitation with these filters since the loss increases rapidly with wavelength. The metal filter is limited to wavelengths near the present design peak.

Efficiency Calculations

The efficiency was computed by calculating the radiation from the emitting surface, assumed at a constant temperature. The optical response of the filter was calculated including energy transmitted through the filter and reflected from the filter. The reflected energy was assumed to be re-absorbed by the emitter and the energy transmitted was assumed absorbed by the PV cell. The power output from the PV cell was calculated by assuming a constant open circuit voltage (V_{OC}) and fill factor (FF) and a wavelength dependent quantum efficiency.

The present approach models all elements separately, with the radiation efficiency defined as the total transmitted energy divided by the energy required to maintain the blackbody at a set temperature.

$$\eta = \frac{\text{Radiated}}{\text{Radiated} - \text{Reabsorbed}} = \frac{E_R}{E_R(1-R)} \text{ where R is reflectivity of the filter.}$$

This is considered slightly more pessimistic but more accurate in modeling the layers and their interactions with the PV cell. The disadvantage is that it does not allow for treatment of the selective emitter. The net power into the emitting surface is the product of the emitter reflectivity and the reflected energy from the filter. Any decrease in emissivity (increase in reflection) of the emitter simply bounces the radiation back to the filter. The effect of changing the reflectivity at the longer wavelengths (a selective emitter) are dominated by the optical properties of the filter. Although an increase in performance is still expected with a selective emitter, we were not able to model it and it was not treated.

Using the filter design code, the efficiency was calculated for various Au thickness for emitter temperatures varying between 1100 and 1800°C. For these curves, the dielectric layers were varied in thickness for each thickness of Au. The power output from the PV cell was then calculated using the transmission of the filter and the collection efficiency of the CuInSe$_2$ cell, with set values of FF and V$_{OC}$. The energy to the burner required to maintain the same temperature, was equal to the radiated energy minus the reflected energy. The efficiency was defined as: $\dfrac{E_{out}}{E_{out} + E_{losses}}$. These calculations were done for each thickness of Au with the dielectric thickness and the maximum efficiency selected for each Au thickness, yielding a different filter design for each temperature.

The efficiency versus Au thickness is shown in Figure 8 at temperatures of 1100, 1300 and 1500°C for the filter designs shown in Table 2. The resulting short circuit current (J$_{SC}$) spectra for these different filter designs are shown in Figure 9 (these efficiencies include optical losses only, no burner or thermal losses are included.). The efficiency is also increasing with emitter temperature, as shown in Figure 10.

TABLE 2. Selected Filter Designs

Au thickness (nm)	Dielectric thickness (nm)
25	120
30	140
40	140
50	100

The relatively small change in efficiency for increasing Au thickness is explained by the increase in reflected power for the increased Au thickness. The transmission at the peak is decreasing but the total power to the emitter is also decreasing since more power is reflected back to the source. Figure 11 which plots the power generated by the PV cell (P$_{OUT}$) versus Au thickness. Figure 12 shows the decrease in power input (P$_{IN}$) with increasing Au thickness. The ratio of these is defined as the efficiency ($\eta = 100\ P_{OUT}/P_{IN}$).

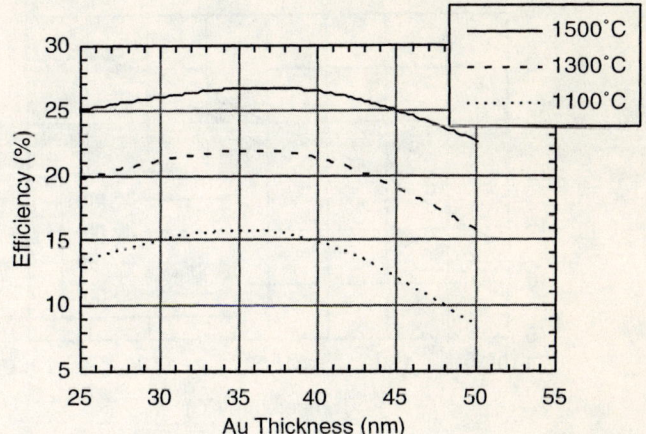

FIGURE 8. Efficiency versus Au thickness at temperatures of 1100, 1300 and 1500°C for the filter designs shown in Table 2.

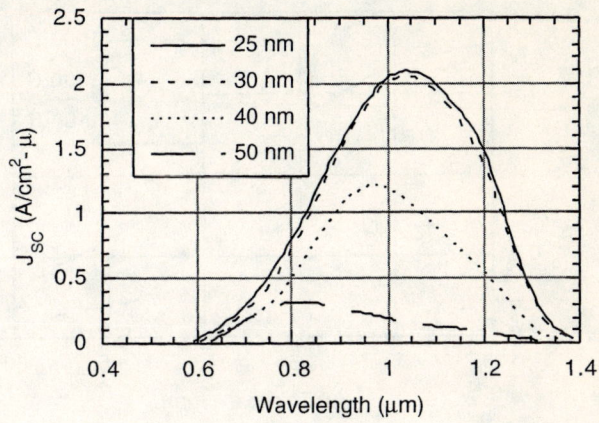

FIGURE 9. Short circuit current versus wavelength for the different filter designs shown in Table 2. All data is for a temperature of 1200°C

453

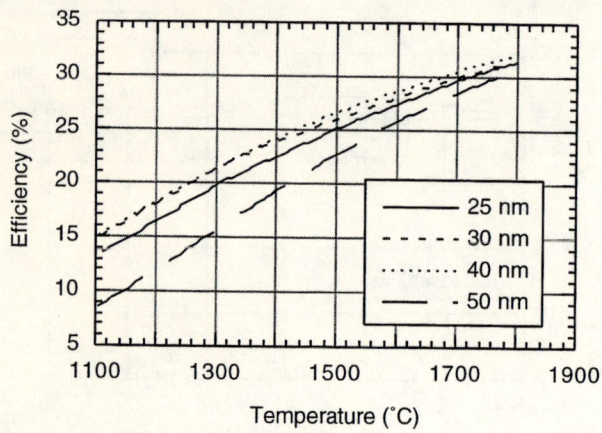

FIGURE 10. Efficiency versus emitter temperature for different filter designs shown in Table 2

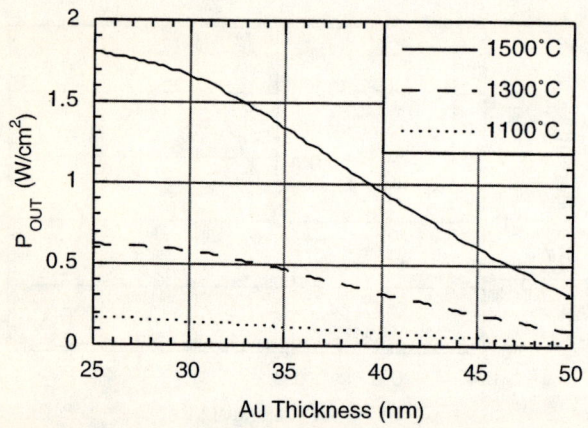

FIGURE 11. Power out versus Au thickness for temperatures of 1100, 1300 and 1500°C

The final filter design, listed in Table 1, uses a Au thickness of 31 nm and an a-Si thickness of 39 nm, which was the design fabricated. With this filter, a calculated efficiency of 18.2% is predicted at a temperature of 1200°C. The thickness of the Au film is expected to be advantageous compared to the much thinner Au used for transparent conductors. A thin film does not attain bulk values for its optical constant until it reaches a certain thickness. In thin layers, the deposition results in islands that coalesce together to form a continuous film. Even after the film is continuous, the optical constants will remain low for some time. The optical properties may not attain bulk values until a relatively thick layer is reached.

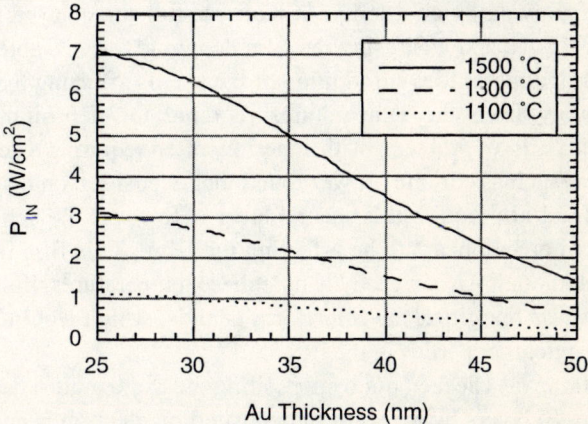

FIGURE 12. Power in versus Au thickness at temperatures of 1100, 1300, and 1500°C

Thin Gold Films

Au films can be deposited by a number of techniques. The most common technique is DC sputtering from a metal target. Films as thin as 5 nm may be electrically continuous, but they still consist of isolated islands. They nucleate into islands that coalesce together to form a continuous layer. Before this coalescence, their optical properties will be a mixture of the substrate and the film. This continuous layer usually forms below 10 nm and depends on the technique as well as the substrates. Techniques such as ion bombardment of the growing films are often used to assist the quality of the film.

Electrodepositon has been used to deposit very high quality Au films. Epner Technology (6), has reported very high reflectivity in the IR (>99.5) with Au films. These films are thicker, typically 1 µm. Epner Technology can electrodeposit films below 100 nm. in thickness. This approach needs a conductive substrate. Strike layers of chrome and/or Au, below 10 nm thick, may provide sufficient conductivity for the electrodeposition with only a nominal effect on the optical properties.

EXPERIMENTAL RESULTS

Initially, IEC deposited Au films onto glass and glass coated with a-Si. The thickness ranged from 15 to 25 nm. The reflectivity of the Au films measured ~99 % for wave numbers between 3000 and 1000 (3.3 to 10 μm).

This was followed by an attempt to make a full three layer structure that consists of a-Si, Au and a-Si using the film design given in Table 1. Based on these samples, it appears that the addition of the a-Si is affecting the properties of the Au, possibly from the temperatures required for deposition. Au films deposited on glass have problems with adhesion, often requiring a deposition of an adhesion layer, such as chrome. It was found that deposition onto a-Si gave good adhesion and did not require this second layer. However the deposition of the second layer of a-Si appears to be affecting the film. This film is deposited at 200-300°C. Although Au can easily handle these temperatures, if the adhesion to the first layer was marginal, the films form islands, which would have a strong effect on the optical properties in the IR.

The reflection was checked out by depositing the same multi-layer structure by replacing the a-Si layers by layers of Si deposited by electron beam evaporation. These results are shown in Figure 13. The reflectivity in the passband has increased slightly and this is assumed due to the lower dielectric constants of the deposited film. The measured IR reflectivity is shown in Figure 14. Measurements of absorption of this film predict a conversion efficiency of 5.4% at an emitter temperature of 1200°C.

SYSTEM DESIGN

Since the PV cells are assumed to be relatively inexpensive, the design attempted to minimize thermal losses with a minimal cost approach. This is shown in Figure 15. It consists of a burner with an inner diameter of approximately 2 inches at a surface temperature of 1200°C and an IR filter with an outer diameter of 12 inches at ambient. The overall length is 12 inches with a generated power of 150 watts. This design keeps the short circuit current from the PV cells to about 3 times the current generated at AM1 illumination. It also minimizes power lost via thermal conduction. The outer surface of the cylinder is at ambient temperature so that any power absorbed by the cells, either by radiation or conduction through the air, is lost. Increasing the power level from the PV cell can shrink the size although a smaller size will increase losses by conduction. The exact design will depend strongly on the size and cost of the PV cells available at that time.

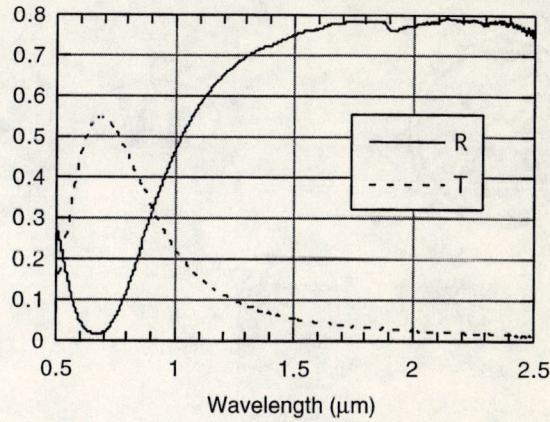

FIGURE 13. Transmission (T) and Reflection (R) measured on a Si/Au/Si filter design with thicknesses 63/31.5/63 nm on 7059 glass

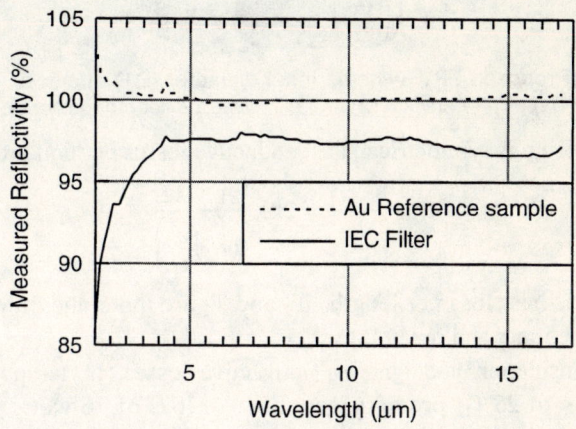

FIGURE 14. IR reflection measured on a Si/Au/Si filter design with thicknesses 63/31.5/63 nm on 7059 glass

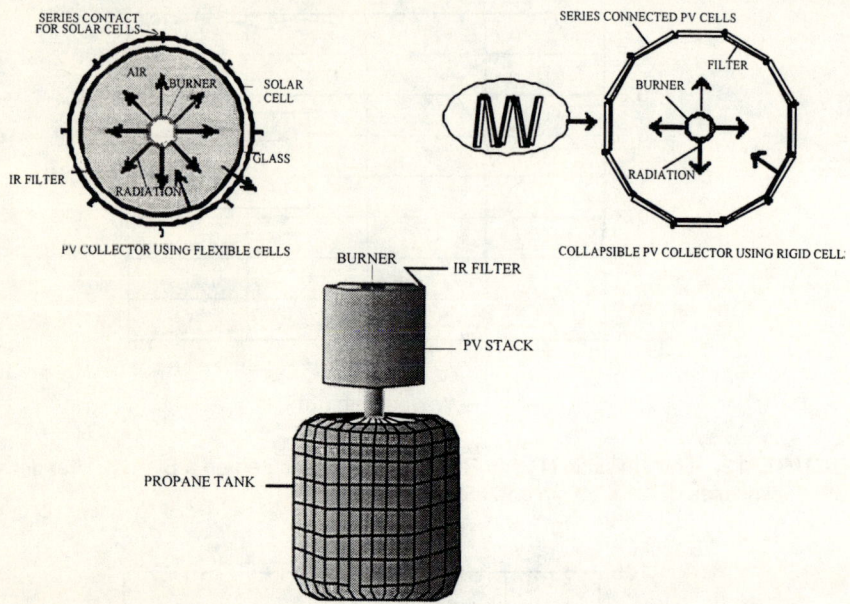

FIGURE 15. Proposed TPV system. Inner diameter (D1) of burner ≈2" at 1200°C. Outer diameter (D2) of TPV stack ≈12" at 23°C. With 12" length, power out is 150 Watts.

Since the design is symmetrical, the conductive losses per unit are:

$$\frac{q_{cond}}{L} = 2\pi k \frac{(T_1 - T_2)}{\ln\left(\frac{r_2}{r_1}\right)}$$

where q_{cond}/L is heat loss per length. T_1 and T_2 are inner and outer temperature and r_1 and r_2 are inner and outer radius.

Using air insulation and ignoring convective losses, at temperatures T_1 of 1200°C and T_2 of 23°C, predicts a conductive loss of 76 watts. This emitter design was also put into a finite element analysis program and the results show good agreement at ~77 watts. When these losses are considered, the overall system efficiency drops to 16.7%

CONCLUSIONS AND SUMMARY

A high efficiency TPV generator is proposed using low cost components, mostly off-the-shelf, including the burner and, most important, the solar cells. The filter will be custom designed but is a simple three layer structure requiring minimal control on the processing. The resulting system, requires no breakthrough technology. All the proposed components, including the PV cell, , exist and the calculations on the efficiency have been made using either measured or published data. The approach involves only a minimal amount of unproved technology

Potential markets include high end specialized applications This includes co-generation from waste heat but primarily will target portable liquid fuel powered generators at low power levels. The largest market is a low cost system using a propane powered generator. Small tanks of propane could generate a few hundred watts of power for a few hours. In these consumer applications, recreational boats and campers, the competition is are gasoline powered generators. The cost, weight and size of the TPV generator will be significantly better than the gasoline powered generators at power levels up to a few hundred watts.

ACKNOWLEDGMENTS

Part of this work was supported by the U. S. Army Research Office under Contract # DAAJ04-96-C-011; CLIN 0002AB.

REFERENCES

1. Schock, A., Or, C., and Mukunda, M., "Effect of Expanded Integration Limits and of Infrared Filter Improvements on Performance of RTPV System", presented at the 2nd NREL Conf. on TPV Generation of Electricity, Colorado Springs, CO, July, 1995, pp. 55-80
2. Wu, X., Mulligan, W. P., Webb, J. D., and Coutts, T. J., "TPV Plasma Filters Based on Cadmium Stannate", presented at the 2nd NREL Conf. On TPV Generation of Electricity, Colorado Springs, CO, July, 1995, pp. 329-338
3. Murthy, S. Dakshina, Langlois, E., Bhat, I., Gutman, R., Brown, E., Dzeindziel, R., Freeman, M., and Choudhury, N., "Characteristics of Degenerately Doped Silicon for Spectral Control In Thermophotovoltaic Systems", presented at the 2nd NREL Conf. On TPV Generation of Electricity, Colorado Springs, CO, July, 1995, pp. 290-311
4. Lissberger, P., "Coatings with Induced Transmission" Appl. Opt. **20**, 1995, p. 14
5. Horne, W E , Morgan, M. D. and Sundaram V. S.,"IR Filters for TPV Converters", presented at the 2nd NREL Conf. On TPV Generation of Electricity, Colorado Springs, CO, July, 1995, pp. 35-51
6. Epner Technology, 25 Division Place, Brooklyn, NY 11222

SESSION 8:
MODELING AND CHARACTERIZATION
OF TPV SYSTEMS

Efficiency measurements of TPV cells

Lars Broman*, Kenneth Jarefors*, Jörgen Marks**, and Mark Wanlass***

*Solar Energy Research Center, Högskolan Dalarna
S-781 88 Borlänge, Sweden
e-mail lbr@du.se, kja@du.se

**Department of Operational Efficiency, Swedish University of Agricultural Sciences
Box 7060, S-750 07 Uppsala, Sweden
e-mail jorgen.marks@stek.slu.se

*** National Renewable Energy Laboratory
1617 Cole Boulevard, Golden, CO 80401-3393, USA
e-mail mw@br4130mail.nrel.gov

Abstract: An apparatus for measuring TPV cell efficiencies at different radiation intensities and for different graybody emitter temperatures has been constructed. The apparatus has been used for measuring V-I characteristics, efficiencies and fill factors for several InGaAs TPV cells. Measured results are used to determine how cells may function together with edge filters, and those results are compared with theory.

INTRODUCTION

A joint project between US National Renewable Energy Laboratory NREL, Swedish University of Agricultural Sciences SLU, and Solar Energy Research Center SERC at Högskolan Dalarna, aims at building a wood powder fuelled TPV generator. Our work started with a thorough literature search [1]. Wood powder has a high energy density, 20 MJ/kg dry matter, and it is combustible in existing oil furnaces (in the range 1-15 MW) with little alteration [2]. Much of our initial work has dealt with the problem of combusting wood powder in such a way that a stable flame temperature of 1500 K is maintained [3,4]. This goal was achieved in a prototype pilot-scale burner that includes feeder [5] and combustion chamber.

Simultaneously, development work is being carried out on other parts of the TPV generator [6]. Four converters for testing have been built at NREL. They are 0.6 eV $In_{0.65}Ga_{0.35}As$ epitaxially grown lattice mismatched on an InP substrate using MOVPE. The metallization and sawing were done at ASEC and the 2-layer antireflection coatings were applied at NREL. The converters are mounted onto $50 \times 50 \times 2$ mm^3 copper plates (electroplated with gold) by means of electrically conductive epoxy. Modeling is in progress; internal quantum efficiency data have been measured at NREL and reflection losses have been measured at SERC, typically being about 20%.

CP401, *Thermophotovoltaic Generation of Electricity: Third NREL Conference,*
edited by Benner/Coutts

A Swedish TPV R&D program is being planned, aiming at co-generation of electricity and heat from refined wood fuel. Many components of the TPV converter; combustion chamber, emitter, optical path design, dichroic filter, and TPV cell mounting and cooling, are considered. Close co-operation with researchers at NREL, working on thin film TPV cell development, is an important aspect of the program. The program is however not adequately financed.

A more thorough background account is presented in another paper at this conference [7]. The present paper deals with our attempts to design equipment for reliable cell efficiency measurements, to measure cell efficiencies, and to apply the results to a possible TPV generator design.

EXPERIMENTAL SETUP

A set-up has been constructed for measuring TPV cell characteristics. It uses a halogen lamp filament as an approximate greybody emitter, the temperature of which was varied between 1500 K and 2600 K by varying the supply voltage.

The radiation intensity at the investigated cell is varied between minimum 0.3 W/cm^2 and maximum 10 W/cm^2 by means of changing the distance between emitter and cell. The total radiated power is assumed to be well approximated by the electric power supplied to the emitter, since only a small fraction is lost through radiation and convection from the bulb. The bulb is highly transparent quartz glass that will hardly absorb more than 1 % of the radiation (an extinction coefficient of 4 m^{-1} and glass thickness of 1 mm give transmission by absorptance $\tau_a = 0.996$ for normal incidence).

The radiative power per sterradian varies however more than 20 % between the most and the least favored direction as measured by us (using a grease spot bolometer). This and the varying optical geometry when changing the distance between the emitter and the test cells has been taken into account when determining the radiation intensity at the cell surface. Thus, the only significant error in this intensity comes from the measurement of the distance between emitter and cell, which is determined ± 0.25 mm. The distance between the emitter and the cell varies between 8 mm (for low temperature and high intensity) and 75 mm, creating a max. 6 % (for the most unfavorable cases) and min. 0.7 % error in the calculated intensity.

Emitter temperature was determined in two ways. Firstly, the dimensions of the tungsten filament were measured (by means of projecting a glowing filament on a distant wall). Since the filament consists of a fine wire (with diameter $d = 0.37$ mm) wound into a spiral (with height $h = 2.78$ mm and length $l = 8.30$ mm), the effective radiating surface is estimated to 72 ± 6 mm^2. Knowing the electric input to the emitter and assuming its emissivity being close to unity, Stefan-Boltzmann's law then gives the temperature. Secondly, the temperature was measured for several different electric inputs using a calibrated Minolta/Land Cyclops 52 infrared pyrometer, giving an effective radiating area equal to 71.8 ± 1.2 mm^2. We would have prefered a determination based on measuring the emission spectrum, which should give a more exact temperature.

464

The largest temperature errors are due to our uncertainty of the emissivity ε of a tungsten wire enclosed in a glass bulb. All temperatures given in the following chapters are given for ε = 0.8 ± 0.1, giving an error in the estimated emitter temperature ranging from ± 50 K for the lowest temperatures to ± 80 K for the highest.

The cell is air cooled and its temperature monitored; a set-up with cell temperature regulation is being planned. The cell is connected in series with an ampèremeter, a 100 W resistance of of 10-47 ohms, and a voltage supply. The V-I characteristic is achieved by varying the voltage in the circuit manually while the current through and the voltage over the cell are registered evenly along the characteristic curve and stored in a computer.

MEASUREMENTS AND RESULTS

The first four cells that are being characterized are the 0.6 eV InGaAs thin film cells grown at NREL. They each have an area of 1 cm^2. The top contact is 2 mm wide, so the irradiated area equals 0.8 cm^2. This is the area used in the cell efficiency calculations. One of the cells (W446) has cracked (mechanical causes), and has a V-I characteristic that is more or less a straight line. The three remaining test cells have been thoroughly measured; they are labelled W445, W451, and W452, respectively. In the figures we normally show results from just one cell.

In order to determine cell parameters for the standard two-diode solar cell model (see e. g. Araújo et. al [8]), their characteristics were measured with no radiation. Figure 1 a shows the V-I characteristic for cell W445 in linear scales, from which the series resistance r_s is determined. Figure 1 b is an enlargement of the characteristic nearest origo, from which the shunt resistance r_{sh} is determined. Figure 1 c is the characteristic in lin-log scales, from which the two diode currents i_{01} and i_{02} are determined. Values for all four cells are summarized in Table 1.

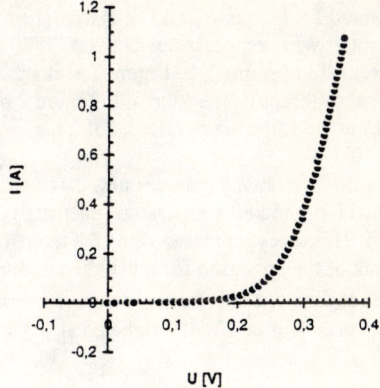

FIGURE 1 a. Non-radiation V-I characteristic of InGaAs TPV cell W 445.

465

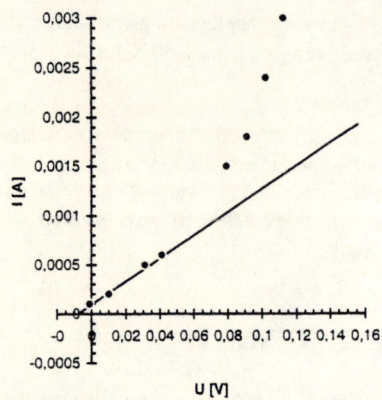 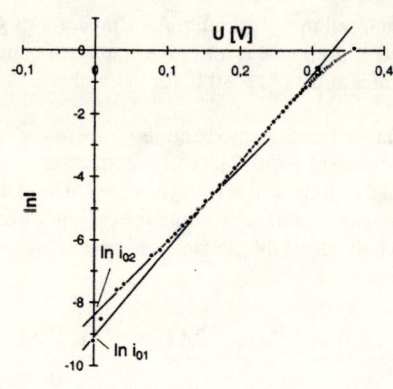

FIGURE 1 b. Part of non-radiation V-I characteristic of InGaAs TPV cell W 445 near origo.

FIGURE 1 c. Non-radiation V-I characteristic of InGaAs TPV cell W 445 in lin-log scales.

TABLE 1. Two-diode parameters measured for four InGaAs cells

Cell	W445	W446	W451	W452
R_s [Ω]	0.027	0.025	0.046	0.035
R_{sh} [Ω]	85	50	220	150
i_{01} [mA]	0.112	0.335	0.0555	0.0613
i_{02} [mA]	0.225	0.335	0.106	0.175

V-I characteristics were measured for the three good cells irradiated with intensities between 0.3 and 10 W/cm² from an emitter with temperatures between 1580 and 2640 K. Some typical results for cell W452 are presented in Figure 2. In Figure 2 a, characteristics are shown for the intensity 1.0 W/cm² and several different temperatures. In Figure 2 b, characteristics are shown for several different intensities at the same temperature, 2380 K.

From the characteristics in Figure 2 and many similar ones, maximum power point efficiencies and fill factors were determined for some 80 temperature-intensity combinations for each of the three good cells. Generally, efficiency increases when the temperature increases. This is of course expected, since the peak of the spectrum for all studied temperatures is at a longer wavelength λ than that corresponding to the bandgap, λ_1. An example of measured efficiency as a function of emitter temperature (for cell W452 radiated at 1.0 W/cm²) is shown in Figure 3 a below.

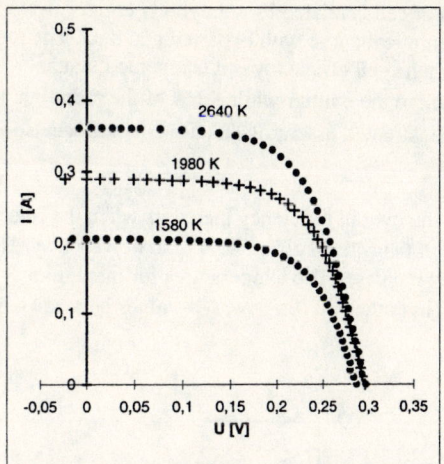 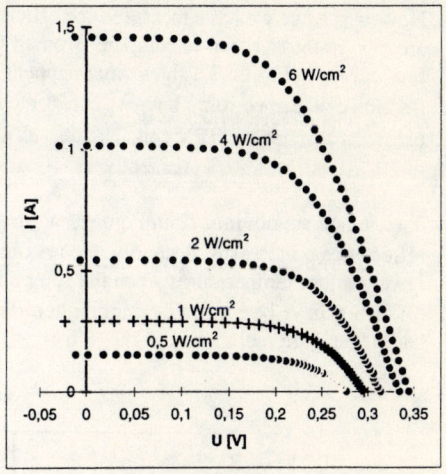

FIGURE 2 a. Typical characteristics for a TPV cell (W452) radiated from an emitter at different temperatures at 1.0 W/cm^2.

FIGURE 2 b. Typical characteristics for a TPV cell (W452) radiated from an emitter at 1980 K at different intensities.

The efficiency is instead fairly independent of the intensity of the radiation. Typically, our measurements show a very broad maximum around 1-2 W/cm^2, becoming a few percent lower only at the highest intensities. Presently, we are not sure whether this depends on the slightly increased cell temperature at the higher intensities or something else. Efficiency measurements at monitored cell temperatures are planned.

As long as the radiation intensity is below 2 W/cm^2, fill factors are fairly independent of emitter temperature, being typically between 61 and 64 %. For higher intensities, the fill factor decreases both with increased emitter temperature and increased intensity.

DISCUSSION

All the cells studied have bandgaps $E_g = 0.6$ eV, and it is obvious from our measurements that if they are irradiated from a greybody emitter, efficiency is quite low at temperatures that can be anticipated in most applications using a fuel heated emitter, i. e. 1500 - 1600 K. It has however theoretically been shown by Broman [9] that higher efficiency is reached at lower emitter temperature when an edge filter between the emitter and the TPV cell reflects back radiation with wavelengths above λ_1. The solid line in Figure 3 b is taken from this work, and shows how the best value for $\lambda_1 T$ varies with filter quality; the variable ρ equals the fraction of radiation above λ_1 that is reflected back to the emitter.

However, since we have measured the efficiency of cells radiated by a greybody emitter, we are now in the position to compare Broman's theoretical curve with experimental data. The top four curves in Figure 3 a show what happens with the cell efficiency if it is assumed that a fraction of the spectrum above λ_1 is reflected back to the emitter while 90 % of the radiation below λ_1 reaches the TPV cell. The curves are constructed assuming an above-λ_1 reflectance of 60, 70, 80, and 90 %, respectively.

Two things are obvious from Figure 3 a: Firstly, the overall efficiency increases when the filter reflectance ρ increases. Secondly, the maximum of the filtered efficiency curve moves towards lower emitter temperatures when the filter quality increases.The temperatures for maximum efficiency have been estimated for eight different hypothetical filters with ρ-values between 0.5 and 0.9 and are included in Figure 3 b.

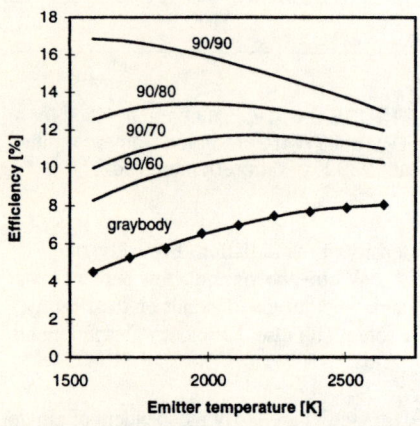

FIGURE 3 a. Bottom curve shows measured efficiency of cell 452 as a function of graybody emitter temperature. The four top curves shows efficiencies if edge filters are inserted between emitter and cell.

FIGURE 3 b. Optimum value of $\lambda_1 T$ as a function of edge filter reflectance ρ. Bars are calculated from measurements, the theoretical curve is taken from ref [9].

Each experimental point in Figure 3 b is represented by a bar due to our uncertainty regarding the emissivity of the tungsten filament in a bulb as mentioned above; the top $\lambda_1 T$ of each bar is calculated for emissivity $\varepsilon = 0.7$ and the bottom $\lambda_1 T$ for $\varepsilon = 0.9$. It may be noted that the general variation of best $\lambda_1 T$ varies exactly as theory predicts, and also that the absolute theoretical and experimental values are in reasonably good agreement. (The factor λ_1 here of course equals $1.24/E_g = 2.07$ μm for $E_g = 0.6$ eV.)

Finally, it should be pointed out that a reasonably good filter also gives a reasonably good result at moderate emitter temperatures. This is obvious from the fact that the curves in Figure 3 a are quite level and have very broad maxima.

468

REFERENCES

1. Broman, L., Thermophotovoltaics bibliography. *Progress in Photovoltaics* 3(1995)65-74.

2. Marks, J., Wood powder: an upgraded wood fuel. *Forest Prod. J.* 42(1992)52-56.

3. Broman, L. and Marks, J., Co-generation of electricity and heat from combustion of wood powder utilizing thermophotovoltaic conversion. *Proc. First NREL Conf. on Thermophotovoltaic Generation of Electricity*, pp 133-138 (1995).

4. Broman, L. and Marks, J. Development of a TPV converter for co-generation of electricity and heat from combustion of wood powder. *Proc. First World Conference on Photovoltaic Energy Conversion* (1995).

5. Marks, J. *Device for feeding of powder*. Swedish Patent Application No. 9701518-4 (1997).

6. Broman, L. Jarefors, K., Marks, J., and Wanlass, M., Electricity from wood powder. Report on a TPV generator in progress. *Proc. Second NREL Conf. on Thermophotovoltaic Generation of Electricity*, pp 177-180 (1996).

7. Schubnell, M., Gabler, H., and Broman, L., Overview on European activities in thermophotovoltaics. *Proc. Third NREL Conf. on Thermophotovoltaic Generation of Electricity* (1997).

8. Araújo, R. G., Krauter, S., Onneken, K., and Quaschning, V., *Photovoltaic energy systems*, http://emsolar.ee.tu-berlin.de/lehre/english/pv1/ (1997).

9. Broman, L., *Optimum bandgaps of TPV devices for blackbody emitters and selective filters* (1997; to be published).

TPV Efficiency Predictions and Measurements for a Closed Cavity Geometry

C. K. Gethers, C. T. Ballinger, M. A. Postlethwait, D. M. DePoy and P. F. Baldasaro

Lockheed Martin Corporation, Schenectady, New York

Abstract: A thermophotovoltaic (TPV) efficiency measurement, within a closed cavity, is an integrated test which incorporates four fundamental parameters of TPV direct energy conversion. These are: (1) the TPV devices, (2) spectral control, (3) a radiation/photon source, and (4) closed cavity geometry effects. The overall efficiency of the TPV device is controlled by the TPV cell performance, the spectral control characteristics, the radiator temperature and the geometric arrangement. Controlled efficiency measurements and predictions provide valuable feedback on all four. This paper describes and compares two computer codes developed to model 16, 1 cm^2 TPV cells (in a 4x4 configuration) in a cavity geometry. The first code, subdivides the infrared spectrum into several bands and then numerically integrates over the spectrum to provide absorbed heat flux and cell electrical output performance predictions (assuming infinite parallel plates). The second code, utilizes a Monte Carlo Photon Transport code that tracks photons, from birth at the radiation source, until they either escape or are absorbed. Absorption depends upon energy dependent reflection probabilities assigned to every geometrical surface within the cavity The model also has the capability of tallying above and below bandgap absorptions (as a function of location) and can support various radiator temperature profiles. The arrays were fabricated using 0.55 eV InGaAs cells with Si/SiO interference filters for spectral control and at steady state conditions, array efficiency was calculated as the ratio of the load matched power to its absorbed heat flux. Preliminary experimental results are also compared with predictions.

Introduction

In a thermophotovoltaic (TPV) generator, four fundamental systems are integrated to generate useful electrical power from infrared radiation:

- a radiator/photon source
- a semiconductor diode
- a spectral control system/filters
- cavity geometry effects

The primary function of the radiator is to provide infrared energy for TPV conversion. The TPV cell can convert only a fraction of the infrared energy emitted by the radiator into electrical power (~28% of the blackbody spectrum for a 1090°C radi-

CP401, *Thermophotovoltaic Generation of Electricity: Third NREL Conference,*
edited by Benner/Coutts
© 1997 The American Institute of Physics 1-56396-734-0/97/$10.00

ator with 0.55 eV TPV cells). Photons with lower energy (below bandgap photons) contribute to the temperature rise of the cell unless they are reflected back to the radiator. The spectral control system reflects these below bandgap photons back to the radiator and transmits above bandgap photons, which can be converted to electricity by the cell. Also, the system's geometry affects the associated photons, creating polarization, view factor, multiple reflections and angle of incidence concerns that must be accounted for in system models. By performing controlled efficiency measurements and predictions, valuable feedback can be provided for cell, filter and radiator development. This testing also provides feedback on the integration of these three fundamental systems into a TPV system by incorporating closed cavity geometry effects.

Reference [1] describes tow methods for determining the single cell TPV conversion efficiency. In one method, the single cell efficiency is calculated from measurements of the quantum efficiency and reflection versus wavelength and measured cell electrical characteristics (ideality factor, series resistance, and dark current). For the second method, the efficiency is measured using a calibrated blackbody source and copper block transient calorimetry technique. Excellent agreement was observed between the measured and predicted single cell efficiencies of 0.55 eV indium gallium arsenide (InGaAs) TPV cells which demonstrated approximately 13.6±10.0% at a radiator temperature of ~1090°C and a cell temperature of ~25°C.

Array (16 cells for this study) efficiency measurements are more complex than those for a single cell. This type of experiment integrates all thermal, electrical and system geometry issues into one measurement. For an array efficiency measurement, a cavity is formed between the radiator and TPV cell surface. The cavity introduces geometry dependent concerns which include polarization, view factor, multiple reflections, radiator temperature non-uniformities and angle of incidence effects. In addition, there are first order effects such as diode performance, diode matching, illumination uniformity and spectral control that play a role.

Two computer models have been developed to quantify these effects. The first code, TPVCalc, subdivides the infrared spectrum into several bands and then numerically integrates over the spectrum to provide absorbed heat flux and cell performance predictions. The second code, utilizes a Monte Carlo code that tracks photons, from birth at the radiation source, until they either escape the cavity or are absorbed. Absorption depends upon energy dependent reflection probabilities assigned to every geometrical surface within the cavity. The code also has the capability of tallying above and below bandgap absorptions as a function of location, and can support various radiator temperature profiles.

The 16 cell array efficiency, η, is experimentally determined by taking the ratio of the array's maximum power output to its total absorbed radiative heat flux,

$$\eta = \frac{P_{OUT}}{P_{ABS}} = \frac{V_{OC}I_{SC}ff}{\dot{m}C_P\Delta T} \qquad \text{Eqn. 1}$$

where,

   P_{OUT} = maximum power output by the array,

   P_{ABS} = the array's total absorbed radiative heat flux,

   V_{oc} = array open circuit voltage,

   I_{sc} = array short circuit current,

   ff = array fill factor,

   $\dot{m}$ = mass flow rate of the coolant,

   C_p = specific heat of coolant and

   ΔT = coolant temperature rise between inlet and outlet of the array.

This paper describes and compares two computer codes that were developed to model 16, 1 cm^2 TPV cells (in a 4x4 configuration) in a cavity geometry. Experimental results are also compared with predictions. The arrays are fabricated using 0.55 eV InGaAs cells with Si/SiO interference filters for spectral control.

Experimental Setup

Figure 1 provides a top-down view of the heater cavity with the lid and radiator material removed showing the TPV array below the window. Figure 2 shows a schematic of the heater cavity test stand. The internal walls and window of the heater cavity are polished to minimize photon losses in the wall and to promote uniform illumination incident upon the array. A square 0.08 inch thick piece of high emissivity Poco Graphite functions as the radiator material and rests over the window. The radiator was instrumented with two C-type thermocouples and its surface is approximately 0.300 inches from the surface of the TPV device. The thermocouples have the flexibility of being placed at either the radiator's center or edge.

In all measurements, the radiator was raised to a target temperature and the system allowed to reach thermal equilibrium (a thermal gradient of ~40°C was measured across the radiator.). The radiative absorbed heat flux and maximum output power of the array were measured and used to calculate array efficiency. All absorbed heat flux measurements were taken with the array in open circuit mode to ensure that the heat flux was transferred to the coolant, and not lost to joule heating of an external load.

The first computer code (TPVCalc) used to model TPV efficiency assumes an infinite parallel flat plate geometry. This model requires radiator temperature as an input and assumes that the radiator is perfectly isothermal. For modeling of the cavity test, an average of the measured edge temperatures was used as the radiator temperature. The code is based on a derivation of the model presented by Baldasaro, *et al* [2], where the efficiency, η, is given by:

$$\eta = F_0 \times \frac{V_{oc}}{E_g} \times \overline{QE} \times ff \times fu_{cell} \times fu_{module} \times f_{unknown} \quad \text{Eqn. 2}$$

where,

F_0 = usable fraction of thermal radiation with energy > E_g,

E_g = cell bandgap energy,

$\overline{QE}$ = average energy weighted charge carrier flux/photon flux of energy > Eg,

fu_{cell} = fraction of total absorbed radiation with energy > E_g,

fu_{module} = fraction of absorbed radiation due to module inactive area,

$f_{unknown}$ = fraction of performance decrement due to non-uniform illumination and cell mismatching effects

Equation 2 is analogous to equation 1 except that it is based on energy ratio parameters. This equation was initially developed to model single cell efficiency and has been modified to account for the additional variables incorporated in 16 cell array efficiency predictions. Specifically, the array equation accounts for fabrication (fu_{module}) and other unknown effects ($f_{unknown}$) such as those caused by illumination and variations in cell matching. Explicit knowledge of the variables in equation 2 is required to accurately predict and measure array efficiency.

In the TPVCalc model, the infrared spectrum is divided into intervals and efficiency is determined by numerically integrating the absorbed heat flux and array output over the spectrum. To calculate the absorbed heat flux in the array, the spectral characteristics of the array and radiator are measured with an Fourier Transform Infrared Spectrometer (FTIR) and numerically integrated to evaluate the spectral utilization of the system. The model also assumes that the only mode of heat transfer is through radiation. This method of calculating absorbed heat flux and array efficiency was compared to the classical grey body solution for radiative exchange between two infinite parallel flat plates. The absorbed heat flux calculated from the model and the heat flux from the classical solutions were identical. Modeling of the array electrical characteristics is based on measured spectral quantum efficiency and an analytically determined over-excitation factor, F_0, which is coupled with the spectral characteristics of the radiator and array (Figures 3-5). The fill factor and open circuit voltage of the array is based on the array cell characteristics (bandgap, dark current, series resistance and ideality factor) and the array temperature. The spectral utilization of the module is also dependent on the inactive area of the array which is not present for a single cell efficiency measurement.

Monte Carlo Photon Transport Code

A Monte Carlo code was also developed to perform photon transport calculations for this specific thermophotovoltaic application. This new Monte Carlo code was used to calculate the heat fluxes, spectral utilization, view factors, and even the photon-induced short circuit current of a 16 cell array of TPV cells. The complex geometric arrangement of the cavity make predicting the efficiency of such an arrangement difficult without this multidimensional computer code.

In general, the Monte Carlo method does not solve an explicit transport equation, but rather simulates individual particle histories and records some aspects of their average behavior. The individual probabilistic events that comprise a history are simulated sequentially by sampling physically-based probability distributions that describe the event. A Monte Carlo calculation consists of tracking millions of photon histories from their birth at the source to death at a terminal event (absorption or escape). This data is then tallied by geometric location and post-processed to determine the heat flux absorption and electrical performance of the array.

Assigning Photon Attributes

The Monte Carlo Photon Transport code simulates individual photon histories from birth to termination. Location, direction, polarization, and energy of each photon must be determined at birth to completely characterize the individual photons. A temperature distribution across the radiator was measured for the cavity experiments which complicates the Monte Carlo simulation, since the photon production rate and energy spectrum become location dependent due to the temperature distribution. Nevertheless, an analytic description of the photon birth location from the radiator surface can be derived. The number of photons emitted by a blackbody per unit volume, N, is given by [3],

$$N(\vec{r}) = \int_0^\infty n(\vec{r}, E)dE = \frac{8\pi\varepsilon}{(hc)^3}(kT(\vec{r}))^3 \int_0^\infty \frac{\alpha^2 d\alpha}{e^\alpha - 1}. \qquad \text{Eqn. 3}$$

where,
   k = Boltzmann's constant,
   h = Planck's constant,
   ε = radiator emissivity,

c = the speed of light,

$n(\vec{r}, E)dE$ = the blackbody energy spectrum (number of photons of energy dE about E per unit volume per unit energy),

$T(\vec{r})$ = location dependent temperature (K) and

$$\int_0^\infty \frac{\alpha^2 d\alpha}{e^\alpha - 1} \approx 2.4011 .$$

A parabolic temperature profile across the radiator is assumed in these calculations based on measured center and edge temperatures. The photon density can be calculated based on the average radiator temperature distribution and converted into a probability distribution describing the photon birth location.

Like the birth location, an analytic solution for the photon directional probability distribution is possible. Integration over solid angles is necessary when calculating a directional probability distribution. Figure 6 shows a schematic representation of solid angle integration where,

$$\int_{4\pi} d\Omega = \int_{-\frac{\pi}{2}}^{\frac{\pi}{2}} \int_0^{2\pi} \sin(\theta) d\phi d\theta = 4\pi . \qquad \text{Eqn. 4}$$

Hence, the average angle for emitted photons from a radiator $\bar{\theta}$ is calculated by integrating over a half space as,

$$\bar{\theta} = \int_0^{\frac{\pi}{2}} \int_0^{2\pi} \theta p(\theta, \phi) \sin(\theta) d\phi d\theta \qquad \text{Eqn. 5}$$

The angular distribution appropriate for a diffuse emitter [4], like graphite, is given by,

$$p(\theta, \phi) = \cos(\theta)/(2\pi) , \qquad \text{Eqn. 6}$$

which results in an average angle of 45° from the surface normal. This average angle was used when designing the filters since it is the most probable photon angle of incidence on the filters in the cavity experiments. Notice that the diffuse scattering angle probability distribution (Eqn. 6) is independent of azimuthal angle, ϕ. The azimuthal angle probability distribution is therefore uniformly distributed between 0 and 2π, which is easily sampled in the Monte Carlo Photon Transport code by generating a random number, $\xi \in (0, 1)$, and defining,

$$\phi_{sampled} = 2\pi\xi . \qquad \text{Eqn. 7}$$

The scattering angle, θ, can be sampled from the probability distribution given in Equation 6 by inverting to define a scattering angle, $\theta_{sampled}$, given a random number, $\xi' \in (0, 1)$, by,

$$\theta_{sampled} = acos(2\xi' - 1) \qquad \text{Eqn.8}$$

These two angles, $\theta_{sampled}$ and $\phi_{sampled}$, completely describe the photon direction in the Monte Carlo simulation.

A blackbody spectrum is used to describe the photon energy distribution [3]. However, the temperature of the radiator surface is location dependent. Hence, the photon birth location is first determined then a blackbody spectrum is constructed based on the local temperature. The photon production density, $n(E)$, that is emitted from a blackbody radiator at a temperature T_o (K) with energy in dE about E is given by

$$n(E)dE = \frac{8\pi\varepsilon E^2}{(hc)^3} \frac{1}{e^{E/(kT_o)} - 1} dE \qquad \text{Eqn. 9}$$

where,

$hc = 1.240 \times 10^3$ eVnm and

k = Boltzmann's constant.

In the Monte Carlo Photon Transport code, the source photon energies are sampled from the corresponding probability distribution based on

$$p(E)dE = \frac{\dfrac{8\pi E^2}{(hc)^3} \dfrac{1}{e^{E/(kT_o)} - 1} dE}{\displaystyle\int_0^{E_{max}} \dfrac{8\pi E^2}{(hc)^3} \dfrac{1}{e^{E/(kT_o)} - 1} dE} \qquad \text{Eqn. 10}$$

where E_{max} is set to an arbitrarily large value of $10kT_o$.

Photon Transport

Once the location, direction and energy of the source photons are determined, the Monte Carlo transport simulation can begin. Millions of photon histories are simulated in typical Monte Carlo calculations. Scattering events from the reflective walls and filters are common during a photon's life cycle. An accurate model of these scattering events is of paramount importance in a Monte Carlo calculation since these scattering events occur frequently. Figure 7 shows the computational model of the cavity experiment with some possible photon histories superimposed. In the Monte Carlo Photon Transport code, materials are considered either opaque or transparent to these photons and every material interface is assigned a reflection probability. Surface reflectance values for the important materials have been obtained as a function of photon wavelength, polarization, and angle of incidence. The reflectance data for the materials used in the cavity experiments has been introduced into Monte Carlo in the form of boundary conditions. These boundary conditions are determined for individual photons each time a photon strikes a boundary. A photon of energy, E_γ, and polarization, ρ_γ, that hits a surface at an angle, θ_γ, that has an energy/angle/polarization dependent reflection probability, $P(E,\theta,\rho)$, is reflected if

$$\xi \leq P(E_\gamma, \theta_\gamma, \rho_\gamma) \qquad \text{Eqn. 11}$$

where ξ = a random number $\in (0, 1)$. Otherwise, the photon enters the material where it is absorbed or transmitted depending on the material opacity. Hence, the accuracy of the Monte Carlo Photon Transport code results are strongly dependent upon the accuracy of the surface reflectivities and material opacities.

The code can be used to calculate the heat flux absorbed in any region of interest regardless of model geometry. Analytic solutions for the heat flux are generally only possible for simple geometries, $e.g.$, parallel and perpendicular rectangular plates or simple cavities. View factors and heat transfer between infinite parallel plates were calculated as part of the Monte Carlo validation procedure. Analytic view factors become very difficult to define for all, but the most simple geometries. The Monte Carlo Photon Transport code absorption edits are equivalent to view factors, by definition, and are calculated for any geometry of interest. Figure 8 shows the geometry for a view factor calculation for two parallel χ-by-χ square plates with a separation distance of Υ and the analytic solution [4] for the view factor is given by

$$F_{1 \rightarrow 2} = \frac{2}{\pi \chi \Upsilon}\left[\ln\left(\frac{(1+\chi^2)(1+\Upsilon^2)}{1+\chi^2+\Upsilon^2}\right)^{1/2} + \chi\sqrt{1+\Upsilon^2}\,\text{atan}\frac{\chi}{\sqrt{1+\Upsilon^2}} + \Upsilon\sqrt{1+\chi^2}\,\text{atan}\frac{\Upsilon}{\sqrt{1+\chi^2}} + (-\chi)\,\text{atan}\chi + (-\Upsilon)\,\text{atan}\Upsilon \right]$$

$$\text{Eqn. 12}$$

The analytic solution for this geometry is $F_{1 \rightarrow 2} = 0.589$ while Monte Carlo calculates the view factor to be $F_{1 \rightarrow 2} = 0.590$. Other view factors were calculated as part of the validation with similar good agreement. In addition, a set of radiative heat transfer calculations between infinite parallel plates was performed and the results from the Monte Carlo Photon Transport code were in good agreement with the analytic solution. The Monte Carlo code only predicts the I_{sc} output of each cell or column of cells. Its data has to be post processed using a SABER circuit simulation to predict I_{sc} for the entire array. To calculate an overall efficiency, Eqn. 1, requires the open circuit voltage and fill factor. The Monte Carlo Photon Transport code does not calculate these parameters and was used primarily for absorbed heat flux predictions.

Discussion of Results

Results from both measurements and computational predictions were generated in the course of performing the cavity studies. In addition, absorbed heat fluxes and short circuit currents became the focus of both the measurements and the computational methods. As a first step in analyzing the results, heat flux calculations from the two computer codes were compared. Table 1 lists the results from the testing of various grey absorber materials using the infinite parallel plate geometry (graphite, gallium antimonide wafer and gold foil). There is good agreement (12% or less) between the predicted absorbed heat flux of TPVCalc and the Monte Carlo Photon Transport code. These results were not expected to be identical since the Monte Carlo code filter reflectivities are represented by 100 evenly spaced energy points while the TPVCalc solution is based on reflectivities of variable energy resolution. TPVCalc also assumes that all energy emitted from the radiator is incident upon the cell; whereas, the Monte Carlo Photon Transport code accounts for absorptions in materials other than the array and incorporates cavity end losses into its predictions. In this series of comparisons, the heat flux predictions from the TPVcalc and the Monte Carlo Ray-Tracing methods are in good agreement.

One of the goals of these cavity studies was to determine if the measured device efficiency could be predicted. Table 2 shows the data predicted and measured for a 16 cell array. For an average radiator temperature of $1090°C$ and cell temperature of $25°C$, the array's efficiency was predicted to be approximately 7.0% using TPV-Calc based upon electrical characterization of the array. Testing of 16, 1 cm^2 TPV cells in a 4x4 configuration (array) resulted in an array efficiency of 6.3%. Although the relative absorbed heat flux and short circuit current measurements are higher than predictions, as will be discussed, the overall efficiency of the device is only affected slightly. This is due to the increase of both the numerator and denominator of equation 1. The measured efficiency of 6.3% is lower than the predicted efficiency of the infinite parallel plate model of 7.0%.

Significant differences were observed between the measured and calculated electrical current and heat fluxes. The absorbed heat flux of the array was predicted to be 6.32 W/cm^2 and 5.34 W/cm^2 using TPVCalc and Monte Carlo, respectively. The measurements show that 7.8 W/cm^2 was absorbed in the TPV array for a radiator and cell temperature of $~1090°C$ and $~25°C$, respectively. In addition, measurements indicate that the array produced approximately 0.324 volts/row and 3.45 amps/row for a total 8.62 watts at a the same temperatures. Both computer codes predicted a short circuit current (I_{sc}) that is $~20\%$ lower than the measured value. Hence, the computer codes under-predict both the heat flux and short circuit current when compared to measured values.

Analysis illustrates that an average radiator temperature of approximately $65°C$

higher than the measured temperature could account for both of these discrepancies. This disagreement could also be explained by low measurements in spectral quantum efficiency and errors in filter reflectivity measurements. Modeling assumptions (the Monte Carlo Photon Transport code assumes purely specular reflectivity within the cavity) and/or radiator emissivity at temperature could also play a role in the observed discrepancies. The predicted absorbed heat flux is also strongly dependent on the radiator surface temperature distribution. An estimated 10% of this discrepancy may be attributable to the uncertainty in our calorimetry measurements as well.

Table 3 summarizes the various parameters in equation 2 and the effects of the various measured and calculated electrical and optical properties on array efficiency. This table also provides a comparison of efficiency between the measured 16 cell array and efficiency for a high performing 0.55 eV InGaAs cell. The data in this table suggests that array efficiency should approach 13.6% for an array constructed of cells with characteristics identical to that of the single cell under uniform illumination. This assumes that the higher than predicted heat flux and short circuit current are both caused by a photon flux which does not agree with measured radiator temperature.

Conclusions

In conclusion, two computer codes (TPVCalc and a Monte Carlo Photon Transport code) were developed to predict 16 cell array efficiency in a cavity geometry. The results obtained from these codes agreed well with analytic solutions for various, simple problems. The codes were also in good agreement with the absorbed heat flux measurements of known grey absorbers. TPVCalc was used to predict the efficiency of the array used in this initial test to be ~7.0%. Actual measurements of the array's efficiency was ~6.3%, showing good agreement with the codes predictions. However, the code predicted significantly lower absorbed heat flux and short circuit current from the array. The Monte Carlo Photon Transport code was used to calculate a best estimate of the heat flux using the exact geometry. The heat absorbed calculated in the Monte Carlo Photon Transport code was similar to that of the TPVCalc results. This adds credibility to the calculated heat flux over the measured heat flux values given the modeling input.

Efficiency was not significantly affected by this discrepancy because the two measurements were ratioed to quantify efficiency. Possible reasons for this discrepancy include inaccurate radiator temperature measurements and other modeling inputs. The array used in this testing was comprised of poor performing cells and was purposely selected for this testing. Use of better cells in this type of array configuration, a more uniform radiator, and a reduction in series resistance would result in efficiencies considerably higher (~13.6%). Table 3 shows the improvements in cell, spectral control and manufacturing required to improve array efficiency from 6.3% to approach that of the aforementioned prediction.

References

1) Charache, G. W. et. al., "Measurement of Conversion Efficiency of Thermophotovoltaic Devices", Proc. of the 2nd NREL Conf. on TPV Gen. of Elec., 351 (1995).

2) Baldasaro, P. F. et. al., "Experimental Assessment of Low Temperature Voltaic Energy Conversion", Proc. of the 1st NREL Conf. on TPV Gen. of Elec., 29 (1994).

3) Krane, Kenneth, *Modern Physics*, John Wiley & Sons, New York, 1983.

4) Modest, M.A., Radiative Heat Transfer, McGraw-Hill, New York, 1993.

5) OptiLayer Code Ver. 3.06, Debell Design, Los Altos, CA, 1996.

Acknowledgments

The authors would like to acknowledge Dr. J. L. Egley, Dr. L. R. Danielson, Dr. J. R. Parrington and E. J. Brown for testing and analysis support. We would also like to thank Dr. G. W. Charache, and Dr. S. R. Sreepada for their valuable insight and suggestions.

481

Table 1: Predicted heat flux comparisons for various greybody absorbers

	Radiator Temperature (°C)	*TPVCalc Heat Flux (W/cm^2)	**Monte Carlo Heat Flux (W/cm^2)	% Difference
Graphite	1055	14.4	14.3	0.9
GaSb Wafer	1055	10.31	10.30	0.1
Gold Foil	1096	0.34	0.30	11.7

*Predictions based on average edge radiator temperature
** Predictions based on isothermal radiator , 100% reflective walls and no loss paths

Table 2: 16 cell array predictions and measurements

	Radiator Temperature (°C)	Absorbed Heat Flux (W/cm^2)	Short Circuit Current (A)	Power out (W)	Efficiency (%)
*TPVCalc	1108	6.38	12.57	7.73	7.0
Monte Carlo	**1109	5.345	12.0	NA	NA
Measurement	1108	7.9	13.88	8.62	6.3

*Predictions based on average edge radiator temperature
**Predictions based on an average radiator temperature of 1069C assuming a parabolic temperature profile

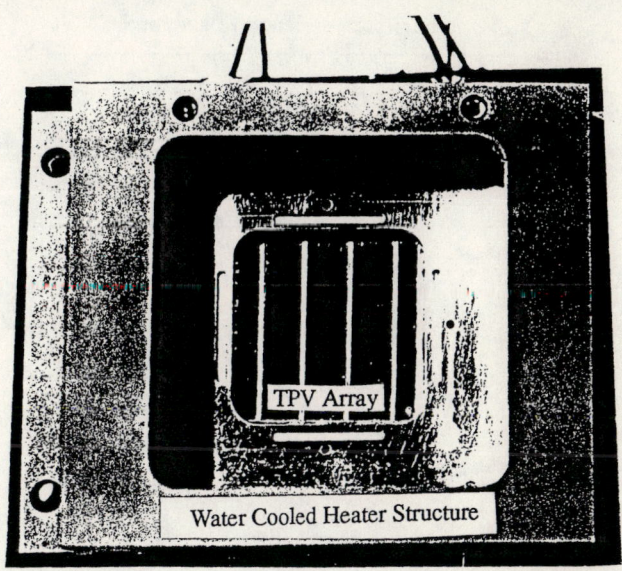

Figure 1. Top view of the heater cavity with the radiator material removed, showing the window for the 16 cell array.

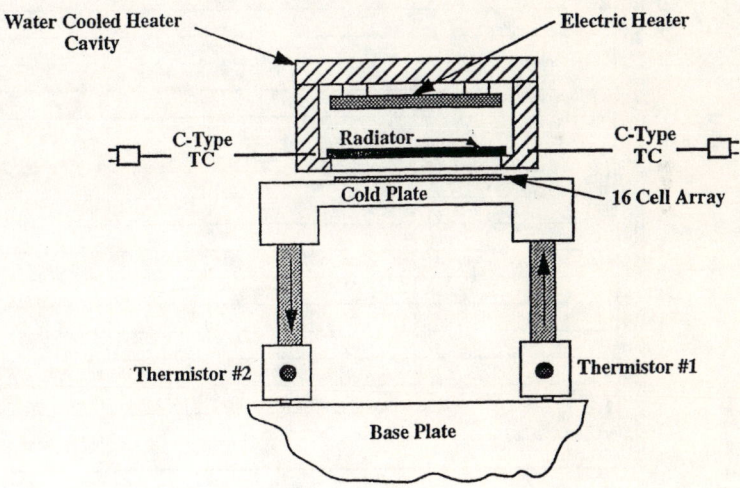

Figure 2. Schematic representation of the heater cavity test stand.

483

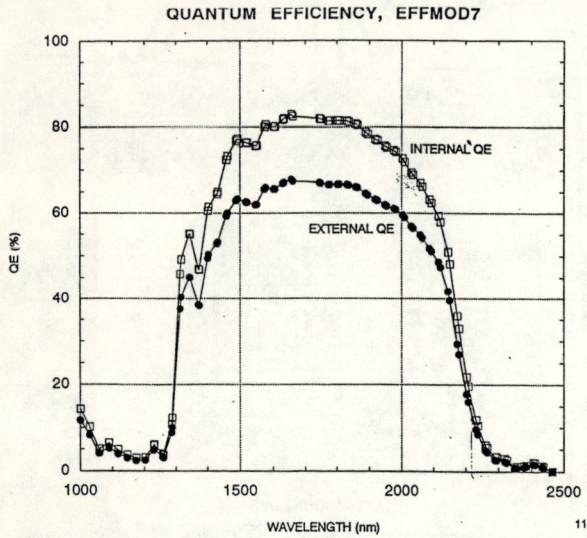

Figure 3: Internal and external quantum efficiency of a typical cell on the sixteen cell array tested in the efficiency test.

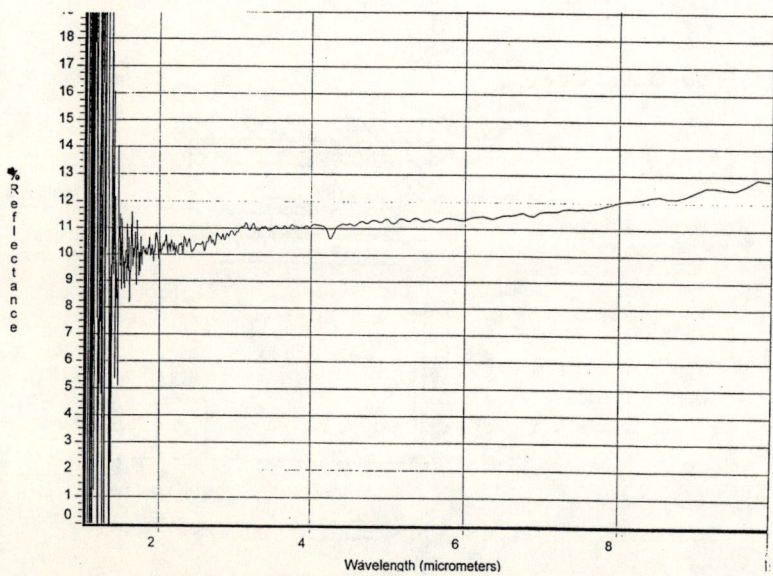

Figure 4: Measured reflectivity of the Poco Graphite radiator used in the efficiency test. The measured data was taken with an integrating sphere/FTIR at room temperature.

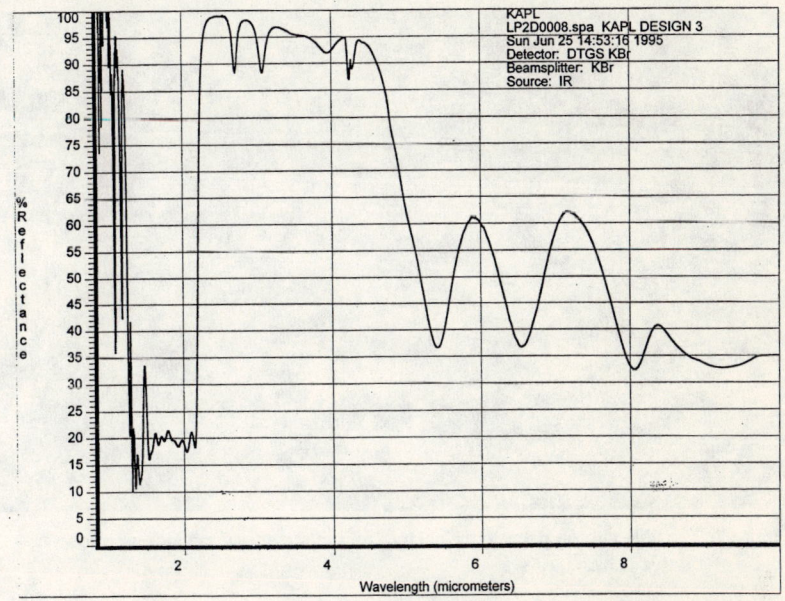

Figure 5: Typical spectral reflection characteristics of an interference filter for the array used in the efficiency test. The measurement was taken using an FTIR spectrometer.

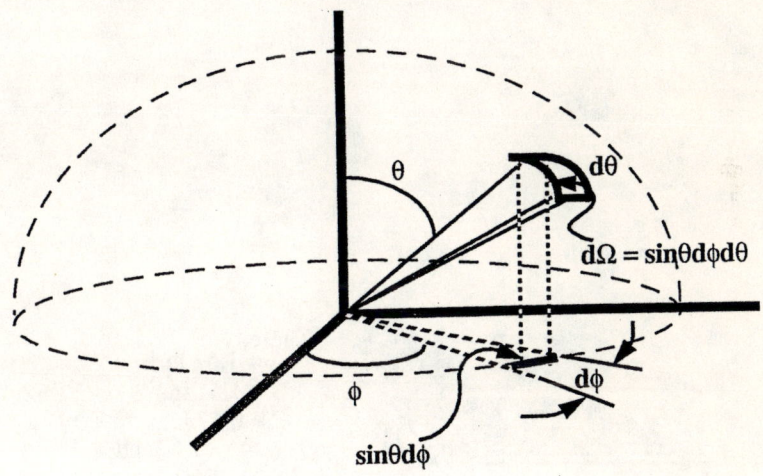

Figure 6. Solid angle integration.

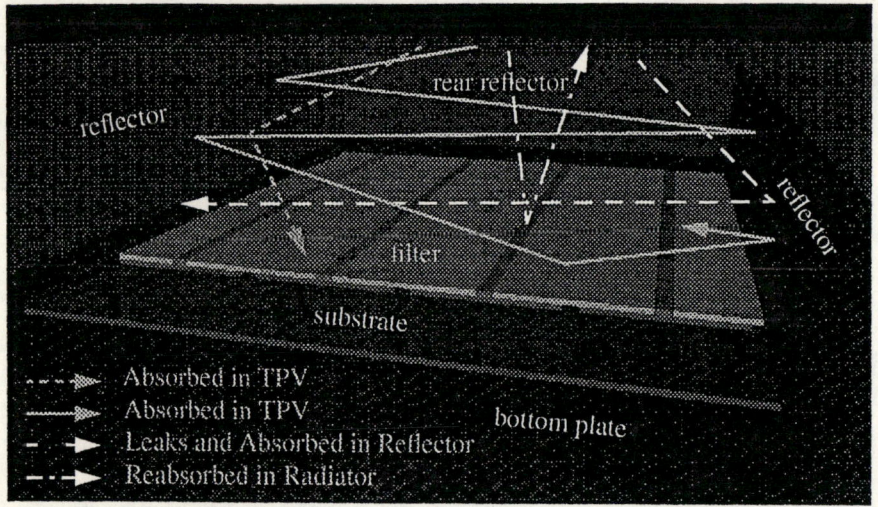

Figure 7. Possible photon paths in the cavity, front reflector and radiator are shown in this view.

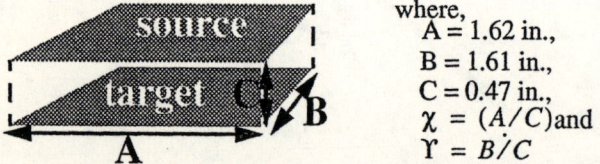

where,
A = 1.62 in.,
B = 1.61 in.,
C = 0.47 in.,
$\chi = (A/C)$ and
$\Upsilon = B/C$

Figure 8: Geometry for the view factor calculation

Effects of Geometry on the Efficiency
of TPV Energy Conversion

Brian S. Good and Donald L. Chubb
NASA Lewis Research Center
Cleveland, Ohio

Abstract.   The importance of view factors in determining the efficiency of
a thermophotovoltaic (TPV) energy conversion system was pointed out at the first
NREL TPV meeting by D. C. White and H. C. Hottel. In this paper we consider the
effect of view factors on efficiency for both planar- and cylindrical-geometry TPV
systems. One method for reducing the view factor radiation loss is to introduce
reflecting surfaces at locations where the radiation loss occurs. We also consider the
effect of these additional reflectors on efficiency.

To determine the cavity efficiency (efficiency of emitter, window, filter and lateral
reflector in combination) a set of nine algebraic equations for the radiation fluxes
incident on each of the cavity components must be solved. The solution depends
on the wavelength-dependent optical properties of the components, as well as the
view factors between components. Once the fluxes are obtained as functions of
wavelength, the cavity efficiency can be calculated.

Results to be presented will include a comparison of the view factor effect for planar
and cylindrical geometries. The effect of including lateral reflectors in the cavity
will be discussed. Finally, the effects of the interaction of the filter with the lateral
reflectors will be considered.

INTRODUCTION

Selective-emitter thermophotovoltaic systems are potentially attractive
as power sources for both space and terrestrial applications. Such systems are
being investigated at several laboratories, including The Auburn Space Power
Institute (1), ThermoLyte Corporation (2), Quantum Group (3), McDonnell
Douglas (4-5), JX Crystals (6), Lockheed-Martin (7), Westinghouse (8) and
NASA Lewis Research Center (9-12). Previously we have described work
using a computer modeling code that predicted the performance of a TPV
system, based on simplified models for the individual system components.
In this work, we consider in more detail the spectral conditioning portion

CP401, *Thermophotovoltaic Generation of Electricity: Third NREL Conference,*
edited by Benner/Coutts
© 1997 The American Institute of Physics 1-56396-734-0/97/$10.00

of the system, that is, the thermal cavity consisting of a thermal source (a selective or gray-body emitter), a protective window, a filter (included only if the emitter is a gray body), and lateral reflectors. A new computer code predicts the thermal performance (that is, the output power, steady-state component temperatures, and cavity efficiency) of this cavity, using component models that are more detailed than those used in previous work.

MODEL OVERVIEW

The TPV system configurations considered here are shown schematically in Fig. 1. The selective emitter system consists of a selective emitter, a protective window, lateral reflectors, and a PV cell array. The filter system consists of a gray-body emitter, a protective window, a bandpass filter, lateral reflectors, and a PV cell array. For both types of system, we consider two system geometries–planar, in which the emitter, window, filter and cell array are all parallel and of the same area, while the lateral reflector is a cylinder which encloses the other components, and cylindrical, in which the emitter, window, filter and cell array are concentric cylinders, and the reflectors are the annular pieces at the ends of the cylinders.

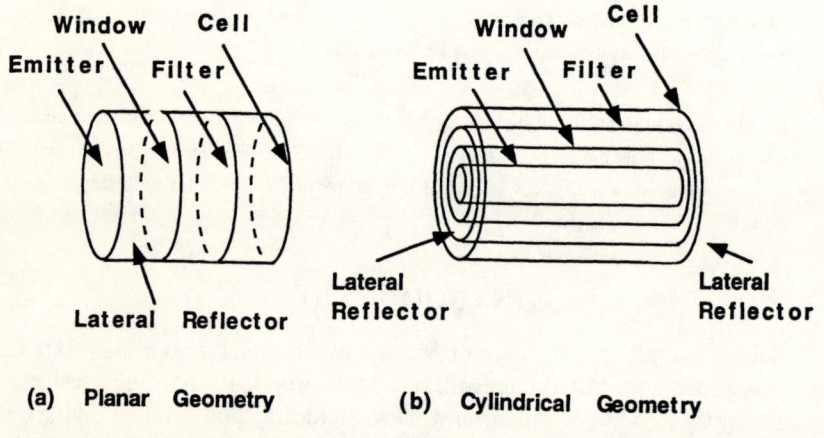

FIGURE 1. Schematic diagram of model system components. (a.) Planar geometry. (b.) Cylindrical geometry.

Radiation is emitted by the thermal source and reaches the cell array after passing through any intervening components, or after reflection by one or more components. The radiation may also be absorbed and re-emitted by any component having nonzero emittance.

In our earlier modeling work, we considered planar TPV systems where the radiative viewfactors between pairs of components were taken to be unity. This corresponds to a best case situation where the components are nearly in physical contact. Real-world materials considerations may render this impractical, at least for the emitter and window (or filter, if one is present). The components must then be separated to some degree, and when this is done, radiative viewfactors must be computed and included in the computer model. The relative sizes of the viewfactors, and their variation with cavity dimensions, will to a great degree determine the cavity thermal performance.

Fundamentally, the cavity model incorporates a set of nine coupled equations for the steady-state radiation fluxes incident on all internal surfaces within the cavity: the inner face of the emitter, both faces of the window and filter, the inner face of the PV cell array, and the inner faces of the lateral reflectors, of which there are three distinct regions. These equations are solved for the steady state fluxes at the nine surfaces. However, because many of the surfaces have non-zero emittances, the fluxes will be temperature dependent. For this study, the emitter and cell array temperatures are fixed at 1500K and 300K, respectively. For the other components, however, we compute the component temperatures iteratively so as to guarantee that no energy is gained or lost within each component, i.e. the radiation incident on each component surface is equal to the sum of the reflected and emitted radiation at the surface. In addition, the computer code permits this iterative temperature calculation to be selectively disabled for any combination of window, filter and lateral reflectors, so that their temperatures remain fixed at specified values. Any component held at fixed temperature will not satisfy the energy balance condition at its surfaces. The net flux into or out of the component will then determine the amount of heating or cooling needed to maintain the component operating at the specified temperature.

THEORETICAL DEVELOPMENT

There are nine distinct surfaces involved in the radiative heat transfer within the cavity. These surfaces (as shown in Fig. 1) are the in-cavity face of the emitter, both faces of the window and filter, the in-cavity face of the cell array, and the in-cavity reflector faces. For the planar geometry it can be seen that the window and filter divide the reflector faces into three regions. Because radiation reaching the different reflector regions will pass through different combinations of components, the three reflector regions must be treated separately, even if their optical properties are identical. The situation is similar in the cylindrical case, where there are two reflectors at opposite ends of the cylinder, each of which is divided into three regions by the window and filter.

The system of equations for the fluxes at these nine faces is derived by computing the incoming and outgoing fluxes at each face, and requiring that energy balance be maintained at each face. The net incoming flux at face l is specified as the sum of the fluxes incident on l from all other component faces:

$$A_l q_{il} = \sum_m \tau_{ml} A_m F_{ml} q_{om} \qquad (1)$$

where q_{il} is the net incoming flux incident on face l, q_{om} is the outgoing flux from face m, A_l and A_m are the areas of faces l and m, F_{ml} is the appropriate radiative viewfactor, and τ_{ml} is the total transmittance of all components separating m and l. It is assumed that all fluxes are uniform across the corresponding component surfaces, and that all surfaces behave in a diffuse manner.

The radiative viewfactors F_{ml} between two finite areas A_m and A_l are given by (13)

$$F_{ml} = \frac{1}{A_m} \int_{A_m} \int_{A_l} \frac{cos\theta_m cos\theta_l}{\pi S^2} dA_m dA_l \qquad (2)$$

where S is the magnitude of the vector $\vec{S}$ connecting differential elements dA_m and dA_l, and $cos\theta_m$ and $cos\theta_l$ are the angles between S and the surface normals at dA_m and dA_l.

The viewfactors are governed by a reciprocity relation:

$$A_l F_{lm} = A_m F_{ml} \qquad (3)$$

The required condition of energy balance at each face is given by

$$q_{ol} = \rho_l q_{il} + q_l \qquad (4)$$

where ρ_l is the reflectance at face l, and q_l is the emissive power emitted by face l, given by

$$q_l = \epsilon_l e_b = \epsilon_l 2\pi k c_o^2 \lambda^{-5}[exp(hc_o/kT\lambda) - 1]^{-1} \qquad (5)$$

where ϵ_l is the emittance at face l, e_b is the blackbody function, k is Boltzmann's constant, h is Planck's constant, c_o is the speed of light, T is the temperature of the emitting surface, and λ is the wavelength. Because these fluxes are functions of wavelength (as are the component optical properties, as discussed later), the above equations, and the following ones which derive from them, will be wavelength dependent as well. By applying Eqs. (1) and (3)-(5) to each of the nine surfaces, the following set of equations for the radiative fluxes are derived.

$$g(1 - \rho_a F_{aa})q_{oa} = \rho_a F_{ae'}q'_{oe} + \rho_a \tau_w \tau_f F_{ac}^{fw} q_{oc} + \rho_a \tau_w F_{af}^{w} q_{of}$$
$$+\rho_a F_{aw}q_{ow} + \rho_a \tau_w F_{ab}^{w} q_{ob} + \rho_a \tau_w \tau_f F_{ad}^{fw} q_{od} + q_a \qquad (6)$$

$$g(1 - \rho_b F_{bb} - \tau_w^2 \rho_b F_{bb}^{w})q_{ob} = \rho_b \tau_w F_{be'}^{w} q'_{oe} + \rho_b F_{bw'} q'_{ow}$$
$$+\rho_b \tau_f (F_{bc}^{f} + \tau_w^2 F_{bc}^{fw})q_{oc} + \rho_b (F_{bf} + \tau_w^2 F_{bf}^{w})q_{of} + \rho_b \tau_w F_{bw}^{w} q_{ow}$$
$$+\rho_b \tau_w F_{ba}^{w} q_{oa} + \rho_b \tau_f (F_{bd}^{f} + \tau_w^2 F_{bd}^{fw})q_{od} + q_b \qquad (7)$$

$$g(1 - \rho_d F_{dd} - \tau_f^2 \rho_d (F_{dd}^{f} + \tau_w^2 F_{dd}^{fw}))q_{od} = \rho_d \tau_f \tau_w F_{de'}^{fw} q'_{oe}$$
$$+\rho_d \tau_f F_{dw'}^{f} q'_{ow} + \rho_d F_{df'} q'_{of} + \rho_d (F_{dc} + \tau_f^2 (F_{dc}^{f} + \tau_w^2 F_{dc}^{fw}))q_{oc}$$
$$+\rho_d \tau_f (F_{df}^{f} + \tau_w^2 F_{df}^{fw})q_{of} + \rho_d \tau_w \tau_f F_{dw}^{fw} q_{ow} + \rho_d \tau_w \tau_f F_{da}^{fw} q_{oa}$$
$$+\rho_d \tau_f (F_{db}^{f} + \tau_w^2 F_{db}^{fw})q_{ob} + q_d \qquad (8)$$

$$(1 - \rho_w F_{ww})q_{ow} = \rho_w F_{we'} q'_{0e} + \rho_w \tau_w \tau_f F_{wc}^{fw} q_{oc} + \rho_w \tau_w F_{wf}^{w} q_{of}$$
$$+\rho_w F_{wa}q_{oa} + \rho_w \tau_w F_{wb}^{w} q_{ob} + \rho_w \tau_w \tau_f F_{wd}^{fw} q_{od} + q_w \qquad (9)$$

$$(1 - \rho_f (F_{ff} + \tau_w^2 F_{ff}^{w}))q_{of} = \rho_f \tau_w F_{fe'} q'_{oe} + \rho_f F_{fw'} q'_{ow}$$
$$+\rho_f \tau_f (F_{fc}^{f} + \tau_w^2 F_{fc}^{fw})q_{oc} + \rho_f \tau_w F_{fw}^{w} q_{ow} + \rho_f \tau_w F_{fa}^{w} q_{oa}$$
$$+\rho_f (F_{fb} + \tau_w^2 F_{fb}^{w})q_{ob} + \rho_f \tau_f (F_{fd}^{f} + \tau_w^2 F_{fd}^{fw})q_{od} + q_f \qquad (10)$$

$$(1 - \rho_c(F_{cc} + \tau_f^2 F_{cc}^f + \tau_f^2 \tau_w^2 F_{cc}^{fw}))q_{oc} = \rho_c \tau_w \tau_f F_{ce'} q_{oe}'$$
$$+\rho_c \tau_f F_{cw'} q_{ow}' + \rho_c F_{cf'} q_{of}' + \rho_c \tau_f (F_{cf}^f + \tau_w^2 F_{cf}^{fw})q_{of}$$
$$+\rho_c \tau_w \tau_f F_{cw}^{fw} q_{ow} + \rho_c \tau_w \tau_f F_{ca}^{fw} q_{oa} + \rho_c \tau_f (F_{cb}^f + \tau_w^2 F_{cb}^{fw})q_{ob}$$
$$+\rho_c (F_{cd} + \tau_f^2 (F_{cd}^f + \tau_w^2 F_{cd}^{fw}))q_{od} + q_c \qquad (11)$$

$$q_{of}' = \rho_f' F_{f'c} q_{oc} + \rho_f' F_{f'd} q_{od} + q_f' \qquad (12)$$

$$q_{ow}' = \rho_w' \tau_f F_{w'c} q_{oc} + \rho_w' F_{w'f} q_{of} + \rho_w' F_{w'b} q_{ob} + \rho_w' \tau_f F_{w'd}^f q_{od} + q_w' \qquad (13)$$

$$q_{oe}' = \rho_e' \tau_w \tau_f F_{e'c} q_{oc} + \rho_e' \tau_w F_{e'f} q_{of} + \rho_e' F_{e'w} q_{ow} + \rho_e' F_{e'a} q_{oa}$$
$$+\rho_e' \tau_w F_{e'b}^w q_{ob} + \rho_e' \tau_w \tau_f F_{e'd}^{fw} q_{od} + q_e' \qquad (14)$$

In the above, the following notation conventions are used.

The subscripts are defined as follows:

e'	radiating face of emitter
w	face of window closest to emitter
w'	face of window closest to cells
f	face of filter closest to emitter
f'	face of filter closest to cells
c	absorbing face of cells
a	portion of reflector(s) bounded by emitter and window
b	portion of reflector(s) bounded by window and filter
d	portion of reflector(s) bounded by filter and cells

Primed fluxes or optical properties indicate that the face is on the downstream (or "cell-facing") side of the component, while unprimed fluxes indicate the upstream (or "emitter-facing") side. Reflector fluxes are unprimed.

ρ_l, ϵ_l and τ_l indicate the reflectance, emittance and transmittance of component l. It is assumed that ρ_l and ϵ_l are identical for both sides of the window and filter, and that both sides of these components are at the same temperature. These optical properties are subject to the constraint that $\rho_l + \epsilon_l + \tau_l = 1$.

492

For all viewfactors, the subscripts indicate the two components exchanging radiation. Superscripts, if present, indicate any transmitting components located between m and l.

Note that in the cylindrical case, because certain components have nonzero transmittances, it is possible to have mutually visible concave faces, e.g. the interior surfaces of the window and filter. The viewfactors for these components will be nonzero, while for the planar case the same viewfactors will be zero. Therefore all viewfactors with two unprimed subscripts, both of which are one of $\{w, f, c\}$, are zero for the planar geometry.

In addition, the quantity g in Eqs. (6)-(8) has the value 1.0 for planar geometry, and 2.0 for cylindrical. This factor takes into account the fact that there are two separate sets of reflectors in the cylindrical geometry.

COMPONENT OPTICAL PROPERTIES

The optical properties, i.e. emittance ϵ, reflectance ρ, and transmittance τ, of the components are assumed to vary with wavelength. For some of the components, e.g. the emitter (12), the optical properties may vary rapidly with wavelength in certain regions of the spectrum, but may vary only slowly in other regions. We therefore divide the spectrum into a number of constant bands, over which the optical properties do not vary with wavelength, and a number of variable bands, in which significant variation with wavelength is allowed. Note that because the emissive power terms in Eqs. (6)-(14) are wavelength-dependent, these equations must be solved at all points (including those in the constant bands) on a wavelength mesh that spans the spectral region of interest (here 1 nm $\leq \lambda \leq 10000$ nm). The use of constant bands does, however, allow for a great reduction in the required input data.

For this preliminary work, the spectrum is divided into five "constant" bands; the detailed structure of the selective emitter emission band is ignored. While this is a simplification, it will not significantly affect the results of the effects of geometry on cavity performance. The band limits are chosen to be consistent with experimental data, as described below. The component optical properties are summarized in Table 1.

The emitter may be a black or gray body, or a narrow-band selective emitter. The black or gray-body emitter is characterized only by its emittance and reflectance; its transmittance is assumed to be zero. For all filter systems considered here, the emitter is a nearly-black body, with $\rho_e = 0.99$ and $\tau_e = 0.01$. The selective emitter may be characterized by three features: a low, essentially constant background emittance attributable to the emitter's metal substrate, a relatively narrow band of high emittance (due to the rare earth dopant) which ideally should correspond to radiation of energy just above the cell bandgap, and a rise in emittance at long wavelength due to the host material. The limits of the three lower bands are chosen to be consistent with experimental data from the rare-earth selective emitters developed in our laboratory (12), optimized for a cell bandgap energy of 0.75 eV, with a dimensionless bandwidth (10) of 0.15. The in-band and out-of-band emittances are 0.8 and 0.1.

The protective window should be as transparent as possible for radiation of energy near the cell bandgap. The upper two band limits are chosen to be consistent with the behavior of a quartz window, which is transparent for wavelengths less than about 3000 nm, and highly absorbing for longer wavelengths.

We consider three distinct filters. All filters are assumed to have been optimized to work with the same PV cell bandgap energy as the selective emitter, and are assumed to have a dimensionless bandwidth of 0.4.

Filter 1 is a nearly-ideal filter with reflectances of 0.01 in the passband, and 0.95 outside the band; the emittance is 0.02 across the whole spectrum. All calculations except the last set use the Filter 1 parameters. Filters 2 and 3 are less optimal; the emittances of these filters are 0.05 and 0.1, respectively. It should be pointed out that these filters are not intended to closely represent any particular real-world filter, but rather to allow investigation of the generic dependence of cavity performance on filter optical properties.

The cell is assumed to be highly absorbing ($\epsilon_c = 0.98$) for photon energies above the bandgap energy, and moderately reflecting ($\rho_c = 0.3$) for sub-bandgap energies.

The lateral reflectors are characterized primarily by their reflectance, which ranges from 0.5 to 0.99 for the various cases considered. It is assumed that the transmittance of the reflectors will be made as small as possible, although we consider some configurations whose reflectors have been effectively removed ($\tau = 1.0$). All reflectors are assumed to have optical propereties that are independent of wavelength.

TABLE 1. Component Optical Properties

Gray-Body Emitter

Band Limits	ρ	ϵ	τ
All	0.01	0.99	0.0

Selective Emitter

Band Limits, nm	ρ	ϵ	τ
1-1423	0.9	0.1	0.0
1424-1653	0.2	0.8	0.0
1654-3000	0.9	0.1	0.0
3001-5000	0.9	0.1	0.0
5001-10000	0.2	0.8	0.0

Window

Band Limits, nm	ρ	ϵ	τ
1-3000	0.03	0.02	0.95
3001-10000	0.03	0.95	0.02

PV Cell

Band Limits, nm	ρ	ϵ	τ
1-1653	0.02	0.98	0.0
1654-10000	0.3	0.7	0.0

Lateral Reflectors

Band Limits	ρ	ϵ	τ
Reflector 1 (All)	0.99	0.01	0.0
Reflector 2 (All)	0.9	0.1	0.0
Reflector 3 (All)	0.8	0.2	0.0
Reflector 4 (All)	0.7	0.3	0.0
Reflector 5 (All)	0.6	0.4	0.0
Reflector 6 (All)	0.5	0.5	0.0

TABLE 1. Component Optical Properties (continued)

Filters

Band Limits, nm	ρ	ϵ	τ
Filter 1			
1-1102	0.95	0.02	0.03
1103-1653	0.01	0.02	0.97
1654-3000	0.95	0.02	0.03
3001-5000	0.95	0.02	0.03
5001-10000	0.95	0.02	0.03
Filter 2			
1-1102	0.92	0.05	0.03
1103-1653	0.01	0.05	0.94
1654-3000	0.92	0.05	0.03
3001-5000	0.92	0.05	0.03
5001-10000	0.92	0.05	0.03
Filter 3			
1-1102	0.87	0.1	0.03
1103-1653	0.01	0.1	0.89
1654-3000	0.87	0.1	0.03
3001-5000	0.87	0.1	0.03
5001-10000	0.87	0.1	0.03

RESULTS AND DISCUSSION

The two basic geometrical configurations of the cavity, denoted planar and cylindrical, are shown in Fig. 1. For both geometries, the window and filter are located one-third and two-thirds of the way between the emitter and cells, respectively. Because there is a separation between the emitter and cells, the emitter-cell viewfactor is less than unity, which will degrade the cavity efficiency. To investigate this effect in isolation, we eliminate the effect of reflector reflectance by setting the reflector transmittance equal to 1.0, and consider a number of variations of both the planar and cylindrical geometries. We define baseline cases of each geometry, where the emitter areas and the ratios of emitter area to reflector area are identical. For the planar geometry, the cavity length L_p is varied so as to give emitter-cell separations of 0.1, 0.2, 0.5, 1.0, 2.0 and 5.0, normalized to the emitter radius. For the cylindrical geometry, the radius of the cell array R_c is varied so as to give the same emitter-cell separations, again normalized to the emitter radius.

Results are presented in Fig. 2, where the cavity efficiencies for planar and cylindrical selective emitter systems are presented as functions of emitter-cell separation. It should be noted that the cylindrical geometry has, in a sense, an extra degree of freedom. For the planar system, the emitter radius and cavity length, along with the locations of the window and filter, determine the cavity dimensions. For the cylindrical geometry, on the other hand, the emitter radius and cavity length determine the emitter area, but the cell array radius must be specified in order to determine the reflector area. Thus the requirements of equal emitter areas and equal emitter-area-to-reflector-area ratios for the two geometries do not uniquely determine the cylindrical cavity dimensions. We impose the arbitrary restriction that, for the baseline cylindrical geometry, the emitter radius is one-half the cell array radius. Because the cavity efficiency varies with cavity dimensions, a direct comparison of the emitter and cylindrical cavity efficiencies is probably meaningless.

It can be seen that for both geometries there is a very rapid decrease in cavity efficiency as the emitter-cell separation increases. This is unsurprising, given that increasing this distance decreases the emitter-cell viewfactor, and increases the emitter-reflector viewfactors; because the reflectors here are completely transparent, increasing the emitter-reflector viewfactors simply increases the area from which the emitter radiation can leak.

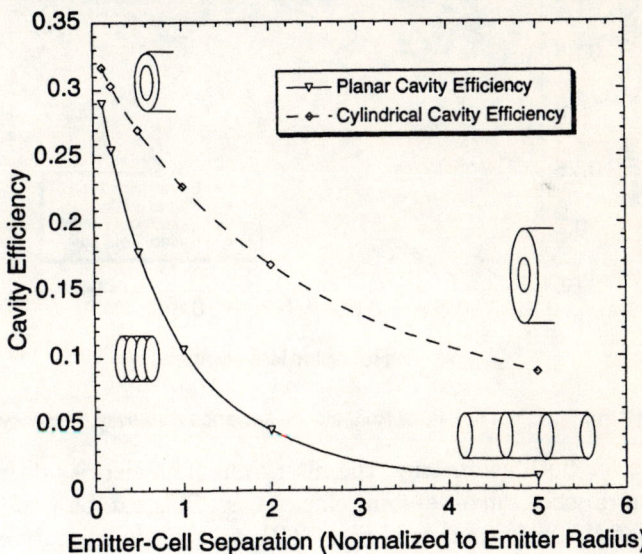

FIGURE 2. Effect of Emitter-Cell Separation on Cavity Efficiency

Given this large deterioration, the inclusion of some kind of reflector is mandatory for systems where the emitter and cell array are separated. In Fig. 3 we present plots of the normalized cavity efficiencies for the two baseline geometries (equal emitter areas, equal emitter area to reflector area ratios, and equally-spaced components) as functions of reflector reflectance. For each geometry, we include data from both selective emitter and gray-body/filter systems; the gray-body emittance is taken to be 0.99, and the filter optical properties are given in Table 1 under the heading "Filter 1." In all cases, the reflector transmittance is zero, so that as the reflectance decreases, the emittance increases correspondingly. Two trends are evident. First, for both the filter and selective emitter systems, the cylindrical geometry shows a slower degradation of efficiency with decreasing reflector ρ than does the planar geometry. Given the additional degree of freedom present in the cylindrical geometry, it is at present unclear whether this is a characteristic of the cylindrical geometry, or of the particular cavity dimensions used here.

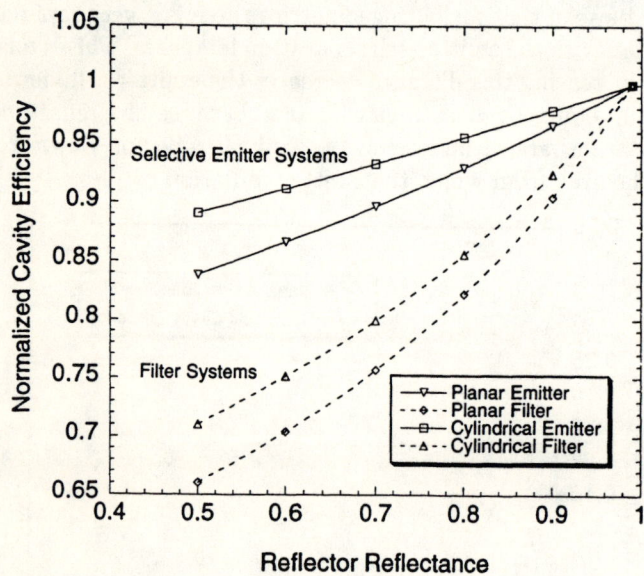

FIGURE 3. Effect of Lateral Reflector Reflectance on Cavity Efficiency

Second, for both geometries, the filter system efficiencies degrade more rapidly than do the emitter system efficiencies. This can be understood by recalling that the performance of filter TPV systems is strongly dependent upon the radiation outside the passband of the filter being reflected back to the emitter. This results in two related effects: first, when the reflectors are transparent, a portion of the out-of-band radiation (determined by the filter-to-reflector viewfactor) is lost, in addition to the portion lost via direct

498

radiation from the emitter. Second, when the reflectors are not transparent, a highly-reflective reflector provides higher efficiency than does a highly-emissive one; because emission is assumed to be diffuse, a significant fraction of the energy reflected from the filter onto the reflector is absorbed, but re-emitted away from the emitter. While there is a similar effect observed in selective emitter systems, it is much less pronounced.

As described above, the viewfactors between components play an important part in determining the system performance. We therefore consider next the effect of the placement of window and filter, when the emitter-cell distance is fixed, so that the emitter-cell viewfactor remains fixed, but all viewfactors involving the window or filter will change. We compare the cavity efficiencies of planar and cylindrical systems in which the window and filter are equally spaced between the emitter and cell array (equal spaced), and ones in which the window is close to the emitter, and the filter is close to the cell array (close spaced). Specifically, we use our baseline optical parameters and geometries for the equal-space cases. For the close-spaced cases, the emitter-window and filter-cell distances are taken to be one-tenth the emitter-cell distance, for both the planar and cylindrical geometries. Again, the optical properties of the filter are given in Table 1 under the "Filter 1" heading.

It is found that the component location is not a major determinant of system performance, as long as the window and filter are lossless, i.e., their temperatures are allowed to float at values that result in energy balance. The results are presented in Table 2. It can be seen that the component spacing makes very little difference in cavity efficiency, and a relatively small difference in the component temperatures.

TABLE 2. Effect of Component Location on Cavity Efficiency

Geometry	Spacing	Emitter/Filter	Efficiency	T_w	T_f	T_s
Planar	Equal	Emitter	0.309	1002		1006
Planar	Close	Emitter	0.308	1016		981
Cylindrical	Equal	Emitter	0.310	917		875
Cylindrical	Close	Emitter	0.313	947		846
Planar	Equal	Filter	0.766	1475	1262	1343
Planar	Close	Filter	0.773	1471	1257	1427
Cylindrical	Equal	Filter	0.707	1461	1238	1282
Cylindrical	Close	Filter	0.681	1460	1230	1376

It should be noted that the cavity efficiencies of the selective emitter systems are approximately half those of the filter systems. This is not a general characteristic of these systems. It has been shown (10) that optimized emitter and filter systems are competitive. However, the emitter systems that yield the highest efficiencies are ones that use back-surface reflectors on the PV cells. Because such reflectors have some of the characteristics of filters, their inclusion here would introduce ambiguity into the results. We have therefore excluded such reflectors on the PV cells modeled here.

As seen above, when the component temperatures are allowed to float, some of the equilibrium values are quite high. This is most likely not a problem for the window, which would probably be constructed of quartz, with a softening point of approximately 1950K (clear fused quartz) (14), or a similar material. The PV cell array, on the other hand, would certainly require cooling, the effects of which have been included in our calculations. The need for cooling of the filter depends in detail on the filter material; without modeling specific filters, we cannot provide definitive information. However, we have performed calculations that provide information on the effects of maintaining the filter at a fixed temperature lower than the equilibrium ("floating") temperature. In Table 3 we present cavity efficiencies for planar equal-spaced systems, where results from floating filter temperature calculations are compared with results of calculations for which the filter temperatures are held fixed at 1000K, 750K and 500K, for systems containing three different filters. The complete filter optical properties are presented in Table 1; most notable are the filter emittances, which are taken to be 0.02, 0.05 and 0.1 for Filters 1,2 and 3, respectively. For each filter, it can be seen that the cavity efficiency exhibits only a small dependence on filter temperature. In the absence of conductive or convective heat losses, the cooling load (and hence the filter waste heat) is simply the difference between the power absorbed (and re-radiated) at the free-floating temperature and that at the fixed temperature. While this radiated power goes as T^4, the small value of the filter emittance means that the magnitude of the waste heat will remain small.

TABLE 3. Effect of Filter Temperature on Cavity Efficiency

Filter Emittance	T_f Floating	$T_f = 1000K$	$T_f = 750K$	$T_f = 500K$
0.02	0.766	0.729	0.715	0.711
0.05	0.580	0.624	0.599	0.591
0.10	0.506	0.498	0.466	0.456

This result also indicates that, even if the window needs to be cooled, its low emittance will mean that the effect of this cooling on cavity performance will be minimal.

Increasing the value of the filter emittance at a given temperature does cause a significant drop in cavity efficiency, as seen by comparing the efficiencies of the Filter 1, 2 and 3 systems at each temperature. This behavior is due primarily to the fact that, as the emittance increases, an increasing fraction of the radiation incident on the filter is absorbed, and approximately half of this fraction is re-radiated towards the cell, thereby impeding the selective function of the filter to some degree.

CONCLUSIONS

1. Unless the emitter, window, filter and cell array can be sandwiched together such that the various viewfactors between these components are nearly unity, lateral reflectors significantly increase the cavity efficiency.

2. The planar geometry is slightly more sensitive to reflector reflectance than is the cylindridcal geometry.

3. Filter systems show markedly greater sensitivity to reflector reflectance than do selective emitter systems.

4. When the emitter-cell distance remains fixed, the locations of the window and filter within the cavity have only a minor effect on cavity performance.

5. Requiring that the filter be cooled results in only a small loss in cavity efficiency.

6. Increasing the filter emittance causes a significant decrease in cavity efficiency.

REFERENCES

1. Adair,P. L. and Rose, M.F.,"Composite Emitters for TPV Systems", in AIP Conference Proceedings 321, The First NREL Conference on Thermophotovoltaic Generation of Electricity, 1994, pp. 245-262.

2. Nelson, R.E., "Grid-Independent Residential Power Systems",in AIP Conference Proceedings 358, The Second NREL Conference on Thermophotovoltaic Generation of Electricity, 1995, pp. 221-237.

3. Holmquist, G.A., Wong, E.M. and Waldman, C.H., "Laboratory Development TPV Generator",in AIP Conference Proceedings 358, The Second NREL Conference on Thermophotovoltaic Generation of Electricity, 1995, pp. 138-161.

4. Stone, K.W., Chubb, D.L., Wilt, D.M., and Wanlass, M.W.,"Testing and Modeling of a Solar Thermophotovoltaic Power System",in AIP Conference Proceedings 358, The Second NREL Conference on Thermophotovoltaic Generation of Electricity, 1995, pp. 199-209.

5. Stone, K. W., Kusek, S. M., Drubka, R. E., and Fay, T. D., "Analysis of Solar Thermophotovoltaic Test Data From Experiments Performed at McDonnell Douglas," in AIP Conference Proceedings 321, The First NREL Conference on Thermophotovoltaic Generation of Electricity, 1994, pp. 153-162.

6. Fraas, L. M., et al, "Development of a Small Air-Cooled 'Midnight Sun' Thermophotovoltaic Generator," in AIP Conference Proceedings 358, The Second NREL Conference on Thermophotovoltaic Generation of Electricity, 1995, pp. 128-133.

7. Baldasaro, P. F., Brown, E. J., DePoy, D. M., Campbell, B. C., and Parrington, J. R., "Experimental Assessment of Low-Temperature Voltaic Energy Conversion," in AIP Conference Proceedings 321, The First NREL Conference on Thermophotovoltaic Generation of Electricity, 1994, pp. 29-43.

8. Wilt, D. M. et al, "Electrical and Optical Performance Characteristics of 0.74eV p/n InGaAs Monolithically Interconnected Modules," The Third NREL Conference on Thermophotovoltaic Generation of Electricity, 1997.

9. Chubb, D.L., Good, B.S., Wilt, D.M., Lowe, R.A., Fatemi, N.S., Hoffman, R.H. and Scheiman, D.,"Review of Recent Thermophotovoltaic (TPV) Research at Lewis Research Center", in Proceedings of the Fourteenth Space Photovoltaic Research and Technology Conference, NASA Conference Publication 10180, pp. 191-207.

10. Good, B.S., Chubb, D.L. and Lowe,R.A., "Comparison of Selective Emitter and Filter Thermophotovoltaic Systems", in AIP Conference Proceedings 358, The Second NREL Conference on Thermophotovoltaic Generation of Electricity, 1995, pp.16-34.

11. Good, Brian S., Chubb, Donald L., and Lowe, Roland A., "Temperature-Dependent Efficiency Calculations for a Thin-film Selective Emitter," in AIP Conference Proceedings 321, The First NREL Conference on Thermophotovoltaic Generation of Electricity, 1994, pp. 263-275.

12. Lowe, Roland A., Chubb, Donald L., and Good, Brian S., "Radiative Performance of Rare Earth Garnet Thin Film Selective Emitters," in AIP Conference Proceedings 321, The First NREL Conference on Thermophotovoltaic Generation of Electricity, 1994, pp. 291-297.

13. Siegel, Robert, and Howell, John, "Thermal Radiation Heat Transfer," Second Edition, McGraw-Hill, 1981, pp. 189-190.

14 Weast, R. C., "Handbook of Chemistry and Physics," 49th Edition, The Chemical Rubber Company, 1968, p. F57.

An Improved Model for TPV Performance Predictions and Optimization

K. L. Schroeder*†, M. F. Rose* and J. E. Burkhalter†

*Space Power Institute and †Department of Aerospace Engineering,
Auburn University, Alabama 36849

Abstract. Previously a model has been presented[1] for calculating the performance of a TPV system. This model has been revised into a general purpose algorithm, improved in fidelity, and is presented here. The basic model is an energy based formulation and evaluates both the radiant and heat source elements of a combustion based system. Improvements in the radiant calculations include the use of ray tracking formulations and view factors for evaluating various flat plate and cylindrical configurations. Calculation of photocell temperature and performance parameters as a function of position and incident power have also been incorporated. Heat source calculations have been fully integrated into the code by the incorporation of a modified version of the NASA Complex Chemical Equilibrium Compositions and Applications (CEA) code[2]. Additionally, coding has been incorporated to allow optimization of various system parameters and configurations. Several examples cases are presented and compared, and an optimum flat plate emitter / filter / photovoltaic configuration is also described.

INTRODUCTION

Research in TPV has been primarily focused at the component level (i.e. a particular type of emitter or photovoltaic). This has lead to the proposal of a wide variety of different system configurations. Unfortunately, the primary basis of these configurations often depends on whichever element is of particular interest to the researcher, which has lead to limited regard for any system effects. While the performance of the element may seem to be promising, it may be degraded or out-weighed by the other components when incorporated into a system. Thus, a method is needed for evaluating these system effects and comparing the performance of the various system types.

The model presented in this paper is for a combustion based TPV system and has been developed as a general purpose code. It allows the user to select the elements in the system from an existing data base as listed in Table 1, or to incorporate ones own data. Utilization of the code requires the user to select the type of system configuration and to define the systems geometry. Note that while provisions have been incorporated in the model for evaluating both flat plate and cylindrical

CP401, *Thermophotovoltaic Generation of Electricity: Third NREL Conference,*
edited by Benner/Coutts

Table 1. System Elements Database

Designator	Type		Reference
Emitters			
1	Gray body		N/A
2	Erbium	(composite)	3
3	Holmium	(composite)	3
4	Neodymium	(composite)	3
5	Ytterbium	(composite)	3
6	Ytterbium	(fibrous)	4, 5
7	Erbium	(solid)	6
8	Neodymium	(solid)	6
9	Ytterbium	(solid)	6
Filters			
1	- None -		N/A
2	Quartz, type 214 fused		7
3	Quartz, type 124 fused		7
4	Edtek T45		8, 9
5	Edtek T48		8, 9
6	Edtek T50		8, 9
Photovoltaics			
1	Silicon, myticrystaline		10
2	InGaAs, Spire		11, 12
3	InGaAs, NLRC		13, 14
4	GaSb		15, 16

configurations, only flate plate configurations will be presented here. In addition to defining the radiant configuration, the user must specify the values of various heat source parameters such as the percentage of heat loss and recuperation. A database of fuel properties is also included providing the user with a wide selection of gaseous and liquid fuel types.

Thus, the model allows the user to analytically predict system performance for a given system configuration, using the best available experimental data. Additional coding allows for comparing and optimizing the effect of various component types and system configurations.

SYSTEM MODEL IMPROVEMENTS

The analytical model is energy based and represents the TPV system as two principle components composed of a radiant transfer element and a heat source, as described in reference 1. Significant improvements have been made in the original model in an effort to better represent a practical system. The principle revisions are described in the following sections.

Radiant Transfer Element

Radiant Cavity

The radiant transfer element considers the interaction between the emitting source and the photovoltaics. It includes the emitter, photovoltaic cells and any semitransparent barrier separating them. Previously, the amount of spectral power incident on the photocell was optimistically calculated assuming no losses due to the geometry of the optical cavity. Functionally this is equivalent to representing the emitter / filter / pv cell configuration as infinite parallel flat plates. With this assumption the fraction of energy leaving the emitter, that is intercepted by the cell, F_{E-C}, is equal to one times the transmittance of the filter.

In an actual system the components of the radiant transfer element form a cavity of finite geometry. Interior to this cavity, multiple reflections of radiation occur between the components until the energy is absorbed by an element or lost by the system. An example of one possible flat plat configuration is shown in Figure 1

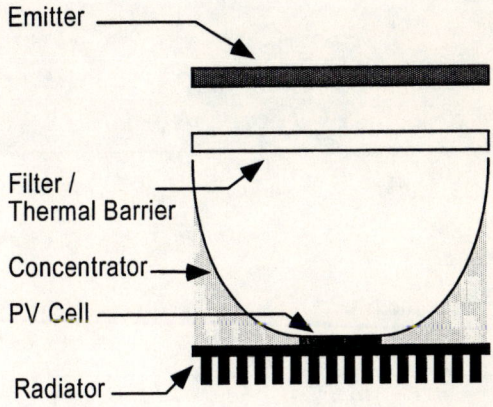

Emitter

Filter /
Thermal Barrier

Concentrator

PV Cell

Radiator

Figure 1. Example Flat Plate Configuration

In this type of configuration the effects of the finite dimensions of the cavity can significantly reduce the fraction of incident power, decreasing the overall performance of the system. The incorporation of spectral reflectors and concentrators is one method for reducing the potential losses and countering the geometric effects. These elements increase the amount of incident radiation on the photocell as reflected by an apparent increase in its view factor..

To account for the cavity effects of the flat plate configuration, the amount of energy incident on or absorbed in the elements of the system can be derived using a ray tracking method. This has been done by Hottel[17], assuming a cell-to-shield view factor of one, and identical emitter-to-shield and shield-to-emitter view factors. These equations have been reformulated and incorporated into the model. They have been derived without the view factor assumptions and include provisions for end effects[18]. However, care must be used in the evaluation of the view factor terms due to the presence of the filter. It has been assumed that the emitter is both diffuse and radiating uniformly, yielding a uniform radiation distribution on the surface of the photocell. For our purpose it is also assumed that optical properties are directionally independent and that Kircholioff's law holds as applied to the filter properties.

The resulting equations are functions of the wavelength λ and are expressed on a per unit of black body radiant power basis. Dropping the λ notation for simplicity, these equations are as follows:

energy fraction reabsorbed by the emitter,

$$G = \frac{\varepsilon^2 F_{ES} R F_{SE} B}{1 - (\rho_C F_{CS})(\rho_S F_{SC})\left[1 - F_{ES} R F_{SE} B(1-\varepsilon)\right]} \qquad (1)$$

energy fraction transmitted through the emitter,

$$G = \frac{\varepsilon F_{ES} R F_{SE}(1 - B)}{1 - (\rho_C F_{CS})(\rho_S F_{SC})\left[1 - F_{ES} R F_{SE} B(1-\varepsilon)\right]} \qquad (2)$$

energy fraction absorbed in the filter,

$$G = \frac{\varepsilon F_{ES} A}{1 - (\rho_C F_{CS})(\rho_S F_{SC})\left[1 - F_{ES} R F_{SE} B(1-\varepsilon)\right]} \qquad (3)$$

508

energy fraction incident on the photocell,

$$G = \frac{\varepsilon F_{ES} I}{1 - F_{ES} R F_{SE} B(1-\varepsilon)} \qquad (4)$$

The R, A, and I terms are written as

$$R = \rho_S F_{SE} + \frac{[F_{SE}(\rho_C F_{CS}) + F_{SS}](1 - \alpha_S - \rho_S)^2}{1 - (\rho_C F_{CS})(\rho_S F_{SC})} + \frac{F_{EE}}{F_{ES}} \quad , \qquad (5)$$

$$A = \alpha_S \left(1 + \frac{(1 - \alpha_S - \rho_S)[F_{SE}(\rho_C F_{CS}) + F_{SS}]}{1 - (\rho_C F_{CS})(\rho_S F_{SC})} \right) \quad , \qquad (6)$$

and

$$I = \frac{F_{SC}(1 - \alpha_S - \rho_S)}{1 - (\rho_C F_{CS})(\rho_S F_{SC})} \quad . \qquad (7)$$

The "F" terms in the above equations represent the apparent view factors between the designated components, with the subscripts E, S and C, refering to the emitter, shield or barrier and photocell respectively. The B term represents the fraction of the emitter surface area that is radiating or intercepting any reflected energy (i.e. for a solid emitter $B=1$).

The power terms which have been formulated on a per unit of emitter area basis are then

$$P_\lambda'' = G_\lambda \times P_{\lambda,b}'' \qquad (8)$$

The term $P_{\lambda,b}$ is the spectral distribution of black body emissive power at the emitter radiation temperature, T_{RAD}. Integration of Equation 8 over all possible wavelengths yields the total power terms associated with Equations 1 through 4. Previously only the amount of power incident on the photocells was evaluated. This integration was performed numerically using a spread sheet with the spectral distribution being discretized into 12.5 nanometer intervals. The evaluation of all the power terms associated with Equations 1 - 7 have now been incorporated as a subroutine into the main program.

509

The effectiveness of a radiant configuration is evaluated in terms two parameters; a cavity transfer coefficient, C_T, and a recovery factor, R. The cavity transfer coefficient represents the ratio of the incident power on the photocell, P_{INC}, to the total amount of power radiated by the emitter, $P_{E,TOT}$.

$$C_T \equiv \frac{P_{INC}}{P_{E,TOT}} \tag{9}$$

The recovery factor accounts for the amount of heat that is optically recuperated by the emitter. It is defined as,

$$R \equiv \frac{P_E}{P_E - P_{E,ABS}} \tag{10}$$

where $P_{E,ABS}$ is the amount of power that has been reflected back to the emitter and reabsorbed as evaluated from Equation 1. Since, this recuperation results in an improvement in the systems efficiency, it is included as a multiplier in Equation 21.

Photocell Parameters

Given the spectral distribution of power incident on the photocells and cell response parameters, the cell efficiency can be derived. By definition, the efficiency of a photocell is the ratio of the amount of electrical power generated to the amount of incident optical power and is calculated from the expression,

$$\eta_{PV} \equiv \frac{P_M}{P_{INC}} = \frac{V_{OC} I_{SC} FF}{incident\ power} \tag{11}$$

where P_M is the maximum obtainable power and the parameters V_{OC}, I_{SC} and FF are the open circuit voltage, short circuit current and fill factor respectively. Previously, these parameters where estimated from experimental data which was typically taken under AM0 conditions at a cell temperature of 25 °C. These parameters however are a function of the incident photon flux and cell temperature. Thus, several provisions have been incorporated to calculate the cell temperature, T_C, and to "correct" the parameters for the illumination conditions of the radiant cavity.

a) The user has the option of defining the cell temperature, the cell position or both conditions. If only the position is specified an iterative process is used to evaluate the temperature. This process requires the following energy balance to be satisfied,

$$P_{INC} = P_M + Q_{LOSS} \tag{12}$$

The Q_{LOSS} term represents the amount of energy convected or radiated away from the back side of the cells. It is express in terms of an overall heat transfer coefficient, U, as

$$Q_{LOSS} = UA_C(T_{C,BACK} - T_{SUR}) \tag{13}$$

where $T_{C,BACK}$ is the back surface temperature of the cell, and T_{SUR} is the surrounding temperature.

Initially a cell temperature is assumed and the photovoltaic efficiency and maximum power are calculated from Equation 11 given the incident power. The amount of heat loss is then determined using Equation 13, for the assumed temperature given the cell properties and transfer coefficient. The sum of these values are then compared with the left side of the equality in Equation 12. If the equality is not met, the assume value of T_C is incrementally increased until it is satisfied. Similarly, provision have been incorporated to calculate the position of the cells given a specified cell operating temperature. In this case, the distance from the cell to the emitter is incrementally increased until the energy balance is satisfied.

b) The user now has the option of specifying a fixed value of V_{OC} or letting the program derive its value from the cell data. In the case where the cell data is used, it is evaluated and corrected as follows: First, the short circuit current density is calculated as the integrated product of the spectral distribution of incident power on the cell and it responsivity. The program then determines the V_{OC} value from the data set which most closely corresponds to the calculated value of J_{SC}. This value, $V_{OC,J}$, is then corrected to match the cell temperature by the formulation

$$V'_{OC} = V_{OC,J} + \frac{dV_{OC,J}}{dT}(T_C - T_J) \tag{14}$$

where the $dV_{OC,J}/dT$ term is also evaluated from the data set. Secondly, the V'_{OC} value is then corrected to match the illumination level using

$$V_{OC} \approx V'_{OC} + \frac{kT_C}{q} \ln\left(\frac{J_{SC}}{J_{SC,J}}\right) \tag{15}$$

where the kT_C/q term is the associated thermal voltage of the cell.

c) If the user chooses to let the model evaluate the value of the open circuit voltage, the program will also determine the fill factor term. Ideally, the fill factor is only a function of V_{OC}, and is modeled using an empirical expression from Green[19].

$$FF_0 = \frac{v_{OC} - \ln(v_{OC} + 0.72)}{(v_{OC} + 1)} \tag{16}$$

This expression is accurate to four decimal places provided $v_{OC} > 10$, where v_{OC} is a normalized voltage defined as $V_{OC}/(kT_C/q)$.

Since pv cells normally have some parasitic losses due to series and shunt resistances the resulting fill factor will be less than ideal. A measure of this loss is evaluated from the experimental data, as the difference (A_1) between the data set value corresponding to $V_{OC, J}$ and the value calculated from Equation 16. The corrected FF value is then calculated using the corrected V_{OC} from Equation 15 as

$$FF = A_1 FF_0 \tag{17}$$

Having evaluated the effectiveness of the cavity transfer and the terms in Equation 11, both the maximum power output and the radiant transfer efficiency can be calculated. The radiant transfer efficiency has been previously defined as the ratio of output power to the total power emitted, such that for a flat plate,

$$\eta_{RT} \equiv \frac{P_M}{P_{E,TOT}} = C_T \eta_{PV} \tag{18}$$

Since the emitter is flat, it is radiating in two directions; away from the photocell and toward the photocell, such that $2 P_E = P_{E,TOT}$. In the case where it is assumed that all of the energy radiated away from the photocell is recovered, P_E is used in place of $P_{E,TOT}$ in Equation 9, and the resulting calculation is referred to as the "one sided" efficiency η^1_{RT}.

Heat Source

The heat source portion of the model considers the amount of available heat, q_{34}, from a hydrocarbon fueled recuperative type burner. As described in reference 1, the principal elements of the combustor model include the inlet, air preheat, combustion, energy extraction, exhaust and ambient states. Previously, the STANJAN[20] program was used to determine the temperature and enthalpy at the six states in the model. In doing this, fuel properties inputs had to be provided by the user at each step and the program had to be run outside of the model used for calculating the radiant transfer.

The primary improvement in the heat source portion of the model was the full integration of the calculations into code. This negates the need to employ STANJAN[20] as an external code. In its place, a modified version of the NASA Complex Chemical Equilibrium Compositions and Applications (CEA) code was added as a subroutine to the main program. Additionally, since the CEA coding includes a large number of species in its thermodynamics database, the need to input fuel property data was eliminated.

With the inclusion of the modified CEA coding, calculation of the thermodynamic states of the combustor from the user's point of view is now a one step process given the system parameters. Once the temperature and enthalpy of the six states have been evaluated, the amount of energy available for radiation at the emitter temperature is determined. Defining the overall thermal efficiency of the heat source, η_{HS}, as the ratio of the maximum available heat at the radiation temperature, to the heat input of the fuel, yields

$$\eta_{HS} = \frac{q_{34}(1 + r)}{H_C} \tag{19}$$

The r term is the air to fuel mass ratio, and the quantity $(1 + r)$ accounts for the fact that q_{34} is evaluated on a per unit mass or mixture basis. Note that this is a reformulation of the original equation and that it is a function of the radiation temperature, T_4, and independent of the mass flow rate. For any given radiative heat rate, $Q(T_4)$, the required amount of fuel mass flow is

$$\dot{m}_f = \frac{Q}{q_{34}(1 + r)} \tag{20}$$

Several other improvements or revisions have been made as part of the heat source calculations. In addition to evaluating the thermal efficiency, the adiabatic flame temperature and stoichiometric values are evaluated in conjunction with the CEA calculations. The properties of the exhaust products are also evaluated in anticipation of future heat exchanger sizing calculations. The "combustion efficiency" factor, η_{COMB}, which was previously used to indicate incomplete combustion has also dropped.

System Efficiency

As a result of the revisions to the model, the evaluation of the total system efficiency has been reformulated as,

$$\eta_{SYS} = \eta_{HS} R \eta_{RAD} \tag{21}$$

EXAMPLE PERFORMANCE PREDICTIONS

Several example cases have been evaluated and are briefly discussed to demonstrate the effects of various system configurations on system performance. In each of the example cases a GaSb photocell was used and various other parameters where changed.

CASE: P1
Config.:   Infinite parallel flat plates
Emitter: black body η_{HS} = 100.0 %
Filter: none η^1_{RAD} = 14.5 %
Photocell: GaSb, JX Crystals - 25 °C η_{SYS} = **16.2 %**
Radiation Temperature: 1800 °C P_d = 15.2 W/cm^2

Note that the efficiency of the thermal source has been defined to be 100%. This case then represents the maximum possible overall efficiency obtainable for a 1800°C black body / GaSb non-concentrator configuration. Incorporating a filter in the radiant cavity yields,

CASE: P2
Config.:   Infinite parallel flat plates
Emitter: black body η_{HS} = 100.0 %
Filter: quartz, type 214 fused - t = 0.1cm η^1_{RAD} = 13.6 %
Photocell: GaSb, JX Crystals - 25 °C η_{SYS} = **16.2 %**
Radiation Temperature: 1800 °C P_d = 14.2 W/cm^2

In this case, the radiant efficiency has decreased in comparison to P1. Primarily this is due to the absortion of the quartz which reduces the amount of incident power on the photocell. This effect is reflected by a reduction in the transfer coefficient from $C_T = 1.0$ in case P1 to $C_T = 0.8$. Note that the system efficiency has remained the same as in P1 case. This is due to an increased percentage of the emitted energy being reflected toward and reabsorbed in the emitter, increasing the overall recovery factor.

Now consider case P2 given finite geometry and an non-ideal heat source. Assuming dimensions of 5 cm x 5 cm, and the following heat source parameters:

Fuel: Propane
Oxidizer: Air
Theo. air: 110 %
Thermal loss parameter: 10 %
Thermal recuperation: 50%

then,

CASE: P3
Config.:   Parallel flat plates (5 cm x 5 cm)
Emitter: black body η_{HS} = 52.7 %
Filter: quartz, type 214 fused - t = 0.1cm η^1_{RAD} = 0.4 %
Photocell: GaSb, JX Crystals - 25 °C η_{SYS} = **0.2 %**
Radiation Temperature: 1800 °C P_d = 1.9 W/cm^2
Position: r_E = 0.0 cm, r_F = 1.0 cm, r_C = 4.5 cm

The drastic reduction in the system performance from the previous case is the result of the inclusion of the system effects. Going from the previous assumption of an ideal heat source with 100% efficiency (in the P2 case), to a combustion heat source reduced the overall performance by almost 50% even with thermal recuperation. Geometric effects account for an additional 43% performance reduction. The majority of the radiative losses is due to end effects.

One method for counteracting the radiant losses of a finite configuration is the inclusion of spectral reflectors. In addition, the reflectors can be used to concentrate the incident radiation on the photocell, potentially improving cell efficiency. This can be shown by considering the follow case given the same heat source parameter as specified for P3.

CASE: P4
Config.:   Parallel flat plates
Emitter: black body η_{HS} = 52.7 %
              5 cm x 5 cm η^1_{RAD} = 8.3 %
Filter: quartz, type 214 fused - t = 0.1cm η_{SYS} = **5.4 %**
              5 cm x 5 cm P_d = 43.7 W/cm^2
Photocell: GaSb, JX Crystals - 25 °C
              1 cm x 5 cm
Radiation Temperature: 1800 °C
Position: r_E = 0.0 cm, r_F = 1.0 cm, r_C = 4.5 cm

Reflectors:
- flat spectral reflectors on the sides between the emitter and filter
  (2 sides, $\rho = 0.9$)
- flat plate concentration between the sides of the filter and the photocell
  (2 sides, $\rho = 0.9$)
- flat spectral reflectors on the ends between the filter and the photocell
  (2 ends, $\rho = 1.0$)

When comparing case P4 to case P3, there is a notable performance increase both in efficiency and power density. The improvements in radiant efficiency are primarily due to a 270% increase in the transfer coefficient. Additionally, with the radiation now being concentrated on a smaller cell area, the photocell efficiency has increased from 10% to 17%. Both of these factors also contribute to the improved power density for the configuration.

In both the P3 and P4 cases, a cell temperature of 25 °C was assumed. To maintain this assumption would require an additional 180 W/cm^2 of power dissipation from the cell. If this amount of energy can not be removed from the cell its temperature will rise, reducing the efficiency and power density. For example assuming a cell temperature of 100 °C, results in a 25% reduction in cell efficiency.

Finally, note that even with the end losses for case P4, the P_d term is greater than the ideal case (i.e. P1) where there are no end losses. This is the direct result of the optical concentration.

OPTIMIZATION CODING AND RESULTS

In addition to improving the analytical model, provisions have been included in the code to allow for the optimization of various system parameters. Currently, these provisions simply perform multiple runs of the performance code, and allow for the direct comparison of the calculated parameters. These parameters include system efficiency, power density, size and cost. However, system size and cost can not be accurately evaluated given the current data sets.

Several near optimum configurations were evaluated by looping through 28,900 runs of the performance code. In each case an infinite parallel flat plate configuration was chosen. The configurations included all possible emitter / filter / photocell combination available in the data set as listed in Table 1. Emitter temperatures between 1000 °C and 1800 °C where considered in 5 °C increments.

This resulted in the following near optimum configurations:

O1: Maximum Radiant Efficiency - $\eta_{RAD} = 60.4\%$ ($\eta_{SYS} = 33.3\%$)
Emitter: Erbium, composite
Filter: Edtek T45
Photocell: InGaAs, MIM - NLRC
Emitter temperature: 1800 °C

O2: Maximum Power Density - $P_d = 35.1$ W/cm^2 ($\eta_{SYS} = 19.8\%$)
Emitter: black body
Filter: Edtek T45
Photocell: InGaAs, MIM - NLRC
Emitter temperature: 1800 °C

To illustrate an optimum system configuration it is necessary to define the heat source parameters. Otherwise the coding would simply increase the amount of heat recuperation to 100%, and decrease the loss parameter to 0%, until the heat source efficiency reached 100%. Choosing the heat source parameters to be; Propane / Air, 110% theoretical air, a 10% loss parameter, and 50% heat recuperation, yields

O3: Maximum System Efficiency - $\eta_{SYS} = 57.2\%$
Emitter: Erbium, composite
Filter: Edtek T50
Photocell: InGaAs, MIM - NLRC
Emitter temperature: 1595 °C

In this case, is should be observed that the emitter temperature is lower than in the case of the configuration with the maximum radiant efficiency. This is due to opposing effects of the radiant and heat source efficiencies. Typically, the radiant efficiency will increase with emitter temperature. Whereas, the efficiency of the heat source will linearly decrease with radiant temperature.

SUMMARY

In summary, an improved model has been presented for analytically predicting system performance. This model includes provision for evaluating the effects of finite system geometry and non-ideal heat sources which have been show by various authors to drastically reduce the overall performance of the system. This has been illustrated by the presentation of several example cases. Additionally, the model has been used to evaluate the best emitter / filter / photocell combinations for several optimum cases.

ACKNOWLEDGMENTS

The authors would like to take this opportunity to thank the Army Research Office and the Department of Energy for their funding and support under research grants DAALO39260205-1 and DE-FG05-95ER12156. Thanks is also due to Bonnie McBride of NASA Lewis for providing the CEA code incorporated into the model. In addition, we would also like to thank the various researcher including; Peter Adair, Jim Avery, Zheng Chen, Mark Morgan, Robert Nelson, Dave Wilt, and Steve Wojtczuk, for providing the data used in the optimization portion of this work.

REFERENCES

1. Schroeder, K. L., Rose, M. F. and Burkhalter, J. E., "A Parametric Study of TPV Systems and the importance of Thermal Management in System Design and Optimization," Proceedings of the 30th Intersociety Energy Conversion Engineering Conference, Aug. 1995.
2. McBride, B. J. and Gordon, S., "Computer Program for Calculation of Complex Chemical Equilibrium Compositions and Applications," NASA Reference Publication 1311, part II Users Manual and Program Description, June 1996.
3. Adair, P. L. and Chen, Zheng, personal communications, 1997.
4. Parent, C. R., Nelson, R. E. and Olow, J. F. S., "Exploratory Research on Natural Gas-Fired Thermophotovoltic Systems," final report, GRI-88/0138, Gas Research Institute, April 1988.
5. Nelson, R. E., personal communications, March 1997.
6. Guazzoni, G. E. and Shapiro, S. J., "Spectral Emittance of Neodymium, Samarium, Erbium and Ytterbium Oxides at High Temperature," ECOM-3281, May 1970.
7. brochure, "GE Quartz," 295/5M.
8. Horne, W. E., Morgan, M. D. and Sundaram, V. S., "IR Filters for TPV Converter Modules," AIP Conference Proceedings No. 358 - The Second NREL Conference on Thermophotovoltaic Generation of Electricity, American Institute of Physics Press., pp. 35-51, 1995.
9. Morgan, M., personal communications, Feb. 1997.
10. Panton, R. A.- Solar Sales Manager, Kyocera, personal communications Feb. 1997.
11. Wojtczuk, S. and Colter, P., "Production Data on 0.55eV InGaAs Thermophotovoltic Cells," Proceedings of the 25th IEEE PV Specialist Conference, 1996.
12. Wojtczuk, S., personal communications, Oct. 1996.
13. Wilt, D. M., et. al., "Monolithically interconnected InGaAs TPV Module Development," Proceedings of the 25th IEEE PV Specialist Conference, 1996.

14. Wilt, D. M., personal communications, Feb. 1997.
15. Frass, L. M., et. al., "Fundamental Characterization Studies of GaSb Solar Cells," 22nd IEEE PV Specialist Conference, 1991.
16. Avery, J., personal communications, 1997.
17. White, D. C., and Hottel, H. C., "Important Factors in Determining the Efficiency of TPV Systems, Part II - "Radiative Transfer Efficiency of a Flat Plate TPV System: Analytical Model and Numerical Results," AIP Conference Proceedings No. 321 - The Second NREL Conference on Thermophotovoltaic Generation of Electricity, American Institute of Physics Press., pp. 437-454, 1994.
18. Schroeder, K. L., "Performance Optimization of Thermophotovoltic Systems," Ph.d. dissertation, 1997.
19. Green, M. A., <u>Solar Cells Operating Principles, Technology and System Applications</u>, reprinted by The Univ. of New South Wales, 1992.
20. Reynolds, W. C., "STANJAN" computer code, Stanford University.

Practical Development and Thermodynamic Modeling of a Complete TPV Generator

Edward M. West

Vehicle Research Institute
Western Washington University, Bellingham, WA 98225

Abstract. Research on thermophotovoltaic (TPV) generator technology at Western Washington University has been focused on complete systems engineering and development. Descriptions and results of computer models and hardware evaluations are presented, and conclusions about TPV systems are made where appropriate. Specific areas of discussion include; heat exchange from combustion gases to the emitter surface, modeling and evaluation of radiation cavity performance and photovoltaic cell cooling system development. All areas of investigation involve both the construction and testing of prototype hardware and the development of engineering models to predict system performance using actual system parameters.

INTRODUCTION

Thermophotovoltaic (TPV) generator development at Western Washington University (WWU) is progressing along two avenues. The primary emphasis is toward the practical development of application oriented TPV generator systems, while thermodynamic modeling is carried out to support the TPV generator system development. This paper outlines the major areas of work directed toward hardware development. It also describes in detail some of tools developed to model the complex thermodynamic interactions within the TPV system.

PRACTICAL DEVELOPMENT

At WWU, the development of TPV specific hardware is progressing in the following areas:

- Fuel mixing and combustion stability enhancement hardware

CP401, *Thermophotovoltaic Generation of Electricity: Third NREL Conference,*
edited by Benner/Coutts
521

- Matched emitters for GaSb based systems[1]
- Ceramic and stainless steel exhaust to air recuperators
- Compressed natural gas fuel storage and delivery
- Electrically heated receiver test station
- Photovoltaic cell support, wiring and assembly systems
- Photovoltaic cell water cooling systems
- Peak power tracker power converters[2]

Combustion work at WWU is focused on stability enhancement to improve burner start-up and warm-up characteristics. Because of the high preheat temperatures, the fuel must be introduced at the desired combustion point. This constraint demands that the fuel mixing is thorough and rapid. An air diffuser has been developed which provides a low velocity, high turbulence region into which the fuel is injected. This arrangement has allowed for good starting, moderate warm-up emissions and excellent high load performance.

Matched emitters have been developed which have an emission band that is matched to the bandgap of the GaSb photovoltaic (PV) cells that are used in our system (1). These emitters are ceramic / ceramic composite which have been doped to create a matched emittance band. Improved power and efficiency due to the matched emitters has been demonstrated in a 25-watt generator (2). Development is continuing to create larger emitters suitable for use in 500-watt generators.

The TPV system efficiency can be substantially enhanced through the use of heat exchangers to recuperate the exhaust heat into the intake air stream. Stainless steel counterflow heat exchangers have been developed which are more than 65% efficient. These recuperators are capable of handling intake temperatures of 1400°K with good durability. Research is ongoing in the area of high temperature ceramic heat exchangers. Ceramic recuperators have been developed which can tolerate intake temperatures of 1700°K. Further work on the ceramic heat exchanger is needed to achieve the full emitter temperature of 1800°K.

Compressed natural gas storage and delivery systems are under development. The use of carbon fiber wrapped storage tanks provides safe and effective fuel storage up to 3600 psi. Two stage regulators are used to provide a stable supply of fuel to the generator system. Finally, control valves and stainless steel fittings control and deliver the fuel to the burner. Automated burner control systems that monitor the combustion process and control the fan, fuel and igniter are being evaluated.

---

[1] Development in partnership with JX Crystals, Issaquah, WA
[2] Development in partnership with Xantrex, Burnaby, B.C., Canada

An internally developed electrically heated receiver test station is available to evaluate changes made to the radiation cavity system. A programmable process controller, coupled to a silicon-controlled rectifier (SCR) power module, provides accurate temperature control of the emitter. Phase compensated input power measurement hardware provides accurate information about the required emitter power. Array output voltage and current meters provide the capability to produce a system I-V curve as well as measure maximum power output. Low temperature thermocouples are available to measure cell and cooling system temperatures. Finally, high temperature thermocouples are available to measure emitter, shield and insulator temperatures. Continuous water-cooling is available so that any water-cooled array can be tested for extended periods.

PV array systems have been developed in a variety of cylindrical configurations. Cell support systems including circuit boards, jumper leads, board interconnects, and water-cooled receivers have been developed. Circuit and receiver manufacturing equipment has been developed and fabricated. This equipment allows for the manufacture of the complete water-cooled receiver circuits. Complete self-contained water-cooled receiver arrays have been fabricated and tested on the receiver test station.

Internally finned, water-cooled receivers that provide effective cooling of the PV cells, are manufactured in-house. Upper and lower water galleries provide physical support for the receivers and also deliver and remove the water from the receiver array. Water to air radiators are used to cool the water before it is returned to the receiver system.

Peak power trackers (PPT) are being developed which will maximize system output power and efficiency. These trackers provide the ability to load match the array to a wide variety of external loads with precise control. The presence of a PPT in the TPV system also provides a degree of fault tolerance. With a PPT in the system, part of the system could fail, but the balance of the system would continue to operate.

THERMODYNAMIC MODELING

To support the development of TPV system hardware, various thermodynamic models have been developed. These models include, but are not limited to, the following system processes:

- Convection, conduction and radiation transfer from combustion gas to emitter
- Radiation interaction heat transfer for the emitter radiation cavity
- Conduction heat transfer / temperature profile in the receiver cross section

The first two models use an iterative numerical solution methodology, while the third uses Finite Element Analysis (FEA) to arrive at a solution. These models are described in the following sections, with their results compared to the relevant data generated using current hardware.

Combustion Gas to Emitter Heat Transfer

This analysis is carried out to determine the geometric effects of the combustion gas to emitter heat transfer, and provide a design tool for assisting the development of emitter systems that have a uniform temperature distribution. This heat transfer analysis includes combustion gas to ceramic tube convection, tube wall conduction and tube surface radiation.

Convection. The analysis of the heat transfer from the combustion gases to the emitter begins with determination of the Reynolds number. Given the mass flow through the system and the annular geometry of the interior of the emitter heat exchanger, the Reynolds number is calculated in the following form

$$Re = \frac{m \cdot D_h}{\mu \cdot A_f} \tag{1}$$

where m is the mass flow, D_h is the hydraulic diameter[3], μ is the viscosity and A_f is the flow area. For the systems that have been developed, the Reynolds number ranges from $\approx 1,000$ to $\approx 1,750$. Given that the onset of turbulence for internal flow is $\approx 2,300$, we conclude that the flow is in a laminar regime (3).

Given laminar flow conditions and uniform heat flux at both surfaces, the Nusselt number for the two annular surfaces is computed with the following equations

$$Nu_i = \frac{Nu_{ii}}{1 - \frac{q_o}{q_i} \cdot \theta_i} \tag{2a}$$

$$Nu_o = \frac{Nu_{oo}}{1 - \frac{q_i}{q_o} \cdot \theta_o} \tag{2b}$$

where the influence coefficients (Nu_{ii}, Nu_{oo}, θ_i, θ_o) are derived empirically for annular geometries and can be obtained in heat transfer literature (4). With an

---

[3] For annular flow geometry, D_h is defined as $D_o - D_i$ for the give annulus.

expression for the Nusselt number, the convection coefficients for the two surfaces can be evaluated in the following form

$$h = \frac{Nu \cdot k}{D_h} \tag{3}$$

where k is the thermal conductivity of the fluid. Finally, the total heat transfer to each surface can be computed with the expression

$$q_{conv} = h \cdot A \cdot \left(T_m - T_s \right) \tag{4}$$

where T_m and T_s are the mean fluid temperature and surface temperature respectively and A is the surface area. Since the computation of the Nusselt number in equation 2 is dependent on the heat flux ratio (q_i/q_o), the derived solution is recursive and requires an iterative numerical solution.

Conduction. Conduction heat transfer through the walls of the ceramic tubes is simultaneously evaluated during the numerical solution. Since the analysis is for steady state conditions, the conduction heat transfer calculation is carried out by evaluating the following expression

$$q_{cond} = \frac{k \cdot A}{L} \cdot \left(T_i - T_o \right) \tag{5}$$

where k is the thermal conductivity, L is the thickness of the ceramic wall and T_i and T_o are the temperature of the inner and outer surfaces respectively.

Radiation. To complete the analysis, the radiation heat transfer must also be determined. Surface to surface transfer modeling is accomplished through the evaluation of the *Stefan-Boltzmann law* for a gray body. This law states that the total emissive power for a gray body is

$$q_{rad} = A \cdot \varepsilon \cdot \sigma \cdot T_s^{\,4} \tag{6}$$

where ε is total surface emissivity, σ is the *Stefan-Boltzmann* constant and T_s is the surface temperature.

Additionally, the possibility of radiation exchange exists between the high temperature gas stream and the adjoining surfaces. Gaseous radiation exchange can be evaluated using the method developed by Hottel (5). Given the system geometry present inside our burner, the interacting gas inside the flow annulus has a *mean beam length* $(L_e) \approx .005$ m. With L_e and the partial pressures of CO_2 and H_2O in the combustion gases, gas emissivity is estimated to be $\approx .002$ at 2000°K. With such a low gas emissivity, gaseous radiation in this system is negligible.

Energy Balance. Control surface equations are developed for each of the five interacting surfaces in the emitter heat exchanger system. Each of these equations follows the form

$$q_{tot} = q_{conv} - q_{cond} - q_{rad} + q_{abs} \qquad (7)$$

where q_{conv}, q_{cond}, q_{rad} are defined in equations 4, 5 and 6. The q_{abs} term is the absorbed radiant power from other radiating surfaces, considering the view factor.

Numerical Solution. Each of the five control surface equations is evaluated over one-centimeter increments of the emitter heat exchanger system. In addition to the evaluation of the control surface equations, the gas temperature and gas properties are reevaluated to track the cooling effects in the combustion gases. This is necessary at high temperatures because the specific heat of gases changes with temperature in a non-linear manner.

The iterative numerical solution of the control surface and property equations explicitly determines the five interacting surface temperatures, the combustion gas stream temperature and the incremental and cumulative output power. From the model, the five surface temperatures and the gas stream temperature are plotted versus emitter position in Figure 1.

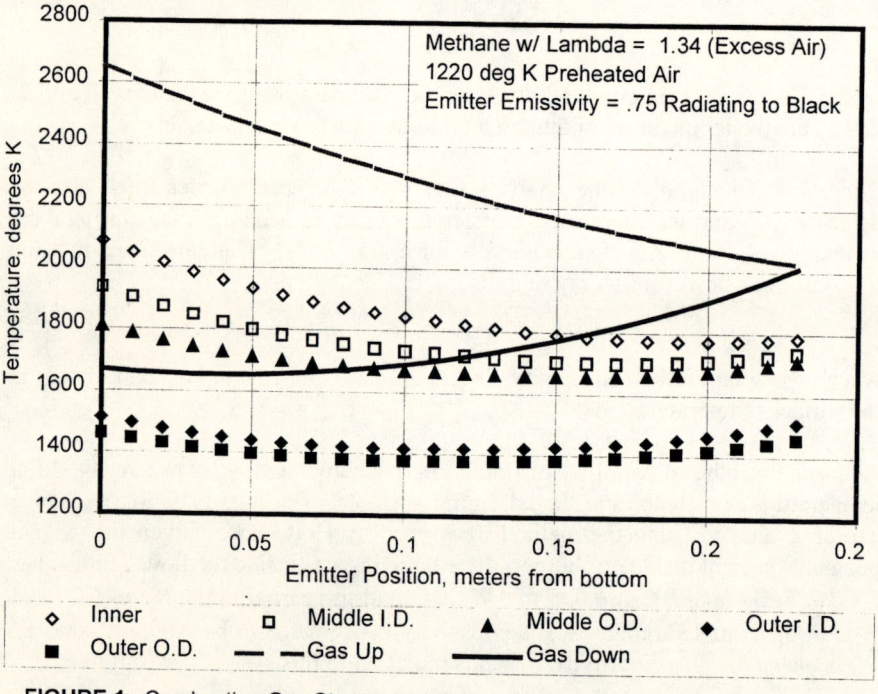

FIGURE 1. Combustion Gas Stream and Emitter Surface Temperatures vs. Position

This data indicates that the cumulative effect of the three modes of heat transfer with this geometric configuration can allow for a uniform emitter temperature distribution.

Comparative Results. The geometry used in this model is the current intermediate size emitter with a nominal outside diameter of 5.3 cm. The fuel flow rates and air preheat temperature used in the model are measured data. As can be seen in Figure 2, the results from the thermodynamic model agree well with the measured temperature distribution for an emitter of the same geometry. The maximum error of prediction for the model is four percent at the top of the emitter. It is believed this discrepancy is caused by heat transfer enhancement from the induced turbulence at the turn-around at the top of the emitter.

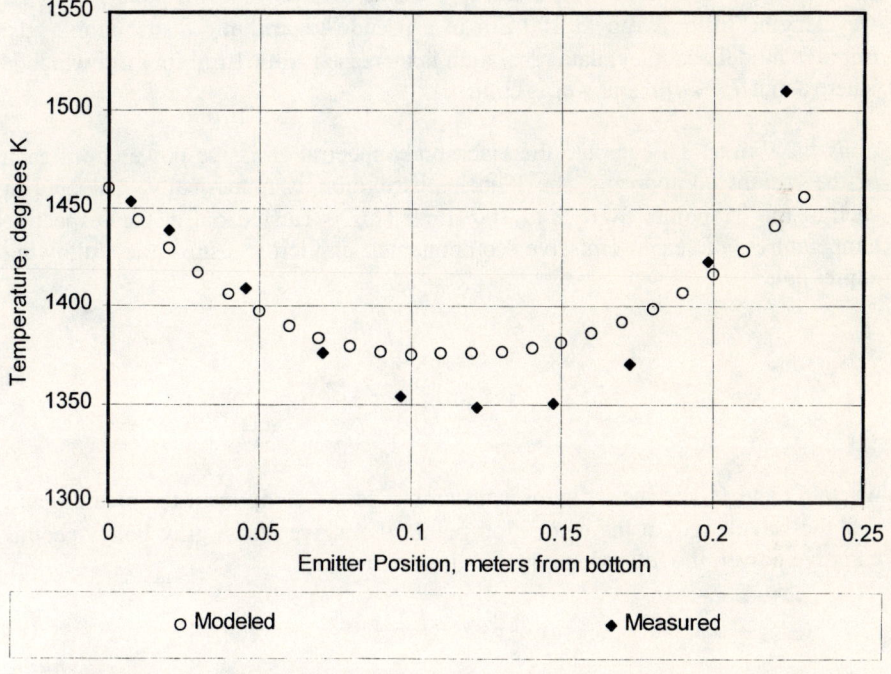

FIGURE 2. Modeled and Measured Emitter Surface Temperature vs. Position

This model is now being used to maximize heat transfer and enhance temperature uniformity in larger emitter systems. These larger emitters are being developed to improve the emitter to receiver array view factor. This model also provides a means to evaluate potential system improvements, such as material substitutions, without having to fabricate and test the resulting effects.

The emitter to receiver thermodynamic model is a radiation interaction model developed to investigate the effects of geometry and material properties on system performance. This model accounts for the cylindrical geometry of the emitter, the quartz shields and the receiver array. The receiver array geometry data includes the percentage of mirror, parasitic gap and cell grid lines.

Radiation Properties. To address the effects of the material properties, each of the interacting materials' radiation properties are provided in tabular form, either directly measured or from the literature (6). The tabular data for each of the materials includes relevant aspects of the materials emissivity, absorptivity, transmissivity or reflectivity. Each of the data sets provides 31 data points for wavelengths from .2 µm to 100 µm in a pseudo-logarithmic distribution. To improve model accuracy, data resolution is increased from 1 µm to 4 µm which is where about 75% of the power is emitted.

Emissive Power. To compute the black body spectral emissive power from each of the radiant components, the Planck distribution is numerically evaluated at each of the 31 points from .2 to 100 µm. This is carried out at the respective temperatures of each emissive components, explicitly using the following expression

$$E_b = \frac{C_1}{\lambda^5 \cdot \left(exp\left(\frac{C_2}{\lambda \cdot T} \right) - 1 \right)} \tag{8}$$

where C_1 and C_2 are the radiation constants, T is the component temperature and λ is the wavelength at the evaluation point. To arrive at the gray body spectral emissive power, the following equation is employed

$$E_\lambda = E_b \cdot \varepsilon_\lambda \tag{9}$$

where ε_λ is the spectral emissivity and A is the emissive area.

Control Surfaces. To model the current WWU emitter / receiver system, a set of eight control surfaces track the energy flow from the emitter to the receiver array. There are control surfaces for the emitter and receiver array, and two surfaces at each of the three quartz shields. The eight control surfaces each require two equations to account for the energy flow in and out of each surface. The pairs of control surface equations are in the basic form

$$ql_{in} = Gl_\lambda \cdot A_1 + ql_{out} \cdot \left(1 - F_{10} \right) \tag{10a}$$

$$q1_{out} = E_T \cdot \varepsilon_\lambda \cdot A_1 + q1_{in} \cdot \rho_\lambda + q2_{in} \cdot (1 - \rho_\lambda) \cdot \tau_\lambda \tag{10b}$$

These equations are interactive, so an iterative numerical solution is carried out.

Energy Balance. Through the evaluation of the eight control surface equations over the 30 spectral bands, the flow of energy is modeled throughout the system. These results are combined for each component to derive the system temperatures and also arrive at a converged equilibrium solution. This is accomplished by satisfying the energy equilibrium equation

$$\int_0^\infty E_{abs} \, d\lambda - \int_0^\infty E_{rad}(T) \, d\lambda = 0 \tag{11}$$

where E_{abs} is the power absorbed by a component and E_{rad} is the power emitter by a component at its temperature. In the case of the emitter and the receiver array, the user specifies each temperature and equation (11) need not be satisfied.

Electric Power Output. With a converged system solution, the spectral irradiation (G_λ) on the PV cells can be explicitly derived from the control surface equation for the receiver array. Irradiation is then used to calculate a photon flux, which is incident on the junction, as a function of wavelength with the relationship

$$Flux_\lambda = \frac{G_\lambda \cdot \lambda}{h \cdot c} \tag{12}$$

where h and c are the Planck constant and the speed of light respectively. Knowing the photon flux and the spectral quantum efficiency (Q_e) of the PV cell, the output power is evaluated with the expression

$$P_{electric} = Q_e \cdot Flux_\lambda \cdot V \cdot eV \tag{13}$$

where V is the peak power voltage of the PV cell and eV is the electron volt.

System Efficiency. With an evaluation of the output power complete, the system efficiency can now be calculated. To determine system efficiency, the input power must be known. Input power ($P_{emitter}$) is calculated using equation (11) as part of the energy balance evaluation for the system. System efficiency is then calculated using the expression

$$\eta_{sys} = \frac{P_{electric}}{P_{emitter}} \tag{14}$$

Comparative Results. The results of the radiation cavity analysis can then be plotted versus temperature for a given geometry. These results can be seen in Figure 3 where power and efficiency are plotted versus emitter temperature. Figure 3 also contains a set of measured data curves from the actual system to compare with the model results.

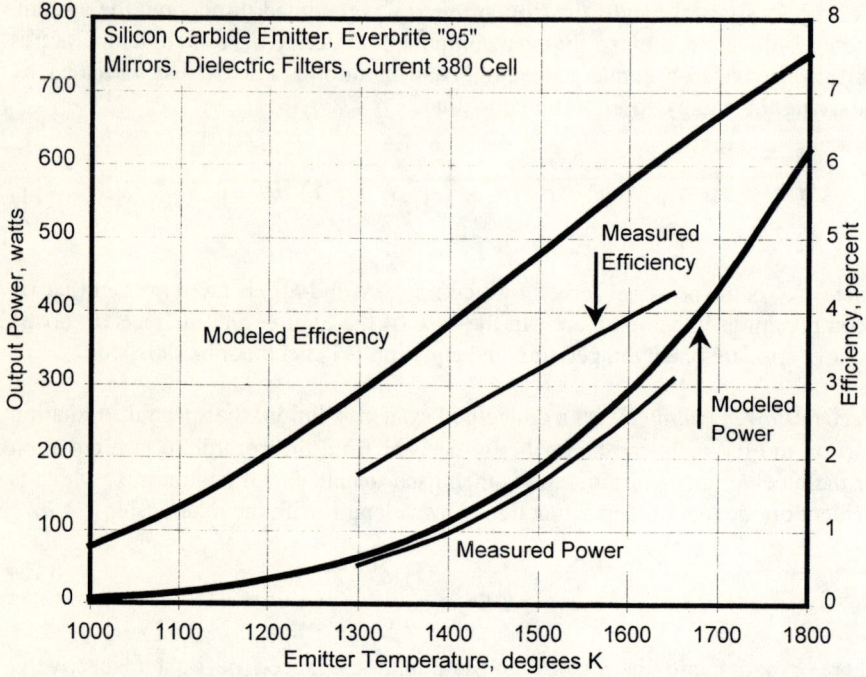

Figure 3. Modeled and Measured Power and Efficiency versus Emitter Temperature

These results clearly indicate that the model can accurately predict the output power of this system. However, the model does consistently over predict the output power by about 15%. By virtue of the accuracy, the model can be effectively used for sensitivity studies when evaluating potential system improvements. Although the absolute predicted efficiency is not accurate, the model results do correlate well with the efficiency trend of the system. The efficiency predictions also correlate well with other published model results (7). The principal error in the prediction of efficiency is believed to arise from ignoring the conduction losses to the array.

Temperature Profile in Receiver

To develop an effective cooling system for the PV cells, an analysis is carried out to evaluate the cell cooling system. The primary performance determinant is the receiver water cooling system. To analyze this system, two modes of heat transfer are modeled; Convection heat transfer and conduction heat transfer.

Convection. Convection is modeled for the internal water cooling passage to determine the heat transfer coefficient. To determine the heat transfer, the first parameter that must be evaluated is the Reynolds number. The Reynolds number for internal flow is calculated using the following expression

$$Re_d = \frac{\rho \cdot V \cdot D_h}{\mu_f} \tag{15}$$

where ρ, V and μ_f are the density, velocity and viscosity of the water respectively. D_h is the hydraulic diameter for the internal flow geometry, which is determined from the following expression

$$D_h = \frac{4 \cdot A_c}{P} \tag{16}$$

where A_c is the flow area and P is the wetted perimeter of the flow area. For the developed geometry used in the WWU receiver, the Reynolds number is about 700. This value of Reynolds number is well below the onset of turbulence which occurs at about 2300 for internal flow.

The heat transfer coefficient is then determined to arrive at the boundary conditions on the inside of the receiver passage. The heat transfer coefficient is calculated from the expression

$$h = \frac{Nu_D \cdot k}{D_h} \tag{17}$$

where k is the thermal conductivity of water and Nu_D is the dimensionless parameter for temperature gradient at the interior surface of the passage. Nu_D is based on the solutions of the differential momentum and energy equations for flow in duct sections of the given geometry, and is found in the literature (8). Once the heat transfer coefficient is derived, the evaluation of the temperature profile can be made.

Geometry Model. To define the receiver heat transfer geometry, a solid model of the receiver is developed. First, a solid model is created for all of the different interacting components including geometry and thermal conductivity. Then, the entire receiver system is assembled from the library of components with all of the thermal interactions defined. Table 1 lists the library of components and their respective thermal conductivity.

Table 1. Receiver components, materials and thermal conductivity.

Component	Material	Thermal Conductivity
Photovoltaic Cell	Gallium Antimonide	40 W/m·K
Electrical Solder	50% Lead / 50%Tin	50 W/m·K
Circuit Trace	Copper	384 W/m·K
Insulating Film	Polyamide	.19 W/m·K
Circuit Board	Copper	384 W/m·K
Circuit Board Adhesive	Aluminum Filled Epoxy	8.65 W/m·K
Receiver Body	Aluminum (6061-T6)	167 W/m·K

Model Solution. The temperature profile in the receiver is determined by finite element analysis (FEA) with an explicit definition of geometry, thermal conductivity and convection coefficient. First, an FEA mesh is developed for the given model assembly which is used to derive the constraining system equations. Then, a numerical solver is used to satisfy the system equations and constraints to arrive at a converged system solution.

This solution is output from the system in the form of a temperature profile image. There is also an explicit database of temperatures throughout the cross-section of the receiver. This information can then be used to evaluate the benefits of any design changes to arrive at a satisfactory system configuration.

Analysis Results. Figure 4 shows the temperature distribution image for one of the current WWU receiver designs.

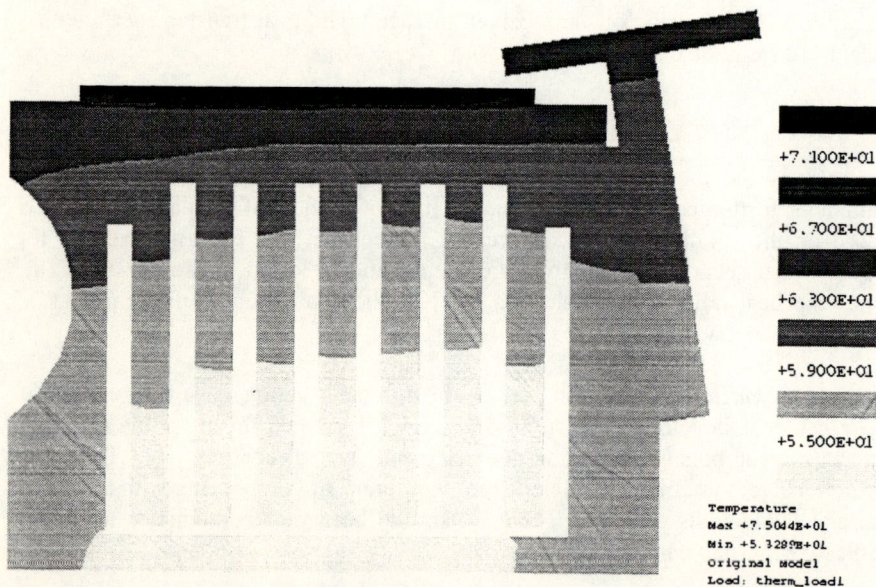

+7.100E+01

+6.700E+01

+6.300E+01

+5.900E+01

+5.500E+01

Temperature
Max +7.5044E+01
Min +5.3289E+01
Original model
Load: therm_load1

Figure 4. Receiver Temperature Profile Generated Using Finite Element Analysis

The temperature distribution shown is for a 25 w/cm^2 heat load on the cell face, and a 4 w/cm^2 load on the mirror surface. This load condition is representative of an 1800°K emitter temperature for the current WWU array geometry. The maximum junction temperature is 75°C, while the cooling water is at 50°C and is flowing at one liter per minute.

The thermal analysis allows for the improvement of receiver cooling design in an efficient manner without having to build multiple evaluation units. Using this technique, the cooling performance of the receiver has been increased from the older design. The temperature drop from the PV junction to the water has been reduced by 70%.

Water Cooling Benefits. The use of water cooling in this type of system has a number of advantages. Water cooling of the PV cell provides a much higher heat transfer coefficient, to the coolant, than does the use of air cooling. Additionally, an automotive type radiator provides a higher available surface area from water to air when compared to air-cooled systems using fins directly contacting the PV cell. The result is a higher heat transfer rate from the junction to the environment for a given temperature difference. Since the water cooling system can hold the junction temperature closer to ambient temperature than air cooling, the system will have a higher efficiency. An added effect of using water-cooling is that the temperature uniformity of the receivers is good due to the high specific heat of the cooling water. Finally, having water behind the cells provides a safety margin if the cooling system fails. In the event of a cooling system failure, the water can boil off while the system is being shut down, thus preventing the PV cells from being damaged.

CONCLUSION

Development of functional TPV generator systems is critical to the future success of TPV technology. To support this, the systematic creation of analysis tools for the development of TPV generators is underway. The thermodynamics of many of the TPV generator sub-systems have been analyzed with good results. This paper has outlined the development and validation of three such analysis tools.

The following analysis models have been developed:

- Convection, conduction and radiation transfer from combustion gas to emitter
- Radiation interaction heat transfer for the emitter radiation cavity
- Conduction heat transfer / temperature profile in the receiver cross section

These three models, combined with other analysis currently under way, should ultimately allow for the successful modeling of the entire TPV generator system. Accurate modeling will lead to further understanding of the intricacies of the TPV generator. Enhanced knowledge of TPV systems will allow for the development of commercially viable production of power using the TPV generator.

ACKNOWLEDGEMENTS

This material is based upon work supported by the U. S. Army Research Office under contract number 33881-CH-MUR and by the U. S. Department of Energy through the Division of Advanced Energy Projects under contract number DE-FG06-94ER1249. The author would like the thank Michael Seal, Gavin Campbell and William Connelly at WWU, and Lewis Fraas and Luke Ferguson at JX Crystals, for all their valuable contributions to this research.

REFERENCES

1. Ferguson, L., et al., "Matched Infrared Emitters for Use with GaSb TPV Cells," presented at The Third NREL Conference on the Thermophotovoltaic Generation of Electricity, Colorado Springs, Colorado, May 18–21, 1997.

2. Williams, D., et. al., "Electric Power Generating Lantern Using Forced-Air-Cooled Low-Bandgap Photovoltaic Cells," presented at The Third NREL Conference on the Thermophotovoltaic Generation of Electricity, Colorado Springs, Colorado, May 18–21, 1997.

3. Incopera, F. P. and DeWitt D. P., *Fundamentals of Heat and Mass Transfer*, 3rd Edition., New York: John Wiley & Sons, 1990, Chap. 8, pp. 469, 502–504.

4. Kays, W. M. and Perkins H. C., in W. M. Rohsenow and J. P. Hartnett, Eds., *Handbook of Heat Transfer*, Chap. 7, McGraw-Hill, New York, 1972.

5. Hottel, H. C., "Radiant-Heat Transmission," in W. H. McAdams, Ed., *Heat Transmission*, 3rd Edition., McGraw-Hill, New York, 1954.

6. Touloukian, Y. S. and DeWitt, D. P., *Thermophysical Properties of Matter*, New York: IFI / Plenum, 1972, Vol. 8: Thermal Radiative Properties, Nonmetallic Solids.

7. Charache, G.W., et. al. "Measurement of Conversion Efficiency of Thermophotovoltaic Devices," in *The Second NREL Conference on the Thermophotovoltaic Generation of Electricity*, 1995, pp. 351–360.

8. Incopera, F. P. and DeWitt D. P., *Fundamentals of Heat and Mass Transfer*, 3rd Edition., New York: John Wiley & Sons, 1990, Chap. 8, pp. 501–502.

Modeling the TPV System Optical Cavity

Dale R. Burger

Consultant
South Pasadena, California 91030

Abstract. Thermophotovoltaic (TPV) system design requires a sophisticated system performance model. A TPV system consists of three major subsystems: photon production from combustion or nuclear heat sources; optical cavity and energy conversion (OCEC); and waste heat recuperation or rejection. This paper covers an OCEC subsystem computer model which is designed to work with the other two subsystem models. The OCEC model utilizes a one-dimensional simplification after calculation of three-dimensional configuration factors. The model requires input of seven optical parameters as a function of wavelength: emitter emission, shield (or filter) transmission; shield (or filter) absorption; photovoltaic (PV) cell reflectivity; PV cell metallization reflectivity; cell-to-cell gap reflectivity; and PV cell spectral response. Other model requirements are: fraction of PV cell plane area taken by the cell, cell metallization, and cell-to-cell gap; a good estimate of end losses and cooling system energy requirements; PV cell voltage and current coefficients as a function of temperature and intensity; and PV cell and wiring resistance losses. Use of the model allows parametric analyses under different system conditions and with different OCEC component specifications. Once parameters showing the most sensitivity to change are determined, then the component specifications relating to those parameters can be reviewed and new or modified components made which will move toward an achievable optimum subsystem design. Another part of the OCEC computer model calculates component figures of merit. This capability provides additional insight toward achieving improved subsystem performance.

INTRODUCTION

The thermophotovoltaic (TPV) generator is an emerging technology which received attention over a 30 year period but has never reached commercialization. Numerous theoretical models have been presented in the past [1-5]. Advances made in the last decade permit the cost effective fabrication of low band gap photovoltaic materials, dichroic filters, resonance filters, and plasma filters. With these new components it is now reasonable to model a TPV system with the expectation that the required components can be customized. Actual design and fabrication of many of the needed TPV system components is already proceeding at a rapid pace so the application of a system optimization model is timely.

CP401, *Thermophotovoltaic Generation of Electricity: Third NREL Conference,*
edited by Benner/Coutts
© 1997 The American Institute of Physics 1-56396-734-0/97/$10.00

This paper starts with an introduction to TPV system elements and their requirements. Following this introduction is the development of a model of the optical cavity and energy conversion (OCEC) subsystem. Areas to be covered are: multiple reflections from emitter, filter, and cell; filter transmittance and absorptance; cell spectral response as a function of temperature and intensity; cell voltage temperature and intensity coefficients; cell current series resistance effects due to intensity; definition and calculation of figures of merit; and use of parametric sensitivity analyses.

DEFINITIONS

It is assumed that the wavelength band is narrow so that properties like emissivity, absorptivity, and reflectance are constants unchanging with multiple reflection. With this caveat in mind the following definitions are offered:

σ	=	Stefan-Boltzmann constant (5.6703×10^{-12} W/cm^2-K^4)
λ	=	wavelength (μm)
$\rho_c(\lambda)$	=	TPV cell reflectance
$\rho_f(\lambda)$	=	filter reflectance
$\rho_g(\lambda)$	=	cell plane gap reflectance
$\alpha_f(\lambda)$	=	filter internal absorptance
$\epsilon(\lambda)$	=	emitter surface emissivity = absorptivity
		$(1-\epsilon)$ = reflectance
γ_i	=	fraction of emitter radiation at emitter temperature T_e which lies in the ith wavelength band (W/cm^2)
c	=	speed of light (2.9979×10^{10} cm/s)
F_{ec}	=	view factor from emitter to cells
F_{ce}	=	view factor from cells to emitter
h	=	Planck constant (6.626196×10^{-34} W-sec/Hz)
k	=	Boltzmann constant (1.380662×10^{-23} W-sec/K)
C	=	TPV cell active area fraction of cell plane area
G	=	cell-to-cell gap fraction of cell plane area
M	=	TPV cell metallization fraction of cell plane area

APPROACH

A simple system model was sought in order to ease the mathematical treatment of the major subsystems: emitter, OCEC, and heat rejection. The optical cavity is viewed as being bounded by the emitter and the heat rejection system both of which are assumed to be planes in space with surface properties but no mass. The configuration factors are calculated using the usual three dimensional algorithms.

System Design Requirements

In a TPV system the emitter, filter, and TPV cell are difficult to specify since they are interrelated and dependent upon required output power density and level of heat input. Other design issues that need to be addressed in a complete model, but are not addressed here, are: parasitic heat losses due to end effects; TPV array losses due to cell mismatch and emitter planar temperature variation; and long term effects due to radiation damage and surface optical degradation.

Some areas where the modeling effort can be relaxed are: effect of load changes on TPV system operation (this can be handled by the power system); effect of heat rejection system efficiency due to distance from sun; and cell and material aging effects. These are real but not pressing issues.

The OCEC subsystem is assumed to consist of an emitter, a cell-mounted filter, and TPV cells with metallization grids on top and gaps in-between. The OCEC parameters of interest are: the emitter spectral emissivity; the filter reflectance and absorptance; emitter/filter/cell multiple reflection effects; TPV cell reflectance; TPV cell spectral response as a function of temperature; TPV cell operating parameters as a function of intensity and temperature; TPV cell metallization and interconnect reflectance; and cell gap reflectance.

Component Requirements

Each of the OCEC subsystem components is defined below in detail to provide a defined starting point for model development.

Emitter

The emitter is located between the heat source and the filter. The emitter is required to transform conducted heat into radiated energy of appropriate spectral character and intensity. The emitter's emission spectrum will be dependent upon temperature as well as material and surface properties and therefore will be modeled either mathematically as a grey body obeying Lambert's law or, preferentially, by interpolation between tabular values derived from measurements made at different temperatures. Temperatures over the heat source and the emitter surfaces are assumed to be constant. Other emitter requirements are adequate mechanical properties, retention of optical properties with time, and low volatility at operating temperature. The space between the emitter and the filter is assumed to be non-absorbing and non-emitting and have an index of refraction of 1.0.

Filter

The filter is located between the emitter and the cell and modifies the photon energy spectrum. The requirements placed upon the filter are thus dependent upon the emitter spectrum and TPV cell spectral response. Filter basic design requirements are the ability to reflect photons with unwanted energy levels and transmit the remaining photons with low absorbance. Filter parameters are assumed to be independent of temperature since they are cell mounted and will only see temperatures of about 30-80 °C. Other filter requirements are adequate mechanical properties and retention of optical properties with time.

TPV Cell

The TPV cell is located between the filter and the heat sink and is viewed as a transducer with a spectral response for conversion of photons of various wavelengths to electrons. The TPV cell spectral response must be based upon measured values at different cell temperatures. Temperature effects are very important with regard to TPV cell conversion efficiency since low band gap cells have large voltage temperature coefficients. The intensity effect is also large since TPV systems may run at an irradiance level of 26.7 W/cm^2 (200 suns) or higher. The temperature drop across the filter-TPV cell-heat rejection system combination is likely to be small in a good design but can not be neglected at the expected operating intensity levels. Cell basic design requirements are: bandgap matched to emitter spectrum, good spectral response, low series resistance, and low dark current. Other design considerations are cell-to-cell interconnection, cell size, and thermal grounding.

Heat Recuperation and Rejection System

The heat rejection system is located between the cell and the outside environment and has to be sized to handle all heat energy not converted to electricity. The optical cavity side of the heat rejection system consists of two different surfaces: those that are covered by the TPV cells and thermally coupled to them; and those gaps between cells which are exposed to radiant energy from the emitter. The outer side of the heat rejection system is dependent upon configuration factors as well as material properties which are subject to other design limitations so a model was not pursued here. Similarly heat recuperation will not be covered here since this is also subject to many other design constraints. Heat sink design requirements are: ability to dissipate heat at a given design temperature; high heat transfer rate; and, in the case of spacecraft, low mass and storage capability. Only the heat dissipation requirement is discussed here.

538

System Efficiency

It is possible to look at the heat producing system or the heat rejecting system apart from the rest of the TPV system only as long as overall system efficiency issues are addressed. The system under consideration starts with the emitter surface and ends with the heat rejection system. Some of the total input energy is converted to electrons by the TPV cell while the energy required to run all of the subsystems required for energy conversion must be subtracted from the converted energy before system efficiency is calculated. Of special interest is the relationship between the heat rejected from the cell, which is the difference between the energy incident upon the cell and the energy converted since a lower conversion efficiency implies a higher heat rejection load and therefore a higher heat rejection subsystem overhead loss. This line of reasoning can be extended to heat rejection systems with no energy conversion subsystem losses (i.e. passive) by utilizing additional mass and additional cost as penalty criteria. The above example shows that simple analyses must be viewed with caution. This is especially true when the system elements are very interactive.

It is not productive to separately analyze the elements of the OCEC subsystem. Here the level of interaction is too high. Hottel, in Part II of Ref. 6, has done a detailed analysis of the optical system for a fibrous emitter, a shield, and a continuous TPV cell plane. The analysis here uses the same approach as Hottel but assumes a solid emitter; a filter (or composite set of filters) instead of a shield; and a discontinuous TPV cell plane with gaps and cell metallization as necessary inclusions. General classes of emitters will first be discussed and then emitters as part of the OCEC subsystem of the TPV system.

Emitters can be separated into three general classes: gray body, selective, and band-pass. A gray body is used for emitters which have an emission spectrum proportional to the Planck black body spectrum. The gray body is used since a large area black body can not be made economically. An emissivity goal of a gray body for a TPV system should be equal to or greater than 0.8.

A selective emitter (usually a rare earth oxide) would seem to be ideal for TPV systems, however, there are some problems. First the peak emissivity of a selective emitter is usually low (<0.7). Second, the total energy under the emissivity curve is usually low due to narrow band width and low emissivity. Low energy density lowers the TPV cell conversion efficiency and may lead to some sort of optical concentration system being required. Third, the rare earth oxides are fragile in bulk and must be made into fibers or mounted on a substrate for application. Since the substrates are also heated their emissions must be masked which imposes design constraints. Fourth, many of the rare earth oxides have secondary emission peaks. Despite these problems, the reduced demands upon subsequent filters may lead to an optimal TPV system using a selective emitter or maybe a hybrid of selective and band-pass emitters.

539

Band-pass emitters have not been discussed much in the recent TPV literature, however, they should not be overlooked. A band-pass emitter can be thought of as any material which exhibits a non-black body spectral distribution. A selective emitter is an extreme example of a band-pass emitter. Some specular metal surfaces, such as polished tungsten, have been shown to have an emission spectrum which is a good match for some TPV cells.

TPV Cell Characterization Baseline

Characterization of TPV cell must cover more variables than that covered in PV cell characterization. As a minimum, cell temperature and spectral response must be measured over the expected range of operating temperatures and intensities.

There is a variety of possible TPV emitter surfaces and operating temperatures which affect the emitter spectral distribution. Since TPV systems actively modify the spectrum presented to the TPV cell and the cells are deliberately designed to use this modified spectrum, then some care must be used in establishing a baseline for cell characterization. In order to establish a baseline some standard conditions must be set:

1. TPV cell characterization temperature (T_0) should be 28 °C since TPV cells must be kept cool, so operating temperatures are not likely to go too far above this baseline;

2. The intensity (I_0) should be 1 "something" since the common logarithm of 1 is zero this simplifies some mathematical statements. However, neither the AM0 spectrum for space nor the AM1 for terrestrial use are a good spectral match for TPV cells. A better intensity standard might be found by using a black body radiating at a temperature equal to the cell designed bandgap and then setting the cell-to-black body aperture distance equal to 2 cm or some other agreed upon value and calling that one "something".

Use of a black body and an aperture distance as an intensity standard means characterization tests can be run at any laboratory and comparable results obtained. The black body temperature for a given cell bandgap can be calculated from the knowledge that wavelength can be determined from the following relationship:

$$E_{bg} = h\frac{c}{\lambda} \tag{1}$$

where E_{bg} is cell bandgap energy (eV). This gives the relationship of wavelength equals 1.2395 μm/E_{bg}. Wien's displacement law for the wavelength corresponding to the maximum energy is $\lambda_{max}T = 2897.8$, where λ_{max} is in microns and T is in Kelvin. This results in the following relation for the temperature of a black body which has its maximum energy at the TPV cell bandgap: $T = 2337.9 \times E_{bg}$.

540

The radiant flux emitted per unit area (W/cm^2) of an ideal black body source can now be simply calculated by:

$$W = \epsilon \sigma T_0^4 \tag{2}$$

where ϵ is the emissivity ($\epsilon = 1$ for a black body) and T_0 is the source or black body emitter temperature as derived above and the sink temperature is assumed to be much smaller than the black body temperature.

The irradiance is the amount of incident energy falling upon the cell per unit area (W/cm^2). In order to calculate the irradiance from the black body the following relationship can be used:

$$H = W \frac{a^2}{4d^2} \tau \tag{3}$$

where a is limiting aperture diameter, d is distance from the aperture to the cell, and τ is transmission of the optical path.

The final problem in cell characterization is the measurement of cell temperature at different intensity levels. One simple means is to use the linear change of V_{oc} with temperature as a temperature sensor. First the V_{oc} cell temperature coefficient must be determined. This can most easily be done by obtaining the V_{oc} at specific temperatures by using a pulsed simulator and a cell heat sink. Next the cell is characterized with intensity at different cell heat sink temperatures. Since the cell surface (junction) temperature rise above the heat sink is essentially linear with intensity the intensity changes to the V_{oc} can now be determined.

Optical Cavity and Energy Conversion Analysis

Using the multi-reflection approach by Hottel [6] and the definitions and components as described above, the resultant irradiance incident upon the TPV cells is:

$$H_c(\lambda) = CF_{ec} \frac{\epsilon \gamma (1 - \alpha_f - \rho_f)}{D} \tag{4}$$

where D is defined to be:

$$D = [1 - \rho_f (C\rho_c + G\rho_g + M\rho_m)] \left[1 - F_{ec}F_{ce}(1-\epsilon) \left(\rho_f + \frac{(C\rho_c + G\rho_g + M\rho_m)(1 - \alpha_f - \rho_f)^2}{1 - \rho_f(C\rho_c + G\rho_g + M\rho_m)} \right) \right] \tag{5}$$

541

Energy incident upon the cell then is:

$$E_i = \sum_0^\infty H_c(\lambda) \qquad\qquad (6)$$

while energy converted in the TPV cell is:

$$E_c = F(H_c)\sum_1^2 H_c(\lambda)S(\lambda_1 - \lambda_2) \qquad\qquad (7)$$

where $S(\lambda_1 - \lambda_2)$ is the measured spectral response of the TPV cell at the operating temperature and $F(H_c)$ is a function dependent upon intensity which changes the conversion efficiency of the TPV cell.

There are four elements involved in the $F(H_c)$ factor: V_{oc} increases logarithmically with intensity; cell temperature (T_c) rises with intensity assuming constant heat sink design, which reduces V_{oc}; cell series resistance (R_s) and wire resistance (R_w) losses increase as the square of current which is linearly proportional to intensity; and cell bandgap decreases as cell temperature rises so the current temperature coefficient increases. The use of these four elements as multiplicative factors or ratios is dependent upon the assumption that a percentage change in V_{oc} or I_{sc} is equivalent to an equal percentage change in cell conversion efficiency, E_c.

If we now look at the logarithmic change of V_{oc} with a change in intensity we can state it as follows:

$$R_{V_{oc}}(I) = 1 + \frac{b_1}{V_{oc1}}\log\left(\frac{H_{c2}}{H_{c1}}\right) \qquad\qquad (8)$$

where b_1 is the intensity slope constant and V_{oc1} is V_{oc} at H_{c1}.

The second element of the intensity factor is the standard V_{oc} temperature correction but must be carefully stated since the heat sink temperature is also intensity dependent. If the heat sink cross-sectional area, length, and thermal conductivity, k_c, are kept constant then the heat sink has the following intensity temperature dependence:

$$\Delta T_s = T_{s1}\left[\left(\frac{H_{c2}}{H_{c1}}\right)^{\frac{1}{4}} - 1\right] \qquad\qquad (9)$$

where T_{s1} is heat sink temperature at H_{c1}. This dependency assumes that the heat sink is radiating into a much colder sink.

Since the heat sink and cell are closely coupled thermally then:

$$\Delta T_c = \Delta T_s + \left(\frac{H_{c2}}{H_{c1}} - 1\right)(T_{c1} - T_{s1}) \tag{10}$$

where T_{c1} is the cell temperature at H_{c1}. The second element can now be stated as:

$$R_{Voc}(T) = 1 + \frac{a_2 \Delta T_c}{V_{oc1}} \tag{11}$$

where a_2 is the usual cell voltage temperature coefficient.

The third element of the intensity factor is simply I^2R losses but has to be related to total cell power in order to create a useable ratio. This element then should be:

$$R_R(I) = \frac{V_{oc2} I_{sc2} FF - I_{sc2}^2 (R_s + R_w)}{V_{oc1} I_{sc1} FF - I_{sc1}^2 (R_s + R_w)} \tag{12}$$

where the V_{oc} and I_{sc} values are open circuit voltage and short circuit current of the cell for the two operating conditions and FF is the TPV cell fill factor.

The fourth element is the current temperature coefficient, which is expressed as a percentage, and must also include the effect of cell temperature increase due to intensity increase. The fourth element should therefore be stated as:

$$R_{Isc}(I,T) = \frac{H_{c2}(1 + a_2 \Delta T_c)}{H_{c1}} \tag{13}$$

where a_2 is the usual cell current temperature coefficient. The intensity factor then is:

$$F(H_c) = R_{Voc}(I,T) R_{Voc}(T) R_R(I) R_{Isc}(I,T) \tag{14}$$

The filter acts to reflect unwanted radiation to the emitter and in the process absorbs some of the radiation from the emitter and reflected from the cell as shown:

$$A_f = F_{ec} \sum_0^\infty \frac{\alpha_f \epsilon(\lambda) \gamma_i \left(1 + \frac{(C\rho_c + G\rho_g + M\rho_m)(1 - \alpha_f - \rho_f)}{1 - \rho_f (C\rho_c + G\rho_g + M\rho_m)}\right)}{D} \tag{15}$$

The energy absorbed by the heat sink due to the gaps in the cell plane and the absorptance of the metallization is:

$$A_{g,m} = \sum_0^\infty H_c(\lambda)[(1-\rho_g)G + (1-\rho_m)M] \tag{16}$$

If end, conduction, and convection losses are all assumed to be small by virtue of good design then they can be considered to be constants at this time for simplicity. The total energy to be handled by the heat rejection subsystem then is:

$$E_r = A_f + A_{g,m} + E_i - E_c \tag{17}$$

The overall OCEC subsystem efficiency then is:

$$\eta_{OCEC} = \frac{E_c}{E_r + E_c} \tag{18}$$

This neglects heat generation efficiency and any support system energy requirement since these occur outside of the scope of this model.

Figures of Merit

Figures of merit are an excellent way to focus materials development effort. In general it is best to have figures of merit that are independent of the system design. This has been attempted here with mixed results.

An emitter figure of merit can be developed for a TPV system by assuming that an ideal emitter would only radiate in the wavelengths where the TPV cell efficiently converts the radiation to electrons. The above assumption also implies that the ideal emitter can not absorb any radiation outside of the desired band. The emitter figure of merit could then be stated as:

$$FM_e(T) = \frac{\sum_1^2 \epsilon(\lambda)\gamma_i}{\sum_0^\infty \epsilon(\lambda)\gamma_i} \tag{19}$$

where the numerator summation limits are the cut-on and cut-off wavelengths for the TPV cell of interest. Since the emissivity is temperature dependent the emitter figure of merit is also temperature dependent.

The figure of merit for a filter is achieved by assuming that the ideal filter would pass all radiation in the wavelengths where the TPV cell efficiently converts the radiation to electrons and would pass no radiation outside of this band. The ideal filter would also have zero absorptance in this band. The filter figure of merit could then be stated as:

$$FM_f(T) = \frac{\left[\Sigma_0^1 \, \epsilon\gamma_i(\rho_f-1) + \Sigma_1^2 \, \epsilon\gamma_i(1-\rho_f) + \Sigma_2^\infty \, \epsilon\gamma_i(\rho_f-1)\right] - A_f}{E_i} \tag{20}$$

The figure of merit for a TPV cell is based upon an ideal cell which converts all incident radiation and has no internal losses. In order to provide a consistent radiation source for the figure of merit a black body emitter at expected system operating temperature is assumed as it was above for cell characterization. Since the emitter and filter are the system components which shape the spectrum, no spectral changes are assumed here and the range of integration is only over the spectral response range.

$$FM_c(I,T) = \frac{F(H_c) \, \Sigma_1^2 \, E_{\lambda b} S(\lambda_1 - \lambda_2) - H_c^2(R_s)}{\Sigma_1^2 \, E_{\lambda b}} \tag{21}$$

where the figure of merit is both temperature and intensity dependent, $F(H_c)$ is used to correct to expected intensity, and $E_{\lambda b}$ (W/cm^2-um) is derived from Planck's law:

$$E_{\lambda b} = \frac{2\pi h c^2 \Delta\lambda}{\lambda^5 e^{\frac{hc}{k\lambda T}} - 1} \tag{22}$$

where $\Delta\lambda$ is wavelength range in microns and T is black body temperature in Kelvin. The choice of the wavelength limits for the TPV cell figure of merit may be subject to some trial and error as the conversion efficiency of a TPV cell is not very high at the smaller wavelength region and the additional heat rejection required from accepting that radiation region may outweigh the energy conversion benefit.

PARAMETRIC ANALYSIS

Parametric analyses using the OCEC subsystem model are done in the usual manner - varying one parameter while holding all independent parameters constant and examining changes in the dependent parameters. The first parameters to be varied probably should be the emitter temperature and emittance versus wavelength since they induce so many other changes. The second parameters to be studied would probably be the filter reflectance and absorptance versus wavelength. These two sets of parametric changes give insight into the importance of shaping the spectrum. The next parameters which could be varied are the various cell optical and photovoltaic properties especially the bandgap voltage and the cell series resistance. These parameters illuminate the importance of cell materials and design. The final set of parameters to be analyzed would be the cell metallization and cell-to-cell gap area fractions. This study gives insight into the importance of control of parasitic losses and proper system design.

RESULTS

Equations were developed to allow sensitivity analyses of components of a small radioisotope TPV system.

REFERENCES

[1] Wedlock, B. D., "Thermo-Photo-Voltaic Energy Conversion," Proc. IEEE, 694, (1963).

[2] Kittl, E. and Guazzoni, G., "Design Analysis of a TPV Generator System," Proc. 25th Annual Power Sources Conf., Atlantic City, NJ, May 1972.

[3] Bass, J. C., et. al., "Nuclear-Thermophotovoltaic Energy Conversion , Final Report," General Atomic Co., GA-A16653, August 1982.

[4] Horne, W. E. and Day, A. C., "Thermal Photovoltaic Space Power System," NAS8-33436, Final Report, February 1987.

[5] Schock, A., et. al., "Radioisotope Thermophotovoltaic (RTPV) Generator and Its Applicability to an Illustrative Space Mission," FSC-ESD-217-93-519A, Fairchild Space and Defense Corp., Germantown, MD, February 14, 1994.

[6] White, David C. and Hottel, Hoyt C., "Important Factors in Determining the Efficiency of TPV Systems, Part II" AIP Conf. Proc. 321, "Proc. of 1st NREL Conf, on Thermophotovoltaic Gen. of Elect., Copper Mtn, CO, July 24-27, 1994," pp 437-454.

[7] Emery, Keith, "Important Factors in Determining the efficiency of TPV Systems," AIP Conf Proc. 321, "Proc. of 1st NREL Conf, on Thermophotovoltaic Gen. of Elect., Copper Mtn, CO, July 24-27, 1994," pp 484-489.

ADDENDUM TO SESSION 1

Measurements and Characterization in Photovoltaics: Lessons Learned for TPV

Lawrence L. Kazmerski

National Renewable Energy Laboratory
1617 Cole Boulevard
Golden, Colorado 80401

Abstract. The NREL measurements and characterization activities, with origins in the late 1970s, have evolved with and within the DOE Photovoltaics Program—specifically to support that effort. These centralized facilities, established for reasons of technical and economical advantages for the program, have included four major functions or approaches: (1) analytical measurement service; (2) standardized evaluations (performance through materials); (3) collaborative research; and, (4) measurement technique development. Each of these are described in terms of their importance and contributions to program and project support. The current facilities and activities are highlighted, and the growth and change of these support efforts are historically delineated. The evolution and contributions of these laboratories to photovoltaics provide some lessons and models for the emerging TPV program. The utility of centralized measurement and characterization functions for technology development is assessed in terms of methods of operation, prioritization and customer satisfaction, program unity and focus, response time, and proprietary data and materials. Specifics relating to materials and measurement standards, centralized data bases, client interactions, program directions, and expectations are cited in terms of both successes and deficiencies for these program-support efforts. The return on investment for estimated in terms of benefits to the program compared to alternative approaches.

INTRODUCTION

Characterization has been essential to the advancement and realization of the photovoltaic technology (1-3). The technology owes its scientific development to an ensemble of creative contributions from its parts—covering thought, planning, realization, evaluation, and verification. Theory, materials science, processing, device development, modeling, testing, and measurement are co-dependent. None of these areas can stand alone; together, they provide for the successes and future of the technology. The U.S. National Program in Photovoltaics recognized the importance of measurement leadership and support at its inception. These support operations were among the earliest investments by the National PV Program, with

CP401, *Thermophotovoltaic Generation of Electricity: Third NREL Conference,*
edited by Benner/Coutts
© 1997 The American Institute of Physics 1-56396-734-0/97/$10.00

origins in the late 1970s (4). This paper documents some of the experiences of this program, and provides some lessons, directions, and cautions about similar approaches for the evolving Thermophotovoltaics Program.

MEASUREMENTS, TESTS, AND TECHNIQUES: OPERATIONS

The range of available measurement and characterization techniques is considerable, covering the expanse of photovoltaic interests from arrays to atoms (4,5). Table 1 presents a portioned summary of current techniques, covering a wide-range of materials (composition, chemistry, and structure) and device (electro-optical, performance, and reliability) methods (6-8). The National Renewable Energy Laboratory offers some 57 identifiable techniques for the analysis of PV materials and devices, and this investment in the test, measurement, and characterization facilities by DOE at its National Laboratories has been substantial (4,5). Multi-technique laboratories have the advantage of being able to apply the correct analysis method for the problem and/or use a number of complementary techniques to unambiguously solve that problem.

The arguments for the establishment of centralized analytical support functions include economics (investment in major capital equipment to effectively serve the entire program), complementary diverse techniques at a single location (multi-technique application to problem solution), a centralized expertise and information archive (for cross fertilization and as a repository for related technology issues), and the establishment of a independent source for standard evaluation of photovoltaic product. These services are utilized, and the NREL measurements and characterization laboratories analyzed more than 16,000 samples with about 65,000 measurements in 1996. The methodology and extent of operations include a number of very important components. The first is the most cited requirement for such a facility: *measurement support and collaborative research*. The support function provides service for the client and should involve extensive interaction between the measurement scientist and the person who knows most about the sample. This focuses the analysis and is one area that distinguishes the established program facility from a commercial analytical laboratory. The analysis is more than just "sample in, data out". A logical extension is the establishment of a collaborative research activity between the characterization and material/device research entities. This project objective would likely include a range of complementary analytical functions to address the problem. This mode of operation enhances "buy in" from both sides, and usually leads to some additional output such as a joint publication in the scientific literature. Not every analysis fits into this category; some materials and device projects need only some chemical or electrical information for establishment of process. However, those that are aimed at extended development or problem solution can benefit from the collaborative approach. In the NREL PV efforts, about 2/3 of the interactions with the client fall into the collaborative category. An extremely important program support function is the second category: *standard test and measurements*. The centralized facility facilitates the independent source for the evaluation of components. None has been more visible or contributory than that for the determination of the performance (efficiency) of cells and modules. This has not only provided the basis for credibility, but has led to a worldwide network of

550

TABLE 1. Compilation of selected characterization techniques for materials and devices.

Cells and Module

- Simulator efficiency (continuous, pulsed)
- Outdoor performance
- Quantum efficiency
- Current-voltage (dark, light)
- Reference cell calibration
- Spectral measurements
- Reliability and durability
- Solar/optical radiation measurements

Electro-Optical Measurements

- Spectrophotometry
- Ellipsometry (fixed wavelength, spectral)
- Photoluminescence
- Photoconductivity
- Fourier transform techniques (FTIR, FTPL)
- Transient spectroscopies
- Capacitance, current-voltage spectroscopy
- Deep-level transient spectroscopy
- Minority-carrier lifetime spectroscopy
- Mobility/resistivity/carrier measurements

Materials Microcharacterization

- Scanning electron microscopy
- Transmission electron microscopy (scanning)
- Associated e-beam analyses (EBIC, EBIV, CL)
- Electron microprobe analysis (EDS, WDS)
- X-ray techniques (diffraction, topography)
- Quartz and beam microbalance
- Atomic absorption, gas chromograph spectroscopy
- High performance liquid chromatography
- Thermogravometric analysis

Surface Analysis

- Auger electron spectroscopy
- Electron energy loss spectroscopy
- X-ray photoelectron spectroscopy
- Ultraviolet photoelectron spectroscopy
- Synchrotron source measurements
- Secondary ion mass spectrometry (dynamic, static)
- Neutral ion and ion scattering spectrometries
- Scanning tunneling and force microscopies
- Ballistic electron energy microscopy
- Near-field scanning optical microscopy/spectroscopy

551

standards laboratories that ensure the independent and fair comparison of devices (within a given technology and between technologies). The centralization of the standard measurement capability has also helped in the dissemination of standard test methods and reporting conditions within the photovoltaic community. It has led to international intercomparisons to make sure that efficiencies measured in India can be expected to track those in Europe and Japan. Almost all these central and standard performance-evaluation laboratories can trace their standards to common sources.

A third component is that of *technique development*. Because of the expertise in the particular technology (e.g., photovoltaics), these measurement scientists and engineers are able to adapt techniques from other electronics technologies and develop new methods in response to the technologies needs. An example of one such method, which has leveraged its development with use in other semiconductor electronic applications, is the minority-carrier lifetime spectrometer developed by Ahrenkiel (9,10). The system, shown schematically in Fig. 1, provides a non-contact and non-destructive means of evaluating the most fundamental measure of the quality of a semiconductor for use in a solar cell—the lifetime of the minority electrons and holes (11). The technique is extremely versatile, as illustrate by the data shown in Fig. 2. It can accurately determine the lifetimes in direct and indirect bandgap semiconductors, bulk and large-area materials and sub-micron thick films, for single crystals and polycrystalline semiconductors, and for bandgaps covering the range from about 0.5 eV to more than 3.0 eV. This versatility certainly has appli-

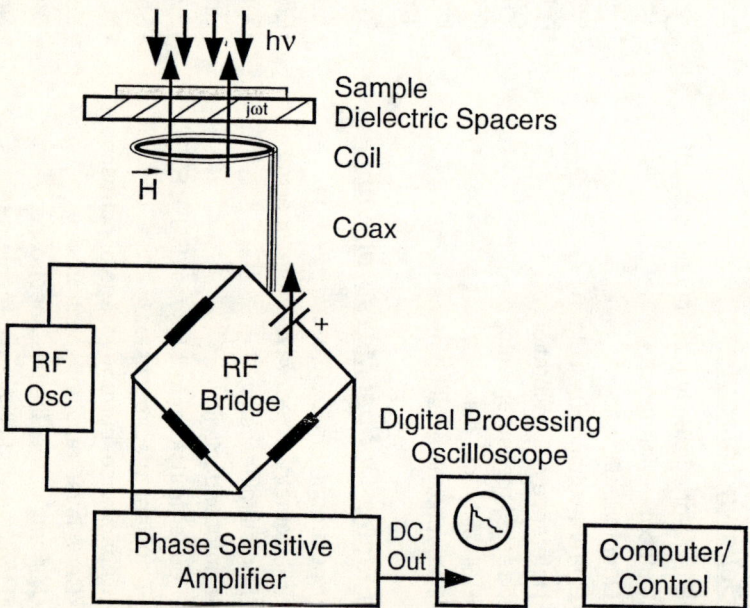

FIGURE 1. Schematic representation of ultrahigh-frequency minority-carrier lifetime spectrometer (ref. 10). Instrument determines the fundamental minority-carrier parameters in semiconductors and devices.

552

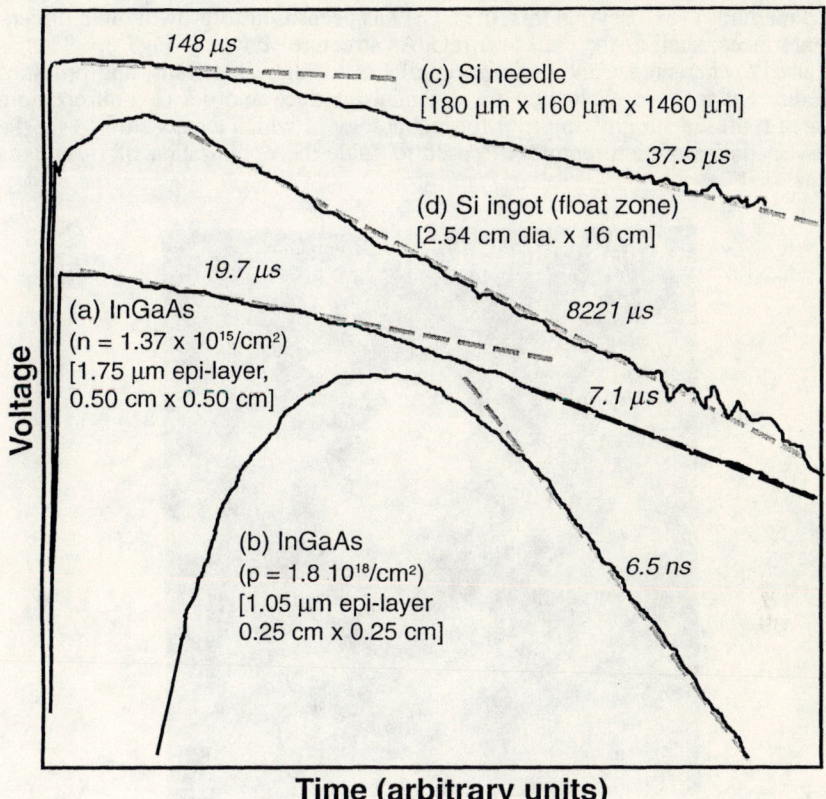

148 μs

(c) Si needle
[180 μm x 160 μm x 1460 μm]

37.5 μs

(d) Si ingot (float zone)
[2.54 cm dia. x 16 cm]

19.7 μs

(a) InGaAs
(n = 1.37 x 10¹⁵/cm²)
[1.75 μm epi-layer,
0.50 cm x 0.50 cm]

8221 μs

7.1 μs

(b) InGaAs
(p = 1.8 10¹⁸/cm²)
[1.05 μm epi-layer
0.25 cm x 0.25 cm]

6.5 ns

Voltage

Time (arbitrary units)

FIGURE 2. Spectroscopic data showing versatility of the minority-carrier spectrometer of Fig. 1: (a) InGaAs film, near intrinsic; (b) InGaAs film, p-type, high doping; (c) small Si needles; and, (d) large Si boule.

cation in PV manufacturing and research settings that deal with a variety of semiconductor products. Additionally, the technique can also be adapted directly into the manufacturing line (see following sections) to provide material evaluation and quality control before or following device processing. Such developments provide a important link, contribution, and technology transfer from the university or research lab to the manufacturing environment. Technique development support for technology advancement can include contributions on the frontiers of science. An example relating to thermophotovoltaics is in the nanoscale science area—using the proximal probe methods (i.e., scanning tunneling microscopy (STM), atomic force microscopy (AFM), and near-field scanning tunneling spectroscopy (NSOM)) to investigate fundamental issues (12-14). For example, the STM has been used to manipulate atoms at InGaAs surfaces to create vacancies (point defect creation) and to directly investigate the electronic levels associated with these using NSOM-related methods (15). Beyond this, the STM has been used to grow atomic dimension interfaces, such as the GaAs on InGaAs structure shown in Fig. 3. This is being used to characterize and optimize interfaces for these materials, and provides some direct diagnostic evaluation for potential interface and device optimization. This area represents a growing tool for technology in which events studied on the highest spatial resolution regimes are used to guide the optimization of macroarea devices.

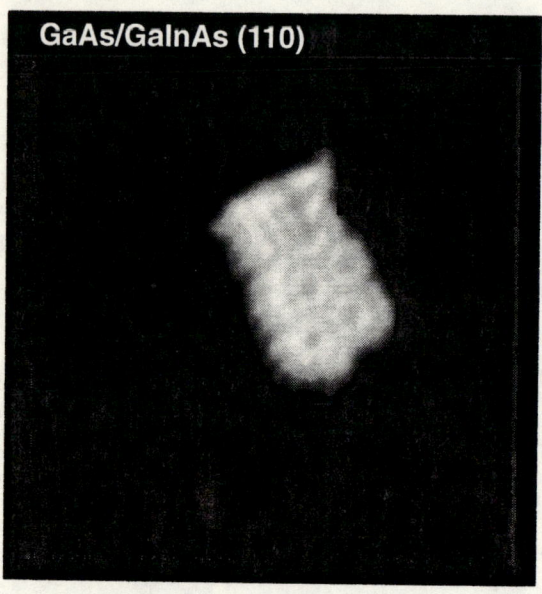

FIGURE 3. Single layer of GaAs deposited atom-by-atom on InGaAs surface using STM for atomic manipulation; intended for study of TPV interfaces on nanoscale.

A <u>fourth</u> and very important aspect is the ability to provide *proprietary measurements* for the clients. The protection of the intellectual property and commercial interests of the client is mandatory. This leads to trust between the parties, and to efficiency and effectiveness in measurement time since information important to the analysis is able to be shared. Obviously, the advancement of any commercial operation is important to the program. At NREL, the National Photovoltaics Program has established the capability to perform protected and proprietary tests and measurements for U.S. industry and other U.S. organizations.

<u>Finally</u>, it is important to *minimize the number of interfaces* between the analysis and the client. Certainly, tracking is an important factor for a program, but the client must be able to work directly with that entity concerned with her/his sample measurements. Because rapid turn around time is essential to the research group or manufacturer, this decreases the chance that samples will be misplaced, left on the desk or mailbox of someone on travel, or that the data will be released incorrectly. Protected electronic transfer and interaction, cited below, are keys to ensuring and enabling rapid response, archiving, and openness of analysis for these operations.

COST AND RETURN ON INVESTMENT

The arguable and perceptible value of the central versus diverse implementation of major characterization capabilities is cost. For more sophisticated and costly equipment, a program cannot afford to provide such facilities for all contractors—especially if the demands for such equipment are intermittent. It should also be noted that the economics of the centralization of major facilities goes beyond the capital investment. Confining the operation of the instrument to a limited number of operators helps minimize maintenance problems and ensures consistency and correctness in data acquisition. Even under the best control, maintenance costs for such major facilities annually runs 7-10% of the equipment cost. This is illustrated in the data for the NREL Measurements and Characterization Laboratories in Fig. 4. There have been clear correlations between the number of operators and the increased cost of maintenance, downtime, and error. In general, the more sophisticated and specialized the technique, the more critical is the attention to controlling the number of operators. This feature is one that a central facility approach has a clear benefit. It is also a principal under which the primate analytical laboratories operate. In their case, profitability is tied to the time that the measurement system operates. Analogously, the "profitability" of the DOE central facility is its ability to contribute to the needs of the client and the progress of the technology.

These centralized facilities have traditionally offered multiple techniques for the microscopic and macroscopic evaluation of components. The range, illustrated in Table I, provides vital information on the chemical, compositional, structural, electrical, and optical properties of surface and intramaterial features, as well at the performance properties of the devices (12). Special strengths include the ability to link nanoscale spectroscopic events with the operation of large-area devices. Two important issues regulate the effectiveness of the central facility approach to characterization support: (1) the ability (range and correct application of techniques and interpretation) to address the problem; and, (2) adequate response time. The former is a measure of the quality of and investment in the analysis program; the latter is an indication of the administration and extent of the workload. Prioritization, scheduling, and adequate communication between the client and the analyst are essential concerns to ensure adequate turn-around time. The ability to access diagnostic results rapidly can greatly assist and direct the advancement of a technology.

From a cost consideration, the cost per sample and the cost per hour for analysis time is comparable to those at commercial facilities. However, the time can be less because of the experiences of these laboratory facilities with the technology. The costs should also be considered with the availability of multiple techniques at a single location, availability of standards, and the less tangible factors (e.g., interest of the staff in the technology).

The potential for leveraging an investment in measurements and characterization for a program like thermophotovoltaics represents the greatest opportunity for TPV. To some extent, some initial leveraging has been accomplished with programs at NREL, JPL, NASA, and some non-federal organizations. No formal agreement-shave been made, and no definitive standards activities have been initiated. Certainly, with the growth in the thermophotovoltaics program, it is time that such investments be considered to ensure independent evaluations for the program's prod-

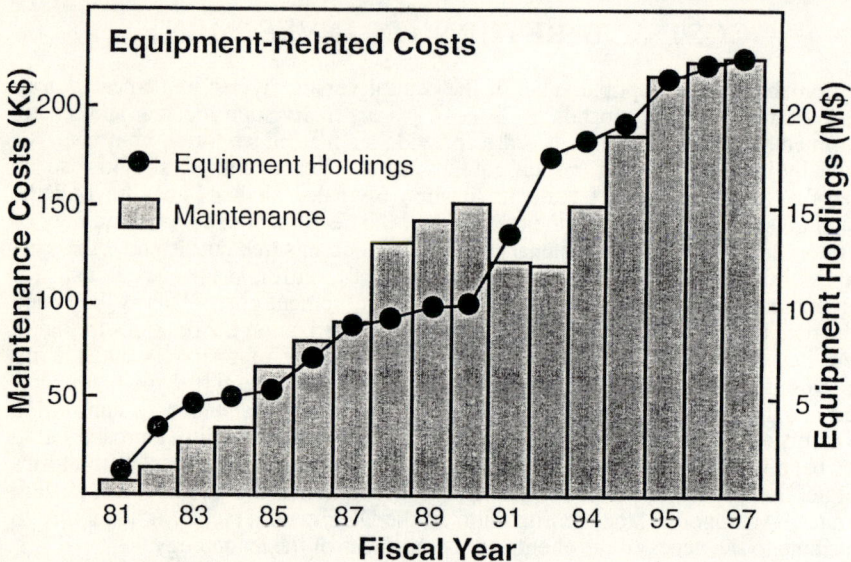

FIGURE 4. Relationships between maintenance costs and capital equipment investment as a function of time for NREL measurements and characterization facility.

ucts—and to guide the critical investments for next-generation technologies. Just as the rebirth of TPV as a technology *leveraged* the advancements in III-V and low-bandgap cell technologies from the PV programs, the TPV programs can *leverage* their position by utilizing the test, measurement, and characterization facilities and expertise that have helped guide and support the advancement of photovoltaics.

REQUIREMENTS, RECOMMENDATIONS, AND TRENDS

The requirements and trends for characterization facilities to assist technology development can be defined in several areas:

- Rapid response. The ability to access data in the shortest possible time is critical to the research or manufacturing program. Central analytical facilities have to work with the client to ensure that the analysis is performed in a time reasonable to the manufacturer client; prioritization and communication are key to ensuring responsiveness. Barriers and interfaces between the analysis and the client have to be minimized. The keys to useful analytical assistance are communication, responsiveness, and turn-around time.

- Interpretation and information. The primary interaction and output metric between the characterization scientist/engineer and the client is the interpretation. Of course, the interpretation depends upon a number of factors from applying the correct technique (analysis end) to the information on the sample (client end). Care to ensure proper interpretation is paramount; attention to the span of information is equally important. Data dumps can delay, hide, and misdirect problem solution. There is a growing realization that simplicity and trend recognition (not detail and complexity) are the measurement areas that provide the manufacturing-line operator with the capabilities to control product quality. Finally, the background and experience of the analytical facility with the technology is crucial to interpretation of results.

- Technique versatility. With the increasing trend in product diversity (e.g., lower and different bandgaps, films and bulk single crystals, operation temperature, concentration) for technology programs, the ability to use a given technique for a variety of device types is desirable. Of equal importance, the capability to perform non-destructive, non-contact measurements greatly enhances importance of the technique for rapidity of analysis and more implementation directly into the production line.

- Electronic communications. The evolution of electronic communications should guide the interaction between the measurement site and the client. The transfer of data should become immediate, archiving should become routine, and protected access to both is already a reality. The future holds the potential for real time interactions between sites and remote user-facility operation by the client.

- Research and technique development. The demands of the technology, whether PV or TPV, continue to expand with the diversity of the product. These demands require research into measurement science and the development of new measurement techniques aimed specifically at the technology. These requirements range from the analytical aspects of data interpretation, through alteration of standard techniques for application, to the development of new characterization methods to meet the needs of the manufacturing environment. Technique development covers the regime from large areas for modules to nanoscale technologies for diagnostic evaluations; from determination of rapid events such as pico- and femtosecond lifetimes to longer-term reliability testing; from more precise measurements of performance through routine evaluations for product classification. Measurement needs and capabilities are expanding with the technology.

- Leveraging investment. The investment into a credible and extensive measurement capability for the thermophotovoltaics can be minimized by leveraging its investment by utilizing existing facilities. The investment is leveraged not only from funding considerations, but also from the experience and background that such facilities can offer thermophotovoltaics.

Certainly, thermophotovoltaic technologies are advancing, and the support of extensive and dedicated measurement and characterization laboratories can be used to advantage by this program. The design, designation, and implementation of these research and support facilities should be carefully considered, leveraging the required investment with existing laboratories and analytical functions with similar missions. Standards, traceability, technique development, rapid response, and protection of intellectual and product property should be central to the thermophotovoltaic efforts.

557

ACKNOWLEDGMENTS

The author expresses his most-sincere gratitude and appreciation to colleagues of the NREL Center for Measurements and Characterization who selflessly helped in reading, reviewing, revising, and adding to this overview. The author would also like to thank the conference Chair for his patience and for the opportunity to present these idea. This paper was prepared partially through the support of NREL and the U.S. Department of Energy under contract No. DE-AC36-83CH10093, and through the support of the U.S. DOE Applied Energy Projects Program for TPV and nanoscale portions of this paper.

REFERENCES

1. See, Proc. IEEE Photovoltaic Specialists Conf.: IEEE, New York; 1971-1996; Also, T.J. Coutts and J.D. Meakin, Eds., *Current Topic in Photovoltaics*:: Academic Press, New York; 1986-1989.

2. S.R. Wenham, M.A. Green, and M.E. Watt, *Applied Photovoltaics*: Centre for Photovoltaic Devices and Systems, University of New South Wales, Australia; 1994, pp. 42-45.

3. L.L. Kazmerski, in K. Boer, Ed., Advances in Solar Energy: Plenum Press, New York; 1986, pp. 1-123.

4. Worldwide web site: www.nrel.gov/measurements

5. See, "From Molecules to Materials", Brochure detailing the capabilities of the Center for Measurements and Characterization: NREL, Golden, Colo.; 1997.

6. C.R. Brundle, C.A. Evans, Jr., and S. Wilson, *Encyclopedia of Materials Characterization* : Butterworth-Heinemann, Boston; 1992.

7. F.A. Abulfotuh and L.L. Kazmerski, in *Handbook of Thin Films*: Plenum Press, New York; 1997 (in-press).

8. L.L. Kazmerski, J. Materials Research: MRS, Pittsburgh, Penn.; 1997, Special issue edited by F. Wald and R. Bell, in-press.

9. R.K. Ahrenkiel, in *Semiconductors and Semimetals*, Vol. 39, R.K. Ahrenkiel and M.S. Lundstrom, Eds.: Academic Press, New York; 1993, pp. 39-150.

10. R.K. Ahrenkiel, Patent in process (June, 1997).

11. J.W. Orton and P. Blood, *The Electrical Characterization of Semiconductors: Measurement of Minority-Carrier Properties* : Academic Press, London; 1990).

12. L.L. Kazmerski, Proc. 23rd IEEE Photovoltaic Specialists Conf. (IEEE, New York; 1993, pp. 1-7. Also, Vacuum 43, 1011 (1992).

13. L.L. Kazmerski, J. Vac. Sci. Technol. B, 9 1549 (1991).

14. L.L. Kazmerski, Proc. European Photovoltaic Solar Energy Conference, Amstermdam (Kluwer Publ., The Netherlands; 1994).

15. L.L. Kazmerski, J. Vac. Sci. Technol. (1997) in-press.

Peter Adair
Space Power Institute
3911 Becora Court
Panama City, FL 32405
Phone: (904) 265-0688

Clint Ballinger
Lockheed Martin
P.O. Box 1072
Schenectady, NY 12301
EMail:
clinton_ballinger@ccgateway.amc.edu

Jochen Bard
Kassel University
Wilhelmshoeher Allec 73
FB 16, IEE
Kassel 34121
Germany
Phone: +49-561-804-6203
Fax: +49-567-804-6434
EMail: bard@re.e-technik.uni-kassel.de

Eric Barringer
Babcock & Wilcox
R&D Division, MC 76
Mt. Athos Road
Lynchburg, VA 24506
Phone: (804) 522-6191
Fax: (804) 522-6980

Frederick Becker
Tecogen
45 First Avenue
Waltham, MA 02254
Phone: 617-622-1059
Fax: 617-622-1025

John Benner
National Renewable Energy Lab
1617 Cole Boulevard
SERF
Golden, CO 80401
Phone: (303) 384-6496

Fax: (303) 384-6481
EMail: john_benner@nrel.gov

Andreas Bett
Fraunhofer Solar Energy Systems
Oltmannsstr 22
Freiburg D-79100
Germany
Phone: +49-764-4588-257
Fax: +49-761-4588-250
EMail: bett@ise.fhg.de

Kurt Betzler
JX Crystals
40 Rockwood
Irvine, CA 92714
Phone: (714) 786-5990

Ishwara Bhat
Rensselaer Polytechnic Inst.
ECSE Department
Troy, NY 12180-3590
Phone: (518) 276-2786
Fax: (518) 276-6261
EMail: bhatl@rpi.edu

William Biter
Sensortex. Inc.
402 D North Mill Road
Kennett Square, PA 19348
Phone: (610) 444-2383
Fax: (610) 444-6193
EMail: wbiter@bellatlantic.net

Stephen Bolger
Independent Inventor
127 W. 79th Street, Apt. 11J
New York, NY 10024
Phone: 212-873-6632
Fax: 212-873-6564

Jose Borrego
Rensselaer Polytechnic Inst.
39 Riverview Road

Brighton, MA 12125-1835
Phone: (617) 782-4947
Fax: (617) 782-4947
EMail: borrego@ecse.rpi.edu

Lars Broman
Dalarna University, SERC
SERC
S-78188
Borlange
Sweden
Phone: +46-2377-8700
Fax: +46-2377-8701
EMail: lbr@du.se

Dale Burger
Consultant
618 Arroyo Drive
So. Pasadena, CA 91030
Phone: (818) 799-5545
Fax: (818) 799-9054
EMail: dburger2@juno.com

Brian Campbell
Knolls Atomic Power Lab.
P. O. Box 1072
Schenectady, NY 12301
Phone: (518) 395-7438
Fax: (518) 395-6136

Jeffrey Carapella
National Renewable Energy Lab
1617 Cole Boulevard
Golden, CO 80401
Phone: (303) 384-6442
Fax: (303) 384-6430
EMail: jeff_carapella@nrel.gov

Zheng Chen
Space Power Institute
231 Leach Center
Auburn University, AL 36849-5320
Phone: (334) 844-5906
Fax: (334) 844-5900
EMail: chenzhe@mail.auburn.edu

David Christensen
National Renewable Energy Lab
1617 Cole Boulevard
Golden, CO 80401
Phone: (303) 275-3015
Fax: (303) 275-3040

Donald L. Chubb
NASA
Cleveland, OH

Eric Clark
Lewis Research Center/NASA
21000 Brookpark Road
MS: 302-1
Cleveland, OH 44135
Phone: (216) 433-3926

George Cody
Exxon Corporate Research
Route 22 East
Clinton Township
Annandale, NJ 08801
Phone: (908) 730-3022
Fax: (908) 730-3031
EMail: gdcody@ere.nj

Miguel Contreras
National Renewable Energy Lab
1617 Cole Boulevard
Golden, CO 80401
Phone: (303) 384-6478
Fax: (303) 384-6430
EMail: miguel_contreras@nrel.gov

Timothy Coutts
National Renewable Energy Lab
1617 Cole Boulevard
Golden, CO 80401
Phone: (303) 384-6561
Fax: (303) 384-6430
EMail: tcoutts@nrel.nrel.gov

Crispin DeBellis
Babcock & Wilcox RDD
1562 Beeson Street
Alliance, OH 44601
Phone: (330) 829-7664
Fax: (330) 829-7362
EMail: crispin.debellis@mcdermott.com

Larry DeShazer
Quantum Group Inc.
11211 Sorrento Valley Road
San Diego, CA 92121
Phone: (619) 457-3048
Fax: (619) 457-3229

Neelkanth Dhere
Florida Solar Energy Center
1679 Clearlake Road
Cocoa, FL 32922-5703
Phone: (407) 638-1442
Fax: (407) 638-1010
EMail: dhere@fsec.ucf.edu

Bob DiMatteo
Draper Laboratory
555 Technology Square
Cambridge, MA 02139
Phone: (617) 258-3052
Fax: (617) 258-2800
EMail: rdimatteo@draper.com

Edward Doyle
Tecogen
45 First Avenue
Waltham, MA 02254
Phone: 617-622-1053
Fax: 617-622-1025

Anna Duda
National Renewable Energy Lab
1617 Cole Boulevard
Golden, CO 80401
Phone: (303) 384-6574
Fax: (303) 384-6430
EMail: aduda@nrel.nrel.gov

Partha Dutta
Rensselaer Polytechnic Institute
Dept. of MEAEM, JEC
Troy, NY 12180
Phone: (518) 276-8842
Fax: (518) 276-6025
EMail: duttap@rpi.edu

Hassan Ehsani
Rensselaer Polytechnic Institute
6034, JEC Bldg.
15th Street
Troy, NY 12180
Phone: (518) 276-6664
Fax: (518) 276-6261
EMail: ehsani@ecse.rpi.edu

Keith Emery
National Renewable Energy Lab
1617 Cole Boulevard
Golden, CO 80401
Phone: (303) 384-6632
Fax: (303) 384-6604

Navid Fatemi
Essential Research, Inc.
23811 Chagrin Blvd.
Suite 220
Cleveland, OH 44122
Phone: (216) 433-5586
Fax: (216) 433-6106
EMail: nfatemi@lerc.nasa.gov

Luke Ferguson
JX Crystals, Inc.
Issaquah, WA 98027

Mark Fitzgerald
Institute for Sustainable Power, Inc.
P. O. Box 4036
Highlands Ranch, CO 80126
Phone: (303) 683-4748
Fax: (303) 470-8239
EMail: markfitz@cyberneering.com

Lewis Fraas
JX Crystals, Inc.
1105 12th Avenue, N.W.
Suite A2
Issaquah, WA 98027
Phone: (206) 392-5237
Fax: (206) 392-7303
EMail: fraas@jxcrystals.com

Alex Freundlich
University of Houston
Science & Research
Building One
Houston, TX 77204-5507
Phone: (713) 747-7724
Fax: (713) 743-3621
EMail: alex@space.svec.uh.edu

Hansjorg Gabler
Fraunhofer Institute
Oltmannsstrasse 5
Freiburg 79100
Germany
Phone: +49-761-4588-229
Fax: +49-761-4588-217
EMail: gabler.ise.fhg.de

Linda Garverick
Essential Research, Inc.
23811 Chagrin Blvd.
Suite 220
Cleveland, OH 44122
Phone: (216) 831-0177
Fax: (216) 831-0113
EMail: garveric@midwest.er.com

Mark Goldstein
Quantum Group Inc.
11211 Sorrento Valley Road
San Diego, CA 92121
Phone: (619) 457-3048
Fax: (619) 457-3229

Brian Good
NASA Lewis Research Center
21000 Brookpark Road

Cleveland, OH 44135
Phone: (216) 433-6296
Fax: (216) 433-6100

Paul Griffin
EXSS Group
London
United Kingdom
Phone: +44-171-594-7583
Fax: +44-171-594-7565
EMail: p.griffin@ic.ac.uk

Guido Guazzoni
CECOM-ASD
Myer Center
Ft. Monmouth, NJ 07712
Phone: (908) 427-4081
Fax: (908) 426-3665
EMail: guazzoni@doim g.monmouth.
army.mil

Ronald Gutmann
Rensselaer Polytechnic Inst.
CIEEM-CII 6129
Troy, NY 12180-3590
Phone: (518) 276-6794
Fax: (518) 276-8761
EMail: rgutmann@unix.cie.rpi.edu

Stephanie Haywood
University of Hull
Cottingham Road
Dept. of Electronic Engineering
Hull, East Yorkshire HU6 7RX
England
Phone: +44-1482-466074
Fax: +44-1482-466664
EMail: s.k.haywood@e-eng.hull.ac.uk

Collin Hitchcock
Rensselaer Polytechnic Inst.
1519 15th Street
Troy, NY 12180
Phone: (518) 271-8133
EMail: hitchc@rpi.edu

Russell Hollingsworth
Materials Research Group, Inc.
12441 W. 49th Avenue
Suite 2
Wheat Ridge, CO 80033
Phone: (303) 425-6688
Fax: (303) 425-6562

William Horne
EDTEK, Inc.
7082 South 220th Street
Kent, WA 98032
Phone: (206) 395-8084
Fax: (206) 395-8086
EMail: edtekinc@cnet.com

Carlos Huggins
EEV Ltd.
106 Waterhouse Lane
Chelmsford, Essex CMI 2QU
England
Phone: +1245-493493
Fax: +1245-492492
EMail: trevor.cross@eev.com

Denis Huguenin
Rhone Poulenc
Sz Rue Haie Cog
Aubervilliers 93308
France

Osamu Ikki
Resources Total System Co.
2 Floor Kariya Bldg.
2-7-11 Shinkawa Chuo-ku
Tokyo 104
Japan
Phone: 81-3-3551-6345
Fax: 81-3-3553-8954
EMail: ged02723@niftyserve.or.jp

James Jaeschke
Eaton Corporation
4201 N. 27th Street
Dept. H503
Milwaukee, WI 53216

Phone: (414) 449-6571
Fax: (414) 449-7084

Raj Jain
NASA Lewis Research Center
MS 302-1
2100 Brookpark Road
Cleveland, OH 44135
Phone: (216) 433-2227
Fax: (216) 433-6106
EMail: raj.k.jain@lerc.nasa.gov

Phillip Jenkins
Essential Research, Inc.
23811 Chagrin Blvd.
Suite 220
Cleveland, OH 44122
Phone: (218) 433-2233
Fax: (216) 433-6106
EMail: phil.jenkins@lerc.nasa.gov

Steven Johnson
Sabrina, Inc.
4545 McIntyre
P. O. Box 4040
Golden, CO 80401
Phone: (303) 279-6543
Fax: (303) 202-0454

Dave Joslin
Spectrolab, Inc.
12500 Gladstone Ave.
Sylmar, CA 91342
Phone: (818) 898-2836
Fax: (818) 361-5102

Nasser Karam
Spectrolab, Inc.
12500 Gladstone Ave.
Sylmar, CA 91342
Phone: (818) 898-7514
Fax: (818) 361-5102

Lawrence Kazmerski
National Renewable Energy Lab
1617 Cole Blvd.

Golden, CO 80401
Phone: (303) 384-6600
Fax: (303) 384-6604
EMail: kaz@nrel.gov

Daniel Krommenhoek
Lockheed Martin
P. O. Box 1072
Schenectady, NY 12301-1072
Phone: (518) 395-4480
Fax: (518) 395-6136

John Kruger
U.S. Army Research Office
P. O. Box 12211
Research Triangle Pk, NC 27709-2211
Phone: (919) 549-4323
Fax: (919) 549-4248
EMail: kruger@aro-em1.army.mil

Aleks Kushch
Quantum Group Inc.
1121 Sorrento Valley Road
San Diego, CA 92121
Phone: (619) 457-3048
Fax: (619) 457-3229

Ekkehard Laqua
Robert Bosch GMBH
P. O. Box 106050
Stuttgart 70049
Germany
Phone: +49-711-811-7660
Fax: +49-711-811-6174
EMail: eklagua@s.0694.am.bosch.de

Keith Lindler
U.S. Naval Academy
590 Holloway Road
Marine Engineering Dept.
Annapolis, MD 21402-5042
Phone: (410) 293-6447
Fax: (410) 293-2219
EMail: lindler@nadn.navy.mil

Kurt Lund
University of California
La Jolla, CA 92093-0411
Phone: (619) 481-8914
Fax: (619) 793-2446
EMail: klund@electriciti.com

Patrick Magari
Creare Incorporated
P. O. Box 71
Hanover, NH 03755
Phone: (603) 643-3800
Fax: (603) 643-4657
EMail: pjm@creare.com

Jorgen Marks
Dalarna University, SERC
S-78188 Borlange
Borlange
Sweden
Phone: +46-23-778700
Fax: +46-23-778701

Charles Martin
U.S. Dept. of Energy/Dept. of Navy
2521 Jefferson-Davis Highway
Arlington, VA 22242-5160
Phone: (703) 603-5505 x6
Fax: (703) 602-5220

Ramon Martinelli
Sarnoff
CN 5300
Princeton, NJ 08543-5300
Phone: (609) 734-2403
Fax: (609) 734-2039
EMail: martinelli@sarnoff.com

Michael Mauk
Astro Power Inc.
Solar Park
Newark, DE 19716-2000
Phone: (302) 366-0400
Fax: (302) 368-6474

Malachy McAlonan
Teledyne Brown Engineering
10707 Gilroy Road
Hunt Valley, MD 21031
Phone: (410) 771-8600
Fax: (410) 771-8619

Robert McConnell
National Renewable Energy Lab
1617 Cole Boulevard
Golden, CO 80401
Phone: (303) 384-6419
Fax: (303) 384-6481
EMail: robert_mcconnell@nrel.gov

Bill Micklethwaite
Firebird Semiconductors
2950 Highway Drive
Trail, BC V1R2T3
Canada
Phone: (250) 364-5605
Fax: (250) 364-5643
EMail: bmicklet@awinc.com

Tom Moriarty
National Renewable Energy Lab
1617 Cole Boulevard
Golden, CO 80401
Phone: (303) 384-6551
Fax: (303) 384-6604
EMail: tom_moriarty@nrel.gov

James Mosquera
Department of the Navy
2521 Jefferson-Davis Hwy
Arlington, VA 22242-5160
Phone: (703) 602-8695
Fax: (703) 602-8893

Bill Mulligan
JX Crystals, Inc.
1105 12th Ave. NW Az
Issaquah, WA 98027
Phone: (425) 392-5237
Fax: (415) 392-7303
EMail: bmulligan@jxcrystals.com

Christopher Murray
Westinghouse Bettis
P. O. Box 79
West Mifflin, PA 15122
Phone: (412) 476-5990
Fax: (412) 476-5151

Mitsuo Nakajima
NEDO
Sunshine 60 29F, Toshima-ku
3-1-1 Higashi-Ikebukuro
Tokyo 170
Japan
Phone: 81-3-3987-9423
Fax: 81-3-5992-6440

Robert E. Nelson
Quantum Group Inc.
1121 Sorrento Valley Road
San Diego, CA 92121
Phone: (619) 457-3048
Fax: (619) 457-3229

Alan Newhouse
Newhouse Consulting
11108 Deborah Drive
Potomac, MD 20854
Phone: (301) 299-4285
Fax: (301) 983-4836
EMail: newhouse@intr.net

Ugur Ortabasi
United Innovations
11585 Sorrento Valley Road
San Diego, CA 92121
Phone: (619) 350-1829
Fax: (619) 794-9609
EMail: ugurO@aol.com

Aleksandar Ostrogorsky
Rensselaer Polytechnic Inst.
TEC 2026
Troy, NY 12180-3590
Phone: (518) 276-6975
Fax: (518) 276-6025
EMail: ostrod@rpi.edu

Jan-Christoph Panitz
Paul Scherrer Institute
OFLC/10 7A
Villigen PSI CH-5232
Switzerland
Phone: +41-56-310-4194
Fax: +41-56-310-2199
EMail: panitz@psw234.psi.ch

Dmitry Paramonov
Westinghouse
1310 Beulah Road
401-4A8
Pittsburgh, PA 15235
Phone: (412) 256-1656
Fax: (412) 256-2444
EMail: paramonov.d.v.@wcsmail.com

James Phillips
Institute of Energy Conversion
University of Delaware
501 Wyoming Road
Newark, DE 19716-3820
Phone: (302) 831-6244
Fax: (302) 831-6226
EMail: jep@udel.edu

Jie Piao
Epitaxial Laboratory, Inc.
25 East Loop Road
Stony Brook, NY 11790-3350
Phone: (516) 444-6114
Fax: (516) 271-6743
EMail: jpeli@worldnet.att.net

Roland Pitts
National Renewable Energy Lab
1617 Cole Boulevard
Golden, CO 80401
Phone: (303) 384-6485
Fax: (303) 384-6655
EMail: roland_pitts@nrel.gov

Stanley Pond
Pond Engineering
2501 So. County Road 21

Berthoud, CO 80513
Phone: (303) 651-1678
Fax: (303) 532-4802

David Riley
Westinghouse Bettis
P. O. Box 79
West Mifflin, PA 15122
Phone: (412) 476-6229
Fax: (412) 476-5151
EMail: driley-2@mail.idt.net

Steven Ringel
The Ohio State University
Dept. of Electrical Engineering
2015 Neil Avenue
Columbus, OH 43210
Phone: (614) 292-6904
Fax: (614) 292-7596

Ronald Roedel
Arizona State University
Dept. of Electrical Engineering
85287-5706, AZ
Phone: (602) 965-6622
Fax: (602) 965-8118
Email: r.roedel@asu.edu

Millard Rose
Space Power Institute
231 Leach Center
Auburn University, AL 36849-5320
Phone: (334) 844-5894
Fax: (334) 844-5900
EMail: rosemil@mail.auburn.edu

Robert Rosenfeld
DARPA
3701 N. Fairfax Dr.
Arlington, VA 22203-2342
Phone: (703) 696-2327
Fax: (703) 696-2204
EMail: rrosenfeld@darpa.mil

Stevan Saban
EDTEK, Inc.
7082 South 220th Street
Kent, WA 98032
Phone: (206) 395-8084
Fax: (206) 395-8086
EMail: edtekinc@cnet.com

Sudesh Saroop
Rensselaer Polytechnic Institute
CIE 6015
Troy, NY 12180
Phone: (518) 270-3657
EMail: saroos@rpi.edu

Alfred Schock
Orbital Sciences Corp.
20301 Century Blvd., A-35
Germantown, MD 20874
Phone: (301) 428-6272
Fax: (301) 353-8619
EMail: or@orbital.fsd.com

Kenneth Schroeder
Space Power Institute
231 Leach Center
Auburn University, AL 36849-5320
Phone: (334) 844-5894
Fax: (334) 844-5900
EMail: schrokl@eng.auburn.edu

Markus Schubnell
Paul Scherrer Institute
Villigen 5232
Switzerland
Phone: 41-56-310-2797
Fax: 41-56-310-4413
EMail: markus.schubnell@psi.ch

Paul Sharps
Research Triangle Institute
P. O. Box 12194
3040 Corwallis Road
Research Triangle Pk, NC 27709-2194

Phone: (919) 541-8859
Fax: (919) 541-6515
EMail: prs@RTI.org

Kailash Shukla
Tecogen
45 First Avenue
Waltham, MA 02254
Phone: (617) 622-1049
Fax: (617) 622-1025

David Spears
MIT Lincoln Laboratory
244 Wood Street
Lexington, MA 02173
Phone: (617) 981-7835
Fax: (617) 981-0122
EMail: spears@ll.mit.edu

Ken Stone
McDonnell Douglas Aerospace
5301 Bolsa Ave.
Huntington Beach, CA 92647
Phone: (714) 896-3311 x9193
Fax: (714) 896-1244

Vasen Sundarum
EDTEK
Kent, WA

Carl Taube
Rhone-Poulenc
One Corporate Drive
Shelton, CT 06484
Phone: (203) 925-3309
Fax: (203) 925-8182

Holly Thomas
National Renewable Energy Lab
1617 Cole Boulevard
Golden, CO 80401
Phone: (303) 384-6400
Fax: (303) 384-6490
EMail: thomash@tcplink.nrel.gov

Vadim Unishkov
Quantum Group
San Diego, CA

Parvez Uppal
Sanders
P. O. Box 868
NHQ6-1517
Nashua, NH 03061-0868
Phone: (410) 204-2361
Fax: (410) 204-2100
EMail: puppall@mailgw.sanders.
lockheed.com

James Vander Meer
Eaton Corporation
Navy Controls Division
4265 N. 30th Street, Dept. N141
Milwaukee, WI 53216
Phone: (414) 449-6518
Fax: (414) 449-6038

Christine Wang
MIT Lincoln Laboratory
244 Wood Street
Lexington, MA 02173-9108
Phone: (617) 981-4466
Fax: (617) 981-0122
EMail: wang@ll.mit.edu

Ted Wangensteen
National Renewable Energy Lab
1617 Cole Boulevard
Golden, CO 80401
Phone: (303) 384-6651
Fax: (303) 384-6604

Mark Wanlass
National Renewable Energy Lab
1617 Cole Boulevard
Golden, CO 80401
Phone: (303) 384-6532
Fax: (303) 384-6430
EMail: mwanlass@nrel.nrel.gov

Scott Ward
National Renewable Energy Lab
1617 Cole Boulevard
Golden, CO 80401
Phone: (303) 384-6529

Ronald Watson
Babcock & Wilcox
P. O. Box 785
Lynchburg, VA 24505-0785
Phone: (804) 522-5623
Fax: (804) 522-6906
EMail: ron.c.watson@mcdermott.com

Suhuai Wei
National Renewable Energy Lab
1617 Cole Boulevard
Golden, CO 80401
Phone: (303) 384-6666
Fax: (303) 384-6531
EMail: sgw@nrel.gov

Edward West
Western Washington University
Vehicle Research Institute
High Street
Bellingham, WA 98225-9086
Phone: (360) 650-3045
Fax: (360) 650-3445
EMail: emwest@henson.cc.wwu.edu

Dave Wilt
NASA Lewis Research Ctr.
21000 Brookpark Road, MS 302-1
Cleveland, OH 44135
Phone: (216) 433-6293
Fax: (216) 433-6106

Ed Witt
National Renewable Energy Lab
1617 Cole Blvd.
Golden, CO 80401
Phone: (303) 384-6402
Fax: (303) 384-6481

Steven Wojtczuk
Spire Corporation
One Patriots Park
Bedford, MA 01730
Phone: (617) 275-6000
Fax: (617) 275-7470

David Woodward
Eaton Corporation
Navy Controls Division
4265 N. 30th Street, Dept. N148
Milwaukee, WI 53216
Phone: (414) 449-7411
Fax: (414) 449-6038

Xuanzhi Wu
National Renewable Energy Lab
1617 Cole Boulevard
Golden, CO 80401
Phone: (303) 384-6552
Fax: (303) 384-6430
EMail: xwu@nrel.nrel.gov

Hiromi Yamaguchi
Institute of Research & Innovation
1-6-8 Yushima
Bunkyo-ku

Tokyo 113
Japan
Phone: 81-3-5689-6359
Fax: 81-3-5689-6360
EMail: qc9n-sru@asahi-net.or.jp

Matthias Zenker
Fraunhofer Institute
Oltmannsstrasse 5
Freiburg D-79100
Germany
Phone: +49-761-4588-179
Fax: 49-761-4588-217
EMail: zenker@ise.fhg.de

Michael Zierak
IBM
Essex Junction, VT

Ken Zweibel
National Renewable Energy Lab
1617 Cole Boulevard
Golden, CO 80401
Phone: (303) 384-6441
Fax: (303) 384-6430
EMail: zweibelk@tcplink.nrel.gov

H

Hitchcock, C., 65, 89
Hoffman, Jr., R. W., 237, 249
Hopkinson, M., 411
Horne, W. E., 105
Huang, H. X., 33
Hui, S., 33

J

Jain, R. K., 237
Jarefors, K., 463
Jenkins, P. P., 237, 249
Johnson, S., xxv

K

Kazmerski, L. L., 549
Keser, S., 41
Ketterl, J. R., 105
Keyes, J., 369
Krommenhoek, D. J., xxiii
Kruger, J. S., xxi, 23
Kushch, A. S., 315, 373

L

Lee, H., 389
Lesko, J. D., 117

M

Manfra, M. J., 75
Marks, J., 463
Martinelli, R. U., 389
Matson, R. J., 227
Mauk, M. G., 117, 129
McAlonan, M., 341
McNeely, J. B., 117, 129
Morgan, M. D., 105
Moriarty, T., 227
Morosini, M. B. Z., 105
Morris, N., 389
Mueller, R. L., 117, 129
Mulligan, B., 369

Murray group

Murray, C. S., 227, 237, 249
Murthy, S. D., 139

N

Nelson, J., 411
Nelson, R. E., xviii, 189

O

Odubanjo, T., 389
Ostrogorsky, A. G., 139, 157

P

Panitz, J.-C., 265
Pate, M., 411
Patel, N. B., 105
Phillips, J. E., 443
Postlethwait, M. A., 471

R

Riley, D. R., xvi, 227, 237, 249
Rose, M. F., xx, 181, 277, 505
Rosenfeld, R., xv

S

Saban, S. B., 105
Samaras, J., 369
Sarmiento, P. A., 373
Saroop, S., 139
Schroeder, K. L., 505
Schubnell, M., 3, 265
Scoles, S. W., 355
Scotto, M. V., 355
Sharps, P. R., 215
Shellenbarger, Z. A., 117, 129
Shukla, K., 329
Sims, P. E., 117
Skinner, S. M., 315, 373
Spears, D. L., 75
Stollwerck, G., 41

Sulima, O. V., 41
Sundaram, V. S., 105

T

Taylor, G. C., 389
Thomas, H., xxviii
Timmons, M. L., 215
Turner, G. W., 75

U

Unishkov, V. A., 397

W

Wang, C. A., 75
Wanlass, M. W., 227, 463
Ward, J. S., 227
Webb, J., 403

Weizer, V. G., 237, 249
West, E. M., 521
Wiesner, H., 403
Williams, D., 369
Wilt, D. M., xiv, 237, 249
Wojtczuk, S., 205
Wu, X., 227

X

Xiang, H. H., 369

Y

Ye, S.-Z., 33

Z

Zachariou, A., 411
Zhang, K., 215
Zierak, M., 55

AIP Conference Proceedings

	Title	L.C. Number	ISBN
No. 372	Beam Dynamics and Technology Issues for + - Colliders 9th Advanced ICFA Beam Dynamics Workshop (Montauk, NY, 1995)	96-84189	1-56396-554-2
No. 373	Stress-Induced Phenomena in Metallization (Palo Alto, CA 1995)	96-84949	1-56396-439-2
No. 374	High Energy Solar Physics (Greenbelt, MD 1995)	96-84513	1-56396-542-9
No. 375	Chaotic, Fractal, and Nonlinear Signal Processing (Mystic, CT 1995)	96-85356	1-56396-443-0
No. 376	Chaos and the Changing Nature of Science and Medicine: An Introduction (Mobile, AL 1995)	96-85220	1-56396-442-2
No. 377	Space Charge Dominated Beams and Applications of High Brightness Beams (Bloomington, IN 1995)	96-85165	1-56396-625-7
No. 378	Surfaces, Vacuum, and Their Applications (Cancun, Mexico 1994)	96-85594	1-56396-418-X
No. 379	Physical Origin of Homochirality in Life (Santa Monica, CA 1995)	96-86631	1-56396-507-0
No. 380	Production and Neutralization of Negative Ions and Beams / Production and Application of Light Negative Ions (Upton, NY 1995)	96-86435	1-56396-565-8
No. 381	Atomic Processes in Plasmas (San Francisco, CA 1996)	96-86304	1-56396-552-6
No. 382	Solar Wind Eight (Dana Point, CA 1995)	96-86447	1-56396-551-8
No. 383	Workshop on the Earth's Trapped Particle Environment (Taos, NM 1994)	96-86619	1-56396-540-2
No. 384	Gamma-Ray Bursts (Huntsville, AL 1995)	96-79458	1-56396-685-9
No. 385	Robotic Exploration Close to the Sun: Scientific Basis (Marlboro, MA 1996)	96-79560	1-56396-618-2
No. 386	Spectral Line Shapes, Volume 9 13th ICSLS (Firenze, Italy 1996)		1-56396-656-5
No. 387	Space Technology and Applications International Forum (Albuquerque, NM 1997)	96-80254	1-56396-679-4 (Case set) 1-56396-691-3 (Paper set)
No. 388	Resonance Ionization Spectroscopy 1996 Eighth International Symposium (State College, PA 1996)	96-80324	1-56396-611-5